T0134262

QUANTUM FIELD THEORY APPROACH TO CONDENSED MATTER PHYSICS

A balanced combination of introductory and advanced topics provides a new and unique perspective on the quantum field theory (QFT) approach to condensed matter physics (CMP). Beginning with the basics of these subjects, such as static and vibrating lattices, independent and interacting electrons, the functional formulation for fields, different generating functionals and their roles, this book presents a unified viewpoint illustrating the connections and relationships among various physical concepts and mechanisms. Advanced and newer topics bring the book up-to-date with current developments and include sections on cuprate and pnictide superconductors, graphene, Weyl semimetals, transition metal dichalcogenides, topological insulators and quantum computation. Finally, well-known subjects such as the quantum Hall effect, superconductivity, Mott and Anderson insulators, spin-glasses, and the Anderson-Higgs mechanism are examined within a unifying QFT-CMP approach. Presenting new insights on traditional topics, this text allows graduate students and researchers to master the proper theoretical tools required in a variety of condensed matter physics systems.

EDUARDO C. MARINO is Professor of Physics at the Federal University of Rio de Janeiro. He was Postdoctoral Fellow at Harvard University and a Visiting Professor at Princeton University. He is an elected member of the Brazilian National Academy of Sciences and in 2005 was awarded the National Order of Scientific Merit by the president of Brazil.

QUANTUM FIELD THEORY APPROACH TO CONDENSED MATTER PHYSICS

EDUARDO C. MARINO

Institute of Physics
Federal University of Rio de Janeiro

CAMBRIDGE
UNIVERSITY PRESS

CAMBRIDGE
UNIVERSITY PRESS

University Printing House, Cambridge CB2 8BS, United Kingdom

One Liberty Plaza, 20th Floor, New York, NY 10006, USA

477 Williamstown Road, Port Melbourne, VIC 3207, Australia

4843/24, 2nd Floor, Ansari Road, Daryaganj, Delhi – 110002, India

79 Anson Road, #06–04/06, Singapore 079906

Cambridge University Press is part of the University of Cambridge.

It furthers the University's mission by disseminating knowledge in the pursuit of
education, learning, and research at the highest international levels of excellence.

www.cambridge.org
Information on this title: www.cambridge.org/9781107074118
DOI: 10.1017/9781139696548

© Eduardo C. Marino 2017

First published 2017

Printed in the United Kingdom by Clays, St Ives plc

A catalogue record for this publication is available from the British Library.

Library of Congress Cataloging-in-Publication Data
Names: Marino, Eduardo C., author.
Title: Quantum field theory approach to condensed matter physics / Eduardo
C. Marino (Universidade Federal do Rio de Janeiro).
Description: Cambridge, United Kingdom ; New York, NY : Cambridge
University Press, 2017. | Includes bibliographical references and index.
Identifiers: LCCN 2017026500| ISBN 9781107074118 (alk. paper) |
ISBN 1107074118 (alk. paper)
Subjects: LCSH: Condensed matter. | Quantum field theory.
Classification: LCC QC173.454 .M38 2017 | DDC 530.4/1–dc23
LC record available at https://lccn.loc.gov/2017026500

ISBN 978-1-107-07411-8 Hardback

Concerning matter, we have been all wrong. What we have been calling matter is actually energy, the vibration of which has been lowered so much as to be perceptible to the senses. There is no matter.

A. Einstein

Contents

Part II   Quantum Field Theory

Preface

The inception of quantum field theory (QFT) occurred in 1905, when Einstein, inspired by the work of Planck, postulated the quantization of the electromagnetic radiation field in terms of photons in order to explain the photoelectric effect. Two years later, Einstein himself made the first application of this incipient QFT in the realm of condensed matter physics (CMP). By extending the idea of quantization to the field of elastic vibrations of a crystal, he used the concept of phonons in order to obtain a successful description of the specific heat of solids, which has become one of the first great achievements of the quantum theory. Since their early days, therefore, we see that CMP and QFT have been evolving together side by side.

In 1926, the quantum theory of the electromagnetic field was formulated according to the principles of quantum mechanics, thereby providing a rational description for the dynamics of photons, which were postulated by Einstein more than 20 years before. QFT soon proved to be the only framework where the two foundations of modern physics, namely, quantum mechanics and the special theory of relativity, could be combined in a sensible way.

From then on, QFTs grew up mainly in the realm of particle physics, until they eventually became some of the most successful theories in physics. Familiar examples are the Standard Model (SM) of fundamental interactions and, more specifically, Quantum Electrodynamics (QED), which exhibits some theoretical predictions that can match the experimental results up to twelve decimal figures. It is difficult to find any other model, ever proposed, possessing such accuracy.

Condensed Matter Physics (CMP), by its turn, has proved to be one of the richest areas of physics, keeping under its focus of investigation an incredible variety of systems and materials. These exhibit a plethora of unsuspected kinds of behavior, frequently associated to different responses to all types of external agents, such as electric and magnetic fields, voltage and temperature gradients, pressure, elastic stress and so on. The understanding of these phenomena is an enterprise that is frequently as interesting as it is challenging. Furthermore, like in no other area of physics, mastering the principles and mechanisms of the phenomena under investigation has produced countless technological by-products. These sometimes have produced such impact on the society that its whole structure has been transformed, and many human habits changed. One such example was the development of the transistor, which ocurred after the physics of doped semiconductors was mastered. The whole revolution of electronics, miniaturization and informatics would have been impossible without it.

For decades, CMP made a description of solids that was based on the concept of independent electrons moving on a crystalline substrate. This picture has worked extremely well due to the peculiar properties of the quantum-mechanical behavior of electrons in a periodic potential and served for understanding an enormous amount of properties of metals, insulators and semiconductors. Adding further elements to this picture has enabled the understanding of magnetic materials. Then superconductivity, one of the most beautiful, interesting and useful phenomena in physics, was understood by including the interaction of independent electrons with the crystal lattice vibrations.

By the 1980s, however, the discovery of the quantum Hall effect and the following efforts employed to understand it brought two important features to the center of attention in the realm of CMP. The first one is the existence of material systems where the electrons, rather than being independent, are strongly correlated due to interactions. The second one is the fact that the physical properties of certain states of matter are determined by sophisticated topological constraints that fix the value of some quantities with an incredible accuracy and guarantee the conservation of others, a fact that would not be otherwise anticipated. Both features usually lead to unsuspected results.

Since that time, a large number of new materials either have been developed or are being designed that present strongly correlated electrons, topological phases or both. For understanding such a large amount of new sophisticated advanced materials, an efficient method, capable of describing the quantum-mechanical properties of a system of interacting many-particle systems and their possibly nontrivial topological aspects, was required. QFT was the natural response to this demand. By then, it had become one of the most powerful theoretical tools available in physics, with applications ranging from particle physics to quantum computation, passing through hadron physics, nuclear physics, quantum optics, cosmology, astrophysics and, most of all, condensed matter physics, which is the subject of this book.

Here I present a QFT approach to many different condensed matter systems that have attracted the interest of the scientific community, always trying to explore the beauty, depth and harmony that are provided by a unified vision of physics in such approaches. This not only fosters a deeper understanding of the subject; it opens new ways of looking at it.

An extremely interesting example of the interplay between CMP and QFT is the Anderson–Higgs mechanism, which plays a central role in the Standard Model, and its relation to the Meissner–Ochsenfeld effect of superconductivity. Here, the Landau–Ginzburg field of the superconducting system plays the role of the Anderson–Higgs field of the SM, the only difference being the gauge group. In both cases, a mass is effectively generated to the gauge field, which causes the corresponding propagators to decay exponentially. In the former case, this exponential decay accounts for the extremely short range of the weak interaction, whereas in the later it leads to an extremely short penetration length for the magnetic field inside the bulk of a superconductor, thereby effectively expelling it from inside superconductors, a phenomenon known as the Meissner–Ochsenfeld effect. The fact that the particle excitations associated to the Landau–Ginzburg field reveal themselves as electron-bound states (Cooper pairs) strongly suggests, both on logical and esthetic grounds, that the Anderson–Higgs boson particle should also be composite. This should be a central issue in the realm of particle physics in the near future.

Another beautiful example that is explored in this book is the equivalence between the Yukawa mechanism of mass generation for leptons and quarks in the SM and the Peierls mechanism of gap generation in polyacetylene. Both involve identical trilinear interactions containing a Dirac field, its conjugate and a scalar field. In the former case, the lepton or quark Dirac fields interact with the Anderson–Higgs field, whereas in the later the electron, which can be shown to be described by a Dirac field, interacts with the elastic vibrations field of the polymer lattice. In both cases, the scalar field acquires a nonzero vacuum expectation value: the first one by a judicious choice of the Anderson–Higgs potential, while the second one by the dimerization of the polyacetylene chain. Therefore, the same mechanism that causes polyacetylene to be an insulator generates the mass of all familiar matter. This amazing unification of phenomena that are separated by more than ten orders of magnitude in energy indicates the existence of a deep, underlying unity in physics. A universal unified vision of this science is, consequently, required nowadays. This book is aimed to provide such a unified picture of CMP and QFT.

Writing a book on applications of QFT in CMP, however, is a formidable challenge, in view of the large number of excellent books that already exist on the subject, some of them listed under Further Reading at the back of this book. One can, indeed, always ask: why another book on QFT in CMP? Nevertheless, because of its peculiar characteristics, which include a balanced combination of introductory, advanced and traditionally known material, I feel that this book has its own place in the literature and will be helpful and useful to a broad group of readers.

The book has been divided into three parts. Part I provides a four-chapter introduction to CMP. Part II contains eight chapters on QFT, including an introduction that starts from the very basic principles of QFT as well as a description of the main features of QFT, which may be relevant for applications in CMP. Part III is made up of eighteen chapters covering different applications of QFT in CMP, which include metals, Fermi liquids, Mott insulators, Anderson insulators, polarons, polyacetylene, materials exhibiting the Kondo effect, quantum magnetic chains, quantum magnetic planar systems, the spin-fermion system, spin glasses, superfluids, conventional superconductors, Dirac superconductors, cuprate superconductors, pnictide superconductors, systems presenting the quantum Hall effect, graphene, silicene, transition metal dichalcogenides, topological insulators, Weyl semimetals and systems that are candidates for topological quantum computation.

This book covers topics for which there is, so far, no complete understanding and, consequently, about which no consensus has been reached in the community. Cuprate and pnictide superconductors, for instance, are examples of such topics. Besides these, the book includes some very recent advanced topics, such as Weyl semimetals, topological insulators and materials potentially relevant for quantum computation. I am aware that the inclusion of such topics in the book is a bold venture; nevertheless, I decided to face it and take the involved risks. I feel the inclusion of these topics has made this work much more interesting and exciting. I hope the reader will understand this point and will share the constructive attitude that stands behind the inclusion of such topics.

The book can be used in many different ways. Chapters 1–7 can be used as an introductory course in CMP and QFT. After this introduction, one can follow the sequence of QFT subjects presented in Chapters 8–12, which comprise classical and quantum topological excitations, order-disorder duality, bosonization and anyons, statistical transmutation and Pseudo Quantum Electrodynamics. Then, after a bridge between QFT and CMP offered in Chapter 13, the reader will find in Chapters 14–30 the QFT approach to a variety of materials and mechanisms of CMP.

Alternatively, the book contains several avenues that will take the reader along certain sequences of QFT procedures, which play an important part in different CMP systems. The first of such avenues starts with symmetries (Chapter 7), and then order-disorder duality and quantum topological excitations (Chapter 9), bosonization and generalized statistics (Chapter 10), bosonization of polarons (Chapter 15), bosonization of quantum magnetic systems in 1d (Chapter 18) and anyons with non-Abelian statistics (Chapter 30).

A second avenue deals with electromagnetic interaction of planar systems. It starts with pseudo quantum electrodynamics (Chapter 12) and then goes to graphene (Chapter 27) and silicene and transition metal dichalcogenides (Chapter 28). A third starts with symmetries (Chapter 7), followed by classical Sine–Gordon solitons (Section 8.3), quantum Sine–Gordon solitons (Section 10.6), 2d

Coulomb gas (Section 18.3), application to copper benzoate (Section 18.4.2) and application to the Kosterlitz–Thouless transition (Section 18.4.3). Then, we have an avenue on superconductivity, which starts in superconductivity (Section 4.4), then goes to electron-phonon interactions (Section 3.7), from this to superconductivity of regular electrons (Sections 23.1–2) and then superconductivity of Dirac electrons (Sections 23.3–6). The reader is kindly invited to find further avenues as such.

The book is mainly meant for researchers, posdocs and graduate students in the areas of CMP, QFT, materials science, statistical mechanics and related areas. Nevertheless, being self-contained in the sense that no previous knowledge of either CMP or QFT is required, the book can also be used by undergraduate students who feel inclined toward QFT and CMP.

I want to express my gratitude to people who contributed in different ways toward the completion of this book. First of all, my editor Simon Capelin, who in the many phases of this work never hesitated to provide his unconditional support. To Roland Köberle, who followed the writing of the book for 3 years, thank you for numerous useful suggestions. Thank you to Curt Callan for taking the time to read the manuscript, to Mucio Continentino for the constructive critical reading of selected chapters, to Hans Hansson for helpful suggestions and to Cristiane de Morais Smith for invaluable comments and remarks. I would also like to thank Vladimir Gritsev, Amir Caldeira, Chico Alcaraz, Nestor Caticha, Luis Agostinho Ferreira and Carlos Aragão for (hopefully) reading the manuscript. Special thanks also go to my collaborators of the most recent years: Van Sérgio Alves, Leandro Oliveira do Nascimento and Lizardo Nunes for the fruitful exchange of points of view. I also take this opportunity to thank my home institution, the Institute of Physics of the Federal University of Rio de Janeiro, for all the support received along many years.

Finally, I would like to thank my family for the many, many, many hours taken from their company in order to keep this project going. Most especially, I thank my wife Norma, without whose love, support and patience everything would have been impossible.

Eduardo C. Marino

Part I
Condensed Matter Physics

1

Independent Electrons and Static Crystals

The expression "condensed matter" refers to materials that are either in a solid or in a liquid state. Soon after the atomic theory was established, the structure of matter in these condensed forms became the object of study under that new perspective. These early investigations already revealed that a large amount of the solids, interestingly, exhibit a peculiar structure, which is called a crystal. These rich forms of matter surprisingly assemble their constituent atoms or molecules in such a way that the most stable configuration has a periodic character, namely, there exists a basic unit that repeats itself along the whole sample. The specific geometric form of the periodic crystalline structure is determined by the spatial orientation of the atomic or molecular valence orbitals of the basic components of each crystal material. The existence of this periodic geometric array exerts a profound influence upon the physical properties of the material. These include the energy spectrum, charge and heat transport, specific heat, magnetic and optical properties. The study of crystal lattices, consequently, is of fundamental importance in the physics of condensed matter.

1.1 Crystal Lattices

The mathematical concept that most closely describes an actual crystal lattice is that of a Bravais lattice, a set of mathematical points corresponding to the discrete positions in space given by

$$\{\mathbf{R}|\ \mathbf{R} = n_1\mathbf{a}_1 + n_2\mathbf{a}_2 + n_3\mathbf{a}_3;\ n_i \in \mathbb{Z}\}, \tag{1.1}$$

where $\mathbf{a}_i$, $i = 1, 2, 3$ are the so-called primitive vectors in three-dimensional space. The corresponding structure in one(two)-dimensional space would be analogous to (1.1), but having only one(two) primitive vector(s). We can see that the points in the Bravais lattice form a pattern that repeats itself periodically. A characteristic feature of this type of mathematical structure is that it looks exactly the same from the perspective of any of its points $\mathbf{R}$.

The Bravais lattice is invariant under the operation

$$\mathbf{R} \to \mathbf{R} + \mathbf{T}, \tag{1.2}$$

where

$$\mathbf{T} = L\mathbf{a}_1 + M\mathbf{a}_2 + N\mathbf{a}_3, \tag{1.3}$$

and L, M, N are arbitrary but fixed integers. Indeed, clearly for any $\mathbf{T}$ we have $\{\mathbf{R}\} \equiv \{\mathbf{R} + \mathbf{T}\}$, hence translations by $\mathbf{T}$ are symmetry operations of the Bravais lattice. Examples of two-dimensional Bravais lattices are the square lattice and the triangular lattice, see Figs. 1.1 and 1.2.

A useful concept related to a Bravais lattice is that of a primitive unit cell. This is a region of space containing a single point of the Bravais lattice that will cover the whole volume (area in two dimensions, length in one dimension) encompassed by the lattice when translated by all the symmetry operations $\mathbf{T}$, in such a way that these translations do not produce any superpositions. There are in general different regions, with many possible shapes, that satisfy the previous definition. Surprisingly, however, the volume (area in two dimensions, length in one dimension) of all primitive unit cells is always the same, irrespective of their

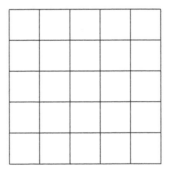

Figure 1.1 Square Lattice: an example of a 2d Bravais lattice

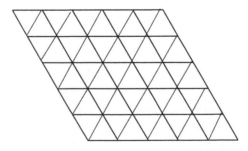

Figure 1.2 Triangular Lattice: an example of a 2d Bravais lattice

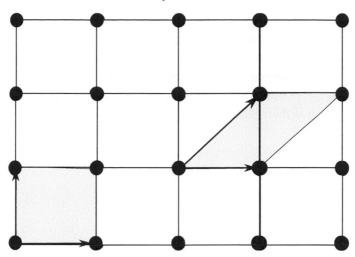

Figure 1.3 Two choices of primitive unit cells for a square lattice, corresponding to different sets of primitive vectors, according to (1.4). Notice that the areas of the two unit cells are, evidently, the same.

specific shape. Evidently, from the definition, the volume V_0 of any primitive unit cell, for a lattice containing N points and a volume V, must be given by $V_0 = \frac{V}{N}$. The mentioned property then follows.

Given a set of primitive vectors, an obvious choice among the many possible primitive unit cells would be

$$\{\mathbf{R}|\ \mathbf{R} = x_1\mathbf{a}_1 + x_2\mathbf{a}_2 + x_3\mathbf{a}_3;\ \ x_i \in [0, 1]\}. \tag{1.4}$$

From this we may infer that the volume of any primitive unit cell is given by

$$V_0 = \mathbf{a}_3 \cdot (\mathbf{a}_1 \times \mathbf{a}_2). \tag{1.5}$$

For two-dimensional lattices, the corresponding area would be

$$A_0 = |\mathbf{a}_1 \times \mathbf{a}_2|, \tag{1.6}$$

whereas for a one-dimensional lattice, we would have the corresponding length

$$L_0 = |\mathbf{a}_1|. \tag{1.7}$$

A crystal structure in general is not just a Bravais lattice; rather it is obtained from the latter by placing what is called a base in each of its points. The base is a finite set of points occupying fixed positions with respect to each of the points of the Bravais lattice.

The so-called honeycomb lattice is an example of a crystal structure, that is not a Bravais lattice. This can be inferred from the fact that points A and B have different perspectives of the lattice, as we can see in Fig. 1.4. This crystal structure is obtained by adding to each point of a triangular Bravais lattice having primitive vectors of length a, a base of two points at $(0, 0)$ and $(h, 0)$, with $h = a/\sqrt{3}$.

The actual crystal material is modeled by placing atoms, ions, molecules or radicals in each of the points of a base $\mathcal{B}$ in a Bravais lattice $\mathcal{BL}$. The crystal mass density distribution is then given by

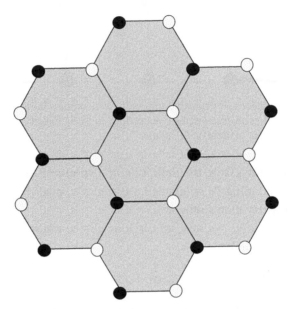

Figure 1.4 Honeycomb crystal structure, showing the two interpenetrating Bravais triangular sublattices A and B, respectively, with black and white dots. Different perspectives of the lattice are clearly obtained from sublattice points A and B.

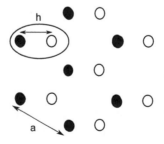

Figure 1.5 Honeycomb crystal structure, showing one Bravais triangular sublattice (black dot with spacing a) and the base (one black and one white dot with spacing $h = a/\sqrt{3}$)

$$\rho(\mathbf{X}) = \sum_{\mathbf{R} \in \mathcal{BL}} \sum_{i \in \mathcal{B}} m_i \, \delta(\mathbf{X} - \mathbf{R} - \mathbf{r}_i), \tag{1.8}$$

where m_i is the mass of the constituent at the point $\mathbf{R} + \mathbf{r}_i$ of the crystal structure. One can verify that the above expression is invariant under the $\mathcal{BL}$ symmetry operations (1.3), namely,

$$\rho(\mathbf{X}) = \rho(\mathbf{X} + \mathbf{T}), \tag{1.9}$$

which follows from the fact that $\sum_{\mathbf{R}} = \sum_{\mathbf{R}-\mathbf{T}}$ for $\mathbf{R} \in \mathcal{BL}$.

In the next section we will study the Fourier expansion of periodic quantities possessing the Bravais lattice symmetry (1.9) and shall explore the important consequences of this condition.

1.2 The Reciprocal Lattice

Let $f(\mathbf{X})$ be a periodic physical quantity exhibiting the same symmetry as a given Bravais lattice, namely

$$f(\mathbf{X}) = f(\mathbf{X} + \mathbf{R}). \tag{1.10}$$

An example of such a quantity is the crystal mass distribution function $\rho(\mathbf{X})$, introduced in (1.8).

The invariance of a function $f(\mathbf{X})$ under translations by Bravais lattice points manifests itself in its Fourier expansion as

$$\begin{aligned} f(\mathbf{X}) &= \sum_{\mathbf{q}} f(\mathbf{q}) \exp\{i\mathbf{q} \cdot \mathbf{X}\} \\ &= \sum_{\mathbf{q}} f(\mathbf{q}) \exp\{i\mathbf{q} \cdot (\mathbf{X} + \mathbf{R})\}, \end{aligned} \tag{1.11}$$

which implies

$$\mathbf{q} \cdot \mathbf{R} = 2\pi n \quad ; \quad n \in \mathbb{Z}. \tag{1.12}$$

This relates the position vectors of a certain Bravais lattice to the argument of the Fourier transform of *any* function having the same symmetry of such a lattice.

Considering that $\mathbf{R} = \sum_i n_i \mathbf{a}_i$, according to (1.1), we see that the above relation is solved by

$$\mathbf{q} = \sum_j l_j \mathbf{b}_j \quad ; \quad l_j \in \mathbb{Z}, \tag{1.13}$$

provided the vectors $\mathbf{b}_j$ satisfy the following relation with the primitive vectors of the Bravais lattice,

$$\mathbf{a}_i \cdot \mathbf{b}_j = 2\pi \delta_{ij}. \tag{1.14}$$

Indeed, this implies

$$q \cdot R = 2\pi \left(\sum_i n_i l_i \right) \quad ; \quad n_i, l_i \in \mathbb{Z}, \tag{1.15}$$

and it is easy to see that the quantity between parentheses above is an integer. We conclude, therefore, that (1.13) with the condition (1.14) satisfies (1.12).

The solution of (1.14) for the vectors $\mathbf{b}_i$ in three dimensions would be

$$\mathbf{b}_1 = \left(\frac{2\pi}{V_0} \right) \mathbf{a}_2 \times \mathbf{a}_3, \tag{1.16}$$

where V_0 is given by (1.5). The vectors $\mathbf{b}_2$ and $\mathbf{b}_3$ are obtained by cyclic permutations. An example in two dimensions would be the square lattice, for which the solution of (1.14) would be

$$\mathbf{b}_i = 2\pi \frac{\mathbf{a}_i}{|\mathbf{a}_i|^2} \quad , \quad i = 1, 2. \tag{1.17}$$

For a one-dimensional lattice, the solution of (1.14) would be

$$\mathbf{b}_1 = 2\pi \frac{\mathbf{a}_1}{|\mathbf{a}_1|^2}. \tag{1.18}$$

The set of vectors $\mathbf{q}$ in (1.13) clearly form themselves a Bravais lattice with primitive vectors $\mathbf{b}_i$, namely

$$\{ \mathbf{Q} | \ \mathbf{Q} = n_1 \mathbf{b}_1 + n_2 \mathbf{b}_2 + n_3 \mathbf{b}_3; \ n_i \in \mathbb{Z} \}. \tag{1.19}$$

This is called "reciprocal lattice," a name derived from the fact that the vectors $\mathbf{b}_i$ have dimension of inverse length, whereas the corresponding vectors of the original lattice, namely $\mathbf{a}_i$ have dimension of length. Notice that there is only one reciprocal lattice associated to a given Bravais lattice and that the latter is the reciprocal of the former.

The Fourier components of a periodic function possessing the same symmetry of a certain Bravais lattice only depend on wave-vectors, which belong to the corresponding reciprocal lattice. This fact has deep consequences, as we shall see. For instance, the Fourier transform of a function satisfying (1.10), for a certain Bravais lattice $\{\mathbf{R}\}$,

$$f(\mathbf{Q}) = \int_V d^3 X f(\mathbf{X}) \exp \{ -i \mathbf{Q} \cdot \mathbf{X} \}, \tag{1.20}$$

becomes after making $\mathbf{X} = \mathbf{r} + \mathbf{R}$, $\int_V = \sum_{\mathbf{R}} \int_{V_0}$

$$f(\mathbf{Q}) = \sum_{\mathbf{R}} \int_{V_0} d^3 r f(\mathbf{r} + \mathbf{R}) \exp \{ -i \mathbf{Q} \cdot (\mathbf{r} + \mathbf{R}) \}$$

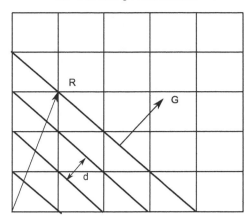

Figure 1.6 Reciprocal lattice vectors **G** are the director-vectors of a family of planes in the original Bravais lattice, spaced by d, according to Eq. (1.22).

$$= N \int_{V_0} d^3 r f(\mathbf{r}) \exp\{-i\mathbf{Q} \cdot \mathbf{r}\}, \tag{1.21}$$

where N is the number of points/cells in the Bravais lattice. The last step follows from the symmetry of the function and the fact that $\mathbf{Q}$ belongs to the reciprocal lattice. We see that the relevant integral sweeps the primitive unit cell only.

The vectors in the set $\{\mathbf{Q}\}$ have an interesting and important feature in connection to its associated Bravais lattice. It is not difficult to see that any Bravais lattice contains different (infinitely many) sets of parallel planes separated by a distance d. The subset of vectors of the Bravais lattice belonging to the nth plane of such set satisfy the relation

$$\mathbf{R} \cdot \frac{\mathbf{G}}{|\mathbf{G}|} = nd \quad ; \quad n \in \mathbb{Z}, \tag{1.22}$$

where $\mathbf{G}$ is a vector perpendicular to this family of planes. By choosing $|\mathbf{G}| = \frac{2\pi}{d}$, we see that (1.22) reduces to (1.12). We then may infer that the director-vectors $\mathbf{G}$ are just the elements of the reciprocal lattice (1.19). Each of the vectors $\mathbf{Q}$ in (1.19), therefore, determines a family of parallel planes in the Bravais lattice, orthogonal to it and such that the basic spacing between adjacent planes is $d = \frac{2\pi}{|\mathbf{Q}|}$.

Let us turn now to a concept that is of foremost importance in the reciprocal lattice. That is the Wigner–Seitz primitive unit cell. This is defined by a peculiar choice of the cell boundary, which is obtained according to the following method. For each of the lattice points, draw lines connecting it to its 1st neighbors, 2nd neighbors... (as much as needed), and then take the set of planes (lines in two dimensions) orthogonal to these lines and intersecting them right at the middle. The resulting cell boundary is the closed surface formed by the union of the

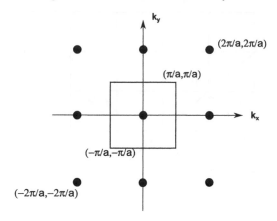

Figure 1.7 The reciprocal lattice of a square lattice with lattice parameter a is a square lattice with spacing $b = \frac{2\pi}{a}$, represented in the figure as black dots. The central square is the First Brillouin zone, the Wigner–Seitz primitive unit cell of the reciprocal lattice.

regions belonging to each of these planes, which form faces. The points of such planes not forming faces of the cell boundary are discarded. The adaptation of this construction to one-dimensional lattices is straightforward.

We have seen that any primitive unit cell has the same volume $\frac{V}{N}$, so this is accordingly the volume of the Wigner–Seitz cell. It can be shown, however, that it is, among all possible primitive unit cells, the one for which the sum of the distances between the cell points and the lattice point it contains is minimal. Another property of the Wigner–Seitz cell is that, by construction, it has the same symmetry as the lattice for which it is defined.

The Wigner–Seitz primitive unit cell of the reciprocal lattice is called the first Brillouin zone. As we shall see, it plays a fundamental role in the quantum-mechanical description of crystalline solids, having profound implications upon the electronic properties of these materials. We shall understand the reason for that in the next section.

The reciprocal lattice also plays an important role in connection with the pattern of x-ray scattering by a crystal. When electromagnetic radiation of wavelength λ falls upon a crystal, the waves reflected by adjacent planes of the Bravais lattice undergo constructive interference whenever the Bragg condition is satisfied, namely

$$2d \sin \theta = n\lambda \quad ; \quad n \in \mathbb{N}, \tag{1.23}$$

where d is the interplane spacing for a family of parallel planes, and θ, the angle between these planes and the direction of the incident radiation (see Fig. 1.9). This,

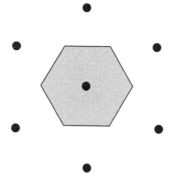

Figure 1.8 Reciprocal lattice of a triangular lattice. The shaded area is the First Brillouin zone.

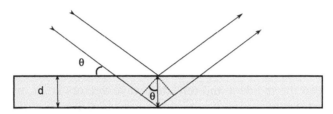

Figure 1.9 Interference between waves scattered from adjacent planes with spacing d, leading to the Bragg condition. The difference in the optical paths of the two beams is $2d \sin \theta$.

of course, will only occur at significant angles when λ is comparable to $d \simeq 0.1$nm, which corresponds to the x-ray region.

The process of reflection of electromagnetic radiation by a crystalline solid may be formulated equivalently as the quantum-mechanical elastic scattering of photons by a periodic potential, which has the same symmetry as the Bravais lattice of the crystal. The probability amplitude for an incident photon with wave-vector $\mathbf{k}_i$ to be scattered into a final state with wave-vector $\mathbf{k}_f$ is given, in first-order Born approximation, by an expression proportional to

$$\langle \mathbf{k}_f | V(\mathbf{r}) | \mathbf{k}_i \rangle = \int d^3 r e^{-i(\mathbf{k}_f - \mathbf{k}_i) \cdot \mathbf{r}} V(\mathbf{r}). \qquad (1.24)$$

This is the Fourier transform of the potential, which can be written as

$$V(\mathbf{r}) = \sum_{\mathbf{R} \in \mathcal{BL}} \sum_{i \in \mathcal{B}} v_i (\mathbf{r} - \mathbf{R} - \mathbf{r}_i). \qquad (1.25)$$

This has the symmetry of the Bravais lattice, Eq. (1.10), and consequently it follows that

$$\mathbf{k}_f - \mathbf{k}_i = \mathbf{Q} \qquad (1.26)$$

is a vector of the reciprocal lattice. Using this fact and making the change of variable $\mathbf{r} \to \mathbf{r} - \mathbf{R} - \mathbf{r}_i$ we see that

$$\langle \mathbf{k}_f | V(\mathbf{r}) | \mathbf{k}_i \rangle = N \sum_{i \in B} v_i(\mathbf{Q}) e^{-i\mathbf{Q} \cdot \mathbf{r}_i}, \tag{1.27}$$

where the sum is over the base points. When this has just a single point, we would have

$$\langle \mathbf{k}_f | V(\mathbf{r}) | \mathbf{k}_i \rangle = N \, v(\mathbf{Q}), \tag{1.28}$$

and when all the base points are occupied by identical constituents,

$$\langle \mathbf{k}_f | V(\mathbf{r}) | \mathbf{k}_i \rangle = N \, v(\mathbf{Q}) \sum_{i \in B} e^{-i\mathbf{Q} \cdot \mathbf{r}_i} \equiv N \, v(\mathbf{Q}) S(\mathbf{Q}). \tag{1.29}$$

Notice the presence of the "geometric form factor," $S(\mathbf{Q})$, whenever the base has more than one point. In the three previous equations, $\mathbf{Q}$ is given by (1.26), a relation known as the von Laue condition. If we square it and use the fact that for elastic scattering $|\mathbf{k}_f| = |\mathbf{k}_i| = \frac{2\pi}{\lambda}$, that $|\mathbf{Q}| = \frac{2\pi}{d}$ for reciprocal lattice vectors, and that the angle between the incident and reflected wave-vectors is 2θ, we can immediately show that (1.26) is just the first Bragg condition for constructive interference. Hence the first-order Born approximation gives the first Bragg peak, which will have an intensity proportional to the squared modulus of the amplitude (1.27).

X-ray spectroscopy constitutes a powerful instrument for the investigation of the structural properties of crystalline solids. In an x-ray experiment, the peaks in the reflected beam will occur right at $\mathbf{k}_i + \mathbf{Q}$, with an intensity proportional to $|v(\mathbf{Q})|^2$, with a possible additional modulation by the geometrical form factor. We conclude that the peaks in the interference spectrum occurring in the x-ray scattering by a crystal provide a direct mapping of the reciprocal lattice of this crystal. The intensity of these peaks will bring information about the local potential v.

In the next section, we explore the consequences of a crystalline structure on the electronic properties of the material.

1.3 Independent Electrons in a Periodic Potential

We will consider here the behavior of electrons in the presence of a periodic potential possessing the same symmetry as a given Bravais lattice. In this first approach we shall neglect the interactions of the electrons among themselves as well as the deviations from an ideal lattice, due, for instance, to thermal and quantum fluctuations. Such periodic potential is created by the basic constituents of the crystal, which are localized at each of the points of the crystalline structure. Its general form is given by (1.25).

The consequences of the presence of a periodic potential in a crystal are of foremost importance for the description of the electronic properties in a crystalline solid. Especially the kinematic properties of the electrons are profoundly modified, as well as the energy spectrum, charge and thermal conductivities, magnetic properties and so on.

We shall explore now the quantum-mechanical properties of an electron (otherwise non-interacting) in the presence of the crystal lattice potential. In order to set the stage for that, we start by considering the cases of a free electron and of a non-periodic potential.

A free electron has Hamiltonian

$$H_0 = \frac{\mathbf{P}^2}{2m} \tag{1.30}$$

and is obviously invariant under arbitrary spatial translations by $\mathbf{a}$. As a consequence, the translation operator $T(\mathbf{a}) = e^{i\frac{\mathbf{P}}{\hbar}\cdot\mathbf{a}}$ commutes with H_0 and its eigenvalues $e^{i\mathbf{q}\cdot\mathbf{a}}$, where $\mathbf{q}$ are arbitrary wave-vectors, are constants of motion.

Adding an arbitrary potential $V(\mathbf{r})$ to H_0 would, in general, break such invariance, causing the wave-vector $\mathbf{q}$ to be no longer a conserved quantity. Nevertheless, when the potential is invariant under translations by Bravais lattice vectors – namely, when $V(\mathbf{r}) = V(\mathbf{r} + \mathbf{R})$ – the Hamiltonian

$$H = \frac{\mathbf{P}^2}{2m} + V(\mathbf{r}) \tag{1.31}$$

becomes also invariant and commutes with $T(\mathbf{R})$. As a consequence, it is possible to find a set of energy eigenfunctions $\psi(\mathbf{r})$ that are also eigenfunctions of $T(\mathbf{R})$ and therefore satisfy

$$T(\mathbf{R})\psi(\mathbf{r}) = e^{i\mathbf{q}\cdot\mathbf{R}}\psi(\mathbf{r}). \tag{1.32}$$

We can always write an arbitrary wave-vector $\mathbf{q}$ as

$$\mathbf{q} = \mathbf{Q} + \mathbf{k}, \tag{1.33}$$

where $\mathbf{Q}$ belongs to the reciprocal lattice and $\mathbf{k}$ is in any of its primitive unit cells. As we shall see, there is a strong reason for choosing the first Brillouin zone, among the many possibilities, as the primitive unit cell of the reciprocal lattice.

Inserting (1.33) in (1.32), and using (1.12), we then realize that

$$T(\mathbf{R})\psi(\mathbf{r}) = \psi(\mathbf{r} + \mathbf{R}) = e^{i\mathbf{k}\cdot\mathbf{R}}\psi(\mathbf{r}). \tag{1.34}$$

We conclude that, for a periodic potential, all the conserved electron wave-vectors belong to the first Brillouin zone. Expression (1.34) is known as Floquet's Theorem.

1.4 Bloch's Theorem

Another important theorem specifies the general form of the energy eigenfunctions $\psi(\mathbf{r})$ for a periodic potential. This is Bloch's Theorem, according to which, for a potential invariant by translations of Bravais lattice vectors $\mathbf{R}$, the energy eigenfunctions are of the form

$$\psi_{\mathbf{k}}(\mathbf{r}) = e^{i\mathbf{k}\cdot\mathbf{r}}u(\mathbf{r}), \tag{1.35}$$

where $\mathbf{k}$ belongs to the first Brillouin zone and $u(\mathbf{r})$ has the same symmetry as the potential, namely $u(\mathbf{r}) = u(\mathbf{r}+\mathbf{R})$. Observe that (1.35) satisfies (1.34).

In order to demonstrate Bloch's Theorem, let us consider the energy eigenvalue problem for the Hamiltonian (1.31), when the potential is invariant under Bravais lattice translations,

$$\left[-\frac{\hbar^2}{2m}\nabla^2 + V(\mathbf{r})\right]\psi(\mathbf{r}) = E\psi(\mathbf{r}). \tag{1.36}$$

Let us take the Fourier transform of the above equation. For that, consider the expansions

$$\psi(\mathbf{r}) = \sum_{\mathbf{q}} \psi(\mathbf{q})e^{i\mathbf{q}\cdot\mathbf{r}} \tag{1.37}$$

and

$$V(\mathbf{r}) = \sum_{\mathbf{Q}} V(\mathbf{Q})e^{i\mathbf{Q}\cdot\mathbf{r}}, \tag{1.38}$$

where the last sum only sweeps vectors of the reciprocal lattice because of the symmetry of $V(\mathbf{r})$. Using the fact that the Fourier transform of a product is a convolution, we get

$$\frac{\hbar^2\mathbf{q}^2}{2m}\psi(\mathbf{q}) + \sum_{\mathbf{K}} V(\mathbf{K})\psi(\mathbf{q}-\mathbf{K}) = E\psi(\mathbf{q}), \tag{1.39}$$

where $\mathbf{K}$ is in the reciprocal lattice. Now, considering that an arbitrary wave-vector $\mathbf{q}$ can be written as in (1.33), we obtain the following set of coupled algebraic equations for the energy eigenvalue problem

$$\left[\frac{\hbar^2}{2m}|\mathbf{k}+\mathbf{Q}|^2 - E\right]\psi(\mathbf{k}+\mathbf{Q}) + \sum_{\mathbf{K}} V(\mathbf{K})\psi(\mathbf{k}+\mathbf{Q}-\mathbf{K}) = 0, \tag{1.40}$$

where $\mathbf{k}$ is in the first Brillouin zone, and $\mathbf{K}$ and $\mathbf{Q}$ in the reciprocal lattice.

Several conclusions can be drawn from the above set of equations. First of all, as expected, we see that $\mathbf{k}$ is a constant of motion. The interaction only couples to $\mathbf{k}$ wave-vectors that differ by reciprocal lattice translations and therefore are outside the first Brillouin zone. This can be clearly seen by taking the $\mathbf{Q} = 0$ component of (1.40).

It immediately follows that the Fourier components $\psi(\mathbf{k}+\mathbf{Q})$ correspond to a fixed $\mathbf{k}$ and must be summed over reciprocal lattice vectors $\mathbf{Q}$. This implies that expansion (1.37) for the energy eigenfunctions is, actually,

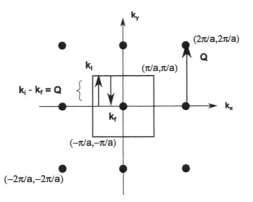

Figure 1.10 The "umklapp" process. By construction, the first Brillouin zone, being the Wigner–Seitz primitive unit cell of the reciprocal lattice, is such that any vector $\mathbf{k}_i$ at its boundaries will satisfy the von Laue condition with $\mathbf{k}_f = -\mathbf{k}_i$. Such backscattering processes open a gap precisely at the zone boundary.

$$\psi(\mathbf{r}) = \psi_{\mathbf{k}}(\mathbf{r}) = \sum_{\mathbf{Q}} \psi(\mathbf{k} + \mathbf{Q}) e^{i(\mathbf{k}+\mathbf{Q})\cdot\mathbf{r}} \tag{1.41}$$

$$= e^{i\mathbf{k}\cdot\mathbf{r}} u_{\mathbf{k}}(\mathbf{r}),$$

where $u_{\mathbf{k}}(\mathbf{r})$ is given by the sum of the $\mathbf{Q}$-dependent terms in the first line. Since this only includes the reciprocal lattice vectors, it follows that $u_{\mathbf{k}}(\mathbf{r})$ has the symmetry of the Bravais lattice. This completes the demonstration of Bloch's Theorem, Eq. (1.35), concerning the energy eigenfunctions of an electron in a periodic potential.

The second conclusion we can draw from (1.40) is that the energy eigenvalues are not changed by translating $\mathbf{k}$ by reciprocal lattice vectors. This means

$$E(\mathbf{k} + \mathbf{Q}) = E(\mathbf{k}), \tag{1.42}$$

namely, the energy eigenvalues are periodic functions of the reciprocal lattice vectors.

A third conclusion can be drawn from (1.40). This will make us realize why the first Brillouin zone (FBZ) is so important and is the chosen primitive unit cell of the reciprocal lattice. From the very construction of the Wigner–Seitz cell, it follows that the von Laue condition (1.26) for constructive interference is satisfied at the first Brillouin zone boundaries by the electron wave-vectors

$$\mathbf{k}_i = \frac{\mathbf{Q}_0}{2} \qquad \mathbf{k}_f = -\frac{\mathbf{Q}_0}{2}$$

$$\mathbf{k}_i - \mathbf{k}_f = \mathbf{Q}_0, \tag{1.43}$$

where $\mathbf{Q}_0$ is any reciprocal lattice vector pointing from a point to each of its neighbors.

This result means the electrons do not move away from the first Brillouin zone; when they approach any of its boundaries they are reflected back, with probability one, in a process that is known as "umklapp," a German word that means "to flip over." The FBZ is the only primitive unit cell of the reciprocal lattice that has this property.

The periodicity and continuity of the energy eigenvalues given by (1.42) imply that these must be bounded from above and from below. This observation, combined with the umklapp mechanism, leads to the conclusion that the possible energy eigenvalues are arranged in the form of allowed energy bands, which are separated by prohibited energy bands, called "gaps." These gaps, by virtue of (1.43), form right at the first Brillouin zone boundaries. It can be shown that each of the allowed energy bands contains as many eigenstates as the number of points in the corresponding Bravais lattice. Choosing the FBZ as the primitive unit cell of the reciprocal lattice has the enormous advantage that, because of the umklapp mechanism, the allowed bands correspond to the different Brillouin zones.

An important consequence of Bloch's Theorem is that, for any periodic potential, the energy eigenstates are extended wave-functions resembling plane waves. As a consequence of this fact, a solid with a partially filled energy band will be able to conduct electric charge when subjected to an external electric field. This type of solid will possess available extended states that can be occupied by the electrons upon application of the external field, thereby establishing an electric current. Conversely, a solid with a completely filled energy band has all the states of the first Brillouin zone occupied and, consequently, despite the fact that these are extended states, there will be no electric current because of the umklapp mechanism and Fermi–Dirac statistics.

These observations allow us to understand the behavior of metals, insulators and semiconductors, even within the independent electron approximation. The first ones have a partially filled energy band, and hence, available unoccupied states. These are separated from the occupied states by the so-called Fermi surface. Insulators and semiconductors, conversely, have a completely filled band, which is separated from the next allowed states by an energy gap. For semiconductors, this gap is of the order of $k_B T$ in such a way that the upper band can be populated by thermal activation. For insulators, the gap is much larger than $k_B T$, and this is not possible.

1.4.1 The Tight-Binding Approach

We turn now to the explicit determination of the energy eigenstates and the corresponding bands. We envisage a situation where a potential $v(\mathbf{r} - \mathbf{R}_i)$ is created at each site $\mathbf{R}_i$ of the Bravais lattice by a local atomic kernel (we consider, for simplicity the situation when the base has just one atom; the total potential is then

$V(\mathbf{r}) = \sum_{\mathbf{R}_i} v(\mathbf{r} - \mathbf{R}_i))$. The local potential has bound states $|n; \mathbf{R}_i\rangle$ with energies ϵ_n, $n = 1, 2, 3, \ldots$ An electron from the outermost shell is subject to this potential and, depending on the overlap of the local wave function $\Psi_n(\mathbf{r} - \mathbf{R}_i) = \langle \mathbf{r}|n; \mathbf{R}_i\rangle$ with its nearest neighbors, it can hop to neighboring atoms and thereby visit the whole crystal.

We may describe this system in two regimes. If the overlap, and consequently the hopping, is small, each of the allowed energy bands will be formed out of the same local eigenstates (same n), thus establishing a one-to-one correspondence among bands and localized states. This situation corresponds to the so-called tight-binding regime. The opposite situation, where there is a strong overlap and intense hopping, is called the "weak periodic potential" regime.

Let us start with the tight-binding approach. The following Hamiltonian captures the relevant features of the electronic properties for an electron in the nth bound state in the tight-binding regime:

$$H_{TB} = \epsilon_n \sum_{\mathbf{R}_i} |n; \mathbf{R}_i\rangle\langle n; \mathbf{R}_i| - t \sum_{\langle \mathbf{R}_i \mathbf{R}_j \rangle} |n; \mathbf{R}_i\rangle\langle n; \mathbf{R}_j|. \tag{1.44}$$

The second sum above runs over nearest neighbors of the Bravais lattice, and t is the hopping parameter, which strongly depends on the overlap of neighboring wave-functions.

The tight-binding Hamiltonian (1.44) remains unchanged by translating each of the Bravais lattice vectors $\mathbf{R}_i$ by an arbitrary vector of the same lattice; hence, it commutes with the operators implementing such translations,

$$[H_{TB}, T(\mathbf{R})] = 0 \quad ; \quad T(\mathbf{R}) = \exp\left\{\frac{i}{\hbar}\mathbf{P} \cdot \mathbf{R}\right\}. \tag{1.45}$$

We can, therefore, seek for common eigenstates of H_{TB} and $T(\mathbf{R})$.

Using the fact that

$$T(\mathbf{R})|n; \mathbf{R}_i\rangle = |n; \mathbf{R}_i - \mathbf{R}\rangle, \tag{1.46}$$

we can see that, for $\mathbf{k}$ in the first Brillouin zone, the states

$$|n; \mathbf{k}\rangle = \frac{1}{\sqrt{N}} \sum_{\mathbf{R}_i} e^{i\mathbf{k}\cdot\mathbf{R}_i} |n; \mathbf{R}_i\rangle \tag{1.47}$$

are eigenstates of the translation operator (N is the number of points in the Bravais lattice). Indeed,

$$T(\mathbf{R})|n; \mathbf{k}\rangle = e^{i\mathbf{k}\cdot\mathbf{R}}|n; \mathbf{k}\rangle, \tag{1.48}$$

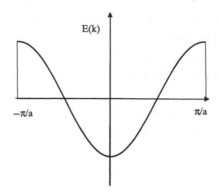

Figure 1.11 The tight-binding eigen-energy for a one-dimensional lattice with spacing a

and therefore $|n; \mathbf{k}\rangle$ are the candidates for energy eigenstates of the system. Then, applying the Hamiltonian (1.44) on the state (1.47), we get

$$H_{TB}|n; \mathbf{k}\rangle = \epsilon_n |n; \mathbf{k}\rangle - \frac{t}{\sqrt{N}} \sum_{\langle \mathbf{R}_i \mathbf{R}_j \rangle} e^{i\mathbf{k} \cdot \mathbf{R}_i} |n; \mathbf{R}_j\rangle. \qquad (1.49)$$

By changing the summation variable as $\mathbf{R}_i \to \mathbf{R}_i' = \mathbf{R}_i - \mathbf{R}_j$ we conclude that

$$H_{TB}|n; \mathbf{k}\rangle = E_n(\mathbf{k})|n; \mathbf{k}\rangle. \qquad (1.50)$$

The energy eigenvalues are given by

$$E_n(\mathbf{k}) = \epsilon_n - t \sum_{\mathbf{R}_i \in \mathbf{R}} e^{i\mathbf{k} \cdot \mathbf{R}_i}, \qquad (1.51)$$

where the sum runs over all nearest neighbors of a given Bravais lattice point $\mathbf{R}$. Note that the above expression is independent of $\mathbf{R}$ and is determined by the form of the Bravais lattice alone. For a one-dimensional lattice with spacing a, (1.51) yields

$$E_n(\mathbf{k}) = \epsilon_n - 2t \cos ka. \qquad (1.52)$$

This is depicted in Fig. 1.11.

For the two-dimensional square lattice we would have

$$E_n(\mathbf{k}) = \epsilon_n - 2t \left[\cos k_x a + \cos k_y a \right], \qquad (1.53)$$

with an obvious generalization for the three-dimensional cubic lattice.

It is instructive to explore the consequences of the results we just found for the energy eigenvalues within the tight-binding approach. Take as an example the square lattice, supposing there is one electron per site and N sites. Since each of

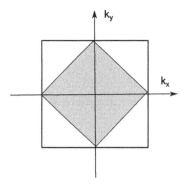

Figure 1.12 Fermi "surface" (line) separating the occupied from non-occupied states of the first Brillouin zone of a square lattice, at half-filling, in the tight-binding approach

the N available states can be occupied by two electrons with different spin orientations, we conclude that the band is half-filled. An important task, then, is to find what is the occupied part of the first Brillouin zone or, equivalently, what is the Fermi "surface" (line) separating the occupied states from the empty ones. Since the states in (1.53) are distributed symmetrically around ϵ_n, we conclude that the Fermi energy must correspond to $\epsilon_F = \epsilon_n$, or

$$\cos k_x a + \cos k_y a = 2\cos\left[(k_x + k_y)\frac{a}{2}\right]\cos\left[(k_x - k_y)\frac{a}{2}\right] = 0$$

$$k_y = \pm k_x \pm \frac{\pi}{a}. \tag{1.54}$$

The four straight lines in (1.54), therefore, determine the Fermi line, which separates the occupied half of the first Brillouin zone from the unoccupied one (see Fig. 1.12). The system, in this case will clearly be a metal.

The result of the previous analysis goes beyond the example chosen. Indeed, we may conclude in general that materials with one active electron per site are metals. This is the case, for instance, of *Au, Ag, Cu* and *Na*.

Observe that, as we increase the hopping parameter t, the band becomes wider and eventually overlaps the next band. It follows that the tight-binding approach is no longer valid in this case and we must resort to a different method for determining the energy bands of a crystal. The appropriate alternative approach is the so-called weak periodic potential.

1.4.2 The Weak Periodic Potential (Large Hopping) Approach

We will consider, for the sake of simplicity, the one-dimensional lattice with spacing a. We look for solutions of the energy eigenvalue equation in wave-vector space, namely (1.40). In the free case, $V = 0$, using (1.33) we would get from (1.40),

$$E_0 = \frac{\hbar^2}{2m}|\mathbf{q}|^2. \tag{1.55}$$

The Fermi surface would then be a spherical surface (circumference in two dimensions).

The weak periodic potential approach is based on the assumption that, because of the strong hopping, the electron is quasi-free, hence the only effect of the potential is felt at the boundaries of the first Brillouin zone, where the von Laue condition applies.

Then, the only Fourier components contributing to the expansion of the eigenfunctions, at $k = \frac{\pi}{a}$ for instance, are precisely the ones evaluated at the boundaries of the first Brillouin zone, namely

$$\alpha_1 \equiv \psi\left(\frac{\pi}{a}\right) = \psi(k) \qquad \alpha_2 \equiv \psi\left(-\frac{\pi}{a}\right) = \psi\left(k - \frac{2\pi}{a}\right). \tag{1.56}$$

Then, making, respectively, $Q = 0$, $K = \frac{2\pi}{a}$ and $Q = K = -\frac{2\pi}{a}$ in (1.40), we obtain the following set of coupled algebraic equations:

$$\left[\frac{\hbar^2 k^2}{2m} - E\right]\psi(k) + V\left(\frac{2\pi}{a}\right)\psi\left(k - \frac{2\pi}{a}\right) = 0 \tag{1.57}$$

and

$$\left[\frac{\hbar^2\left(k - \frac{2\pi}{a}\right)^2}{2m} - E\right]\psi\left(k - \frac{2\pi}{a}\right) + V\left(-\frac{2\pi}{a}\right)\psi(k) = 0. \tag{1.58}$$

Calling $V_0 = V\left(\frac{2\pi}{a}\right)$ and observing that $V\left(-\frac{2\pi}{a}\right) = V_0^*$, because V is real, we can write the two equations above as

$$\begin{pmatrix} E_0(k) - E & V_0 \\ V_0^* & E_0(k - \frac{2\pi}{a}) - E \end{pmatrix} \begin{pmatrix} \alpha_1 \\ \alpha_2 \end{pmatrix} = 0. \tag{1.59}$$

The above equation will have nontrivial solutions, provided the matrix has the determinant equal to zero. This condition yields

$$E_\pm(k) = \frac{E_0(k) + E_0(k - \frac{2\pi}{a})}{2} \pm \left[\frac{\left[E_0(k) - E_0(k - \frac{2\pi}{a})\right]^2}{4} + |V_0|^2\right]^{1/2}. \tag{1.60}$$

For $k = \frac{\pi}{a}$, right at the first Brillouin zone boundary, we have

$$E_\pm\left(\frac{\pi}{a}\right) = E_0\left(\frac{\pi}{a}\right) \pm |V_0| \tag{1.61}$$

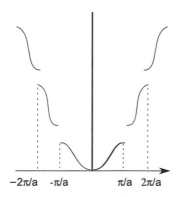

$-2\pi/a$ $-\pi/a$ π/a $2\pi/a$

Figure 1.13 The band structure obtained with the weak periodical potential approach for a one-dimensional lattice. Notice the (free) parabolic shape away from the zone boundaries.

and we see that a gap of width $2|V_0|$ opens at the zone boundary. This should be expected because of the umklapp mechanism. The resulting band structure is depicted in Fig. 1.13. There we can see the next two Brillouin zones for a one-dimensional lattice.

From (1.60), we may infer the general form of the energy eigenvalues close to the boundaries of the first Brillouin zone, for an arbitrary dimension. These are given by

$$E_\pm(\mathbf{k}) = \frac{1}{2}\left[E_0(\mathbf{k}) + E_0(\mathbf{k} - \mathbf{Q}_0)\right] \pm |V_0|. \tag{1.62}$$

Away from the zone boundaries, the energy will be given by the same expression as in the free case.

It is interesting to find out what would be the form of the Fermi surface within the present approach. This can be defined as the surface with a constant energy $E(\mathbf{k}) = E_F$. It follows that $\nabla_\mathbf{k} E(\mathbf{k})$ must be always perpendicular to the Fermi surface. Now, near the zone boundaries, according to (1.62),

$$\nabla_\mathbf{k} E_\pm(\mathbf{k}) = \frac{\hbar^2}{m}\left[\mathbf{k} - \frac{\mathbf{Q}_0}{2}\right]. \tag{1.63}$$

It is not difficult to see that this vector belongs to the boundary planes of the first Brillouin zone whenever $\mathbf{k}$ is in one of these planes. We conclude therefore that the Fermi surface is always perpendicular to the first Brillouin zone boundary planes. The only effect of the lattice on the Fermi surface, within the present approach, is to bend it in such a way that it becomes perpendicular to the zone boundaries. This observation will allow us to draw important conclusions about general electronic properties of solids.

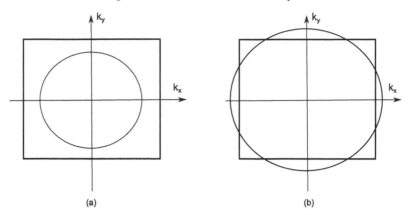

Figure 1.14 The free Fermi "surface" (line): (a) one electron per site, (b) two electrons per site

To give an example, we will consider the two-dimensional square lattice, which is easier to visualize. Let us assume firstly that the system possesses one electron per site. In this case, the area of the free Fermi circle, which gives the number of occupied states, must equal one-half of the first Brillouin zone area. In this case, the radius of the Fermi circle would be $\sqrt{\frac{2}{\pi}}\frac{\pi}{2a}$ and the Fermi circumference would never cross the zone boundaries, remaining therefore unaffected by the lattice (see Fig. 1.14a). The system is a metal.

For a system in the same lattice, but now having two active electrons per site, the area of the free Fermi circle must coincide with the area of the first Brillouin zone. Now, the Fermi circle radius would become larger than $\frac{\pi}{2a}$, namely, $\sqrt{\frac{4}{\pi}}\frac{\pi}{2a}$. Consequently, the Fermi circumference will trespass the first zone boundaries and there will be occupied states in the so-called second Brillouin zone, see Fig. 1.14b. The lattice effect, as we saw, will be to bend and fragment the Fermi circumference in such a way that it becomes perpendicular to the boundaries. This produces a fragmented Fermi line, parts of which are in the first Brillouin zone and parts of which are in the second, see Fig. 1.16. These "pockets" formed in the second zone, however, can be brought back to the first zone by performing translations with reciprocal lattice vectors. In this way, a second band will be formed within the first Brillouin zone.

The presence of a Fermi line in both bands indicates that, in spite of the fact that there are two electrons per site, the system is not an insulator. These materials are usually called semi-metals. The mechanism described here qualitatively allows one to understand, for instance, why the elements in the second column of the periodic table are not insulators in spite of their having two active electrons: they are semi-metals.

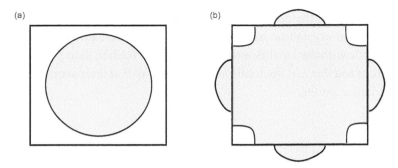

Figure 1.15 Fermi "surface" (line) separating occupied from non-occupied states of a square lattice, for an occupancy of (a) one electron per site; (b) two electrons per site. Both results are obtained in the weak periodic potential approach.

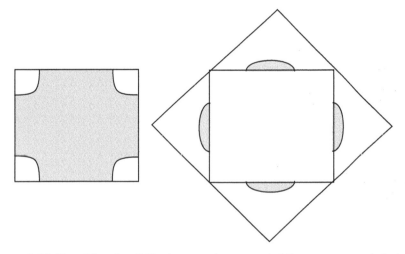

Figure 1.16 Fermi "surface" (line) separating occupied from non-occupied states of a square lattice, for an occupancy of two electrons per site in the weak periodic potential approach. On the left is the first Brillouin zone and on the right is the second Brillouin zone.

The size of the pockets would become smaller for systems with a larger $|V_0|$, thus implying the corresponding materials would present a poorer metallic character. In the limiting situation, the second band would disappear, the whole first Brillouin zone would be occupied, and the system would become an insulator. A system that qualitatively would fit this situation is $NaCl$, which has a cubic Bravais lattice with a base of two electrons per site and is an insulator.

We conclude this chapter by remarking how deep are the effects produced on the electronic properties of crystalline solids by the existence in such materials of a periodic potential, created by constituents that are assembled in a basic structure that repeats itself. In the next chapter we shall explore the fact that the basic constituents of a real crystalline material actually do not remain in the positions that would coincide with the ideal Bravais lattice points. Rather, their positions undergo both quantum and thermal fluctuations in such a way that their average corresponds to Bravais lattice points.

2

Vibrating Crystals

Real crystalline material systems are made of fundamental building blocks, which are typically "atomic kernels" – namely, atoms or ions deprived of their outermost electrons, which are subject to a mutual interaction potential. Depending on the range of external parameters such as temperature and pressure, these basic pieces will assemble themselves in different ways, such that in any case the energetically most favorable situation is achieved. Surprisingly, for a vast amount of materials at room temperature and pressure, the most stable configurations would be such that the average position of the "atomic kernels" coincides precisely with the ideal points of a Bravais lattice, possibly with a certain base. The actual location of the solid's basic constituents nevertheless oscillates around these equilibrium positions because of thermal and quantum fluctuations. Hence, since the physical properties of crystal materials are strongly influenced by the symmetry of the underlying lattice structure, one should expect that such oscillations would have a profound impact on the physical properties of the solid.

We shall consider in this chapter the dynamics of the oscillatory motion of the basic constituents of a material crystal, both from the classical and quantum-mechanical points of view.

2.1 The Harmonic Approximation

For the sake of clarity, in this first approach we will consider crystal structures with a base containing just one kernel or, in other words, just a Bravais lattice. Then we can specify the position of each of the atomic kernels by

$$\mathbf{X}(\mathbf{R}) = \mathbf{R} + \mathbf{r}(\mathbf{R}). \tag{2.1}$$

Classically, $\mathbf{r}(\mathbf{R})$, the relative position with respect to the closest Bravais lattice point $\mathbf{R}$, is a dynamical variable describing the position of an atomic kernel of mass M and momentum $\mathbf{P}(\mathbf{R}) = M\frac{d\mathbf{r}}{dt}(\mathbf{R})$. Notice that we must have $\langle \mathbf{r} \rangle = 0$ so that the average position coincides with the Bravais lattice points, namely $\langle \mathbf{X} \rangle = \mathbf{R}$.

A pair of atomic kernels does interact through a potential energy that depends on the mutual separation $\mathbf{X}(\mathbf{R}) - \mathbf{X}(\mathbf{R}')$. We assume only nearest-neighbors interactions will be relevant. Then, the total Hamiltonian describing the mechanical energy of the material lattice is therefore

$$H = \sum_{\mathbf{R}} \frac{\mathbf{P}^2(\mathbf{R})}{2M} + \sum_{\langle \mathbf{R}\mathbf{R}' \rangle} V\left(\mathbf{R} - \mathbf{R}' + \mathbf{r}(\mathbf{R}) - \mathbf{r}(\mathbf{R}')\right). \tag{2.2}$$

We assume the equilibrium configuration of the material lattice is the Bravais lattice, hence $\mathbf{R} - \mathbf{R}'$ must be a stable equilibrium point of the potential, and therefore

$$\left(\frac{\partial V(\mathbf{X})}{\partial X_i}\right)_{\mathbf{X}=\mathbf{R}-\mathbf{R}'} = 0 \quad i = 1, 2, 3 \tag{2.3}$$

and the Hessian matrix

$$K_{ij} = \left(\frac{\partial^2 V(\mathbf{X})}{\partial X_i \partial X_j}\right)_{\mathbf{X}=\mathbf{R}-\mathbf{R}'} \quad i = 1, 2, 3 \tag{2.4}$$

must have only positive eigenvalues.

We envisage a situation where the system is not too far from the equilibrium configuration, hence we may expand the potential energy (2.2) in $\mathbf{r}(\mathbf{R}) - \mathbf{r}(\mathbf{R}')$ around the equilibrium point $\mathbf{R} - \mathbf{R}'$. Going up to the second order, namely, the first yielding a nontrivial result, we obtain, up to a constant,

$$H = \sum_{\mathbf{R}} \frac{\mathbf{P}^2(\mathbf{R})}{2M} + \frac{1}{2} \sum_{\langle \mathbf{R}\mathbf{R}' \rangle} \left[\mathbf{r}(\mathbf{R}) - \mathbf{r}(\mathbf{R}')\right]_i K_{ij} \left[\mathbf{r}(\mathbf{R}) - \mathbf{r}(\mathbf{R}')\right]_j + \dots \tag{2.5}$$

This is known as the harmonic approximation. It always appears as the first term whenever we expand any potential about a stable equilibrium point. We shall see that for a great amount of applications it will provide a good description of the oscillatory motion of a crystal. For some specific purposes, however, we must go beyond it.

One familiar physical process that requires the inclusion of higher-than-harmonic terms is the thermal expansion. Indeed, a harmonic potential is symmetric about the origin; hence, the average positions of the lattice basic constituents are fixed at the Bravais lattice points, irrespective of the total energy. Consequently, as the temperature increases and the total energy becomes larger, the average positions of the lattice constituents would remain the same. The overall volume of the sample, therefore, would not exhibit any dependence on the temperature. Conversely, by adding anharmonic terms, such as the trilinear term of the expansion above, we would break the symmetry of the potential about the Bravais lattice points. Even though these would continue as the stable equilibrium positions, now the average relative distance of the lattice constituents would in general increase as the temperature and, correspondingly the average energy, are raised. This would provide an explanation for the phenomenon of thermal expansion. Any anomalous behavior, where the volume decreases with the temperature, can also be described by choosing the anharmonic terms in such a way that the average relative distance decreases with the temperature.

2.2 Classical Description of Crystal Oscillations

We now use the harmonic approximation to study the classical dynamics of a particular material lattice. For simplicity, we shall consider a three-dimensional cubic lattice with spacing a, having just one atomic kernel of mass M at each site. In this case we have in (2.4), $K_{ij} = K\delta_{ij}$.

The Hamiltonian (2.5), then, can be written as

$$H = \sum_{\mathbf{R}} \frac{\mathbf{P}^2(\mathbf{R})}{2M} + \frac{1}{2}K \sum_{\langle \mathbf{RR'} \rangle} \left| \mathbf{r}(\mathbf{R}) - \mathbf{r}(\mathbf{R'}) \right|^2. \tag{2.6}$$

The classical equation of motion corresponding to the Hamiltonian (2.6) is

$$M\frac{d^2 \mathbf{r}_i(\mathbf{R})}{dt^2} = -\frac{\partial H}{\partial \mathbf{r}_i(\mathbf{R})} = -K\left[2\mathbf{r}_i(\mathbf{R}) - \sum_{\mathbf{R'} \in \mathbf{R}} \mathbf{r}_i(\mathbf{R'}) \right], \tag{2.7}$$

where the sum runs over the nearest neighbors of the site $\mathbf{R}$ of a cubic lattice.

The solution has three modes, $s = x, y, z$,

$$\mathbf{r}(\mathbf{R}) = C\mathbf{e}_s \, e^{i(\mathbf{k}\cdot\mathbf{R} - \omega_s(\mathbf{k})t)}, \tag{2.8}$$

where $\mathbf{e}$ is a unit vector indicating the oscillation direction. Each mode has an angular frequency given by

$$\omega_s(\mathbf{k}) = 2\sqrt{\frac{K}{M}} \left| \sin\left(\frac{k_s a}{2} \right) \right|. \tag{2.9}$$

Notice that the solution (2.8) remains unaltered by shifting $\mathbf{k} \to \mathbf{k} + \mathbf{Q}$, where the last vector belongs to the reciprocal lattice. For a cubic lattice, this is equivalent to shifting each component as $k_i \to k_i + \frac{2n\pi}{a}$ for $n \in \mathbb{Z}$ and we see that the frequency is accordingly invariant under this shift, namely $\omega(\mathbf{k}) = \omega(\mathbf{k} + \mathbf{Q})$. These considerations lead us to a surprising conclusion: the vector $\mathbf{k}$ in the two previous equations is in the first Brillouin zone. By imposing periodic boundary conditions, we can see that this contains N different values of $\mathbf{k}$, for a Bravais lattice with N points.

For small $|\mathbf{k}|$, the angular frequency modes (2.9) reduce to $\omega_i(\mathbf{k}) = \sqrt{\frac{K}{M}}|k_i|a$, $i = x, y, z$, and vanish in the limit $|\mathbf{k}| \to 0$. These are called acoustic modes. It can be verified that when the crystal structure contains a base, in general there are also solutions with a frequency such that $\omega(\mathbf{k} = 0) = \omega_0 \neq 0$. These are the so-called optical modes.

The physical position of the kernel at $\mathbf{R}$ is actually given by the real part of the solution (2.8). We see that it oscillates about $\mathbf{R}$ in such a way that, on the average, the kernel position coincides with the Bravais lattice points, or equivalently, $\langle \mathbf{r} \rangle = 0$. Deviations from the average might be important, however a quantum-mechanical treatment thereof is most certainly needed. This is the purpose of the next section.

2.3 Quantum Description of Crystal Oscillations

Let us consider here the quantum-mechanical description of the material crystal associated to the Hamiltonian (2.6), within the harmonic approximation. According to the postulates of quantum mechanics, the dynamical coordinates and momenta become operators acting on a Hilbert space labeled by the Bravais lattice points $\mathbf{R}$: respectively, $\mathbf{r}_{op}(\mathbf{R})$ and $\mathbf{P}_{op}(\mathbf{R})$. We then have the canonical commutation rules (omitting henceforth the subscript "op"),

$$[r_i(\mathbf{R}), P_j(\mathbf{R}')] = i\hbar\delta_{\mathbf{R}\mathbf{R}'}\delta_{ij} \quad [r_i(\mathbf{R}), r_j(\mathbf{R}')] = [P_i(\mathbf{R}), P_j(\mathbf{R}')] = 0. \quad (2.10)$$

From (2.6) we see that, within the harmonic approximation, each atomic kernel behaves as a harmonic oscillator subject to the elastic forces exerted by the nearest neighbors. For a harmonic oscillator, it is convenient to introduce the operators

$$a_s(\mathbf{R}) = \sqrt{\frac{M\omega_s(\mathbf{k})}{2\hbar}}\mathbf{r}(\mathbf{R}) \cdot \mathbf{e} + i\sqrt{\frac{1}{2\hbar M\omega_s(\mathbf{k})}}\mathbf{P}(\mathbf{R}) \cdot \mathbf{e} \quad (2.11)$$

and

$$a_s(\mathbf{k}) = \frac{1}{\sqrt{N}}\sum_{\mathbf{R}}e^{-i\mathbf{k}\cdot\mathbf{R}}a_s(\mathbf{R}). \quad (2.12)$$

In the above expressions, $\mathbf{k}$ is in the first Brillouin zone, $\mathbf{R}$ are Bravais lattice points and the subscript s denotes the oscillation mode.

From (2.12), we have

$$\mathbf{r}(\mathbf{R}) = \frac{1}{\sqrt{N}}\sum_{\mathbf{k},s}e^{i\mathbf{k}\cdot\mathbf{R}}\sqrt{\frac{\hbar}{2M\omega_s(\mathbf{k})}}\left[a_s(\mathbf{k}) + a_s^\dagger(-\mathbf{k})\right]\mathbf{e}_s \quad (2.13)$$

and

$$\mathbf{P}(\mathbf{R}) = \frac{1}{\sqrt{N}} \sum_{\mathbf{k},s} e^{i\mathbf{k}\cdot\mathbf{R}} \sqrt{\frac{\hbar M \omega_s(\mathbf{k})}{2}} i \left[a_s(\mathbf{k}) - a_s^\dagger(-\mathbf{k}) \right] \mathbf{e}^s, \qquad (2.14)$$

which satisfy commutation rules (2.10).

These imply the following relations:

$$[a_r(\mathbf{k}), a_s^\dagger(\mathbf{k}')] = \delta_{\mathbf{k}\mathbf{k}'}\delta_{rs} \quad [a_r(\mathbf{k}), a_s(\mathbf{k}')] = [a_r^\dagger(\mathbf{k}), a_s^\dagger(\mathbf{k}')] = 0. \qquad (2.15)$$

The harmonic Hamiltonian (2.6) can be expressed in terms of the above operators as

$$H = \sum_{\mathbf{k},s} \hbar\omega_s(\mathbf{k}) \left[a_s^\dagger(\mathbf{k})a_s(\mathbf{k}) + \frac{1}{2} \right] \equiv \sum_{\mathbf{k},s} H_{\mathbf{k},s}. \qquad (2.16)$$

The above sum runs over the first Brillouin zone, which in the present case contains N different wave-vectors $\mathbf{k}_i$, $i = 1, \ldots, N$ and over the different modes s.

In order to solve the eigenvalue problem for H, let us define

$$|n(\mathbf{k}, s)\rangle = \frac{[a_s^\dagger(\mathbf{k})]^{n(\mathbf{k},s)}}{\sqrt{n(\mathbf{k}, s)!}} |0(\mathbf{k}, s)\rangle \qquad n(\mathbf{k}, s) = 0, 1, 2, \ldots \qquad (2.17)$$

Then, we have

$$H_{\mathbf{k}_i,s} |n(\mathbf{k}_i, s)\rangle = \left[n(\mathbf{k}_i, s) + \frac{1}{2} \right] \hbar\omega_s(\mathbf{k}_i) |n(\mathbf{k}_i, s)\rangle, \qquad (2.18)$$

where $H_{\mathbf{k}_i,s}$ is implicitly defined in (2.16). Then, it immediately follows that the direct product of states

$$|\{n(\mathbf{k}, s)\}\rangle = \prod_s \prod_{\mathbf{k}_1}^{\mathbf{k}_N} |n(\mathbf{k}, s)\rangle \qquad (2.19)$$

is an eigenstate of the total Hamiltonian, namely

$$H|\{n(\mathbf{k}, s)\}\rangle = E(\{n(\mathbf{k}, s)\})|\{n(\mathbf{k}, s)\}\rangle, \qquad (2.20)$$

with energy eigenvalues given by

$$E(\{n(\mathbf{k}, s)\}) = \sum_{i=1}^N \sum_s \left[n(\mathbf{k}_i, s) + \frac{1}{2} \right] \hbar\omega_s(\mathbf{k}_i). \qquad (2.21)$$

The eigenstates of the above Hamiltonian, therefore, can be labeled by a set of N natural numbers, each one corresponding to a different $\mathbf{k}_i$ in the first Brillouin zone, for each mode s:

$$\{n(\mathbf{k}, s)\} = \{n(\mathbf{k}_1, s) \ldots n(\mathbf{k}_N, s)\}. \qquad (2.22)$$

According to (2.21), the number $n(\mathbf{k}_i, s)$ represents the quantity of quantum excitations with a wave-vector $\mathbf{k}_i$, $i = 1, \ldots, N$, each one with an energy $E_{i,s} = \hbar\omega_s(\mathbf{k}_i)$, for each of the frequency modes. These quantum excitations are called "phonons" and (2.17) represents a quantum state with $n(\mathbf{k}_i, s)$ of them. The additive character of the corresponding energy eigenvalues, according to (2.21), indicates that the phonons are non-interacting. We shall see that this is a consequence of the harmonic approximation. Should we go beyond it by including higher order terms in the potential energy expansion (2.5), the phonons would then interact among themselves.

From (2.17), we see that

$$a_s^\dagger(\mathbf{k})|n(\mathbf{k}, s)\rangle = \sqrt{[n(\mathbf{k}, s) + 1]}|n(\mathbf{k}, s) + 1\rangle. \qquad (2.23)$$

Also, from (2.15), we have

$$a_s(\mathbf{k})|n(\mathbf{k}, s)\rangle = \sqrt{n(\mathbf{k}, s)}|n(\mathbf{k}, s) - 1\rangle. \qquad (2.24)$$

We see that the action of the $a^\dagger$ and a operators on the n-phonon state (2.17) has the effect of either increasing or reducing the number of phonons in that state by one unit. For this reason, they are respectively called phonon creation and annihilation operators. $|0(\mathbf{k}, s)\rangle$ is the state with no phonons with wave-vector $\mathbf{k}$ and mode s. The ground state is the one with no phonons for any $\mathbf{k}$. For that reason, it is usually called the vacuum.

The quantum-mechanical description of the crystalline vibrations of a solid is effectively done in terms of phonon elementary excitations. These are non-interacting within the harmonic approximation but become mutually interacting when we add higher-than-quadratic terms to the harmonic Hamiltonian. As an example of the use of phonons for describing quantum elementary processes in a solid, let us consider the neutron scattering experiments. In these, a beam of neutrons is reflected by the multiple planes of the crystal and the resulting interference spectrum of the reflected beam is measured. The neutrons strongly interact with the nuclei of the atomic kernels of the solid. The interaction potential must depend on $\mathbf{X}_N - \mathbf{X}(\mathbf{R})$, which are, respectively, the position operators of the neutron and of the atomic kernels, the last one given by (2.1). The complete Hamiltonian will be

$$H_T = \frac{\mathbf{P}_N^2}{2M_N} + \sum_{\mathbf{R}} V(\mathbf{X}_N - (\mathbf{R} + \mathbf{r}(\mathbf{R}))) + H, \qquad (2.25)$$

where $\mathbf{P}_N$ and M_N are the neutron momentum and mass, the sum is over the Bravais lattice and H is the harmonic Hamiltonian, (2.6).

The total Hamiltonian in (2.25) is invariant under the quantum-mechanical simultaneous translation operations

$$T^\dagger(\mathbf{R}_0)\mathbf{X}_N T(\mathbf{R}_0) = \mathbf{X}_N + \mathbf{R}_0 \tag{2.26}$$

and

$$\tau^\dagger(\mathbf{R}_0)\mathbf{r}(\mathbf{R})\tau(\mathbf{R}_0) = \mathbf{r}(\mathbf{R} - \mathbf{R}_0). \tag{2.27}$$

The translation operator in the neutron sector is clearly given by

$$T(\mathbf{R}_0) = \exp\left\{-\frac{i}{\hbar}\mathbf{P}_N \cdot \mathbf{R}_0\right\}. \tag{2.28}$$

The crystal translation operator, conversely, can be shown to be

$$\tau(\mathbf{R}_0) = \exp\{-i\,\mathbf{K}_l \cdot \mathbf{R}_0\}, \tag{2.29}$$

where

$$\mathbf{K}_l = \sum_{\mathbf{k},s} \mathbf{k}\, a_s^\dagger(\mathbf{k})a_s(\mathbf{k}). \tag{2.30}$$

This satisfies the eigenvalue equation

$$\mathbf{K}_l|\{n(\mathbf{k}, s)\}\rangle = \sum_{\mathbf{k},s} n(\mathbf{k}, s)\mathbf{k}|\{n(\mathbf{k}_s)\}\rangle, \tag{2.31}$$

where the eigenstates $|\{n(\mathbf{k}, s)\}\rangle$ are given by (2.19). The combined translations $T(\mathbf{R}_0)\tau(\mathbf{R}_0)$ must commute with the total Hamiltonian implying that their composite eigenvalues must be conserved in any process. In a neutron scattering experiment with initial and final neutron momentum, given, respectively, by $\mathbf{p}_i$ and $\mathbf{p}_f$, therefore, we would have

$$\exp\left\{-i\left(\frac{\mathbf{p}_i}{\hbar} + \sum_{\mathbf{k},s} n_i(\mathbf{k}, s)\mathbf{k}\right) \cdot \mathbf{R}_0\right\} = \exp\left\{-i\left(\frac{\mathbf{p}_f}{\hbar} + \sum_{\mathbf{k},s} n_f(\mathbf{k}, s)\mathbf{k}\right) \cdot \mathbf{R}_0\right\}. \tag{2.32}$$

It follows from the above equation that

$$\frac{\mathbf{p}_f}{\hbar} - \frac{\mathbf{p}_i}{\hbar} = \mathbf{Q} - \sum_{\mathbf{k},s} \Delta n(\mathbf{k}, s)\mathbf{k}, \tag{2.33}$$

where $\mathbf{Q}$ belongs to the reciprocal lattice. This expression reduces to the von Laue condition, Eq. (1.26) when $\Delta n(\mathbf{k}) = 0$; then phonons are neither created nor destroyed and the neutron scattering process is, therefore, elastic. When $\Delta n(\mathbf{k}, s) \neq 0$, however, phonons are either created or destroyed and the scattering process is, then, inelastic. One phonon inelastic scattering is a very useful method for the experimental determination of the phonon frequency spectrum. Indeed, in this case from (2.33), $\omega(\mathbf{k}) = \omega(\mathbf{k} + \mathbf{Q}) = \omega(\frac{\mathbf{p}_f - \mathbf{p}_i}{\hbar})$. Since we must also have conservation of energy, then

$$\hbar\omega\left(\frac{\mathbf{p}_f - \mathbf{p}_i}{\hbar}\right) = \pm\frac{\mathbf{p}_f^2 - \mathbf{p}_i^2}{2M_N} \qquad (2.34)$$

where the two signs correspond, respectively, to one phonon absorption or emission. By measuring the peaks in the reflected neutron momentum $\mathbf{p}_f$, for a given incident $\mathbf{p}_i$, one is able to determine the phonon frequency for all points of the first Brillouin zone.

An analogous experimental technique is the inelastic photon scattering, for which we would replace the neutron energy with the photon energy, $\hbar\omega$. In the case of optical phonons, this is called Raman scattering.

2.4 Thermodynamics of Phonons

The quantum-mechanical treatment of the oscillatory behavior of a system has a profound impact upon its thermodynamical properties. This was first shown by Planck in 1900, in his treatment of the blackbody radiation. Inspired by Planck's work, Einstein then proposed, in 1905, for the first time a full quantum description of the electromagnetic radiation field and thus explained the photoelectric effect. In this way, Einstein effectively created the embryo of quantum field theory, around 20 years before its actual birth. Two years later, in 1907, he applied the same quantum-mechanical principle to the elastic oscillatory motion of a crystal, effectively introducing the concept of phonons, and thereby determined the specific heat of a crystal [1]. This was the first application of (embryonic) quantum field theory to a condensed matter system. It was also one of the first great successes of quantum theory. We conclude this chapter by examining the thermodynamics of phonons, in a tribute to that magnificent work.

One of the challenges in the early days of the physics of solids was to understand the experimental behavior of the elastic specific heat of a solid. This was in complete disagreement with the classical treatment, based, for instance, on (2.6), which predicted a constant result: $c_V = \frac{N}{V}k_B$, per crystal vibration mode, the so-called Dulong–Petit law. The experimental results, conversely, indicated that c_V would vanish in the regime of low temperatures. The starting point for determining the thermodynamical properties of a system is the partition function. Within a quantum-mechanical treatment, this is given by

$$Z = \text{Tr}\,e^{-\beta H} = \sum_m \langle m|e^{-\beta H}|m\rangle = \sum_m g(E_m)e^{-\beta E_m}, \qquad (2.35)$$

where $\beta = 1/k_B T$, the states $|m\rangle$ form a complete set of energy eigenstates with eigenvalues E_m and degeneracy $g(E_m)$. The sums above run over the states in the complete set (not over the eigenvalues!). For the harmonic Hamiltonian (2.6) or (2.16), the energy eigenstates and eigenvalues are given respectively by (2.19) and (2.21). Since all phonons with a given wave-vector are identical, the phonon states are completely characterized by the set of numbers (2.22); hence, the sum over phonon states consists in summing over each one of the phonon numbers $n(\mathbf{k}_i, s)$, for every mode s and $i = 1, \ldots, N$. The partition function then becomes

$$Z = \prod_{\mathbf{k}_i} \prod_s \sum_{n(\mathbf{k}_i,s)=0}^{\infty} \exp\left\{-\beta\left[n(\mathbf{k}_i, s) + \frac{1}{2}\right]\hbar\omega_s(\mathbf{k}_i)\right\}$$

$$Z = \frac{1}{2}\prod_{\mathbf{k}_i}\prod_s\left[\frac{1}{\sinh\left(\frac{\beta\hbar\omega_s(\mathbf{k}_i)}{2}\right)}\right]. \tag{2.36}$$

From the partition function we can get the internal energy, U:

$$U = -\frac{\partial}{\partial\beta}\ln Z = \sum_{\mathbf{k}_i,s}\hbar\omega_s(\mathbf{k}_i)\left[\frac{1}{e^{\beta\hbar\omega_s(\mathbf{k}_i)} - 1} + \frac{1}{2}\right]. \tag{2.37}$$

In the above equation, we recognize the Bose–Einstein average phonon number as the first term in the expression between brackets, so the sum gives the average total energy of the phonons. The second term is a temperature-independent ground-state energy.

Einstein modeled the phonon frequency spectrum by assuming a constant frequency: $\omega_s(\mathbf{k}_i) = \omega_E$, called the Einstein frequency, an approach that is appropriate for the optical modes.

In this case, (2.37) yields (assuming there are three modes)

$$U_E = 3N\hbar\omega_E\left[\frac{1}{e^{\beta\hbar\omega_E} - 1} + \frac{1}{2}\right], \tag{2.38}$$

whereupon we obtain the specific heat in the Einstein model,

$$c_V = \frac{1}{V}\frac{\partial U_E}{\partial T} = 3\frac{N}{V}k_B\left[\frac{\beta\hbar\omega_E}{2\sinh\frac{\beta\hbar\omega_E}{2}}\right]^2. \tag{2.39}$$

This provides a correct qualitative description of the elastic specific heat of a solid. The low-temperature behavior of the specific heat, however, is dominated by the contribution of the acoustic modes.

Following Einstein's work, Debye proposed in 1912 a model for the acoustic modes where $\omega_s(\mathbf{k}_i) = c_s|\mathbf{k}_i|$. As we saw before, the possible $\mathbf{k}_i$-values are contained in the first Brillouin zone, hence their number is determined by the number of sites N in such a way that the above linear interpolation has an upper bound at the zone boundary. This maximal phonon frequency is known as the Debye frequency, ω_D. Using this model, Debye re-calculated the specific heat c_V. This is in excellent agreement with the experiments, down to low temperatures. At high temperatures, both Einstein and Debye models reduce to the classical result: the law of Dulong–Petit.

Lattice vibrations, especially in the quantum-mechanical form of phonon quantum excitations, play a fundamental role in the physics of crystalline solids. We will see a variety of systems in which the quantum-mechanical treatment of the electron-phonon interactions reveals unsuspected properties of otherwise metallic electrons. These range from the formation of an insulating gap to superconductivity, besides the specific features of the resistivity itself. We shall approach the topic of electronic interactions in the next chapter.

3

Interacting Electrons

In our description of solids so far, we have neglected the electronic interactions. Even though this has produced a surprisingly good picture of crystalline solids, there are many features of these materials for which the inclusion of interactions is essential and may even radically change this picture. For the proper quantum-mechanical description of an interacting system with many electrons, it is convenient to employ a formalism similar to the one used for phonons in the previous chapter.

3.1 Quantum Theory of Many-Electron Systems

Let us consider an electron in a state $|\alpha\rangle \otimes |\chi\rangle$, which is in the space given by the direct product of the orbital and spin Hilbert spaces, within the Schrödinger picture. We assume that $\{|\alpha\rangle\}$ and $\{|\chi\rangle\}$ form complete sets in the corresponding Hilbert spaces. The associated spinor wave-function in the representation $|\mathbf{r}\rangle \otimes |\sigma\rangle$ of position and spin z-component $\sigma = \uparrow, \downarrow$ is

$$\phi_{\alpha\chi}(\mathbf{r}; \sigma) = \langle \mathbf{r}|\alpha\rangle\langle\sigma|\chi\rangle. \tag{3.1}$$

In a way analogous to our early approach with phonons, we introduce an operator that creates an electron in the state $|\alpha\rangle \otimes |\chi\rangle$,

$$|\alpha\rangle \otimes |\chi\rangle = c^\dagger_{\alpha\chi}|0\rangle, \tag{3.2}$$

where $|0\rangle$ is the state with no electrons that we call "the vacuum." Then, defining the operator

$$\psi_\sigma(\mathbf{r}) = \sum_{\alpha\chi} \phi_{\alpha\chi}(\mathbf{r}; \sigma)c_{\alpha\chi}, \tag{3.3}$$

we have

$$\psi_\sigma^\dagger(\mathbf{r})|0\rangle = \sum_{\alpha\chi} \phi_{\alpha\chi}^*(\mathbf{r};\sigma)c_{\alpha\chi}^\dagger|0\rangle$$

$$= \sum_{\alpha\chi} |\alpha\rangle\langle\alpha|\mathbf{r}\rangle \otimes |\chi\rangle\langle\chi|\sigma\rangle = |\mathbf{r}\rangle \otimes |\sigma\rangle \equiv |\mathbf{r};\sigma\rangle. \quad (3.4)$$

We see that the operator $\psi_\sigma^\dagger(\mathbf{r})$, acting on the vacuum, creates a position eigenstate for an electron with the z-component of its spin equal to σ. In case the $|\chi\rangle$ states themselves are eigenstates of S_z, the wave-function (3.1) would be proportional to $\delta_{\chi,\sigma}$ and consequently Eq. (3.3) would reduce to

$$\psi_\sigma(\mathbf{r}) = \sum_\alpha \langle\mathbf{r}|\alpha\rangle c_{\alpha\sigma}. \quad (3.5)$$

The normalized N-electrons' state with definite positions and spins, then, would be

$$|\mathbf{r}_1\sigma_1;\ldots;\mathbf{r}_N\sigma_N\rangle = \frac{1}{\sqrt{N!}}\psi_{\sigma_1}^\dagger(\mathbf{r}_1)\ldots\psi_{\sigma_N}^\dagger(\mathbf{r}_N)|0\rangle. \quad (3.6)$$

The electrons, having spin $1/2$, must be in a completely anti-symmetric state according to the spin-statistics theorem, hence the electron creation operators, unlike the phonons, must satisfy anti-commutation rules, namely

$$\{\psi_r(\mathbf{r}),\psi_s^\dagger(\mathbf{r}')\} = \delta_{\mathbf{rr}'}\delta_{rs} \quad \{\psi_r(\mathbf{r}),\psi_s(\mathbf{r}')\} = \{\psi_r^\dagger(\mathbf{r}),\psi_s^\dagger(\mathbf{r}')\} = 0, \quad (3.7)$$

and for a given set of electronic states $|\alpha_k\rangle$, $k = 1,2,3,\ldots$,

$$\{c_r(\alpha_i),c_s^\dagger(\alpha_j)\} = \delta_{ij}\delta_{rs} \quad \{c_r(\alpha_i),c_s(\alpha_j)\} = \{c_r^\dagger(\alpha_i),c_s^\dagger(\alpha_j)\} = 0. \quad (3.8)$$

The latter implies $(c_s^\dagger(\alpha_i))^2 = 0$, which means there can be no more than one electron with the same spin component in a given state, thereby enforcing Fermi–Dirac statistics.

The analog of the number eigenstates (2.17) and (2.19) is

$$|n_{1,s};n_{2,s};n_{3,s};\ldots\rangle = \prod_{s=\uparrow,\downarrow} \prod_k |n_{k,s}\rangle, \quad (3.9)$$

where

$$|n_{k,s}\rangle = [c_s^\dagger(\alpha_k)]^{n_{k,s}}|0_{k,s}\rangle \quad n_{k,s} = 0,1. \quad (3.10)$$

These are eigenstates of the number operator $N_{k,s} = c_s^\dagger(\alpha_k)c_s(\alpha_k)$, with eigenvalue $n_{k,s} = 0,1$, which therefore indicates the number of electrons with spin s in the state $|\alpha_k\rangle$, $k = 1,2,3,\ldots$

It can easily be verified that the operator $c_s^\dagger(\alpha_k)$ acting on a state with $n_{k,s} = 0$ yields the state with $n_{k,s} = 1$, whereas the operator $c_s(\alpha_k)$ acting on a state with

$n_{k,s} = 1$ yields the state with $n_{k,s} = 0$. For that reason, they are called electron creation and annihilation operators.

Let us consider the electrons to be inside a cubic box of volume V, with periodic boundary conditions. For these, we take the states $|\alpha_k\rangle$ as momentum eigenstates: $|\mathbf{k}\rangle$, with $\mathbf{p} = \hbar\mathbf{k}$ and

$$\langle \mathbf{r}|\alpha\rangle = \langle \mathbf{r}|\mathbf{k}\rangle = \frac{e^{-i\mathbf{k}\cdot\mathbf{r}}}{V^{1/2}}. \tag{3.11}$$

From (3.3), then, we get

$$\psi_\sigma(\mathbf{r}) = \sum_{\mathbf{k}} \frac{e^{-i\mathbf{k}\cdot\mathbf{r}}}{V^{1/2}} c_\sigma(\mathbf{k}),$$

$$c_\sigma(\mathbf{k}) = \int_V d^3r \frac{e^{i\mathbf{k}\cdot\mathbf{r}}}{V^{1/2}} \psi_\sigma(\mathbf{r}). \tag{3.12}$$

Starting from the corresponding classical expressions, we can use the formalism introduced above to obtain the operators associated to charge and spin densities and currents. For the first one, we have

$$\rho(\mathbf{r}) = \sum_\sigma \psi_\sigma^\dagger(\mathbf{r})\psi_\sigma(\mathbf{r}), \tag{3.13}$$

or, in momentum space, using (3.12) and the fact that the Fourier transform of a product is a convolution,

$$\rho(\mathbf{q}) = \sum_{\mathbf{p},\sigma} c_\sigma^\dagger(\mathbf{p})c_\sigma(\mathbf{p}+\mathbf{q})$$

$$\sum_{\mathbf{r}} \rho(\mathbf{r}) = \rho(\mathbf{q}=0) = \sum_{\mathbf{p},\sigma} c_\sigma^\dagger(\mathbf{p})c_\sigma(\mathbf{p}). \tag{3.14}$$

The associated current operator is

$$\mathbf{j} = -\frac{i\hbar}{2m} \sum_\sigma \left[\psi_\sigma^\dagger(\mathbf{r})\nabla\psi_\sigma(\mathbf{r}) - \nabla\psi_\sigma^\dagger(\mathbf{r})\psi_\sigma(\mathbf{r}) \right]. \tag{3.15}$$

For a U(1) invariant Hamiltonian, it satisfies a continuity equation: $\frac{\partial\rho}{\partial t} + \nabla\cdot\mathbf{j} = 0$.

The i-component of the spin density, accordingly, is given by

$$\rho_S^i(\mathbf{r}) = \frac{\hbar}{2} \sum_{\alpha\beta} \psi_\alpha^\dagger(\mathbf{r})\sigma_{\alpha\beta}^i\psi_\beta(\mathbf{r}), \tag{3.16}$$

whereas the associated spin current is

$$\mathbf{j}_S^i = -\frac{i\hbar}{2m}\left(\frac{\hbar}{2}\right) \sum_{\alpha\beta} \left[\psi_\alpha^\dagger(\mathbf{r})\sigma_{\alpha\beta}^i\nabla\psi_\beta(\mathbf{r}) - \nabla\psi_\alpha^\dagger(\mathbf{r})\sigma_{\alpha\beta}^i\psi_\beta(\mathbf{r}) \right]. \tag{3.17}$$

For a rotational invariant Hamiltonian, the spin current and density satisfy a continuity equation, which is similar to the one of charge.

3.2 Non-Interacting Electrons

3.2.1 Free Electrons

We start by applying the previous formalism in the situation where the electronic interaction is neglected. We will consider both the case of free electrons and that of non-interacting electrons in a crystal lattice in the tight-binding regime.

Considering non-interacting electrons with energy $\epsilon(\mathbf{k})$, as well as the number operator, it is not difficult to realize that the total energy operator must be

$$H_0 = \sum_{\mathbf{k},\sigma} \epsilon(\mathbf{k}) c_\sigma^\dagger(\mathbf{k}) c_\sigma(\mathbf{k}), \tag{3.18}$$

where, for free electrons, we have $\epsilon(\mathbf{k}) = \frac{\hbar^2 \mathbf{k}^2}{2m}$.

Using the relation

$$\frac{1}{V} \sum_{\mathbf{k}} \leftrightarrow \int \frac{d^3 k}{(2\pi)^3}, \tag{3.19}$$

resulting from the periodic boundary conditions, and (3.12), we can cast the free Hamiltonian in the form

$$H_0 = \sum_\sigma \int d^3 r d^3 r' \psi_\sigma^\dagger(\mathbf{r}) \left(-\frac{\hbar^2}{2m} \nabla^2 \right) \delta(\mathbf{r} - \mathbf{r}') \psi_\sigma(\mathbf{r}'). \tag{3.20}$$

It is easy to see that the energy eigenstates are given by (3.9) and the corresponding eigenvalues by

$$E_0 = \sum_{\mathbf{k},\sigma} \frac{\hbar^2 \mathbf{k}^2}{2m} n_{\mathbf{k},\sigma}. \tag{3.21}$$

3.2.2 The Tight-Binding Model

Another nice application of the present formalism is found in the tight-binding approach for electrons in a crystal. Indeed we can rewrite the Hamiltonian (1.44) in terms of electron creation and annihilation operators in the state $|n\rangle$ with energy ϵ_n, on the sites of a Bravais lattice.

Let us choose the sum in (3.5) in such a way that $\langle \mathbf{r}|\alpha\rangle \equiv \langle \mathbf{r}|i\rangle = \phi(\mathbf{r} - \mathbf{R}_i)$ is an atomic wave-function localized on the i-site of the Bravais lattice, namely

$$\psi_\sigma(\mathbf{r}) = \sum_i \phi(\mathbf{r} - \mathbf{R}_i) c_{i,\sigma}. \tag{3.22}$$

Writing

$$H_{TB} = \epsilon_n \int d\mathbf{r} \sum_\sigma \psi_\sigma^\dagger(\mathbf{r})\psi_\sigma(\mathbf{r}),$$

$$H_{TB} = \epsilon_n \sum_{ij,\sigma} \int d\mathbf{r}\phi^*(\mathbf{r} - \mathbf{R}_i)\phi(\mathbf{r} - \mathbf{R}_j)c_{i,\sigma}^\dagger c_{j,\sigma} + HC. \tag{3.23}$$

Now, the integral above is equal to one for $i = j$, for normalized atomic wave-functions, while the overlap integral for $i \neq j$ is only appreciable for nearest neighbors. Hence we can write

$$H_{TB} = \epsilon_n \sum_{i,\sigma} c_{i,\sigma}^\dagger c_{i,\sigma} - t \sum_{\langle ij\rangle,\sigma} \left[c_{i,\sigma}^\dagger c_{j,\sigma} + c_{j,\sigma}^\dagger c_{i,\sigma} \right], \tag{3.24}$$

where the hopping parameter is proportional to the overlap integral, for neighboring sites i, j:

$$t = -\epsilon_n \int d\mathbf{r}\phi^*(\mathbf{r})\phi(\mathbf{r} - (\mathbf{R}_j - \mathbf{R}_i)). \tag{3.25}$$

Then, using (3.12), we can put (3.26) in the form

$$H_{TB} = \sum_{\mathbf{k},\sigma} E_n(\mathbf{k})c_\sigma^\dagger(\mathbf{k})c_\sigma(\mathbf{k})$$

$$E_n(\mathbf{k}) = \epsilon_n - 2t \sum_{i=x,y,z} \cos k_i a, \tag{3.26}$$

where the last expression is meant for a cubic lattice. This was the previous result for the tight-binding energy in this case and should be compared with (3.18).

3.3 Electron-Electron Interactions: the Coulomb Interaction

Electrons, being electrically charged particles, will interact among themselves through the electromagnetic field. This is represented by a scalar potential φ and a vector potential $\mathbf{A}$, which are U(1) gauge fields. The interaction is introduced by imposing the principle of gauge invariance under the local U(1) group. This implies the replacement of regular time and space derivatives with the corresponding gauge-covariant ones, namely

$$\frac{\partial}{\partial t} \rightarrow \frac{\partial}{\partial t} + i\frac{e}{\hbar c}\varphi \quad ; \quad \nabla \rightarrow \nabla + i\frac{e}{\hbar c}\mathbf{A}. \tag{3.27}$$

The included electromagnetic field can be either external or generated by the electrons themselves. In the latter case it produces an effective interaction among the electrons. Frequently, only the static limit of this interaction is needed, since the speed of an electron in a solid is much less than the speed of light. In this case, one can show that the net effect of replacing (3.27) into (3.20), in the absence of external fields, is to add to the Hamiltonian H_0 a term

$$H_C = e \int d^3 r \rho(\mathbf{r}) \varphi(\mathbf{r}).$$

Considering that

$$-\nabla^2 \varphi(\mathbf{r}) = e\rho(\mathbf{r}),$$

we can write

$$H_C = \frac{e^2}{2} \sum_{\sigma,\sigma'} \int d^3 r d^3 r' \psi_\sigma^\dagger(\mathbf{r}) \psi_\sigma(\mathbf{r}) \left[\frac{1}{4\pi |\mathbf{r} - \mathbf{r'}|} \right] \psi_{\sigma'}^\dagger(\mathbf{r'}) \psi_{\sigma'}(\mathbf{r'}), \qquad (3.28)$$

where, in the last step, we used the Green function of the Laplacian operator to express the scalar potential $\varphi(\mathbf{r})$ in terms of the charge density. We see that in the static case the electromagnetic interaction just reduces to the familiar Coulomb interaction. In the expression above, $\psi_\sigma(\mathbf{r})$ is given by (3.5).

Let us consider a situation where the electrons undergoing the Coulomb interaction belong to two neighbor atoms, labeled $i = 1, 2$. In this case,

$$\psi_\sigma(\mathbf{r}) = \sum_{i=1,2} \phi(\mathbf{r} - \mathbf{R}_i) c_{i,\sigma}, \qquad (3.29)$$

where $\phi(\mathbf{r} - \mathbf{R}_i)$, $i = 1, 2$ are the corresponding atomic wave-functions.

Inserting (3.29) in (3.28), we see that 16 terms are generated. Out of these, four terms are proportional to the occupation number operator $n_{i,\sigma} = c_{i,\sigma}^\dagger c_{i,\sigma}$, four are proportional to $c_{1,\sigma}^\dagger c_{2,\sigma}$ and $c_{2,\sigma}^\dagger c_{1,\sigma}$ and eight either vanish or yield subleading contributions.

3.4 The Hubbard Model

The subset of terms of the Coulomb interaction provenient from $n_{i,\sigma} = c_{i,\sigma}^\dagger c_{i,\sigma}$ is given by

$$H_C' = U \sum_i n_{i,\uparrow} n_{i,\downarrow} + \frac{U}{2} \sum_{i,\sigma} n_{i,\sigma}, \qquad (3.30)$$

where

$$U = e^2 \int d^3 r d^3 r' |\phi(\mathbf{r})|^2 \left[\frac{1}{4\pi |\mathbf{r} - \mathbf{r'}|} \right] |\phi(\mathbf{r'})|^2 \qquad (3.31)$$

and we used the fact that $n_{i,\sigma}^2 = n_{i,\sigma}$, for fermions. Here, we neglected the nearest-neighbor density interaction terms, which would be proportional to

$$U_{ij} = e^2 \int d^3 r d^3 r' |\phi(\mathbf{r} + [\mathbf{R}_j - \mathbf{R}_i])|^2 \left[\frac{1}{4\pi |\mathbf{r} - \mathbf{r'}|} \right] |\phi(\mathbf{r'})|^2. \qquad (3.32)$$

These will clearly be subleading because of the fast decay of the atomic wave-functions $\phi(\mathbf{r})$.

In undoped systems, one frequently finds the situation where there is precisely one electron per site, either with spin up or down, the so-called half-filling. In this case we have, for each site, the constraint

$$\sum_{\sigma} n_{i,\sigma} = 1, \tag{3.33}$$

which is known as Gutzwiller projection. In this case, the last term in (3.30) becomes just a constant, being therefore trivial.

The Hubbard model Hamiltonian is constructed by adding the electronic interaction H_C', for all sites of the lattice at half-filling, to the tight-binding Hamiltonian, namely

$$H_U = -t \sum_{\langle ij \rangle, \sigma} \left[c_{i,\sigma}^\dagger c_{j,\sigma} + c_{j,\sigma}^\dagger c_{i,\sigma} \right] + U \sum_i n_{i,\uparrow} n_{i,\downarrow}. \tag{3.34}$$

The Hubbard model plays an important role in the description of the electronic interactions in condensed matter systems such as the high-Tc cuprate superconductors, for instance.

3.5 Exchange Interactions and Magnetism

Introduction

Magnetism at a macroscopic scale is one of the most spectacular phenomena in physics, known from ancient times as a property that some materials have to attract each other. We shall see that, surprisingly, it is the result of a quantum mechanical effect, which, because of a collective behavior, is amplified from the atomic to the macroscopic scale.

The very roots of magnetism stem from the fact that a charged particle such as an electron or a proton will produce a magnetic field whenever it is in a state of nonzero angular momentum. Mathematically, we express this result as the relation between the magnetic dipole moment and the angular momentum. An electron, for instance, will have an intrinsic magnetic dipole moment given by

$$\vec{\mu} = -g \frac{\mu_B}{\hbar} \mathbf{S}, \tag{3.35}$$

where $\mathbf{S}$ is the electron spin, $\mu_B = e/2mc$ is the Bohr magneton and g, the gyromagnetic factor. This is a pure number that is specific for the electron. For a proton, for instance, we would have a different g-factor, a positive sign and the proton mass replacing the electron mass m in the Bohr magneton. Curiously, the neutron in spite of being neutral and because of its internal structure exhibits a nonzero magnetic dipole moment with its own g-factor.

The Classic Magnetic Dipole Interaction

Consider an electron belonging to an atom placed on a site on a crystalline lattice. Let us investigate the kind of interaction with neighboring atoms that will cause the magnetic dipole moments of these individual electrons to cooperate in order to form a collective state with a macroscopic magnetic field.

Classically, a pair of magnetic dipole moments in positions separated by a vector **r** has an interaction energy given by

$$U_{12} = \frac{1}{r^3} \left[\vec{\mu}_1 \cdot \vec{\mu}_2 - 3(\vec{\mu}_1 \cdot \hat{r})(\vec{\mu}_2 \cdot \hat{r}) \right]. \tag{3.36}$$

This is minimized when the two moments are aligned with the vector **r**, with an energy $U_{12} = -\mu_1 \mu_2 / r^3$. This could in principle, depending on the crystal structure, provide the basic interaction responsible for the alignment of all magnetic moments in the crystal; however, a simple analysis reveals it simply cannot be. The above alignment energy for a typical crystal is on the order of 10^{-4}eV, which corresponds to a temperature on the order of 1K. This means the macroscopic magnetic order would be destroyed by thermal fluctuations above 1K, whereas it is known to exist well above room temperature. We shall see below that, actually, the Coulomb interaction and quantum-mechanical principles are responsible for the existence of macroscopic magnetic order.

The Exchange Interaction

Let us consider once again the Coulomb interaction (3.28) envisaged in a situation where the two interacting electrons belong to neighboring atoms, as described by (3.29). In the previous section we showed that, considering the terms proportional to $n_{i,\sigma} = c^{\dagger}_{i,\sigma} c_{i,\sigma}$, we were led to the Hubbard model. We take now the terms proportional to $c^{\dagger}_{1,\sigma} c_{2,\sigma}$ and $c^{\dagger}_{2,\sigma} c_{1,\sigma}$. These yield the so-called exchange interaction

$$H_J = J \sum_{\sigma,\sigma'} c^{\dagger}_{1,\sigma} c_{2,\sigma} c^{\dagger}_{2,\sigma'} c_{1,\sigma'}, \tag{3.37}$$

where

$$J = e^2 \int d^3r d^3r' \phi^*(\mathbf{r} - \mathbf{R}_1)\phi(\mathbf{r} - \mathbf{R}_2) \left[\frac{1}{4\pi |\mathbf{r} - \mathbf{r}'|} \right] \phi^*(\mathbf{r}' - \mathbf{R}_2)\phi(\mathbf{r}' - \mathbf{R}_1)$$

$$\tag{3.38}$$

is the exchange integral. Notice that this does not appear in the expression of the classical Coulomb interaction; rather, it is a consequence of the anti-symmetric nature of the 2-electron state-vector

$$|\mathbf{r}\sigma; \mathbf{r}'\sigma'\rangle = \frac{1}{\sqrt{2}} \psi_\sigma^\dagger(\mathbf{r}) \psi_{\sigma'}^\dagger(\mathbf{r}')|0\rangle. \tag{3.39}$$

Observe that J will increase as the overlap between the neighboring atomic wavefunctions is enlarged.

Using Eq. (3.16) for the spin density operator, we obtain the spin operator

$$\mathbf{S} = \frac{\hbar}{2} \sum_{\alpha\beta} \int d^3r \, \psi_\alpha^\dagger(\mathbf{r}) \vec{\sigma}_{\alpha\beta} \psi_\beta(\mathbf{r}),$$

$$\mathbf{S}_i = \frac{\hbar}{2} \sum_{\alpha\beta} c_{i,\alpha}^\dagger \vec{\sigma}_{\alpha\beta} c_{i,\beta} \qquad i = 1, 2. \tag{3.40}$$

Now, using the identity

$$\sum_{a=x,y,z} \sigma_{\alpha\beta}^a \sigma_{\mu\nu}^a = 2\delta_{\alpha\nu}\delta_{\beta\mu} - \delta_{\alpha\beta}\delta_{\mu\nu} \tag{3.41}$$

and Eq. (3.37), we see that the exchange interaction energy is given by

$$H_J = -\frac{2J}{\hbar^2} \mathbf{S}_1 \cdot \mathbf{S}_2 + \frac{4J}{\hbar^2} \left(\sum_\alpha n_{1,\alpha} \right) \left(\sum_\alpha n_{2,\alpha} \right). \tag{3.42}$$

We will consider here the situation where the constraint (3.33) applies, known as "localized magnetism" and appropriate for magnetic insulators. In this case, the exchange interaction

$$H_J = -\frac{2J}{\hbar^2} \mathbf{S}_1 \cdot \mathbf{S}_2 \tag{3.43}$$

is responsible for the magnetic interaction. Notice that for $J > 0$ it favors the parallel alignment of adjacent spins, whereas for $J < 0$, it would favor the antiparallel alignment thereof.

3.6 The Heisenberg Model

Isotropic Case

The extension of the previous expression to all sites of a crystal lattice leads to the celebrated Heisenberg model,

$$H_H = -\frac{2J}{\hbar^2} \sum_{ij} \mathbf{S}_i \cdot \mathbf{S}_j. \tag{3.44}$$

Evidently, the terms corresponding to the nearest neighbors will dominate the sum because the overlap integral in J will be maximal. For $J > 0$, the Heisenberg Hamiltonian will lead to ferromagnetic order, and for $J < 0$, to antiferromagnetic order. Since J is typically of the order of 1eV, these magnetically ordered states will survive thermal fluctuations up to temperatures way above room temperature.

Anisotropy

The Heisenberg Hamiltonian is rotationally invariant. Many systems, however, possess anisotropic magnetic properties and therefore cannot be described by this model. The main sources of anisotropy are the spin-orbit interaction and the crystal geometry itself. The former corresponds to an additional Hamiltonian $H_{SO} = \lambda \sum_i \mathbf{L}_i \cdot \mathbf{S}_i$, where $\mathbf{L}$ is the orbital angular momentum and $\mathbf{S}$ is the spin. H_{SO} is usually treated perturbatively in the orbital sector of the Hilbert space, hence, to first order $\Delta H_{(1)} = \lambda \sum_i \langle \Psi | \mathbf{L}_i | \Psi \rangle \cdot \mathbf{S}_i$.

In most materials, the crystal geometry is such that the angular part of the ground-state vector $|\Psi\rangle$, for $l = 1$, is one of the orbitals $|p_x\rangle = \frac{1}{\sqrt{2}}[|11\rangle + |1-1\rangle]$, $|p_y\rangle = \frac{1}{\sqrt{2}}[|11\rangle - |1-1\rangle]$ and $|p_z\rangle = |10\rangle$. The corresponding expressions for $l = 2, \ldots$ can be found in textbooks on atomic physics. It is not difficult to see that for these states $\langle \Psi | \mathbf{L}_i | \Psi \rangle = 0$, hence $\Delta H_{(1)} = 0$, and we must proceed to the second order, where the correction to the Hamiltonian is

$$\Delta H_{(2)} = \lambda^2 \sum_{ij} \sum_{n \neq 0} \frac{\langle \Psi | L_i^\alpha | n \rangle \langle n | L_j^\beta | \Psi \rangle}{\epsilon_0 - \epsilon_n} S_i^\alpha S_j^\beta$$
$$= \sum_{ij} \tilde{J}^{\alpha\beta} S_i^\alpha S_j^\beta. \tag{3.45}$$

Adding the second-order spin-orbit correction to the Heisenberg Hamiltonian, we get a quadratic form in $S_i^\alpha S_j^\beta$ with coupling $J^{\alpha\beta} = \tilde{J}\delta^{\alpha\beta} + \tilde{J}^{\alpha\beta}$. Diagonalizing the quadratic form, we get the XYZ-Hamiltonian

$$H_{XYZ} = \sum_{ij} \left[J_X S_i^X S_j^X + J_Y S_i^Y S_j^Y + J_Z S_i^Z S_j^Z \right], \tag{3.46}$$

which provides a general description of localized-moments magnetism. Important special cases are the Ising model, for which $J_X = J_Y = 0$, the XY-Model, for which $J_X = J_Y \neq 0$, and $J_Z = 0$ and the Heisenberg model, where $J_X = J_Y = J_Z$.

3.7 Electron-Phonon Interactions

Electrons naturally interact with the crystal lattice because the kernels occupying the lattice sites are usually non-neutral: either ions or charged radicals. The electron-lattice interaction is of fundamental importance in the description of transport, structural and thermodynamical properties of crystalline solids. It is also the one responsible for the phenomenon of superconductivity.

The general expression for the electron-lattice interaction energy is

$$H_{el} = \int d^3x \frac{1}{N} \sum_{\mathbf{R}_i} \rho(\mathbf{x}) V(\mathbf{x} - \mathbf{X}_i), \tag{3.47}$$

where $\rho(\mathbf{x})$ is the electronic density at $\mathbf{x}$ and $\mathbf{X}_i = \mathbf{R}_i + \mathbf{r}_i$ are the positions of the lattice kernels (we assume there is just one per site): $\mathbf{R}_i$ are the equilibrium positions, which coincide with the Bravais lattice sites, and $\mathbf{r}_i$ are the deviations with respect to these. $V(\mathbf{x} - \mathbf{X}_i)$ is the interaction potential between the electron and the crystal material constituents.

Making a Taylor series expansion in $\mathbf{r}_i \equiv \mathbf{r}(\mathbf{R}_i)$ about the Bravais lattice points $\mathbf{R}_i$, we get

$$H_{el} = \int d^3x \frac{1}{N} \sum_{\mathbf{R}_i} \rho(\mathbf{x}) \left[V(\mathbf{x} - \mathbf{R}_i) + \mathbf{r}(\mathbf{R}_i) \cdot \nabla_{\mathbf{R}_i} V(\mathbf{x} - \mathbf{R}_i) + \ldots \right]. \tag{3.48}$$

The first term corresponds to an electron in an ideal Bravais lattice. Using (3.12), (3.13) and (3.14), we can write it as

$$H_{el}^{(0)} = \sum_{\mathbf{k},\mathbf{k}',\sigma} c_{\mathbf{k},\sigma}^{\dagger} c_{\mathbf{k}',\sigma} \int d^3x \, e^{i(\mathbf{k}'-\mathbf{k})\cdot\mathbf{x}} \frac{1}{N} \sum_{\mathbf{R}_i} V(\mathbf{x} - \mathbf{R}_i). \tag{3.49}$$

The integral above is the Fourier transform of a function with the periodicity of the Bravais lattice, and therefore we must have

$$\mathbf{k}' - \mathbf{k} = \mathbf{Q}, \tag{3.50}$$

where $\mathbf{Q}$ is a vector of the reciprocal lattice. Consequently,

$$H_{el}^{(0)} = \sum_{\mathbf{k},\sigma} V(\mathbf{Q}) c_{\mathbf{k},\sigma}^{\dagger} c_{\mathbf{k}+\mathbf{Q},\sigma}. \tag{3.51}$$

Since $E(\mathbf{k}) = E(\mathbf{k} + \mathbf{Q})$, according to the Bloch theorem, this means the scattered electron will have the same energy it has before the scattering. On the other hand, physical momenta corresponding to wave-vectors satisfying the relation (3.50) are precisely the same. Hence we can now completely understand why an electron in a Bravais lattice propagates without being scattered: its momentum and energy simply are not changed as it propagates.

Now consider the first-order term. Using the quantum-mechanical expression for the lattice displacement in terms of phonon creation and annihilation operators, given by (2.13), we obtain

$$H_{el}^{(1)} = \frac{1}{\sqrt{N}} \sum_{\mathbf{k},\mathbf{k}',\mathbf{q},s,\sigma} c_{\mathbf{k},\sigma}^{\dagger} c_{\mathbf{k}',\sigma} \frac{1}{N} \sum_{\mathbf{R}_i} e^{i(\mathbf{k}'-\mathbf{k}+\mathbf{q})\cdot\mathbf{R}_i} \sqrt{\frac{\hbar}{2M\omega_s(\mathbf{q})}} \left[a_s(\mathbf{q}) + a_s^{\dagger}(-\mathbf{q}) \right]$$

$$\times \int d^3x \, e^{i(\mathbf{k}'-\mathbf{k})\cdot\mathbf{x}} \mathbf{e}_s \cdot \nabla_{\mathbf{R}_i} V(\mathbf{x} - \mathbf{R}_i). \tag{3.52}$$

Now, using the fact that

$$\int d^3x \, e^{i(\mathbf{k}'-\mathbf{k})\cdot\mathbf{x}} \nabla_{\mathbf{R}_i} V(\mathbf{x} - \mathbf{R}_i) = \int d^3x \, e^{-i(\mathbf{k}'-\mathbf{k})\cdot(\mathbf{x}+\mathbf{R}_i)} \nabla_{\mathbf{x}} V(\mathbf{x}) \tag{3.53}$$

and the periodicity of $\nabla_{\mathbf{x}} V(\mathbf{x})$, which implies $\mathbf{k}' - \mathbf{k} = \mathbf{Q}$ belongs to the reciprocal lattice, we conclude that the $\mathbf{R}_i$ sum reduces to

$$\sum_{\mathbf{R}_i} e^{i(\mathbf{k}' - \mathbf{k} + \mathbf{q} - \mathbf{Q})\cdot\mathbf{R}_i} = N\delta_{\mathbf{k}' - \mathbf{k} + \mathbf{q}, \mathbf{Q}} \tag{3.54}$$

The electron-phonon interaction Hamiltonian, therefore, is given by

$$H_{e-ph} = \sum_{\mathbf{k},\mathbf{q},s,\sigma} W_{\mathbf{q},s} \, c^{\dagger}_{\mathbf{k}+\mathbf{q},\sigma} c_{\mathbf{k},\sigma} \left[a_s(\mathbf{q}) + a_s^{\dagger}(-\mathbf{q}) \right]. \tag{3.55}$$

where

$$W_{\mathbf{q},s} = \sqrt{\frac{\hbar}{2MN\omega_s(\mathbf{q})}} \int d^3x \, e^{-i\mathbf{q}\cdot\mathbf{x}} \mathbf{e}_s \cdot \nabla_{\mathbf{x}} V(\mathbf{x}). \tag{3.56}$$

The electron-phonon Hamiltonian describes a process where an electron in an initial state is scattered into a final state, either by emitting or absorbing a phonon with energy $\hbar\omega_s(\mathbf{q})$ and wave-vector $\mathbf{q}$, which satisfies the relation

$$\mathbf{k}' - \mathbf{k} + \mathbf{q} = \mathbf{Q}. \tag{3.57}$$

The electron energy gain or loss is $\Delta_E = \pm\hbar\omega_s(\mathbf{q})$.

The full electron-phonon Hamiltonian is

$$H_T = H_e + H_{ph} + H_{e-ph}, \tag{3.58}$$

where H_e is given by (3.18), H_{ph} by (2.16) and H_{e-ph} by (3.55).

Let us explore the effective electron-electron interaction that is induced by the electron-phonon coupling. For this purpose, let us treat the above Hamiltonian in Rayleigh–Schrödinger perturbation theory. It is easy to see that there will be no correction to the energy in first order, because of the linear phonon term. We must therefore proceed to the second order. Using (3.55), we find that the energy correction due to the interaction in second order corresponds to the effective Hamiltonian

$$H_{e-ph}^{(2)} = \sum_{\mathbf{k},\mathbf{k}'} \sum_{\mathbf{q}} \sum_{\sigma,\sigma'} |W_{\mathbf{q}}|^2 \, c^{\dagger}_{\mathbf{k}',\sigma'} c_{\mathbf{k}'+\mathbf{q},\sigma'} c^{\dagger}_{\mathbf{k}+\mathbf{q},\sigma} c_{\mathbf{k},\sigma}$$

$$\times \sum_n \frac{\langle 0| \left[a(-\mathbf{q}) + a^{\dagger}(\mathbf{q}) \right] |n\rangle \langle n| \left[a(\mathbf{q}) + a^{\dagger}(-\mathbf{q}) \right] |0\rangle}{E_0^{(0)} - E_n^{(0)}}. \tag{3.59}$$

There are two non-vanishing terms in the n-sum, represented by the Feynman graphs of Figure 3.1, which correspond respectively to the emission and absorption of one phonon in each vertex. These can be cast in the form

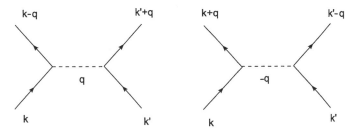

Figure 3.1 The phonon-mediated electronic interaction

$$H^{(2)}_{e-ph} = \frac{1}{2} \sum_{\mathbf{k},\mathbf{k}'} \sum_{\mathbf{q}} \sum_{\sigma,\sigma'} |W_{\mathbf{q}}|^2$$

$$\times \left[\frac{c^{\dagger}_{\mathbf{k}-\mathbf{q},\sigma} c_{\mathbf{k},\sigma} c^{\dagger}_{\mathbf{k}'+\mathbf{q},\sigma'} c_{\mathbf{k}',\sigma'}}{\epsilon_{\mathbf{k}} - \epsilon_{\mathbf{k}-\mathbf{q}} - \hbar\omega} + \frac{c^{\dagger}_{\mathbf{k}'-\mathbf{q},\sigma'} c_{\mathbf{k},\sigma} c^{\dagger}_{\mathbf{k}+\mathbf{q},\sigma} c_{\mathbf{k}',\sigma'}}{\epsilon_{\mathbf{k}} - \epsilon_{\mathbf{k}+\mathbf{q}} + \hbar\omega} \right], \quad (3.60)$$

or, making $\mathbf{q} \to -\mathbf{q}$ in the last term,

$$H^{(2)}_{e-ph} = \frac{1}{2} \sum_{\mathbf{k},\mathbf{k}'} \sum_{\mathbf{q}} \sum_{\sigma,\sigma'} |W_{\mathbf{q}}|^2 \, c^{\dagger}_{\mathbf{k}'+\mathbf{q},\sigma'} c^{\dagger}_{\mathbf{k}-\mathbf{q},\sigma} c_{\mathbf{k},\sigma} c_{\mathbf{k}',\sigma'}$$

$$\times \left[\frac{1}{\epsilon_{\mathbf{k}} - \epsilon_{\mathbf{k}-\mathbf{q}} - \hbar\omega} - \frac{1}{\epsilon_{\mathbf{k}} - \epsilon_{\mathbf{k}-\mathbf{q}} + \hbar\omega} \right], \quad (3.61)$$

where we used the fact that $\omega(\mathbf{q})$ is an even function.

The effective total electron-phonon Hamiltonian, including the phonon-induced effective electronic interaction, is therefore given by

$$H_T = H_0 + \sum_{\mathbf{k},\mathbf{k}',\mathbf{q}} \sum_{\sigma,\sigma'} V(\mathbf{k}, \mathbf{q}) \, c^{\dagger}_{\mathbf{k}'+\mathbf{q},\sigma'} c^{\dagger}_{\mathbf{k}-\mathbf{q},\sigma} c_{\mathbf{k},\sigma} c_{\mathbf{k}',\sigma'}$$

$$V(\mathbf{k}, \mathbf{q}) = |W_{\mathbf{q}}|^2 \left[\frac{\hbar\omega}{\left(\epsilon_{\mathbf{k}} - \epsilon_{\mathbf{k}-\mathbf{q}}\right)^2 - (\hbar\omega)^2} \right]. \quad (3.62)$$

We will see that this Hamiltonian plays a central role in the explanation of superconductivity and forms the basis of the BCS theory, which describes this phenomenon. Curiously, however, it also describes the interaction that is the main cause of a nonzero resistance in the charge transport in solids.

With the above study about the electron-phonon interaction, we conclude the description of the main electronic interactions in a crystal. In the next chapter we study important effects of these interactions.

4

Interactions in Action

The consequences of electronic interactions, lattice oscillations or combinations of both frequently lead to spectacular effects. In this chapter, we focus on some of these, such as long range magnetic order, strong electronic correlations and superconductivity. Electric conductivity/resistivity is also considered.

4.1 Magnetic Order

The Mean Field Approximation

Let us examine here the mechanism by which the magnetic interaction between nearest neighbors can be magnified up to a macroscopic scale, thus exhibiting a collective behavior of all atoms in the crystal. We assume the magnetic properties of the system are described by the Ising model, namely,

$$H_I = -\frac{2J}{\hbar^2} \sum_{\langle ij \rangle} \mathbf{S}_i^Z \mathbf{S}_j^Z, \tag{4.1}$$

where we assume $J > 0$.

We shall use the so-called mean field approximation, where we replace one of the spins above by its average $\mathbf{S}_j^Z \to \langle \mathbf{S}_j^Z \rangle$. Such replacement produces a corresponding system with non-interacting spins in the presence of an effective uniform magnetic field, which conveys all information about the spin interactions:

$$\mathbf{H}_e^Z = \frac{2J_0}{g\mu_B \hbar} \langle \mathbf{S}_i^Z \rangle, \tag{4.2}$$

where $J_0 = zJ$, z being the coordination number (the number of nearest neighbors of a given site). The mean field Hamiltonian then reads

$$H_{MF} = -\frac{g\mu_B}{\hbar} \sum_i \mathbf{S}_i^Z \mathbf{H}_e^Z. \tag{4.3}$$

In the presence of an external magnetic field $\mathbf{H}^Z$, the mean field Hamiltonian would become

$$H_{MF}[\mathbf{H}^Z] = -\frac{g\mu_B}{\hbar} \sum_i \mathbf{S}_i^Z[\mathbf{H}_e^Z + \mathbf{H}^Z]. \tag{4.4}$$

Considering that the eigenvalues of the Z-component of the spins are $\pm\frac{\hbar}{2}$, we readily find the partition function and free energy corresponding to H_{MF} in the presence of an external magnetic field $\mathbf{H}^Z$:

$$Z[\mathbf{H}^Z] = \left\{ 2\cosh\left[\frac{g\mu_B}{2k_BT}[\mathbf{H}_e^Z + \mathbf{H}^Z]\right] \right\}^N$$

$$F[\mathbf{H}^Z] = -Nk_BT \ln 2\cosh\left[\frac{g\mu_B}{2k_BT}[\mathbf{H}_e^Z + \mathbf{H}^Z]\right], \tag{4.5}$$

where N is the number of lattice sites in the crystal.

The Ferromagnetic Order

The magnetization corresponding to this free energy is

$$\mathbf{M}^Z = -\frac{\partial F}{\partial \mathbf{H}^Z}\Big|_{\mathbf{H}^Z=0} = N\frac{g\mu_B}{2}\tanh\left[\frac{g\mu_B}{2k_BT}\mathbf{H}_e^Z\right]. \tag{4.6}$$

Now, observe that the effective magnetic field $\mathbf{H}_e^Z$ is proportional to the magnetization $\mathbf{M}^Z$. Indeed, defining

$$\mathbf{M}^Z = N\frac{g\mu_B}{\hbar}\langle\mathbf{S}_i^Z\rangle, \tag{4.7}$$

we have, according to (4.2),

$$\mathbf{H}_e^Z = \frac{1}{N}\frac{2J_0}{(g\mu_B)^2}\mathbf{M}^Z. \tag{4.8}$$

It follows that expression (4.6) is a transcendental equation for the magnetization:

$$x = \tanh\gamma x, \tag{4.9}$$

where $x = \frac{1}{N}\frac{2}{g\mu_B}\mathbf{M}^Z$ and $\gamma = \frac{zJ}{2k_BT}$.

For $\gamma < 1$ we only have the solution $x_0 = 0$, whereas for $\gamma > 1$ we always have a nontrivial solution $x_0 \neq 0$. From (4.5) we see that this is the solution that minimizes the free energy; therefore, the condition $\gamma = 1$ marks the separation between two different phases of the system: a phase with $x_0 = 0$ and $\mathbf{M}^Z = 0$, which is called paramagnetic, and a phase for which $x_0 \neq 0$ and, consequently, $\mathbf{M}^Z \neq 0$, known as ferromagnetic. In this case the system presents a spectacular effect: a spontaneous magnetization, $\mathbf{M}^Z = N\frac{g\mu_B}{2}x_0$, which, being proportional to N, is macroscopic. This is an example of a quantum effect that can be observed

Interactions in Action

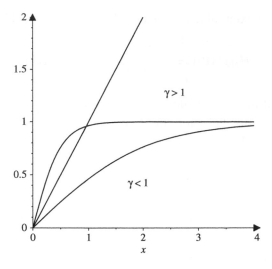

Figure 4.1 The solutions of Eq. (4.9): a nontrivial solution occurs for $\gamma > 1$

at a macroscopic scale. The critical temperature for the onset of the ferromagnetic phase is obtained from the condition $\gamma = 1$, which for a given J implies a critical temperature $T_c = \frac{zJ}{2k_B}$, known as the Curie temperature.

In order to find the temperature dependence of the magnetization below T_c, we expand the hyperbolic function in (4.9) and, observing that $\gamma = T_c/T$, obtain for $T < T_c$

$$x = \frac{T_c}{T}x - \left(\frac{T_c}{T}\right)^3 \frac{x^3}{3} + . \tag{4.10}$$

Considering that $x \neq 0$ and that for $T \simeq T_c$, we may neglect higher-order terms, we obtain the solution $x_0 = \sqrt{3/T_c}(T_c - T)^{1/2}$. This implies the magnetization vanishes as

$$\mathbf{M}^Z(T) = M_0 \left(\frac{T_c - T}{T_c}\right)^{1/2}, \tag{4.11}$$

for $T < T_c$.

Such power-law dependence on the temperature characterizes what is called a critical behavior. An outstanding feature of critical systems is the fact that all its basic components – in this case, localized spins or magnetic moments – are tightly correlated, irrespective of the distance that separates them in the sample, thereby producing a macroscopic effect. This phenomenon looks even more amazing when we recall that the original Hamiltonian only involves interactions between nearest neighbors.

The Spin Correlation Function: Correlation Length

How can we quantitatively determine how correlated a spin system is? This is conveniently done by means of the connected spin correlation function, defined as

$$- \frac{\partial^2 F}{\partial \mathbf{H}_i^Z \partial \mathbf{H}_j^Z}\Big|_{\mathbf{H}^Z=0} = \frac{\lambda^2}{k_B T}\left[\langle \mathbf{S}_i^Z \mathbf{S}_j^Z \rangle - \langle \mathbf{S}_i^Z \rangle \langle \mathbf{S}_j^Z \rangle\right] = \frac{\lambda^2}{k_B T}\langle \mathbf{S}_i^Z \mathbf{S}_j^Z \rangle_C, \quad (4.12)$$

where $\lambda = \frac{g\mu_B}{\hbar}$.

In order to extract the physical meaning of the correlation function, let us use (4.6) and (4.12) to express the free energy as a Taylor series in the external field:

$$F[\mathbf{H}^Z] = F_0 - \sum_{\mathbf{R}_i} \mathbf{M}_i^Z \mathbf{H}_i^Z + \frac{\lambda^2}{2k_B T}\sum_{\mathbf{R}_i \mathbf{R}_j}\langle \mathbf{S}_i^Z \mathbf{S}_j^Z \rangle_C \mathbf{H}_i^Z \mathbf{H}_j^Z + \cdots \quad (4.13)$$

The stability condition for the system requires that $\partial F/\partial \mathbf{H}_i^Z = 0$. Assuming that the external field is weak enough for allowing higher-than-quadratic terms to be neglected, we can express this condition in the so-called linear response regime, where

$$\langle \mathbf{S}^Z(\mathbf{R}_i)\rangle = \frac{\lambda}{k_B T}\sum_{\mathbf{R}_j}\langle \mathbf{S}^Z(\mathbf{R}_i)\mathbf{S}^Z(\mathbf{R}_j)\rangle_C \mathbf{H}_j^Z. \quad (4.14)$$

Let us consider now what would be the effect of applying an external magnetic field localized at the site $\mathbf{R}_k$, namely $\mathbf{H}_j^Z = H_0\delta_{jk}$. It follows from the above equation that

$$\langle \mathbf{S}^Z(\mathbf{R}_i)\rangle = \frac{\lambda}{k_B T}\langle \mathbf{S}^Z(\mathbf{R}_i)\mathbf{S}^Z(\mathbf{R}_k)\rangle_C \, H_0. \quad (4.15)$$

We see that the correlation function $\langle \mathbf{S}_i^Z \mathbf{S}_k^Z \rangle_C$ measures how strongly the magnetic field applied at $\mathbf{R}_k$ influences the magnetization at a different site $\mathbf{R}_i$. The efficiency in the process of conveying the influence from one site to another expresses how tightly correlated the spins are.

In the presence of a time-dependent applied magnetic field, the dynamical spin correlation function is the proper instrument for determining how efficiently correlated the spins are:

$$- \frac{\partial^2 F}{\partial \mathbf{H}_i^Z(t) \partial \mathbf{H}_j^Z(0)}\Big|_{\mathbf{H}^Z=0} = \frac{\lambda^2}{k_B T}\langle \mathbf{S}_C^Z(\mathbf{R}_i, t)\mathbf{S}_C^Z(\mathbf{R}_j, 0)\rangle$$

$$\mathbf{S}_C^Z(\mathbf{R}_i, t) = \mathbf{S}^Z(\mathbf{R}_i, t) - \langle \mathbf{S}^Z(\mathbf{R}_i, t)\rangle. \quad (4.16)$$

The generalization of (4.14) then reads

$$\langle \mathbf{S}^Z(\mathbf{R}_i, t)\rangle = \frac{\lambda}{k_B T}\sum_{\mathbf{R}_j}\int_{-\infty}^{t} dt' \langle \mathbf{S}^Z(\mathbf{R}_i, t)\mathbf{S}^Z(\mathbf{R}_j, t')\rangle_C \mathbf{H}^Z(\mathbf{R}_j, t'). \quad (4.17)$$

Using the fact that the Fourier transform of a convolution is a product, we can express the above equation in a very simple form in terms of $(\mathbf{k}, \omega)$-dependent Fourier transforms, namely

$$\langle S^Z(\mathbf{k}, \omega) \rangle = \frac{\lambda}{k_B T} \langle S^Z S^Z \rangle_C(\mathbf{k}, \omega) H^Z(\mathbf{k}, \omega). \tag{4.18}$$

The proportionality function,

$$\chi_{ZZ}(\mathbf{k}, \omega) = \frac{\lambda}{k_B T} \langle S^Z S^Z \rangle_C(\mathbf{k}, \omega), \tag{4.19}$$

is known as the magnetic susceptibility. It is in general a 3×3 matrix, but only the ZZ component appears in the Ising model.

We are going to show in Chapter 19, using a quantum field theory approach for a $d = 2$ dimensional ferromagnetic crystal, that the spin correlation function, in the continuum limit, is given, in frequency-momentum space, by

$$\langle S^Z S^Z \rangle_C(\mathbf{k}, \omega) = \frac{1}{\omega - ic^2|\mathbf{k} - \mathbf{Q}|^2 + \frac{i}{\xi^2}}. \tag{4.20}$$

In this expression, ξ is the correlation length, a quantity that determines the distance within which the correlation function is significantly different from zero or, equivalently, the size of the region where the spins are tightly correlated.

It can be shown that, close to the Curie temperature, T_c, the correlation length diverges as

$$\xi(T) \propto (T - T_c)^{-1}. \tag{4.21}$$

This fact, implies that, in the ferromagnetic phase, all spins in the crystal are tightly correlated, thereby allowing the manifestation of this quantum phenomenon at a macroscopic scale.

Notice that in the above equation we have shifted $\mathbf{k}$ by a reciprocal lattice vector $\mathbf{Q}$ to comply with the fact that the $\mathbf{R}$-integral is an approximation for $\sum_{\mathbf{R}}$, which only sweeps the Bravais lattice points.

We can write (4.20), for $\omega \to 0$, which represents the response to a static field, as

$$\langle S^Z S^Z \rangle_C(\mathbf{k}, \omega = 0) \propto \left[\frac{1/\xi}{(\mathbf{k} - \mathbf{Q})^2 + \frac{1}{\xi^2}} \right] \xi. \tag{4.22}$$

At the onset of the ferromagnetic phase, for $T \to T_c$, ξ diverges and, using the Lorentzian representation of the Dirac δ, we find for the static magnetic spin correlation function:

$$\langle S^Z S^Z \rangle_C(\mathbf{k}, \omega = 0) \stackrel{T \to T_c}{\sim} \delta(|\mathbf{k} - \mathbf{Q}|)\xi. \tag{4.23}$$

From the result above we conclude that the Fourier transform of the connected correlation function only contains wave-vectors belonging to the reciprocal lattice for $T \to T_c$, because of the Dirac delta in (4.23). This means that, in the ferromagnetic phase, such correlation function has the same symmetries as the Bravais lattice, thus implying a full ordering of the spins according to the Bravais lattice pattern.

Then it follows that the static magnetic susceptibility, which is proportional to $S(\mathbf{k}, \omega = 0)$, behaves as

$$\chi(T) \propto \xi \propto (T - T_c)^{-1} \tag{4.24}$$

for $T \to T_c$. This is again an indication that the local spins of the crystal are stiffly correlated along the whole sample.

An Order Parameter

The magnetization, given by (4.7), according to the results of the preceding analysis, acts ultimately as a measure of how much collective the system behaves or, in other words, how much ordered the system is. Magnetization, therefore, may be considered as an order parameter: namely, a quantity that expresses the amount of ordering in a system. It would vanish in a paramagnetic phase, where each atom shows and independent behavior (thus reflecting a complete absence of ordering) and, conversely, would be nonzero in a ferromagnetic phase, where the atoms show an organized collective behavior. Furthermore, even inside this phase, the value of the order parameter would increase as we would lower the temperature, or equivalently, increase the degree of organization in the collective behavior of the system.

The concept of an order parameter was put forward by Landau as a useful tool in the description of phase transitions in many different systems. He showed in particular, how the free energy could be expanded in terms of the order parameter. We will see that this concept is particularly useful in the theory of superconductivity.

Neutron Scattering

The ordering of the spins in a magnetic crystal can be conveniently probed by neutron scattering, because neutrons, having zero charge, interact with the crystal essentially through their magnetic dipole moment. It can be shown [19] that for a neutron scattered from a state $(\mathbf{k}_i, E_i)$ onto a state $(\mathbf{k}_f, E_f)$, the differential cross section satisfies

$$\frac{d^2\sigma}{dEd\Omega}(\mathbf{k}, E) \propto S(\mathbf{k}, \omega) \tag{4.25}$$

in such a way that $\mathbf{k} = \mathbf{k}_f - \mathbf{k}_i$ and $E = \hbar\omega = E_f - E_i$.

Using expressions (4.22) and (4.23) for the Fourier transform of the spin correlation function, we conclude that in the paramagnetic phase, for $T > T_c$, because
of the Lorentzian distribution functions, neutron scattering will produce a diffuse
image of the reciprocal lattice, whereas in the ferromagnetic phase, where $T < T_c$,
the image will become sharp because of the deltas. In this phase, scattering will
be predominantly elastic ($E = 0$), while in the paramagnetic phase it will be
quasi-elastic.

Antiferromagnetic Order

Another interesting class of magnetic order is the one presented by the so-called
antiferromagnetic systems. These are characterized by the presence of an exchange
coupling, which is negative. Let us consider, for instance, the antiferromagnetic
Ising model, described by (4.1) with a coupling $J < 0$ between nearest neighbors.
We assume the crystal lattice to be bipartite, namely, one for which neighboring
sites belong to different sublattices A and B. A negative J will favor the opposite
orientation of the spins belonging to the different sublattices: $\langle S_i^Z \rangle_B = -\langle S_i^Z \rangle_A$.
The ordered state, known as Néel state, is characterized by a nonzero sublattice
magnetization

$$\mathbf{M}_A^Z = -\mathbf{M}_B^Z = \frac{N}{2} \frac{g\mu_B}{\hbar} \langle S_i^Z \rangle_A. \tag{4.26}$$

It is not difficult to see that the effective mean field acting on the spins of sublattice
A will be the one corresponding to the sublattice magnetization $\mathbf{M}_B^Z = -\mathbf{M}_A^Z$.
Hence, we will have the same transcendental equation (4.9) for $\mathbf{M}_A^Z$, with $\gamma = \frac{-zJ}{2k_BT}$.
We immediately conclude that the ordered antiferromagnetic state will set in below
a critical temperature $T_N = \frac{z|J|}{2k_B}$, the Néel temperature.

We will show in Chapter 19, by means of a quantum field theory approach for a
two-dimensional magnetic system described by the antiferromagnetic Heisenberg
model on a square lattice, that in the continuum limit, the sublattice spin correlation
function is given by

$$\langle S_A^Z(\mathbf{R}, t)S_A^Z(0, 0)\rangle_C = \frac{e^{-\frac{\sqrt{|c^2t^2 - R^2|}}{\xi}}}{4\pi\sqrt{|c^2t^2 - R^2|}}, \tag{4.27}$$

where c is a characteristic velocity and ξ, the correlation length.

The Fourier transform of the above correlation function now reads

$$S_A(\mathbf{k}, \omega) = \frac{1}{\omega^2 - c^2\left[(\mathbf{k} - \frac{\mathbf{Q}}{2})^2 + \frac{1}{\xi^2}\right]}. \tag{4.28}$$

Observe that in the case of antiferromagnetic order, the spin susceptibility, which is proportional to $S_A(\mathbf{k}, \omega)$, behaves as

$$\langle S^Z S^Z \rangle_{C}^{A}(\mathbf{k}, \omega = 0) \overset{T \to T_N}{\sim} \delta\left(|\mathbf{k} - \frac{\mathbf{Q}}{2}|\right) \xi \qquad (4.29)$$

at $T \to T_N$.

Now, the Fourier transform of the spin correlation function contains a Dirac delta forcing the wave-vectors to coincide with half of the reciprocal lattice vectors, thus reflecting the fact that in the Néel state the spins order in a pattern that doubles the corresponding primitive cell of the Bravais lattice.

Magnetism of Itinerant Electrons

The examples of magnetic order examined above apply to systems of localized electrons: namely, insulators. In order to understand the magnetism of metals, let us investigate the mechanism behind the magnetic ordering of itinerant electrons. For this purpose, consider the generalization of the Hubbard Hamiltonian (3.34), where we relax the constraint (3.33). Applying a mean field approximation, where $n_{i,\sigma} \to \langle n_{i,\sigma} \rangle = \langle n_{\sigma} \rangle$, we get

$$H_U = -t \sum_{\langle ij \rangle, \sigma} \left[c_{i,\sigma}^{\dagger} c_{j,\sigma} + c_{j,\sigma}^{\dagger} c_{i,\sigma} \right] + U \sum_{i} \sum_{\sigma} \langle n_{\sigma} \rangle c_{i,\sigma}^{\dagger} c_{i,\sigma}. \qquad (4.30)$$

Using (3.13) and (3.26), we can re-write the previous Hamiltonian in diagonal form,

$$H_U = \sum_{\mathbf{k}} \sum_{\sigma} E_{\sigma}(\mathbf{k}) c_{\mathbf{k},\sigma}^{\dagger} c_{\mathbf{k},\sigma}, \qquad (4.31)$$

where the energy eigenvalues are

$$E_{\sigma}(\mathbf{k}) = E_n(\mathbf{k}) + U \langle n_{\sigma} \rangle. \qquad (4.32)$$

Defining

$$m = \langle n_{\uparrow} \rangle - \langle n_{\downarrow} \rangle \qquad n = \langle n_{\uparrow} \rangle + \langle n_{\downarrow} \rangle,$$
$$\langle n_{\uparrow} \rangle = \frac{1}{2}(n + m) \qquad \langle n_{\downarrow} \rangle = \frac{1}{2}(n - m), \qquad (4.33)$$

we have

$$E_{\pm}(\mathbf{k}) = E_n(\mathbf{k}) + \frac{U}{2}n \pm \frac{m}{2}. \qquad (4.34)$$

Now,

$$m = \langle n(E_+(\mathbf{k})) \rangle - \langle n(E_-(\mathbf{k})) \rangle, \qquad (4.35)$$

where $n(E_\sigma(\mathbf{k}))$ is the Fermi–Dirac distribution. The two previous equations form a self-consistent pair that has a nonzero solution for m at temperatures below a critical temperature T_c, given by

$$\frac{k_B T_c}{E_F} = \frac{\pi}{2}\sqrt{\frac{1}{3}\left(\frac{U}{N}g(E_F) - 1\right)}, \tag{4.36}$$

where N is the number of itinerant electrons, E_F is the Fermi energy and $g(E)$ is the conduction band density of states at energy E.

Observe that for a metal to exhibit a nonzero magnetization at some finite temperature, it must satisfy the condition

$$\frac{U}{N}g(E_F) > 1, \tag{4.37}$$

known as Stoner criterion.

Metals with s or p conduction bands usually have a very small density of states because of the large width of such bands. These metals, therefore, do not satisfy the Stoner criterion and, consequently, do not present a magnetically ordered phase. The d-bands, conversely, are much narrower, thus possessing some of the largest densities of states. For that reason, they are capable of satisfying that criterion. This explains why transition metals such as Fe, Co and Ni are ferromagnetic.

4.2 Strongly Correlated Systems

As an example of a strongly correlated system, let us consider a two-dimensional crystal structure assembled in a square lattice with one atom per site, each of these atoms providing one active electron. Within the tight-binding approach, we have seen that a square Fermi surface exists, separating the occupied states, which fill one-half of the first Brillouin zone, from the unoccupied ones, which will form the second half of this zone. Within this approximation, which neglects the electronic interactions, therefore, the system would be a metal. Let us see how this picture is profoundly modified when the interactions are taken into account.

We assume the interactions are described by the Hubbard model, which is given by (3.34), namely $H_H = H_U + H_t$, in the strong coupling regime where $U \gg t$. This may be treated by considering the U-term as the unperturbed Hamiltonian and using Rayleigh–Schrödinger perturbation theory in H_t.

We start from the ground state, defined by

$$\sum_i \left(n_{i,\uparrow} + n_{i,\downarrow}\right)|0\rangle = |0\rangle. \tag{4.38}$$

This is clearly a state with one electron per site, either with spin up or down; in other words, the exactly half-filled band regime.

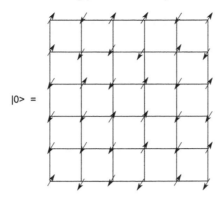

$|0> =$

Figure 4.2 The ground state at half-filling

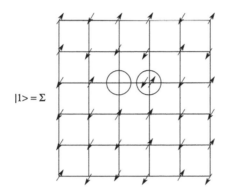

$|1> = \Sigma$

Figure 4.3 The first excited state above half-filling

It is easy to show that $H_U|0\rangle = 0$, because on $|0\rangle$ either $n_{i,\uparrow} = 0$ or $n_{i,\downarrow} = 0$, hence $E_0^{(0)} = 0$.

The first-order correction to the ground-state energy is

$$E_0^{(1)} = \langle 0|H_t|0\rangle = 0. \tag{4.39}$$

This vanishes because $H_t|0\rangle = |1\rangle$, where the state $|1\rangle$ contains at least one empty and one doubly occupied site, being, therefore, orthogonal to the ground state: $\langle 0|1\rangle = 0$. We must, consequently, proceed to the second-order correction.

This is given by

$$E_0^{(2)} = \sum_{n\neq 0} \frac{\langle 0|H_t|n\rangle\langle n|H_t|0\rangle}{E_0 - E_n}, \tag{4.40}$$

where $\{|n\rangle\}$ is a complete set of the unperturbed Hamiltonian H_U.

It is not difficult to see that only $|n\rangle = |1\rangle$ will give a nonzero contribution to the above expression. All other contributions would vanish for reasons similar to the one leading to (4.39). Using the fact that $E_0 = 0$ and $E_1 = U$, since $H_U|1\rangle = U|1\rangle$, we can write

$$
\begin{aligned}
E_0^{(2)} &= -\frac{1}{U} \sum_{n \neq 0} \langle 0|H_t|n\rangle \langle n|H_t|0\rangle \\
&= -\frac{1}{U} \sum_n \langle 0|H_t|n\rangle \langle n|H_t|0\rangle \\
&= -\frac{1}{U} \langle 0|H_t\, H_t|0\rangle,
\end{aligned} \tag{4.41}
$$

where we used the completeness of $\{|n\rangle\}$.

Inserting H_t in the latter expression, we may infer that the only non-vanishing terms are

$$
E_0^{(2)} = -\frac{2t^2}{U} \sum_{\langle ij\rangle} \sum_{\sigma,\sigma'} \langle 0|c_{i,\sigma}^\dagger c_{j,\sigma} c_{j,\sigma'}^\dagger c_{i,\sigma'}|0\rangle. \tag{4.42}
$$

Now, using (3.37) and (3.43) in the half-filled situation when $\sum_\sigma n_{i,\sigma} = 1$, we can cast the above equation in the form (using $\hbar = 1$)

$$
E_0^{(2)} = \langle 0|\frac{4t^2}{U} \sum_{\langle ij\rangle} \mathbf{S}_i \cdot \mathbf{S}_j|0\rangle. \tag{4.43}
$$

Since $U > 0$, this result shows that in the strong coupling regime $U \gg t$, at half-filling, the dynamics of the electrons described by the Hubbard model becomes effectively governed by the antiferromagnetic Heisenberg Hamiltonian with coupling $J = -2t^2/U$:

$$
H_H^{(2)} = \frac{4t^2}{U} \sum_{\langle ij\rangle} \mathbf{S}_i \cdot \mathbf{S}_j. \tag{4.44}
$$

Consequently, below the Néel temperature, the electronic system at half-filling, which has precisely one electron per site, must be in an ordered antiferromagnetic state.

Using the mean field approach and (3.40), we can write

$$
H_H^{(2)} = \frac{2t^2}{U} \sum_{\langle ij\rangle} \sum_{\alpha\beta} (c_{i,\alpha}^\dagger \vec{\sigma}_{\alpha\beta} c_{i,\beta}) \cdot \langle \mathbf{S}_j\rangle. \tag{4.45}
$$

Notice that the above Hamiltonian implies that in the presence of the antiferromagnetic background, the effective lattice determining the electronic properties has the modulus of its primitive vectors doubled. This means the corresponding effective primitive reciprocal lattice vectors are changed accordingly:

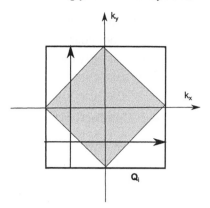

Figure 4.4 The reciprocal primitive vectors of the square lattice, the first Brillouin zone and the occupied states at half-filling (shaded area)

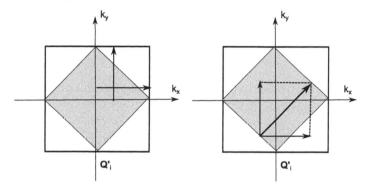

Figure 4.5 The reciprocal primitive vectors of the square lattice after the doubling produced by the magnetic ordered background. The region delimited by the Fermi surface (line) becomes the new first Brillouin zone, which therefore is completely filled. The system, consequently, becomes an insulator.

$$\mathbf{Q} = \left(\frac{2\pi}{a} n, \frac{2\pi}{a} m \right) \longrightarrow \mathbf{Q}' = \left(\frac{\pi}{a} n, \frac{\pi}{a} m \right) \qquad n, m \in \mathbb{Z}. \qquad (4.46)$$

This change in the reciprocal lattice exerts a deep influence on the electronic properties of this system. The most spectacular effect is the generation of a gap right on the Fermi surface (line), thus transforming the metal into an insulator. How does this happen?

Observe that the Fermi surface precisely fits into the Brillouin zone, touching it in four points and occupying exactly one-half of it. This is the so-called nesting property. Because of this property, when the reciprocal lattice is modified by the Néel transition, there are four reciprocal lattice vectors ($n, m = \pm 1$) that connect opposite sides of the Fermi surface, just as the previous reciprocal lattice vectors would connect opposite sides of the Brillouin zone. The opening of a gap at the

Fermi surface then follows naturally in the same way the umklapp mechanism creates a gap at the first Brillouin zone boundaries.

The study of this system shows how deeply the electronic interactions can modify the properties of a system, in this case transforming a metal into an insulator.

4.3 Conductivity

We have studied in Section 4.1 how a material system responds to the application of an external magnetic field. Let us consider now what is the response to the application of an external electric field $\mathbf{E}$. We assume this electric field to be associated to a vector potential $\mathbf{A}$ in a gauge such that $\mathbf{E} = -\frac{\partial \mathbf{A}}{\partial t}$.

Starting from the free electron system described by (3.20), we introduce the electric field according to the usual, minimal coupling prescription (3.27), to obtain the Hamiltonian of the electron system in the presence of an electric field:

$$H[\mathbf{A}] = -\frac{\hbar^2}{2m} \sum_\sigma \int d^3r \, \psi_\sigma^\dagger(\mathbf{r}) \left(\nabla + i\frac{e}{\hbar c}\mathbf{A} \right)^2 \psi_\sigma(\mathbf{r}). \tag{4.47}$$

We use the same prescription in order to derive the electric current operator out of the free-electron expression (3.15), namely,

$$\mathbf{J} = -\frac{ie\hbar}{2m} \sum_\sigma \left[\psi_\sigma^\dagger(\mathbf{r})\mathbf{D}\psi_\sigma(\mathbf{r}) - [\mathbf{D}\psi_\sigma(\mathbf{r})]^\dagger \psi_\sigma(\mathbf{r}) \right], \tag{4.48}$$

where $\mathbf{D} = \nabla + i\frac{e}{\hbar c}\mathbf{A}$ is the so-called covariant derivative. We can write the charge current operator in the above equation as

$$\mathbf{J} = e\mathbf{j} + \frac{e^2}{mc}\rho\,\mathbf{A}, \tag{4.49}$$

where $\mathbf{j}$ is given by (3.15) and ρ is the electronic density, given by (3.13).

Notice that $H[\mathbf{A}]$ is a functional of the $\mathbf{A}$-field. The electric current, accordingly, is the functional derivative of $H[\mathbf{A}]$ with respect to the field $\mathbf{A}$, namely (we refer the reader to Section 5.1 for an introduction to functionals and their differentiation and integration),

$$J^i(\mathbf{r}) = \frac{\delta H[\mathbf{A}]}{\delta A_i(\mathbf{r})}. \tag{4.50}$$

Our main goal here is to obtain an expression for the average current $\langle \mathbf{J} \rangle$ as a function of the external electric field. For this purpose, we shift to the Heisenberg picture and introduce the partition functional and the free energy, respectively, as follows:

$$Z[\mathbf{A}_{\text{ext}}] = \text{Tr} e^{-\beta H[\mathbf{A}_{\text{ext}}]}$$

$$F[\mathbf{A}_{\text{ext}}] = -\frac{1}{\beta} \ln Z[\mathbf{A}_{\text{ext}}], \tag{4.51}$$

where $\beta = 1/k_B T$.

We have assumed that the external field was absent before $t = 0$ and that it starts to act after this instant. The interaction Hamiltonian is, therefore

$$H[\mathbf{A}_{\text{ext}}] = H_0[\mathbf{A}] + \theta(t) \int d^3x \mathbf{J}^i \, A^i_{\text{ext}}. \tag{4.52}$$

It is then clear that

$$\frac{\delta F[\mathbf{A}_{\text{ext}}]}{\delta A^i_{\text{ext}}}] \Big|_{\mathbf{A}_{\text{ext}}=0} = \langle \mathbf{J}^i \rangle \tag{4.53}$$

and

$$-\frac{\delta^2 F[\mathbf{A}_{\text{ext}}]}{\delta A^i_{\text{ext}}(\mathbf{r}, t)\delta A^j_{\text{ext}}(\mathbf{r}', t')} \Big|_{\mathbf{A}_{\text{ext}}=0} = \langle \mathbf{J}^i \mathbf{J}^j \rangle_C = \langle \mathbf{J}^i \mathbf{J}^j \rangle - \langle \mathbf{J}^i \rangle \langle \mathbf{J}^j \rangle. \tag{4.54}$$

In the expression above,

$$\langle \mathbf{J}^i(\mathbf{r}, t)\mathbf{J}^j(\mathbf{r}', t')\rangle_C = \langle \mathbf{j}^i(\mathbf{r}, t)\mathbf{j}^j(\mathbf{r}', t')\rangle_C + \frac{ne^2}{mc}\delta^{ij}\delta(\mathbf{r} - \mathbf{r}')\delta(t - t'), \tag{4.55}$$

where $n = \langle \rho(\mathbf{r}, t)\rangle$.

Now, using (4.53) and (4.55), we can derive the functional Taylor series for the free energy functional, assuming the external field is weak enough for the higher-than-quadratic terms to be neglected:

$$F[\mathbf{A}_{\text{ext}}] = F_0 + \int d^3x dt \langle \mathbf{J}^i(\mathbf{r}, t)\rangle A^i_{\text{ext}}(\mathbf{r}, t) - \frac{1}{2} \int d^3x dt \int d^3x' dt'$$
$$\times \left[\langle \mathbf{j}^i(\mathbf{r}, t)\mathbf{j}^j(\mathbf{r}', t')\rangle_C + \frac{e^2}{mc}\delta^{ij} \langle \rho(\mathbf{r}, t)\rangle\delta(\mathbf{r} - \mathbf{r}')\delta(t - t') \right]$$
$$\times A^i_{\text{ext}}(\mathbf{r}, t)A^j_{\text{ext}}(\mathbf{r}', t') + \dots \tag{4.56}$$

We may obtain a relation between the average current and the external applied field by imposing the stability of the system. This implies

$$\frac{\delta F[\mathbf{A}_{\text{ext}}]}{\delta A^i_{\text{ext}}(\mathbf{r}, t)} = 0. \tag{4.57}$$

According to (4.56), therefore, this condition leads to the following relation between the current and vector potential, up to the first order in the external field (linear response):

$$\langle \mathbf{J}^i(\mathbf{r}, t)\rangle = \int d^3x' dt'$$
$$\times \left[\langle \mathbf{j}^i(\mathbf{r}, t)\mathbf{j}^j(\mathbf{r}', t')\rangle_C + \frac{e^2}{mc}\delta^{ij} \langle \rho(\mathbf{r}, t)\rangle\delta(\mathbf{r} - \mathbf{r}')\delta(t - t') \right] A^j_{\text{ext}}(\mathbf{r}', t'). \tag{4.58}$$

It is convenient to consider the Fourier transform of the above equation. Using the fact that the transform of a convolution is a product, and replacing the average electronic density with $n = N/V$, where N is the number of electrons in the conduction band and V is the system's volume, we get

$$\langle \mathbf{J}^i(\mathbf{k}, \omega) \rangle = \left[\langle \mathbf{j}^i \mathbf{j}^j \rangle_C(\mathbf{k}, \omega) + \frac{ne^2}{mc} \delta^{ij} \right] A^j_{\text{ext}}(\mathbf{k}, \omega). \qquad (4.59)$$

We want to determine the response of the electronic system (more specifically, of the mobile electrons of the conduction band) to an external electric field given by $\mathbf{E}(\omega) = -i\omega \mathbf{A}_{\text{ext}}(\omega)$ in frequency space (for a uniform field $\mathbf{k} = 0$). The response we seek is the 3×3 matrix, called electric conductivity, which, according to (4.59), relates the average total current to the external electric field and is given by

$$\langle \mathbf{J}^i \rangle = \sigma^{ij} \mathbf{E}^j, \qquad (4.60)$$

where

$$\sigma^{ij} = \frac{i}{\omega} \left[\langle \mathbf{j}^i \mathbf{j}^j \rangle_C(\mathbf{k}, \omega) + \delta^{ij} \frac{ne^2}{mc} \right]$$

$$\sigma^{ij} = \frac{i}{\omega} \langle \mathbf{J}^i \mathbf{J}^j \rangle_C(\mathbf{k}, \omega). \qquad (4.61)$$

The equation above is known as Kubo formula. It expresses the response to an applied electric field of frequency ω and wave-number k. For a uniform field, we should take the limit $\mathbf{k} \to 0$. The DC-conductivity, conversely, would be obtained by taking the limit $\omega \to 0$.

There are different electron scattering mechanisms that lead to a finite conductivity or, equivalently, to a nonzero resistivity. Among these, the most important are thermal and quantum fluctuations of the ion positions (phonons) around the equilibrium Bravais lattice sites. This interaction was studied in detail in Section 3.4 and is described by the Hamiltonian (3.62). Defects and impurities as well are a source of electron scatterings. The whole process is controlled by the average time elapsed between two electron scattering events. This characteristic time interval is called "relaxation time": τ ($\tau \simeq 10^{-14}$s).

Now, concerning the different causes of electron scattering, which produce a nonzero resistance, it is natural to expect that different scattering processes, say impurities and phonons, will lead to different relaxation times. How do we determine the total effective relaxation time? Matthiesen's rule states that they add as

$$\frac{1}{\tau} = \frac{1}{\tau_1} + \frac{1}{\tau_2} + \frac{1}{\tau_3} + \dots \qquad (4.62)$$

The inverse conductivity matrix is the resistivity matrix. We conclude, according to Matthiesen's rule, that the resistivities produced by different sources just add.

An important factor that strongly influences the conductivity is the interaction among the electrons themselves. The Kubo formula is sensitive to electronic interactions and will correctly capture and describe their effect.

In Section 14.4, we determine the current correlation functions in a system of electrons, which is subject to random scatterings from localized centers, being otherwise free. For this calculation, we resort to quantum field theory methods. These yield

$$\langle \mathbf{j}^i \mathbf{j}^j \rangle (\mathbf{k} = 0, \omega) = \delta^{ij} \left(\frac{ne^2}{m} \right) \left[\frac{-i/\tau}{\omega + i/\tau} \right]$$

$$\langle \mathbf{J}^i \mathbf{J}^j \rangle (\mathbf{k} = 0, \omega) = \delta^{ij} \left(\frac{ne^2}{m} \right) \left[\frac{\omega}{\omega + i/\tau} \right], \qquad (4.63)$$

and from this we derive an expression for the conductivity, namely, Eq. (14.42). We also obtain, for this system, the average current remaining after the application of an external pulse of electric field, given by (14.48).

This issue is closely related to superconductivity and to the screening of a magnetic field inside a superconductor, known as the Meissner effect. Interestingly and beautifully, this phenomenon is also closely related to the mechanism of mass generation of the particles associated with the gauge fields that mediate the weak interaction in the Standard Model. Indeed, we will see in the next section that the essence of the Anderson–Higgs mechanism, which is behind both the Meissner effect and the mass generation for a gauge field, is the fact that, in the presence of an incompressible density of charge carriers, n, it is possible to prevent the build-up of the second term in (14.48), thus securing a non-vanishing current, even in the absence of an applied electric field.

Let us now turn to the frequency-dependent conductivity, which is given by (14.42). It expresses the way the system would react in the presence of an external AC-field of frequency ω. The DC-limit is obtained by just taking the limit $\omega \to 0$ in that expression.

The existence of a finite conductivity in a metal has profound consequences on the propagation of electromagnetic waves in these materials. Replacing the current density term on the right-hand side of the Ampère–Maxwell equation with (4.60) produces a conductivity-dependent refraction index that determines the speed of light inside the metal:

$$v = \frac{c}{n(\omega)} \qquad n(\omega) = \sqrt{\epsilon(\omega)} \qquad \epsilon(\omega) = 1 + i \frac{4\pi \sigma(\omega)}{\omega}, \qquad (4.64)$$

where $\epsilon(\omega)$ is the dielectric function. Using the expression for the optical conductivity $\sigma(\omega)$, (14.42), obtained within the non-interacting electron model described above, we can cast it in the form

$$\epsilon(\omega) = 1 - \frac{\omega_P^2}{\omega^2}\left(1 - \frac{i}{\omega\tau}\right), \tag{4.65}$$

where $\omega_P^2 = \frac{4\pi n e^2}{m}$ is the so-called plasma frequency ($\omega_P \simeq 10^{16}$Hz).

Now, observe that the electromagnetic wave-vector modulus satisfies

$$k = \frac{\omega}{c}\sqrt{\epsilon(\omega)}, \tag{4.66}$$

hence for this wave to propagate through the metal, a real positive dielectric function is required. This condition is fulfilled for $\omega > \omega_P$ (notice that $\omega_P\tau \simeq 10^2 \gg 1$); otherwise k will have an imaginary part and the wave will be damped. Consequently, metals are transparent for EM waves with a frequency higher than the plasma frequency, which is above the visible region. Now, for EM waves in the visible, we have $\omega < \omega_P$, but still $\omega\tau \gg 1$. This means the refraction index $n(\omega) = \sqrt{\epsilon(\omega)}$ in this case would be purely imaginary and so would be the wave-vector k, thereby precluding the propagation of EM waves on the visible range in the metal.

Energy conservation forces all the visible light incident on a metal to be reflected, since it cannot propagate inside it. Reflectance is the quantity that measures the reflected fraction of an EM wave inciding onto a metal. For perpendicular incidence, the reflectance of a metal is

$$R = \left|\frac{1 - n(\omega)}{1 + n(\omega)}\right| = \left|\frac{1 - \sqrt{\epsilon(\omega)}}{1 + \sqrt{\epsilon(\omega)}}\right|. \tag{4.67}$$

From our previous analysis, therefore, it becomes clear that for visible light in a metal, the dielectric function turns out to be negative. The refraction index, consequently, is purely imaginary, and according to the above equation we get $R = 1$, implying that all the incident light is completely reflected. This explains, for instance, why the metals shine so much, a quality that is certainly responsible for making some of them so valuable, and consequently for so many historic implications.

In this section, we have studied the ability of a system to transport charge in a dissipative regime corresponding to an average rate of scatterings $1/\tau$, where τ is the relaxation time. In the next section, we examine the phenomenon that makes $\tau \to \infty$, thereby eliminating dissipation in the transport of charge and consequently producing zero resistivity.

4.4 Superconductivity

Introduction

Superconductivity is certainly one of the most interesting, beautiful and useful phenomena in physics. It was discovered by Kamerlingh-Onnes, in 1911, after he had developed the technology of helium liquefaction a few years before. He observed that the electrical resistance of mercury, a metal usually presenting a finite resistivity, would suddenly drop to zero when this material was cooled below a temperature of approximately $4K$, by contact with a liquid helium bath. This behavior would allow the existence of persistent electrical currents, even in the absence of an applied electric field, for as long as we keep the material cooled below that temperature. The phenomenon was called superconductivity. The same behavior was subsequently found in different materials at temperatures ranging up to the order of $20K$. Nevertheless, it would take almost 50 years for the mechanism producing this phenomenon to be properly understood. In the meantime the whole quantum theory itself, an unavoidable instrument for its comprehension, had to be built.

Even though a thorough understanding of what became known as conventional superconductivity had been achieved by the 1960s, a new class of superconducting synthetic materials was found by Bednorz and Müller in 1986, [2] involving a new mechanism that could not be explained in the same way as the conventional superconductivity. These are the so-called High-Tc cuprates, such as $La_{2-x}Sr_xCuO_4$ (LSCO) and $YBa_2Cu_3O_{6+x}$ (YBCO). The critical temperature of the former is about $40K$, whereas that of YBCO is roughly of the order of $90K$. This is above the boiling temperature of nitrogen; hence, for the first time the use of liquid nitrogen as a cooler medium was allowed. Because nitrogen is much more abundant and therefore much cheaper than helium, this opened a vast field of potential technological applications for the new superconductors.

In 2008, an even newer class of unconventional superconducting materials was synthesized, with critical temperatures of the order of $30K$ to $60K$ [4]. These are the iron-based pnictides, such as $Sr_{1-x}K_xFe_2As_2$. Both the cuprates and the pnictides have extremely rich phase diagrams, which should provide many clues for understanding the underlying physical processes in these materials. So far, however, the mechanism or mechanisms responsible for the new forms of superconductivity exhibited by cuprates and pnictides is not completely understood, thus opening a fascinating challenge in this field of research. Cuprate and pnictide superconductors are the subject of Chapters 24 and 25, respectively, whereas a detailed quantum field theory approach to superconductivity is presented in Chapter 23.

In 2015, superconductivity was observed at a temperature of $203.5 K$ in hydrogen sulfide, $H_2 S$, at a pressure of $150 G Pa$ [3]. This is claimed to be conventional superconductivity and has the highest transition temperature so far.

An Order Parameter

The most distinguished feature of superconductivity is the existence of a persistent current, namely, a current that would not decay even in the absence of an applied electric field. In (14.48) we will show that when the electric field is removed, after the application of a pulse, the current decays because the $\mathbf{j}(t)$ component of the current builds up to cancel the other component. This observation contains the key for producing a persistent current: we must prevent the $\mathbf{j}$-term from canceling the $\mathbf{A}_0$-term.

We start by introducing the order parameter, namely, a complex function of the temperature, such that its modulus squared coincides with the persistent current carriers: $\eta^2 = n_s$. It follows that $\Psi(\mathbf{r}) = \eta e^{i\varphi}$ must be the wave function for these charge carriers, being different from zero, wherever they are. It is therefore an order parameter for the superconducting state.

We will see that whenever η is a non-vanishing constant, the $\mathbf{j}$ component of the current will no longer cancel the other one in (14.48), thereby leading to a superconducting state where a persistent current exists. For this purpose, we first examine how to obtain the value of the order parameter at a given temperature and, from it, the current $\mathbf{j}$.

The Landau–Ginzburg Theory

The Landau–Ginzburg theory, proposed in 1950, was a turning point in physics. Its influence reaches an impressive amount of systems, ranging from condensed matter through cosmology to particle physics. It was awarded the Nobel Prize in 2003.

The theory is centered on the concept of an order parameter, introduced by Landau, and is formulated by expressing the free energy as a functional of this order parameter, which in the case of superconductivity is a complex field $\Psi(\mathbf{r})$, as we saw. Assuming the system is close to a critical point marking a phase transition, we must have a small value for $|\Psi(\mathbf{r})|$. Assuming, furthermore, that this is slowly varying spatially, we can express the free energy for particles of mass M and charge q as

$$F[\mathbf{A}, \Psi] = F_0 + \int d^3 r \left\{ \frac{1}{2M} \left| \left[-i\hbar\nabla - \frac{q}{c}\mathbf{A} \right] \Psi \right|^2 + a(T)|\Psi|^2 + \frac{1}{2}b|\Psi|^4 \right\},$$

(4.68)

where $b > 0$ and $a(T) = a_0(T - T_c)$, T_c being the critical temperature for the onset of superconductivity and $a_0 > 0$.

The field equation obtained from (4.68) is

$$\frac{1}{2M} \left\| \left[-i\hbar\nabla - \frac{q}{c}\mathbf{A} \right] \right\|^2 \Psi + \left[a(T) + b|\Psi|^2 \right] \Psi = 0. \tag{4.69}$$

In the SC phase, the average current $\langle \mathbf{J} \rangle$, given by (4.49) and (4.50) must be identified with the supercurrent $\mathbf{J}_S$. Now, this is obtained from the free energy by

$$\mathbf{J}_S^i(\mathbf{r}) = \frac{\delta F[\mathbf{A}]}{\delta \mathbf{A}_i(\mathbf{r})}. \tag{4.70}$$

Considering the momentum-velocity relation in the presence of a vector potential, we find that the velocity operator of the SC condensate is given by

$$\mathbf{V} = \frac{1}{M} \left[-i\hbar\nabla - \frac{q}{c}\mathbf{A} \right]. \tag{4.71}$$

It is not difficult to realize that the solution of (4.69) that minimizes the free energy for $T < T_c$ has a constant modulus $|\Psi|^2 = n_S$, where $n_S \neq 0$ is the density of SC carriers. This vanishes for $T > T_c$ and increases as we lower the temperature (for $T < T_c$).

For such a condensate, the wave-function has a constant modulus, $\sqrt{n_S}$ ($\Psi = \sqrt{n_S}\, e^{i\varphi}$), given by

$$|\Psi| = \begin{cases} \sqrt{\frac{a_0}{b}(T_c - T)} & T < T_c \\ \\ 0 & T > T_c \end{cases}, \tag{4.72}$$

and consequently the velocity eigenvalue is given by

$$\mathbf{v}_S = \frac{1}{M} \left[\hbar\nabla\varphi - \frac{q}{c}\mathbf{A} \right]. \tag{4.73}$$

The average $\mathbf{j}$ component of the current is

$$\langle \mathbf{j} \rangle(t) = n_S \frac{q}{M}\hbar\nabla\varphi, \tag{4.74}$$

and the full current, corresponding to (4.70), is given by

$$\mathbf{J}_S = \langle \mathbf{J} \rangle(t) = n_S \frac{q}{M} \left[\hbar\nabla\varphi - \frac{q}{c}\mathbf{A} \right] = q\, n_S \mathbf{v}_S. \tag{4.75}$$

Notice that the above current is gauge invariant. Indeed, it remains unchanged under the gauge transformation

$$\mathbf{A} \rightarrow \mathbf{A} + \frac{\hbar c}{q}\nabla\theta \quad ; \quad \varphi \rightarrow \varphi - \theta. \tag{4.76}$$

Now, observe that the current in (4.75) is persistent. Even if we remove the applied electric field, for a constant $\mathbf{A} = \mathbf{A}_0$, as we did in the previous subsection, the current will persist, provided $n_S \neq 0$. The phase φ has a coherent spatial distribution, determined by the velocity of the charge carriers. These, conversely, are correlated throughout the sample, through the phase gradient.

One can understand physically why the phase coherence of the superconducting condensate produces a persistent current. Indeed, the individual random electron scatterings, which are responsible for resistivity, just cannot occur without destroying the condensate phase coherence. By behaving collectively, tied by the correlation imposed by the fact that velocity is the gradient of the wave-function phase, the electrons just cannot be scattered individually, thereby producing a resistanceless electric current.

The London Equation

If we take the rotational of (4.75), the first term on the right-hand side vanishes and we get

$$\nabla \times \mathbf{J}_S = -n_S \frac{q^2}{Mc}\mathbf{B}. \tag{4.77}$$

Now, using the Ampère law, we can write the previous equation as

$$\nabla \times \nabla \times \mathbf{B} = -\frac{4\pi n_S q^2}{Mc^2}\mathbf{B}, \tag{4.78}$$

and considering Gauss' law of the magnetic field, we arrive at

$$\nabla^2\mathbf{B} = \frac{4\pi n_S q^2}{Mc^2}\mathbf{B}. \tag{4.79}$$

This is the equation satisfied by the magnetic field in a superconductor. It was first derived by the London brothers in 1935. An identical equation may be derived for the current. Indeed, taking the rotational of (4.77), using Ampère's Law and the fact that there is no net charge density in the absence of electric fields, we get

$$\nabla^2\mathbf{J} = \frac{4\pi n_S q^2}{Mc^2}\mathbf{J}. \tag{4.80}$$

The Meissner Effect

The Meissner effect, discovered in 1933 by Meissner and Ochsenfeld, consists in the fact that the magnetic field is expelled from a superconductor. It can be conveniently explained by the London equation. Indeed, let us consider London equation for the following geometry: a superconductor material, placed at $z > 0$ and the vacuum, at $z < 0$. We assume the system extends to infinity in the x, y directions. This correctly portrays the situation close to boundary of a superconducting sample. The symmetry implies that the magnetic field in the London equation just

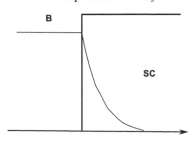

Figure 4.6 Meissner effect: the magnetic field decays exponentially inside a superconductor.

depends on z. One immediately finds, then, from (4.79), that the magnetic field inside the superconductor is given by

$$\mathbf{B}(z) = \mathbf{B}(0) \exp\left(-\frac{z}{\lambda_L}\right) \qquad z > 0, \tag{4.81}$$

where $\lambda_L^2 = \frac{Mc^2}{4\pi n_S q^2}$. The London penetration length $\lambda_L \simeq 10^{-8} m$ determines how deep the magnetic field penetrates the superconductor. Its small value signifies that the magnetic field is essentially zero inside the superconductor, except for a thin, skin-deep region, thus explaining the Meissner effect.

From (4.80) we also see that the current decays exponentially as we recede from the surface into the bulk of a superconductor. Consequently, we come to the surprising conclusion that the persistent current in a superconductor flows along its surface. Notice that this is in agreement with the fact that (4.73), and consequently (4.75), vanish in the bulk, on account of (4.69).

We see that the Landau–Ginzburg theory correctly describes both the existence of a persisting current and the Meissner effect in a superconductor. A key ingredient of this theory is the fact that the modulus of the complex order parameter, which is identified with the wave function of the super-current carriers, acquires a *constant* nonzero value for temperatures below T_c. It is precisely because of this feature that $\langle \mathbf{j} \rangle$ can be written as in (4.74). It follows that this component of the total current no longer cancels the other one, thus allowing the existence of a persistent current.

The Landau–Ginzburg theory describes the onset of this regime, however it does not explain the physical mechanism that produces it. This was an achievement of the BCS theory, proposed by Bardeen, Cooper and Schrieffer in 1957.

The BCS Theory

BCS theory provides the explanation of the mechanism responsible for superconductivity on the basis of the interaction of electrons with the quantized crystalline vibrations of the solid, namely, the phonons. Surprisingly, the same electron-phonon interaction that produces a nonzero resistance increasing with the temperature is, in fact, responsible for the onset of a zero-resistance supercon-ducting state, where the transport of charge occurs without dissipation when the temperature is lowered below a certain critical value T_c in many materials.

A detailed study of the electron-phonon interaction was presented in Section 3.4. Equation (3.62), in particular, expresses the effective electron-electron inter-action, mediated by one-phonon exchange. This expression is remarkable for two reasons. First, this effective interaction becomes negative (that is, attractive) when-ever the energy difference between the two electrons is less than the energy of the exchanged phonon. Secondly, if this energy difference is close to (but less than) the phonon energy, this attractive interaction becomes large and supersedes the Coulomb repulsion between the electrons.

Remember, the phonon frequency is always less than the Debye frequency ω_D, as we saw in Chapter 2. It follows that when the electrons' energy difference $\Delta\epsilon$ is larger than the Debye energy, $\Delta\epsilon > \hbar\omega_D$, the interaction (3.62) will never be attractive. Conversely, for $\Delta\epsilon < \hbar\omega_D$, this will always be attractive for some range of the exchanged phonon energy. The former situation is the one that is most fre-quently found in material systems: the electronic effective interaction is repulsive. This situation is clearly enhanced as we increase the temperature, thereby boosting the electrons' energy differences. Nevertheless, when we lower the temperature in a metal, we reach a situation in which most of the electrons occupy the states below the Fermi surface. Then, a large number of electrons will be close to the Fermi sur-face, and therefore would have small energy differences. Hence, it is likely that a large number of electrons will be found inside a shell of width of the order of $\hbar\omega_D$ around the Fermi surface ($\hbar\omega_D \ll \epsilon_F$). For these electrons, the attractive interaction mediated by phonons (3.62) will be more important than the Coulomb repulsion, and two-electron bound states with opposite spins, called Cooper pairs, will consequently form. These Cooper pairs then form a condensate in which the order parameter has a constant modulus, namely, an incompressible fluid. As we saw above, this is the key condition for the existence of a persistent current. The Cooper pairs therefore are the carriers of the superconducting current.

The net effective electron-electron interaction, which takes into account both the Coulomb repulsion and the phonon-mediated attraction, is described by (3.62), with the interaction potential given by

$$V(\mathbf{k}, \mathbf{q}) = \begin{cases} -\lambda & |\epsilon_{\mathbf{k}} - \epsilon_{\mathbf{k}-\mathbf{q}}| < \hbar\omega_D \\ \\ 0 & |\epsilon_{\mathbf{k}} - \epsilon_{\mathbf{k}-\mathbf{q}}| > \hbar\omega_D \end{cases} \tag{4.82}$$

where $\lambda > 0$ is an effective coupling constant.

In Chapter 23, we use quantum field theory methods to show that this effective, phonon-induced attractive quartic electronic interaction indeed produces an incompressible fluid of Cooper pairs below a certain critical temperature T_c, which is determined as well. This fluid is described by a complex order parameter of constant modulus. As we have seen above, this automatically leads to a superconducting state containing a persistent current.

Magnetic Flux Quantization and Type II Supercondutors

Let us consider expression (4.75) for the current in a superconductor material. We have seen that this current exponentially vanishes inside the material, hence, we must have

$$\hbar \nabla \varphi - \frac{q}{c} \mathbf{A} = 0. \tag{4.83}$$

Integrating this along a closed loop C, we get

$$\oint_C \mathbf{A} \cdot d\mathbf{l} = \frac{hc}{q} \frac{1}{2\pi} \oint_C d\mathbf{l} \cdot \nabla \varphi. \tag{4.84}$$

The left-hand side above is the magnetic flux across the surface bounded by C, namely Φ_C, as implied by the Stokes theorem. Since the Cooper pair wave-function must be single-valued, we have

$$\varphi(2\pi) - \varphi(0) = 2\pi \, n \quad ; \quad n \in \mathbb{Z}. \tag{4.85}$$

We conclude that, for such geometry, the magnetic flux is quantized:

$$\Phi_C = n\Phi_0 \quad , \quad \Phi_0 = \frac{hc}{q}, \tag{4.86}$$

where, in the expression of the flux quantum, we reinstated c and $q = -2e$, corresponding to a Cooper pair.

An example of the geometry used above would be provided by a superconducting ring. In this case, Φ_C would be the quantized magnetic flux across the ring. The most interesting example, however, would be that of a bulk superconductor.

Indeed, by choosing the phase of the Cooper pair wave-function in such a way that

$$\Psi(\mathbf{r}) = \rho(\mathbf{r})e^{i\varphi(\mathbf{r})} = \rho(\mathbf{r})e^{i \, \arg(\mathbf{r})}, \tag{4.87}$$

we immediately comply with (4.85). The function $\arg(\mathbf{r})$, however, is not well defined at the origin, hence, the only possibility to have an acceptable wave-function is that $\rho(\mathbf{r})$ should vanish at the origin. This would open a hole of normal state inside the superconducting bulk, thereby allowing one quantum of magnetic field to pierce through each of these holes. The presence of these quantized magnetic vortices characterizes a class of superconductors called Type II superconductors.

We conclude here our introduction to condensed matter systems. In the second part of this book we present an introduction to quantum field theory, which is an important tool for investigating the former, as we will see in the third part.

Part II
Quantum Field Theory

5

Functional Formulation of Quantum Field Theory

Quantum field theory, from its very inception, emerged as a natural application of the laws of quantum mechanics to systems of fields. This was actually the case already when Einstein applied Planck's concept of quantum to the radiation electromagnetic field in order to explain the photoelectric effect. In its early days, quantum field theory served as a unique framework where the laws of quantum mechanics could be unified with those of the theory of relativity, but soon it had become the main instrument for describing the physics of elementary particles and their interactions. In this area it has produced some of the most accurate theoretical models ever produced in any area of knowledge. Indeed, theoretical predictions in this framework, in some cases, agree with the experiments within up to twelve digits. More recently, the range of applications of quantum field theory was enlarged to include, among other areas, cosmology, astrophysics, hadron physics and condensed matter physics, which is the subject of this book. We start the Second Part, which is about quantum field theory itself, with a chapter providing the basic tools required for operating this powerful theoretical device.

5.1 Functional Integration and Differentiation

5.1.1 Functions and Functionals

Functions and functionals are two important classes of mathematical objects playing a central role in many areas of physics. In both of them, the value of the function or the functional, say, a real number, is determined by a certain input. In the case of functions of a single variable, this input is another real number, called the variable, whereas in the case of functionals the input is in the form of a certain function. Consider the following example.

Let us take the expression

$$y = f(x) \qquad x, y \in \mathbb{R}. \tag{5.1}$$

This may be considered in two different ways. Assuming a fixed functional form f, then y may be considered as a real function of the real variable x, namely, its value is determined by the value of x and in general changes as x is modified. In this case, y is a function of x: $y = f(x)$. Conversely, assuming a fixed value of the argument $x = x_0$, however considering the functional form $f(x_0)$ as arbitrary and subject to change, it follows that the real number y will be determined by the functional form of f. In this case, y will be a functional of f, namely $y = F_{x_0}[f] \equiv f(x_0)$.

Physical quantities in field theory, both at classical and quantum-mechanical levels, are frequently expressed as functionals, which in most cases are of the form

$$F[\varphi] = \int_a^b \mathcal{F}(\varphi(x))dx, \tag{5.2}$$

where $\mathcal{F}$ is a well-known function of φ. In this case, $y = F[\varphi]$ just depends on the functional form of φ. The previous example of a functional, $y = F_{x_0}[f] \equiv f(x_0)$, is a particular case of (5.2), for $\mathcal{F} = f(x)\,\delta(x - x_0)$ and $x_0 \in [a, b]$. Another frequently used functional of the same form is the action functional, where $\mathcal{F}$ is the Lagrangean, and the variable x, the time.

The mathematical formulation of classical and quantum field theory requires the concepts of functional integrals and derivatives. In the next two subsections we shall see how to obtain these natural extensions of the usual derivatives and integrals of functions of one real variable.

5.1.2 Integrals and Derivatives

Let us start by recalling how the integral of a function is defined. Take

$$I(y) = \int_a^y \varphi(x)dx. \tag{5.3}$$

For this purpose, let us discretize the $[a, y]$ interval of the real x-axis in N intervals of uniform width ε, in such a way that $N\varepsilon = y - a$ and

$$\{x_0 = a, x_1, \ldots, x_N = y | x_{i+1} - x_i = \varepsilon\} \overset{\varepsilon \to 0, N \to \infty}{\Longrightarrow} [a, y]. \tag{5.4}$$

It follows that a real function $\varphi(x)$ will become, after such discretization, a collection of N real numbers

$$\{\varphi_1, \varphi_2, \ldots, \varphi_N | \varphi_i = \varphi(x_i) \in \mathbb{R}\} \overset{\varepsilon \to 0, N \to \infty}{\longrightarrow} \varphi(x). \tag{5.5}$$

The Derivative and the Integral

The derivative of the function $\varphi(x)$ at the point x_i is then defined as

$$\frac{\Delta\varphi(x_i)}{\Delta x_i} = \frac{\varphi(x_{i+1}) - \varphi(x_i)}{\varepsilon} \overset{\varepsilon \to 0}{\longrightarrow} \frac{d\varphi(x)}{dx}.$$

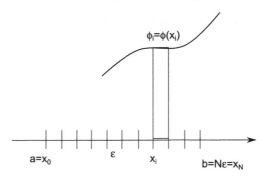

Figure 5.1 The discretization of the interval $[a, b]$

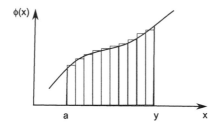

Figure 5.2 The function $F(y) = \int_a^y \phi(x)dx$

Then we have the definite integral of the function $\varphi(x)$ on the interval $[a, y]$, defined by

$$\varepsilon \sum_{i=1}^{N} \varphi_i \xrightarrow{\varepsilon \to 0, N \to \infty} I(a, y) = \int_a^y \varphi(x)dx. \tag{5.6}$$

The Fundamental Theorem of Calculus

It can be verified that $I(a, y)$ is a function of y, and consequently an increment in $I(a, y)$ is clearly given by

$$\Delta I(x_k) = \varepsilon \varphi_k. \tag{5.7}$$

The y-derivative of $I(a, y)$ therefore becomes

$$\frac{dI(y)}{dy} = \lim_{\varepsilon \to 0} \frac{\Delta I(x_k)}{\varepsilon} = \varphi_k \quad ; \quad \frac{dI(a, y)}{dy} = \varphi(y), \tag{5.8}$$

which is a result known as the fundamental theorem of calculus.

5.1.3 Functional Integrals and Functional Derivatives

Functional Integral

Now, consider an arbitrary functional $F[\varphi]$. Upon the discretization described in (5.4), this becomes a *function* of the N real variables defined there, namely

$$F = F(\varphi_1, \varphi_2, \ldots, \varphi_N) \overset{\varepsilon \to 0, N \to \infty}{\longrightarrow} F[\varphi]. \tag{5.9}$$

We therefore define a functional integral over φ, namely an integral sweeping all possible configurations of the function (field) φ as a usual integral over each one of the real variables φ_i, representing the value of the function $\varphi(x)$ at the point x_i:

$$Z = \int D\varphi F[\varphi] = \lim_{\varepsilon \to 0, N \to \infty} \int_{-\infty}^{\infty} d\varphi_1 \ldots \int_{-\infty}^{\infty} d\varphi_N F(\varphi_1, \varphi_2, \ldots, \varphi_N). \tag{5.10}$$

Functional Derivative: Definition

Upon discretization, a functional becomes a function of (infinitely) many variables, as is stated in (5.9). Hence, it follows that the infinitesimal variation of the function F due to a change in the variable φ_i, keeping all the other variables fixed, is the differential

$$\Delta F = \frac{\partial F}{\partial \varphi_i} \Delta \varphi_i. \tag{5.11}$$

The total variation of the multi-variable function $F(\varphi_1, \ldots, \varphi_N)$ is, then,

$$\Delta F = \sum_{i=1}^{N} \frac{\partial F}{\partial \varphi_i} \Delta \varphi_i. \tag{5.12}$$

In the continuum limit, this becomes

$$\Delta F = \lim_{\varepsilon \to 0, N \to \infty} \varepsilon \sum_{i=1}^{N} \frac{\partial F}{\varepsilon \partial \varphi_i} \Delta \varphi_i$$

$$\Delta F = \int_{a}^{b} dx \frac{\delta F}{\delta \varphi}(x) \Delta \varphi(x), \tag{5.13}$$

whence,

$$\frac{\delta F}{\delta \varphi}(x) = \lim_{\varepsilon \to 0, N \to \infty} \frac{\partial F}{\varepsilon \partial \varphi_i}. \tag{5.14}$$

Functional Derivative: Special Cases

Let us focus now on the functional F given by (5.2), which many times appears in the description of physical systems. After the discretization made in (5.4) this becomes the following function of the real variables φ_j:

$$F(\varphi_1, \ldots, \varphi_N) = \varepsilon \sum_{i=1}^{N} \mathcal{F}(\varphi_j)$$

$$F(\varphi_1, \ldots, \varphi_N) = \varepsilon \sum_{i=1}^{N} \mathcal{F}_j, \tag{5.15}$$

where $\mathcal{F}_j \equiv \mathcal{F}(\varphi_j)$.

Now, using the form of F given in (5.15), we find that

$$\frac{\partial F}{\partial \varphi_i} = \frac{\partial \mathcal{F}}{\partial \varphi_i} \varepsilon. \tag{5.16}$$

Hence, in this case, from (5.14) and (5.16),

$$\frac{\delta F}{\delta \varphi}(x) = \lim_{\varepsilon \to 0, N \to \infty} \frac{\partial F}{\varepsilon \partial \varphi_i}$$

$$\frac{\delta F}{\delta \varphi}(x) = \lim_{\varepsilon \to 0, N \to \infty} \frac{\partial \mathcal{F}}{\partial \varphi_i}$$

$$\frac{\delta F}{\delta \varphi}(x) = \frac{\partial \mathcal{F}}{\partial \varphi(x)}. \tag{5.17}$$

The extension of the above formula for the case of a functional where the integrand depends both on the function $\varphi(x)$ and its derivative $\frac{d\varphi(x)}{dx}$, namely

$$G[\varphi] = \int_a^b \mathcal{G}\left(\varphi(x), \frac{d\varphi(x)}{dx}\right) dx, \tag{5.18}$$

can be obtained straightforwardly by replacing (5.16) with

$$\Delta G = \varepsilon \sum_{i=1}^{N} \left[\frac{\partial \mathcal{G}}{\partial \varphi_i} \Delta\varphi_i + \frac{\partial \mathcal{G}}{\partial \varphi_i'} \Delta\varphi_i' \right] \xrightarrow{\varepsilon \to 0, N \to \infty} \int_a^b dx \left[\frac{\partial \mathcal{G}}{\partial \varphi} \Delta\varphi + \frac{\partial \mathcal{G}}{\partial \varphi'} \Delta\varphi' \right]$$

$$\Delta G = \int_a^b dx \left[\frac{\partial \mathcal{G}}{\partial \varphi} - \frac{d}{dx} \frac{\partial \mathcal{G}}{\partial \left(\frac{d\varphi}{dx}\right)} \right] \Delta\varphi, \tag{5.19}$$

where, in the last step, we used the fact that $\Delta\varphi'(x) = \frac{d}{dx}\Delta\varphi(x)$ and integrated by parts the second term, assuming that $\Delta\varphi(a) = \Delta\varphi(b) = 0$. Now, using (5.13), we readily get

$$\frac{\delta G}{\delta \varphi(x)} = \frac{\partial \mathcal{G}}{\partial \varphi} - \frac{d}{dx} \frac{\partial \mathcal{G}}{\partial \left(\frac{d\varphi}{dx}\right)}. \tag{5.20}$$

When G is the action functional, the above expression equated to zero yields the Euler–Lagrange equation that governs the classical behavior of all physical systems. The classical description, therefore, stems from the condition of stationary action known as Hamilton's principle.

5.2 Gaussian Functional Integrals

Let us evaluate here the functional integral

$$S_B = \int D_B\varphi \exp\left\{-\frac{1}{2}\int dxdy[\varphi(x)A(x,y)\varphi(y)]\right\}, \tag{5.21}$$

where the subscript B means we are integrating over bosonic fields. We normalize the above integral, dividing by a constant factor such that, upon discretization, this becomes

$$S_B = \int_{-\infty}^{\infty}\ldots\int_{-\infty}^{\infty}\frac{\varepsilon}{s_0}d\varphi_1\ldots\frac{\varepsilon}{s_0}d\varphi_N \exp\left\{-\frac{1}{2}\varepsilon^2\sum_{ij}\varphi_i A_{ij}\varphi_j\right\}, \tag{5.22}$$

where the $\frac{\varepsilon}{s_0}$-factors correspond to the normalization introduced above and the matrix A_{ij} is supposed to be hermitean. Because of this property, the matrix A may be diagonalized by an orthogonal transformation, namely

$$[OAO^T]_{ij} = a_i\mathbb{I}_{ij} \qquad O^T O = \mathbb{I}, \tag{5.23}$$

where a_i are the eigenvalues of A, and $\mathbb{I}$ is the identity matrix. Then, performing the change of variables

$$\phi_i = O_{ij}\varphi_j \qquad \prod_i d\varphi_i = \det O \prod_i d\phi_i \tag{5.24}$$

and considering the fact that the determinant of an orthogonal matrix is one, the above expression then becomes

$$S_B = \prod_i\int_{-\infty}^{\infty}\frac{\varepsilon}{s_0}d\phi_i \exp\left\{-\frac{1}{2}a_i\varepsilon^2\phi_i^2\right\} = \prod_i\sqrt{\frac{1}{a_i}} = [\text{Det}A]^{-1/2}, \tag{5.25}$$

where we have chosen $s_0 = \sqrt{2\pi}$.

Another extremely useful integral is

$$S_B[J] = \int D_B\varphi \exp\left\{-\frac{1}{2}\int dxdy[\varphi(x)A(x,y)\varphi(y)] + \int dx\varphi(x)J(x)\right\}. \tag{5.26}$$

After discretization it becomes

$$S_B[J] = \int_{-\infty}^{\infty}\ldots\int_{-\infty}^{\infty}\frac{\varepsilon}{s_0}d\varphi_1\ldots\frac{\varepsilon}{s_0}d\varphi_N \exp\left\{-\frac{1}{2}\varepsilon^2\sum_{ij}\varphi_i A_{ij}\varphi_j + \varepsilon\sum_i\varphi_i J_i\right\}, \tag{5.27}$$

Performing the change of variable $\varphi_i \to \eta_i = \varphi_i - A_{ij}^{-1}J_j$, we get

$$S_B[J] = \exp\left\{\frac{1}{2}\varepsilon^2 \sum_{ij} J_i A_{ij}^{-1} J_j\right\}$$

$$\times \int_{-\infty}^{\infty} \cdots \int_{-\infty}^{\infty} \frac{\varepsilon}{s_0}d\eta_1 \cdots \frac{\varepsilon}{s_0}d\eta_N \exp\left\{-\frac{1}{2}\varepsilon^2 \sum_{ij} \eta_i A_{ij}\eta_j\right\}. \qquad (5.28)$$

The last factor above is nothing but the integral (5.25), hence, taking the continuum limit, we find

$$S_B[J] = \exp\left\{\frac{1}{2}\int dxdy[J(x)\Delta(x,y)J(y)]\right\} [\text{Det}A]^{-1/2}, \qquad (5.29)$$

where $\Delta(x,y) = A^{-1}(x,y)$, is the Green function of $A(x,y)$, namely

$$\int dz A(x,z)\Delta(z,y) = \delta(x-y). \qquad (5.30)$$

5.3 Fermion Fields

The path integral formulation of a quantum theory of fermion fields requires the use of the so-called Grassmann variables. These are mathematical objects that generalize real and complex variables in the sense that they anti-commute in spite of not being operators. As a consequence, they have peculiar derivation and integration rules, which turn out to be just the ones needed for the quantum-mechanical path integral formulation of a fermion field.

Given a set of N real Grassmann variables $\{\theta_1, \ldots, \theta_N\}$, then, by definition,

$$\{\theta_i, \theta_j\} = 0 \implies \theta_i^2 = 0. \qquad (5.31)$$

It follows that the most general function of a single Grassmann variable is

$$f(\theta) = a + b\theta, \qquad (5.32)$$

where a and b are real constants.

Differentiation and integration of Grassmann variables are defined, respectively, as

$$\frac{d}{d\theta_i}\theta_j = \delta_{ij}, \quad \int d\theta_i \theta_j = \delta_{ij}. \qquad (5.33)$$

We saw in (5.5) that under the discretization of space, a bosonic field $\varphi(x)$ becomes a collection of N real variables $\{\varphi_1, \ldots, \varphi_N | \varphi_i = \varphi(x_i)\}$. A fermionic field $\psi(x)$, accordingly, under the same discretization, (5.4), will become a collection of Grassmann variables, namely

$$\{\theta_1, \theta_2, \ldots, \theta_N | \theta_i = \psi(x_i)\} \xrightarrow{\varepsilon \to 0, N \to \infty} \psi(x). \qquad (5.34)$$

We can define a functional integral over a fermion field, therefore, by a set of $N \to \infty$ integrals over Grassmann variables. Taking a functional $F[\psi]$ as the integrand, we generalize (5.10) as

$$S_F = \int D\psi\, F[\psi] = \lim_{\varepsilon \to 0, N \to \infty} \int d\theta_1 \dots \int d\theta_N\, F(\theta_1, \theta_2, \dots, \theta_N). \quad (5.35)$$

For a Gaussian fermion integral, we get, instead of (5.21),

$$S_F = \int D_F \psi\, \exp\left\{ -\frac{1}{2} \int dx dy [\psi(x) A(x, y)\psi(y)] \right\}. \quad (5.36)$$

By applying the above discretization, we obtain

$$S_F = \int \frac{d\theta_1}{\varepsilon} \dots \int \frac{d\theta_N}{\varepsilon}\, \exp\left\{ -\frac{1}{2}\varepsilon^2 \sum_{ij} \theta_i A_{ij} \theta_j \right\}, \quad (5.37)$$

where the matrix A_{ij} must be anti-symmetric because of the anti-commuting nature of the θ_i-variables. Also notice that we now have chosen $s_0 = \varepsilon^2$.

According to the Grassmann variables integration rules, only the term of order $N/2$ of the exponential (assume N is even) will be different from zero. This will be

$$S_F = \int \frac{d\theta_1}{\varepsilon} \dots \int \frac{d\theta_N}{\varepsilon} \frac{(-1)^{N/2}}{(N/2)!} \frac{\varepsilon^N}{2^{N/2}} \underbrace{\left[\sum_{ij} \theta_i A_{ij} \theta_j \right] \dots \left[\sum_{ij} \theta_i A_{ij} \theta_j \right]}_{N/2}, \quad (5.38)$$

where we have $N/2$ brackets. Notice that the ε factors cancel out. Straightforward combinatorics then show that

$$S_F = \int d\theta_1 \dots \int d\theta_N \theta_N \dots \theta_2 \theta_1 \sum_{\{P_1, \dots P_N\}} \left[A_{P_1 P_2} \dots A_{P_{N-1} P_N} \right] (-1)^{\epsilon(P)}$$

$$S_F = \sum_{\{P_1, \dots P_N\}} \left[A_{P_1 P_2} \dots A_{P_{N-1} P_N} \right] (-1)^{\epsilon(P)}, \quad (5.39)$$

where we used the fact that the integrals over the Grassmann variables just give one and the sum runs over all permutations in the set $\{P_1, \dots P_N\}$ and $\epsilon(P)$ is the parity of the permutation. The above expression is called the Pfaffian of the matrix A_{ij}, denoted by $\text{Pf}(A)$.

For an $N \times N$ anti-symmetric matrix, as A_{ij}, the determinant vanishes for odd N, whereas for even N we have $det(A) = [\text{Pf}(A)]^2$, or equivalently,

$$\det A_{ij} = \left[\sum_{\{P_1, \dots P_N\}} \left[A_{P_1 P_2} \dots A_{P_{N-1} P_N} \right] (-1)^{\epsilon(P)} \right]^2; \quad (5.40)$$

hence, it follows that

$$S_F = \text{Pf}(A) = [\text{Det}(A)]^{1/2}. \tag{5.41}$$

Following the same steps as before, we can evaluate

$$S_F[\eta] = \int D_F \psi \exp\left\{-\frac{1}{2}\int dxdy[\psi(x)A(x, y)\psi(y)] + \int dx\psi(x)\eta(x)\right\}, \tag{5.42}$$

where ψ and η are fermion fields. By discretizing, making the change of variable $\theta_i \to \xi_i = \theta_i - A_{ij}^{-1}\eta_j$ and going back to the continuum limit, we obtain

$$S_F[\eta] = \exp\left\{\frac{1}{2}\int dxdy[\eta(x)\Delta(x, y)\eta(y)]\right\} [\text{Det}A]^{1/2}, \tag{5.43}$$

where $\Delta(x, y) = A^{-1}(x, y)$, is the Green function of $A(x, y)$.

5.4 Table of Functional Derivatives and Integrals

Here we summarize the results for functional derivative and formulas obtained above.

$$1) \quad \frac{\delta\varphi(y)}{\delta\varphi(x)} = \delta(x - y). \tag{5.44}$$

$$2) \quad \frac{\delta}{\delta\varphi(x)}\int_a^b \varphi(y)J(y)dy = J(x). \tag{5.45}$$

$$3) \quad \frac{\delta}{\delta\varphi(x)}\int_a^b \mathcal{F}(\varphi(y))dy = \frac{\partial\mathcal{F}}{\partial\varphi(x)}. \tag{5.46}$$

$$4) \quad \frac{\delta}{\delta\varphi(x)}\int_a^b \mathcal{G}\left(\varphi(y), \frac{d\varphi(y)}{dy}\right)dy = \frac{\partial\mathcal{G}}{\partial\varphi} - \frac{d}{dx}\frac{\partial\mathcal{G}}{\partial\left(\frac{d\varphi}{dx}\right)}. \tag{5.47}$$

$$5) \quad \int D_B\varphi \exp\left\{-\frac{1}{2}\int dxdy[\varphi(x)A(x, y)\varphi(y)]\right\} = [\text{Det}A]^{-1/2}. \tag{5.48}$$

$$6) \quad \int D_B\varphi \exp\left\{-\frac{1}{2}\int dxdy[\varphi(x)A(x, y)\varphi(y)] + \int dx\varphi(x)J(x)\right\}$$
$$= \exp\left\{\frac{1}{2}\int dxdy[J(x)A^{-1}(x, y)J(y)]\right\} [\text{Det}A]^{-1/2}. \tag{5.49}$$

$$7) \quad \int D_F\psi \exp\left\{-\frac{1}{2}\int dxdy[\psi(x)A(x, y)\psi(y)]\right\} = [\text{Det}A]^{1/2}. \tag{5.50}$$

$$8) \quad \int D_F\psi \exp\left\{-\frac{1}{2}\int dxdy[\psi(x)A(x, y)\psi(y)] + \int dx\psi(x)\eta(x)\right\}$$
$$= \exp\left\{\frac{1}{2}\int dxdy[\eta(x)A^{-1}(x, y)\eta(y)]\right\} [\text{Det}A]^{1/2}. \tag{5.51}$$

5.5 Classical Fields

A classical field is essentially a function of the spatial coordinates, which evolves in time, similarly to any classical variable. Given a field $\varphi(\mathbf{x}, t)$, a fundamental physical quantity, fully determining its classical (and quantum-mechanical) properties is the action $S[\varphi]$, a functional of this field given by

$$S[\varphi] = \int_{-\infty}^{\infty} dt \int d^3x \mathcal{L}\left(\varphi, \partial_\mu \varphi\right). \tag{5.52}$$

In the previous expression, $\mathcal{L}$ is the Lagrangean density, a function of the field and its time and space derivatives that uniquely characterizes a given field theory.

The classical behavior of the field is determined by equations that emerge from the condition that the classical evolution renders the action stationary, namely

$$\frac{\delta S}{\delta \varphi} = 0 \quad \Longrightarrow \quad \frac{\partial \mathcal{L}}{\partial \varphi} - \partial_\mu \frac{\partial \mathcal{L}}{\partial \partial_\mu \varphi} = 0. \tag{5.53}$$

The Euler–Lagrange equation obtained in the last step follows from (5.47) and determines the evolution and all dynamical properties of the classical field. Physical quantities such as the energy, or Hamiltonian, are then usually functionals of the field and its derivatives.

In order to obtain the field theory Hamiltonian, for instance, we start by defining the field-momentum canonically conjugate to φ:

$$\pi(\mathbf{x}, t) = \frac{\partial \mathcal{L}}{\partial \dot{\varphi}}. \tag{5.54}$$

The Hamiltonian and its corresponding density then follow by a Legendre transformation of the Lagrangean density,

$$H = \int d^3x \mathcal{H} \quad ; \quad \mathcal{H} = \pi(\mathbf{x}, t)\dot{\varphi}(\mathbf{x}, t) - \mathcal{L}. \tag{5.55}$$

Notice that the Hamiltonian is a functional $H = H[\pi, \varphi, \nabla\varphi]$.

5.6 Quantum Fields

The universe is quantum-mechanical by nature. As a consequence, the principles of quantum mechanics should be applied to all natural systems including, of course, those in which the basic physical observable is a field. This immediately leads to the concept of a quantum field operator. Consider, for instance, a classical scalar field $\varphi(\mathbf{x}, t)$. It follows that, upon quantization, this must become an operator $\phi(\mathbf{x}, t)$ acting on a Hilbert space. The time dependence normally exhibited by a field hence makes the Heisenberg picture the most natural framework for developing a quantum field theory. Nevertheless, Schrödinger picture field operators have already been considered above in Chapters 2 and 3, for phonons and electrons. The electron field operator was given by (3.3), (3.4), (3.5), (3.7) and (3.8), whereas the phonon field operator was given by (2.13), (2.14) and (2.15). In what follows, we shall use the Feynman formulation as a convenient unifying framework for describing quantum fields.

We can have a clear picture of what a quantum field is by using the discretization of the space coordinate given by (5.4). A classical field $\varphi(\mathbf{x}, t)$ becomes, under such discretization, a collection of real numbers, given by (5.5) (the fact that the space coordinate belongs to $\mathbb{R}^3$ can be easily adapted to the formalism):

$$\{\varphi_1, \varphi_2, \ldots, \varphi_N | \varphi_i = \varphi(\mathbf{x}_i) \in \mathbb{R}\} \xrightarrow{\varepsilon \to 0, N \to \infty} \varphi(\mathbf{x}). \tag{5.56}$$

A quantum field, denoted by $\phi(\mathbf{x}, t)$ in the Heisenberg picture (or $\phi(\mathbf{x})$ in the Schrödinger picture), conversely becomes, under this discretization, the direct product of a set of operators

$$\{\phi_1 \otimes \phi_2, \ldots, \otimes \phi_N\} \xrightarrow{\varepsilon \to 0, N \to \infty} \phi(\mathbf{x}), \tag{5.57}$$

acting on a Hilbert space $\mathcal{H} = \mathcal{H}_1 \otimes \ldots \otimes \mathcal{H}_N$. We have, then, the eigenvalue equation

$$\phi_i | \varphi_i \rangle = \varphi_i | \varphi_i \rangle, \quad i = 1, \ldots, N. \tag{5.58}$$

All the quantum field properties are obtained from the vector state $|\Psi(t)\rangle$, which has its dynamical evolution determined by

$$|\Psi(t)\rangle = e^{-\frac{i}{\hbar}Ht} |\Psi(0)\rangle, \tag{5.59}$$

where H is the quantum Hamiltonian operator derived from (5.55).

5.6.1 Quantum Field Averages

According to the principles of quantum mechanics formulated in the Schrödinger picture, when we measure the field associated to the operator $\phi(\mathbf{x})$ at a time t_i and obtain as the result a certain classical field configuration $\varphi_i(\mathbf{x})$, or, equivalently, the set of discrete real variables $\{\varphi_1^i, \varphi_2^i, \ldots, \varphi_N^i\}$.

It follows that right after this measurement is made, the quantum state of the field must become an eigenstate of the field operator having precisely such field configuration as an eigenvalue, namely

$$|\Psi(t_i)\rangle = |\varphi_i(\mathbf{x})\rangle$$
$$\phi(\mathbf{x})|\varphi_i(\mathbf{x})\rangle = \varphi_i(\mathbf{x})|\varphi_i(\mathbf{x})\rangle. \tag{5.60}$$

Subsequently, the field will evolve according to (5.59) in such a way that, for $t > t_i$,

$$|\Psi(t)\rangle = e^{-\frac{i}{\hbar}H(t-t_i)}|\varphi_i(\mathbf{x})\rangle. \qquad (5.61)$$

Suppose now a second measurement of the field is performed at a later time t_f. Within a quantum-mechanical description, there will be a definite probability amplitude for the subsequent measurement of the field made at such a later time t_f to yield an arbitrary result $\varphi_f(\mathbf{x})$, or, equivalently, $\left\{\varphi_1^f, \varphi_2^f, \ldots, \varphi_N^f\right\}$.

This probability amplitude is given by

$$\langle\varphi_f(\mathbf{x})|\Psi(t_f)\rangle = \langle\varphi_f(\mathbf{x})|e^{-\frac{i}{\hbar}H(t_f-t_i)}|\varphi_i(\mathbf{x})\rangle_S$$
$$= \langle\varphi_f(\mathbf{y}, t_f)|\varphi_i(\mathbf{x}, t_i)\rangle_H, \qquad (5.62)$$

where the last expression is in the Heisenberg picture.

We clearly need a method for calculating the probability amplitude above. This is a central issue in quantum field theory.

The most convenient procedure was provided by Feynman. In a magnificent work, he has shown that this amplitude may be expressed as a functional integral over the classical field φ, weighed by a complex phase, consisting in the action divided by $\hbar$, namely

$$\langle\varphi(\mathbf{y}, t_f)|\varphi(\mathbf{x}, t_i)\rangle_H = \int D\varphi \, \exp\left\{\frac{i}{\hbar}S[\varphi]\right\}\Bigg|_{\varphi(t_i)=\varphi_i(\mathbf{x})}^{\varphi(t_f)=\varphi_f(\mathbf{y})}$$
$$= \int D\varphi \, \exp\left\{\frac{i}{\hbar}\int_{t_i}^{t_f} dt \int d^3x \mathcal{L}\left(\varphi, \partial_\mu\varphi\right)\right\}\Bigg|_{\varphi(t_i)=\varphi_i(\mathbf{x}).}^{\varphi(t_f)=\varphi_f(\mathbf{y})} \qquad (5.63)$$

Notice that the above integral is calculated with the constraint that $\varphi(t_i) = \varphi_i(\mathbf{x})$ and $\varphi(t_f) = \varphi_f(\mathbf{y})$.

The discretization of spatial coordinates introduced at the previous sections of this chapter serves as the operational method for evaluating the functional integral appearing in the expression for the amplitude in (5.63).

Observe that the expression above applies to field operators either in the Heisenberg or in the Schrödinger picture, since

$$\langle\varphi(\mathbf{y}, t_f)|\varphi(\mathbf{x}, t_i)\rangle_H = \langle\varphi(\mathbf{y})|e^{-\frac{i}{\hbar}H(t_f-t_i)}|\varphi(\mathbf{x})\rangle_S. \qquad (5.64)$$

Furthermore, the expectation value of a time-ordered product of Heisenberg operators, in the Feynman formulation, is given by

$$\langle\varphi(\mathbf{y}, t_f)|T\phi(x_1)\ldots\phi(x_N)|\varphi(\mathbf{x}, t_i)\rangle$$
$$= \int D\varphi \, \varphi(x_1)\ldots\varphi(x_N) \exp\left\{\frac{i}{\hbar}\int_{t_i}^{t_f} dt \int d^3x \mathcal{L}\left(\varphi, \partial_\mu\varphi\right)\right\}\Bigg|_{\varphi(t_i)=\varphi_i(\mathbf{x}).}^{\varphi(t_f)=\varphi_f(\mathbf{y})} \qquad (5.65)$$

In the above expression $x_j = (\mathbf{x}_j, t_j)$, $j = 1, \ldots, N$ and $t_i < t_j < t_f$ and T is the time-ordering operator.

A particularly important special case is the one when $t_i \to -\infty$, $t_f \to \infty$. It turns out in this case that for the action to be finite, the field configurations $\varphi_i(\mathbf{x})$ and $\varphi_f(\mathbf{x})$ must reduce to the vacuum values, which are usually zero. Otherwise the action will diverge, thereby giving no contribution to the integral.

We have, therefore, the general quantum average of the time-ordered product of quantum fields:

$$\langle 0|T\phi(x_1)\ldots\phi(x_N)|0\rangle = \frac{1}{\mathcal{N}} \int D\varphi \, \varphi(x_1)\ldots\varphi(x_N) \exp\left\{\frac{i}{\hbar}S[\varphi]\right\}, \qquad (5.66)$$

where

$$S[\varphi] = \int_{-\infty}^{\infty} dt \int d^3x \mathcal{L}\left(\varphi, \partial_\mu\varphi\right)\Big|_{\varphi(-\infty)=0}^{\varphi(\infty)=0} \qquad (5.67)$$

and $\mathcal{N}$ is a normalization factor, guaranteeing that the norm of the vacuum state is normalized to one.

5.7 The Whole Physics in Three Formulas

We can associate to any physical system a certain functional of its dynamical degrees of freedom, called the action, as we saw above. For a system containing a field $\varphi(x)$, in particular, the action is a functional

$$\text{(I)} \quad S = S[\varphi]. \quad \text{(Action)} \qquad (5.68)$$

The whole classical description of the dynamics of an arbitrary system derives from the functional derivative of the action with respect to the dynamical variables. Given the action $S[\varphi]$, we have the whole classical dynamics determined by

$$\text{(II)} \quad \frac{\delta S[\varphi]}{\delta\varphi} = 0. \quad \text{(Classical Physics)} \qquad (5.69)$$

The whole quantum-mechanical description of the dynamics of an arbitrary system, conversely, is obtained from the functional integral of the phase $e^{i\frac{S}{\hbar}}$ over the dynamical field variables. For the specific case of the system containing a field $\varphi(x)$,

$$\text{(III)} \quad \int D\varphi \, e^{\frac{i}{\hbar}S[\varphi]}. \quad \text{(Quantum Physics)} \qquad (5.70)$$

The three equations above contain, in principle, the description of all properties of any physical system, both at the classical and quantum-mechanical levels. The classical description is provided by the functional derivative of the action, while the quantum-mechanical description, by the functional integral of $e^{i\frac{S}{\hbar}}$.

It now becomes clear that for macroscopic systems where $S[\varphi] \gg \hbar$, any trajectories not satisfying (5.69) will be washed out by destructive interference. That is how the macroscopic world seems to behave according to the classical picture. On the other hand, for the microscopic world, we have $S[\varphi] \sim \hbar$, and all trajectories contribute appreciably to the functional integral, thus leading to the well-known quantum effects.

5.8 Finite Temperature

In many applications of quantum field theory in different areas of physics, we must take into account the fact that the system is in contact with a thermal reservoir at a nonzero temperature T. The relevant statistical mechanical quantity to be used in the description of the system is the partition function

$$Z = \text{Tr}\, e^{-\beta H}, \tag{5.71}$$

where $\beta = 1/k_B T$ and H is the Hamiltonian operator. Averages are given by

$$\langle A \rangle = \frac{\text{Tr}\, A e^{-\beta H}}{Z}. \tag{5.72}$$

In order to evaluate the traces, we use a complete set of eigenstates of the field operator $\phi(\mathbf{x}, t = 0)$ now assumed to be in the Schrödinger picture:

$$\phi(\mathbf{x}, t = 0)|\varphi(\mathbf{x})\rangle = \varphi(\mathbf{x})|\varphi(\mathbf{x})\rangle. \tag{5.73}$$

In terms of these, we have

$$Z = \int D\varphi \langle \varphi(\mathbf{x})|e^{-\beta H}|\varphi(\mathbf{x})\rangle. \tag{5.74}$$

Consider now expression (5.63) for $t_i = 0$, which we may write in terms of Schrödinger picture states, namely

$$\langle \varphi_f(\mathbf{x})|e^{-\frac{i}{\hbar} H \Delta t}|\varphi_i(\mathbf{x})\rangle = \int D\varphi \, \exp\left\{ \frac{i}{\hbar} \int_0^{\Delta t} dt \int d^3 x \mathcal{L}\left(\varphi, \partial_\mu \varphi\right) \right\} \Bigg|_{\varphi(0)=\varphi_i}^{\varphi(\Delta t)=\varphi_f}. \tag{5.75}$$

We may cast this in a form similar to (5.74) by means of an analytic continuation into imaginary time, the so-called Wick rotation,

$$\tau = it \qquad \partial_\tau = -i\partial_t \qquad \partial_\mu^E = (\partial_\tau, \partial_i)$$

$$S_E[\varphi] = -\int d\tau \int d^3 x \mathcal{L}_E\left(\varphi, \partial_\mu^E \varphi\right), \tag{5.76}$$

where E is a shorthand for "Euclidean."

Supplementing the Wick rotation by choosing $\Delta t = -i\hbar\beta$, taking periodic boundary conditions $\varphi_i(\mathbf{x}) = \varphi_j(\mathbf{x})$ and integrating on the boundary field configuration $\varphi(\mathbf{x}, \tau = 0) = \varphi(\mathbf{x}, \tau = \hbar\beta)$, we have

$$Z = \int D\varphi \, \exp\left\{-\int_0^{\hbar\beta} d\tau \int d^3x \mathcal{L}_E\left(\varphi, \partial_\mu^E \varphi\right)\right\}\Big|_{\text{periodic.}} \qquad (5.77)$$

The thermal and quantum n-fields' average will be given, accordingly, as

$$\langle T\phi(x_1)\ldots\phi(x_n)\rangle$$
$$= \int D\varphi \, \varphi(\mathbf{x}_1, \tau_1)\ldots\varphi(\mathbf{x}_n, \tau_n) \exp\left\{-\int_0^{\hbar\beta} d\tau \int d^3x \mathcal{L}_E\left(\varphi, \partial_\mu^E \varphi\right)\right\}\Big|_{\text{periodic,}}$$
$$(5.78)$$

where $0 \le \tau_1, \ldots, \tau_n \le \hbar\beta$.

The fact that the temperature is finite imposes severe constraints on the Fourier decomposition of the n-point functions. Let us take, for instance the case where $n = 2$. We can write

$$\Delta(\mathbf{x}, \tau; \mathbf{y}, 0) = \langle T\phi(\mathbf{x}, \tau)\phi(\mathbf{y}, 0)\rangle = \int \frac{d^3k}{(2\pi)^3} \int \frac{d\omega}{2\pi} f(\mathbf{k}, \omega)e^{i\mathbf{k}\cdot(\mathbf{x}-\mathbf{y})}e^{-i\omega\tau}.$$
$$(5.79)$$

Imposing the periodic boundary condition $\Delta(\mathbf{x}, \tau; \mathbf{y}, 0) = \Delta(\mathbf{x}, \tau; \mathbf{y}, \beta)$ implies $e^{-i\omega\beta} = 1$. From this we conclude that $\omega = \omega_n = \frac{2n\pi}{\beta}$.

The result above, however, only applies to the case of bosonic fields. For fermionic fields, we must remember that the T-ordering is defined as

$$T\phi(\mathbf{x}, \tau)\phi(\mathbf{y}, \beta) = \pm\phi(\mathbf{y}, \beta)\phi(\mathbf{x}, \tau), \qquad (5.80)$$

the plus and minus signs applying to bosonic and fermionic fields, respectively. It follows that, in the fermionic case, the boundary condition implies $e^{-i\omega\beta} = -1$, whereupon $\omega = \omega_n = \frac{(2n+1)\pi}{\beta}$ for fermions.

The discrete frequencies ω_n are known as Matsubara frequencies and the Fourier expansion of quantum field averages is modified by replacing the continuous frequency ω by these whenever $T \neq 0$.

For the case of the two-point function, for instance, we would have

$$\langle T\phi(\mathbf{x}, \tau)\phi(0, 0)\rangle = \frac{1}{\beta} \sum_{n=-\infty}^{\infty} \int \frac{d^3k}{(2\pi)^3} f(\mathbf{k}, \omega_n)e^{i\mathbf{k}\cdot\mathbf{x}}e^{-i\omega_n\tau}. \qquad (5.81)$$

We shall use the Matsubara formalism in many instances in order to describe quantum field theory systems at a finite temperature.

5.9 Prescriptions: Meanings and Purposes

It frequently happens that a certain function associated to either quantum or classical fields has an ambiguous mathematical meaning. In such cases a prescription is required in order to make sense out of such functions. Different prescriptions lead to functions exhibiting diverse physical properties, which are used for different practical purposes. Prescriptions, therefore, play a crucial role in determining the mathematical features of a physical system.

Let us take, for instance, Eq. (5.79) for the case of a free, massive scalar field in Euclidean space

$$\Delta_E(\mathbf{x}, \tau; \mathbf{y}, 0) = \int \frac{d^3k}{(2\pi)^3} \int \frac{dk_4}{2\pi} e^{i\mathbf{k}\cdot(\mathbf{x}-\mathbf{y})} e^{-ik_4\tau} \frac{1}{k_4^2 + \mathbf{k}^2 + m^2}.$$

$$(5.82)$$

Undoing the Wick rotation, namely, $\tau = it$, $k_4 = -i\omega$, we see that the integrand becomes singular at $\omega = \pm\omega(\mathbf{k})$, $\omega(\mathbf{k}) = \sqrt{\mathbf{k}^2 + m^2}$. Then, depending on the way we deal with these singularities, the function $\Delta(\mathbf{x}, t; \mathbf{y}, 0)$ will describe different properties of the field and shall be used for different purposes.

5.9.1 Feynman Prescription

The Feynman prescription consists in making

$$\frac{1}{k_4^2 + \mathbf{k}^2 + m^2} \longrightarrow \frac{-i}{\omega^2 - \mathbf{k}^2 - m^2 + i\epsilon}$$

$$\longrightarrow \frac{-i}{[\omega - [\omega(\mathbf{k}) - i\epsilon']] [\omega - [-\omega(\mathbf{k}) + i\epsilon']]},$$

$$(5.83)$$

where $\epsilon' = 2\epsilon\omega(\mathbf{k})$. We see that functions defined through the Feynman prescription allow the Wick rotation to be performed as an analytic continuation in the complex ω-plane. It follows from (5.66) that the function obtained from (5.82) by imposing the Feynman prescription is the time-ordered average of the product of quantum field operators:

$$\Delta_E(\mathbf{x}, \tau; \mathbf{y}, 0) \overset{\omega^2 \to \omega^2 + i\epsilon}{\longrightarrow} \Delta_F(\mathbf{x}, t; \mathbf{y}, 0) = \langle T\phi(\mathbf{x}, t)\phi(\mathbf{y}, 0)\rangle.$$

$$(5.84)$$

Looking at the integral on the complex ω-plane, we see that states with positive frequencies (energies) propagate forward in time, whereas those with negative frequencies (energies) do it backward in time. This is compatible with the picture of positive energy anti-particles as the lack (hole) of a particle with a negative energy. This picture is particularly useful in condensed matter systems, where it expresses the real situation.

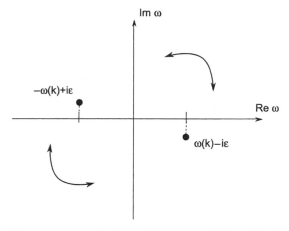

Figure 5.3 The Feynman prescription and its compatibility with the Wick rotation

5.9.2 Retarded and Advanced Prescriptions

Now, suppose we make

$$\frac{1}{k_4^2 + \mathbf{k}^2 + m^2} \longrightarrow \frac{-i}{(\omega + i\epsilon)^2 - \mathbf{k}^2 - m^2}$$
$$= \frac{-i}{[\omega - (\omega(\mathbf{k}) - i\epsilon)][\omega - (-\omega(\mathbf{k}) - i\epsilon)]}, \qquad (5.85)$$

where $\omega(\mathbf{k}) = \sqrt{\mathbf{k}^2 + m^2}$. If we look at the poles in the complex ω-plane, we see they are displaced into the lower half-plane, where the contribution of $t > 0$ comes. There is no contribution from the upper half-plane, which corresponds to $t < 0$. The Euclidean function now becomes the so-called Retarded function, which is proportional to $\theta(t)$.

If we choose, conversely, $\omega \rightarrow (\omega - i\epsilon)$, it is clear we will obtain a function proportional to $\theta(-t)$, the Advanced function.

In summary,

$$\Delta_E(\mathbf{x}, \tau; \mathbf{y}, 0) \xrightarrow{\omega \pm i\epsilon} \Delta_{R,A}(\mathbf{x}, t; \mathbf{y}, 0) = \theta(\pm t)\Delta(\mathbf{x} - \mathbf{y}, t). \qquad (5.86)$$

The Feynman, Retarded and Advanced functions all satisfy

$$(-\Box + m^2)\Delta_{F,R,A}(\mathbf{x}, t; \mathbf{y}, t') = \delta(\mathbf{x} - \mathbf{y})\delta(t - t'), \qquad (5.87)$$

which means they are all Green functions of the Klein–Gordon operator. The Euclidean function Δ_E, also known as Schwinger function, is the Green function of the corresponding Euclidean operator $\Box_E + m^2$.

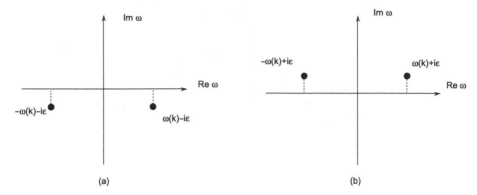

Figure 5.4 The Retarded and Advanced prescriptions, respectively, (a) and (b)

5.9.3 Wightman and Pauli–Jordan Functions

Wightman functions are defined by imposing the so-called spectral condition, $\theta(\pm\omega)\delta(\omega^2 - \mathbf{k}^2 - m^2)$, namely

$$\Delta_{W\pm}(\mathbf{x}, t; \mathbf{y}, 0) = \int \frac{d^3k}{(2\pi)^3} \int \frac{d\omega}{2\pi} e^{i\mathbf{k}\cdot(\mathbf{x}-\mathbf{y})} e^{-i\omega t} \theta(\pm\omega)\delta(\omega^2 - \mathbf{k}^2 - m^2).$$

(5.88)

Using

$$\pi\delta(x) = \lim_{\epsilon \to 0} \left[\frac{i}{x + i\epsilon} - \frac{i}{x - i\epsilon} \right],$$

(5.89)

we may write (5.88) as

$$\Delta_{W\pm}(\mathbf{x}, t; \mathbf{y}, 0) = \int \frac{d^3k}{(2\pi)^3} \int \frac{d\omega}{2\pi} e^{i\mathbf{k}\cdot(\mathbf{x}-\mathbf{y})} e^{-i\omega t} \theta(\pm\omega)$$
$$\times \frac{1}{\pi} \left[\frac{i}{\omega^2 - \mathbf{k}^2 - m^2 + i\epsilon} - \frac{i}{\omega^2 - \mathbf{k}^2 - m^2 - i\epsilon} \right].$$

(5.90)

Performing the ω integration by the method of residues in (5.90) and replacing the last line above with (5.83), we find

$$\Delta_F(\mathbf{x}, t; \mathbf{y}, 0) = \theta(t)\Delta_{W+}(\mathbf{x}, t; \mathbf{y}, 0) + \theta(-t)\Delta_{W-}(\mathbf{x}, t; \mathbf{y}, 0)$$
$$= \langle T\phi(\mathbf{x}, t)\phi(\mathbf{y}, 0)\rangle = \theta(t)\langle \phi(\mathbf{x}, t)\phi(\mathbf{y}, 0)\rangle + \theta(-t)\langle \phi(\mathbf{y}, 0)\phi(\mathbf{x}, t)\rangle. \quad (5.91)$$

We conclude that Wightman functions represent the vacuum expectation value of a simple product of the field operators, namely,

$$\Delta_{W+}(\mathbf{x}, t; \mathbf{y}, 0) = \langle \phi(\mathbf{x}, t)\phi(\mathbf{y}, 0)\rangle$$
$$\Delta_{W-}(\mathbf{x}, t; \mathbf{y}, 0) = \langle \phi(\mathbf{y}, 0)\phi(\mathbf{x}, t)\rangle.$$

(5.92)

From (5.90), we immediately see that Wightman functions satisfy the homogeneous Klein–Gordon field equation

$$(-\Box + m^2)\Delta_{W\pm}(\mathbf{x}, t; \mathbf{y}, 0) = 0, \tag{5.93}$$

hence they are not Green functions.

We finally mention the Pauli–Jordan function,

$$\Delta_{PJ} = \Delta_{W+} - \Delta_{W-} = \langle[\phi(\mathbf{x}, t), \phi(\mathbf{y}, 0)]\rangle, \tag{5.94}$$

which represents the vacuum expectation value of the commutator of quantum field operators.

6

Quantum Fields in Action

The dynamics of a field, either at classical or quantum-mechanical level, is fully determined by the action functional. The classical evolution equations are determined by the condition that the action should be stationary, whereas a functional integral over a phase factor containing the action determines the evaluation of quantum field averages. The physical content of a quantum field theory is completely encoded in the functions $G^{(n)}(x_1, \ldots, x_n) = \langle 0|T\phi(x_1)\ldots\phi(x_n)|0\rangle$. These contain, for instance, all information about scattering cross-sections, energy-spectrum, bound-states of the associated quanta (particles), phase transitions, among others. It can be shown that they are the Green functions of certain appropriate operators, closely related to the proper vertices $\Gamma^{(n)}(x_1, \ldots, x_n)$.

6.1 Green Functions and Their Generating Functionals

6.1.1 The Green Function Generating Functionals

In the previous chapter, we described a detailed method for evaluating each of the functions $G^{(n)}(x_1, \ldots, x_n)$. It would be quite convenient, however, to have available a general method for determining at once all of the $G^{(n)}(x_1, \ldots, x_n)$ functions of a given quantum field theory. For this purpose, we introduce the generating functional

$$Z[J] = \langle 0|T \exp\left\{i \int d^4x\, \phi(x) J(x)\right\}|0\rangle. \tag{6.1}$$

Using (5.45), we quickly conclude that

$$\langle 0|T\phi(x_1)\ldots\phi(x_n)|0\rangle = \frac{\delta^n\, Z[J]}{i^n \delta J(x_1)\ldots\delta J(x_n)}\Big|_{J=0}. \tag{6.2}$$

It follows from this that we may expand $Z[J]$ as a functional Taylor series, given by

$$Z[J] = \sum_{n=0}^{\infty} \frac{i^n}{n!} \int d^4x_1 \dots d^4x_n \langle 0|T\phi(x_1)\dots\phi(x_n)|0\rangle J(x_1)\dots J(x_n). \quad (6.3)$$

6.1.2 The Connected Green Function Generating Functionals

Another quite useful generating functional is $W[J]$, which is introduced by the relation

$$Z[J] = e^{iW[J]} \quad ; \quad W[J] = -i \ln Z[J]. \quad (6.4)$$

Upon functional differentiation, it produces the so-called connected Green functions, namely

$$\langle 0|T\phi(x_1)\dots\phi(x_n)|0\rangle_C = \frac{i \, \delta^n \, W[J]}{i^n \delta J(x_1)\dots\delta J(x_n)}\bigg|_{J=0}. \quad (6.5)$$

These are n-point Green functions that do not contain in their expression any terms corresponding to lower than n Green functions.

Using this, we get, for $n = 1$,

$$\langle 0|\phi(x_1)|0\rangle_C = \frac{\delta \, W[J]}{\delta J(x_1)} = \frac{-i}{Z[J]} \frac{\delta \, Z[J]}{\delta J(x_1)}, \quad (6.6)$$

and, accordingly, for $n = 2$,

$$\begin{aligned}
\langle 0|\phi(x_1)\phi(x_2)|0\rangle_C &= \frac{-i\delta^2 \, W[J]}{\delta J(x_1)\delta J(x_2)} \\
&= -\frac{1}{Z[J]} \frac{\delta^2 \, Z[J]}{\delta J(x_1)\delta J(x_2)} + \frac{1}{Z^2[J]} \frac{\delta \, Z[J]}{\delta J(x_1)} \frac{\delta \, Z[J]}{\delta J(x_2)} \\
&\xrightarrow{J\to 0} \langle 0|\phi(x_1)\phi(x_2)|0\rangle - \langle 0|\phi(x_1)|0\rangle\langle 0|\phi(x_2)|0\rangle. \quad (6.7)
\end{aligned}$$

We see that, indeed, the connected 2-point function is defined by removing from $G^{(2)}$ the $G^{(1)}$ contributions.

The $W[J]$ functional can be expanded in terms of the connected functions as

$$W[J] = \sum_{n=0}^{\infty} \frac{i^{n-1}}{n!} \int d^4x_1 \dots d^4x_n \langle 0|T\phi(x_1)\dots\phi(x_n)|0\rangle_C J(x_1)\dots J(x_n).$$

$$(6.8)$$

6.2 Proper Vertices and Their Generating Functional

A third generating functional, which also turns out to be extremely useful, is defined as a functional Legendre transform of $W[J]$. Indeed, from

$$\frac{\delta\, W[J]}{\delta J(x)} = \langle 0|\phi(x)|0\rangle_C \equiv \varphi_C(x), \tag{6.9}$$

we introduce

$$\Gamma[\varphi_C] = W[J] - \int d^4x\, J(x)\varphi_C(x). \tag{6.10}$$

From (6.9) and (5.45), it becomes clear that

$$\frac{\delta\, \Gamma[\varphi_C]}{\delta J(x)} = 0 \qquad \frac{\delta\, \Gamma[\varphi_C]}{\delta \varphi_C(x)} = -J(x). \tag{6.11}$$

The functional $\Gamma[\varphi_C]$ generates the functions

$$\Gamma^{(n)}(x_1 \ldots x_n) = \left. \frac{\delta^n\, \Gamma[\varphi_C]}{\delta\varphi_C(x_1)\ldots \delta\varphi_C(x_n)} \right|_{\varphi_C=0}, \tag{6.12}$$

which are called proper vertices. In terms of these we may make the functional expansion

$$\Gamma[\varphi_C] = \sum_{n=0}^{\infty} \frac{1}{n!} \int d^4x_1 \ldots d^4x_n \Gamma^{(n)}(x_1 \ldots x_n)\varphi_C(x_1)\ldots \varphi_C(x_n). \tag{6.13}$$

We shall see that $\Gamma[\varphi_C]$ is the functional that generalizes the classical action, but taking into account all quantum effects. The analysis of its behavior, in particular at a finite temperature, allows therefore the obtainment of a realistic phase diagram of the system including the characterization of phase transitions.

We now show that the connected two-point function given by (6.5) is the Green function of the $\Gamma^{(2)}$ proper vertex. Indeed, from (6.11) we have

$$\frac{\delta^2\, \Gamma[\varphi_C]}{\delta\varphi_C(x)\delta\varphi_C(\xi)} = -\frac{\delta\, J(x)}{\delta\varphi_C(\xi)}, \tag{6.14}$$

whereas, from (6.9)

$$\frac{\delta^2\, W[J]}{\delta J(\xi)\delta J(y)} = \frac{\delta\varphi_C(\xi)}{\delta\, J(y)}. \tag{6.15}$$

It follows, by taking the convolution of the above Γ and W second functional derivatives and using (5.44), that

$$-\int d^4\xi \left[\frac{\delta^2\Gamma[\varphi_C]}{\delta\varphi_C(x)\delta\varphi_C(\xi)}\right]\left[\frac{\delta^2 W[J]}{\delta J(\xi)\delta J(y)}\right] = \int d^4\xi\, \frac{\delta J(x)}{\delta\varphi_C(\xi)}\, \frac{\delta\varphi_C(\xi)}{\delta J(y)}$$

$$= \frac{\delta J(x)}{\delta J(y)} = \delta(x-y). \tag{6.16}$$

Considering the above expression at $J = \varphi_C = 0$, we have

$$\int d^4\xi\, \Gamma^{(2)}(x,\xi)\left[-iG_C^{(2)}(\xi,y)\right] = \delta(x-y), \tag{6.17}$$

where $G_C^{(2)}(x, y) = \langle 0|T\phi(x)\phi(y)|0\rangle_C$. Using the fact that the Fourier transform of a convolution is a product, we have, in energy-momentum space

$$\Gamma^{(2)}(p)\left[-iG_C^{(2)}(p)\right] = 1 \quad ; \quad G_C^{(2)}(p) = \frac{i}{\Gamma^{(2)}(p)}. \tag{6.18}$$

These expressions show that the two-point, time-ordered vacuum expectation value of the field operator is the Green function of the proper vertex $\Gamma^{(2)}$. We will see that this property will have far-reaching consequences.

6.3 Free Fields

Let us consider the case of a free field theory, for which the action is a quadratic functional of the field. We have seen examples of such theories in the case of phonons in the harmonic approximation, for instance. The name "free" stems from the fact that the associated quanta or particles do not interact. Here we evaluate the functionals introduced above for the case of a free theory.

Let us assume the action functional is given by

$$S_0[\varphi] = \frac{1}{2}\int d^4x \left[\partial_\mu\varphi\partial^\mu\varphi - m^2\varphi^2\right]. \tag{6.19}$$

The corresponding Hamiltonian is

$$H_0[\varphi] = \frac{1}{2}\int d^3x \left[\pi^2 + \nabla\varphi \cdot \nabla\varphi + m^2\varphi^2\right], \tag{6.20}$$

where the canonically conjugate momentum is given by (5.54).

From (5.66), (5.67) and (6.1), it follows that

$$Z_0[J] = \frac{1}{\mathcal{N}}\int D\varphi \, \exp\left\{i\int d^4x \left[\frac{1}{2}\partial_\mu\varphi\partial^\mu\varphi - \frac{1}{2}m^2\varphi^2 + J\varphi\right]\right\}, \tag{6.21}$$

where $\mathcal{N}$ is chosen in such a way that $Z_0[J = 0] = 1$.

Now, performing the Wick rotation (5.76), we can cast the above expression in the form

$$Z_0[J] = \frac{1}{\mathcal{N}}\int D\varphi \, \exp\left\{-\int d^4x_E \left[\frac{1}{2}\varphi[-\Box_E + m^2]\varphi - J\varphi\right]\right\}. \tag{6.22}$$

From (5.49), it follows that

$$Z_0[J] = \exp\left\{\frac{1}{2}\int d^4x_E d^4y_E \, J(x)\Delta_E(x - y)J(y)\right\}, \tag{6.23}$$

where $\Delta_E(x - y)$ is the Green function of the operator in the quadratic term in (6.22):

$$[-\Box_E + m^2]\Delta_E(x - y) = \delta(x - y)$$

$$\Delta_E(x) = \int \frac{d^4k}{(2\pi)^4} \frac{e^{i[\mathbf{k}\cdot\mathbf{x} - k_4 x_4]}}{k_4^2 + \mathbf{k}^2 + m^2} \tag{6.24}$$

Going back to real time (Minkowski space), we have

$$Z_0[J] = \exp\left\{\frac{i}{2}\int d^4x d^4y\, J(x)\Delta_F(x - y)J(y)\right\}, \tag{6.25}$$

where

$$\Delta_F(\mathbf{x}, t) = \int \frac{d\omega}{2\pi} \int \frac{d^3k}{(2\pi)^3} \frac{e^{i[\mathbf{k}\cdot\mathbf{x} - \omega t]}}{\omega^2 - [\mathbf{k}^2 + m^2] + i\epsilon} \tag{6.26}$$

is the Feynman propagator. Notice the inclusion of the Feynman prescription factor $i\epsilon$, which is required for the Wick rotation to be a genuine analytic continuation, meaning it does not go over any poles.

From (6.25) we can immediately infer that the functional $W_0[J]$ is given by

$$W_0[J] = \frac{1}{2}\int d^4x d^4y\, J(x)\Delta_F(x - y)J(y). \tag{6.27}$$

From (6.9) it follows that

$$\varphi_C(x) = \int d^4y\, \Delta_F(x - y)J(y). \tag{6.28}$$

Inserting (6.28) in (6.27) and using (6.10), we immediately find

$$\Gamma_0[\varphi_C] = \frac{1}{2}\int d^4x d^4y J(x)\delta(x - y)\varphi_C(y)$$

$$\Gamma_0[\varphi_C] = \frac{1}{2}\int d^4x d^4y J(x)\left(-\Box_y - m^2\right)\Delta_F(x - y)\varphi_C(y)$$

$$\Gamma_0[\varphi_C] = \frac{1}{2}\int d^4y \varphi_C(y)\left(-\Box_y - m^2\right)\varphi_C(y), \tag{6.29}$$

where in the last step we used (6.28).

Now, using the expressions just derived for the basic generating functionals, we obtain some fundamental properties of free quantum field theories. From (6.27) we have that the only connected Green function, for a free quantum field theory, is the two-points one, namely

$$G_{0,C}^{(2)}(x, y) = i\Delta_F(x - y). \tag{6.30}$$

Now, the non-connected functions can be obtained from the functional $Z_0[J]$, given by (6.25). The result implies only the n-even functions are non-vanishing and given by products of 2-point functions:

$$G_0^{(n)}(x_1, \ldots, x_n)$$
$$= \sum_{\{P_1 \ldots P_{2n}\}} i \Delta_F(x_{P_1} - x_{P_2}), \ldots, i \Delta_F(x_{P_{2n-1}} - x_{P_{2n}}); n = \text{ even}$$
$$G_0^{(n)}(x_1, \ldots, x_n) = 0; \quad n = \text{odd}, \tag{6.31}$$

where the sum goes over all permutations in the set $\{P_1 \ldots P_{2n}\}$. This result is known as the Wick theorem.

From the effective action generating functional (6.29), we obtain the only proper vertex occurring in the free field case:

$$\Gamma_0^{(2)}(x, y) = -(\Box + m^2)\delta(x - y)$$
$$\Gamma_0^{(2)}(p) = p^2 - m^2. \tag{6.32}$$

Using (6.26) and (6.30), we can verify that indeed the two-point proper vertex $\Gamma^{(2)}$ and the two-point connected Green function $G_C^{(2)}$ satisfy (6.18).

Knowledge of $G_C^{(2)}$ allows us to infer about the energy spectrum of the system. Indeed, performing the ω integral in (6.26) by the method of residues, we conclude that only the poles of the integrand contribute to the dispersion relation $\omega(\mathbf{k})$ of $G_C^{(2)}$. The poles occur at $\omega = \pm\sqrt{\mathbf{k}^2 + m^2}$, which correspond to the energy of a free relativistic particle of momentum $\mathbf{k}$ and mass m, thus confirming that the quanta associated to the field described by the action (6.19) are free relativistic particles of mass m.

6.4 Interacting Fields

Let us consider now a field theory described by the action

$$S[\varphi] = S_0[\varphi] + S_I[\varphi] = \int d^4x \left[\frac{1}{2}\partial_\mu\varphi\partial^\mu\varphi - \frac{1}{2}m^2\varphi^2 - V(\varphi) \right], \tag{6.33}$$

where the field potential $V(\varphi)$ is a function of φ containing higher than quadratic terms.

Now, the $Z[J]$ functional generator, according to (5.66), (5.67) and (6.1), will be given by

$$Z[J] = \frac{1}{N} \int D\varphi \, \exp\left\{ i \int d^4x \left[\frac{1}{2}\partial_\mu\varphi\partial^\mu\varphi - \frac{1}{2}m^2\varphi^2 - V(\varphi) + J\varphi \right] \right\}. \tag{6.34}$$

This functional integral can no longer be evaluated exactly; however, with the help of the identity

$$\varphi(x) \exp\left\{ i \int d^4x \, J\varphi \right\} = -i \frac{\delta}{\delta J(x)} \exp\left\{ i \int d^4x \, J\varphi \right\}, \tag{6.35}$$

we can re-write (6.34) as

$$Z[J] = \exp\left\{ i \int d^4 x \, V\left(-i\frac{\delta}{\delta J(x)}\right)\right\} Z_0[J], \tag{6.36}$$

where $Z_0[J]$ is given by (6.25). Expanding the exponential in the expression above, we can write the exact generating functional of an interacting theory as a sum of terms, each of them containing a finite number of functional derivatives of the free theory generator $Z_0[J]$,

$$Z[J] = \sum_n \frac{(-i)^n}{n!} \int d^4 x_1 \ldots \int d^4 x_n \, V\left(-i\frac{\delta}{\delta J(x_1)}\right) \ldots V\left(-i\frac{\delta}{\delta J(x_n)}\right)$$
$$\times \exp\left\{\frac{i}{2}\int d^4 x d^4 y \, J(x)\Delta_F(x-y)J(y)\right\}. \tag{6.37}$$

This is a useful expression, as it allows us, according to (6.2), to express an arbitrary n-point Green function in terms of a combination of Feynman propagators $\Delta_F(x - y)$. By truncating the series at some finite value of n, we would obtain a perturbative result for the generating functional and associated Green functions.

6.5 Feynman Graphs

An extremely convenient method for calculating each term of the above expansion was devised by Feynman. This consists in providing a graphical representation of each of these terms, according to some "Feynman rules." According to these, each Feynman propagator is represented by a line, whereas an interaction term $V(\varphi) = \lambda\varphi^n$ is represented by a point, out of which n lines emerge, which we call the interaction "vertex." One can obtain all terms appearing in the expansion above by just adding vertices and lines, according to the Feynman rules, very much like assembling the pieces of a puzzle.

6.5.1 The Exact Propagator

Let us consider now the exact two-point Green function in momentum space,

$$G_C^{(2)}(p) = \frac{i}{\Gamma^{(2)}(p)}. \tag{6.38}$$

We write the exact two-point vertex function as

$$\Gamma^{(2)}(p) = \Gamma_0^{(2)}(p) - \Sigma(p)$$
$$\Gamma^{(2)}(p) = p^2 - m^2 - \Sigma(p), \tag{6.39}$$

where $\Sigma(p)$ is called the self-energy. This is frequently expressed as the Schwinger–Dyson equation,

$$[G_C^{(2)}(p)]^{-1} = [G_{0,C}^{(2)}(p)]^{-1} + i\Sigma(p). \tag{6.40}$$

Then, we can write the exact two-point Green function as

$$G_C^{(2)}(p) = \frac{i}{p^2 - m^2 - \Sigma(p)} \tag{6.41}$$

or, equivalently,

$$G_C^{(2)}(p) = G_{0,C}^{(2)}(p) \left[\frac{1}{1 + i\Sigma(p)G_{0,C}^{(2)}(p)} \right]$$
$$G_C^{(2)}(p) = G_{0,C}^{(2)}(p) + G_{0,C}^{(2)}(p)[-i\Sigma(p)]G_{0,C}^{(2)}(p) + \dots \tag{6.42}$$

The corresponding Feynman graphs are shown in Figs. 6.1 and 6.2, for the case of a φ^3 interaction.

Different degrees of approximation can then be used. In the tree level approximation, for instance, only the first term in the expansion in Fig. 6.1 is considered. In the so-called RPA approximation, conversely, the whole series of the expansion in Fig. 6.1 is considered, however the self-energy is approximated by just the first term in the series depicted in Fig. 6.2.

Now, looking at the two-point exact Green function in coordinate space, we have

$$G_C^{(2)}(\mathbf{x}, t) = i \int \frac{d\omega}{2\pi} \int \frac{d^3k}{(2\pi)^3} \frac{e^{i[\mathbf{k}\cdot\mathbf{x}-\omega t]}}{\omega^2 - [\mathbf{k}^2 + m^2 + \Sigma(\omega, \mathbf{k})] + i\epsilon}. \tag{6.43}$$

Observe that we can no longer express the dispersion relation as a free particle Einstein energy-momentum relation, as we did before. This is an evidence of the fact that the quanta associated to the fields described by the action (6.33) are

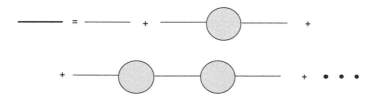

Figure 6.1 The exact propagator for an interacting theory

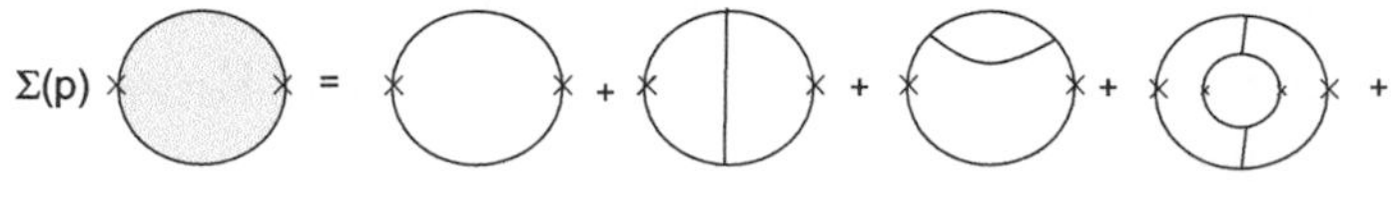

$\Sigma(p)$

Figure 6.2 The Self-Energy expansion for a theory with a φ^3 interaction

interacting particles. Nevertheless, we will see that in the large-distances asymptotic limit they behave as free particles, but with a renormalized mass, which is the physical one.

6.6 The Effective Action and the Effective Potential

Consider the expression for the functional $\Gamma[\varphi_C]$, (6.13). Using (6.39), we can cast this functional in the form

$$\Gamma[\varphi_C] = \frac{1}{2} \int d^4 x_1 \int d^4 x_2 \varphi_C(x_1) \left[(-\Box + m^2)\delta(x_1 - x_2) + \Sigma(x_1 - x_2) \right] \varphi_C(x_2)$$
$$+ \sum_{n=3}^{\infty} \frac{1}{n!} \int d^4 x_1 \dots d^4 x_n \Gamma^{(n)}(x_1 \dots x_n)\varphi_C(x_1) \dots \varphi_C(x_n). \qquad (6.44)$$

Noting that each function $\Gamma^{(n)}$ contains the "contact" terms, which appear in the action, we conclude that the functional $\Gamma[\varphi_C]$ is a generalization of the classical action. This contains, besides the original terms appearing in the classical action, some new terms generated by the interaction. $\Gamma[\varphi_C]$, consequently, is called the effective action.

A useful concept is that of the effective potential. Let us envisage a situation in which the vacuum expectation value φ_C given by (6.9) is a constant. This should occur for $J = 0$. In this case, the functional Γ, (6.13), is given by

$$\Gamma[\varphi_C] = \sum_{n=0}^{\infty} \frac{1}{n!} \int d^4 x_1 \dots d^4 x_n \Gamma^{(n)}(x_1 \dots x_n)\varphi_C^n. \qquad (6.45)$$

Now, the integrals above are nothing but the Fourier transform of the proper vertices at zero energy and momentum:

$$\Gamma^{(n)}(p_1 = 0 \dots p_n = 0) = \int d^4 x_1 \dots d^4 x_n \Gamma^{(n)}(x_1 \dots x_n). \qquad (6.46)$$

Hence, in this case, the $\Gamma[\varphi_C]$ functional becomes a function of φ_C, given by

$$V_{\text{eff}}(\varphi_C) = \sum_{n=0}^{\infty} \frac{1}{n!} \Gamma^{(n)}(p_1 = 0 \dots p_n = 0)\varphi_C^n. \qquad (6.47)$$

Now observe that the proper vertices $\Gamma^{(n)}(p_i = 0)$ have a perturbative expansion starting with the tree level component that is simply the mass, in the case of $\Gamma^{(2)}$, or a coupling parameter, for higher n. When the tree level vertex is multiplied by the corresponding φ_C power, it yields the classical potential $V_{\text{cl}}(\varphi_C)$.

We have, therefore,

$$V_{\text{eff}}(\varphi_C) = V_{\text{cl}}(\varphi_C) + \mathcal{V}(\varphi_C), \qquad (6.48)$$

where $\mathcal{V}(\varphi_C)$ represents the quantum corrections to the classical potential. We conclude that the effective potential is a generalization of the classical potential but with the important difference that it already includes quantum effects. It is, therefore, a powerful tool for studying the phase diagram of the system, both at zero and finite temperatures. In the first case, it is particularly useful for describing the so-called quantum phase transitions, which occur at $T = 0$, due to quantum fluctuations.

6.6.1 The $Z[J]$-Functional of QED: the Fermionic Determinant

Before proceeding to study renormalization in QFT, we introduce Quantum Electrodynamics of Dirac fermions (QED) as an example of an interacting QFT.

QED is the quantum theory of electrons and positrons interacting through the electromagnetic field, which is also quantum-mechanical. Although its precise mathematical formulation was not achieved before the early 1950s, its conception dates back to the pioneering work of Einstein on the photoelectric effect, in 1905, when he proposed the quantization of the electromagnetic radiation field. In modern times, QED has become one of the most successful theoretical models ever created.

Here we present an instructive calculation of the Γ-functional of QED, which will serve to clarify the physical meaning of this important quantity.

QED is defined by the Lagrangean

$$\mathcal{L} = -\frac{1}{4}F_{\mu\nu}F^{\mu\nu} + i\overline{\psi}\not{\partial}\psi - m\overline{\psi}\psi - e\overline{\psi}\gamma^{\mu}\psi A_{\mu}, \tag{6.49}$$

where the four-vector A_{μ} is the photon field and the four-components spinor ψ, the electron-positron field. $F_{\mu\nu} = \partial_{\mu}A_{\nu} - \partial_{\nu}A_{\mu}$ is the electromagnetic field intensity tensor, γ^{μ} are the 4×4 Dirac matrices, $\not{A} = \gamma_{\mu}A^{\mu}$ and e, m are, respectively, the electron charge and mass.

The generating functional of photon correlation functions is

$$Z[J_{\mu}] = e^{iW[J_{\mu}]} = \int DA_{\mu}D\psi D\overline{\psi} \exp\left\{i\int d^4x\left[\mathcal{L} + J_{\mu}A^{\mu}\right]\right\}. \tag{6.50}$$

Performing the integrations on the fermion field in the expression above, we obtain, using the table of functional integrals provided in Chapter 5,

$$Z[J_{\mu}] = \int DA_{\mu}\exp\left\{i\int d^4x\left[\mathcal{L}_M + J_{\mu}A^{\mu}\right]\right\}\frac{\text{Det}[(i\not{\partial} + m + \not{A})]}{\text{Det}[(i\not{\partial} + m)]}$$

$$Z[J_{\mu}] = \int DA_{\mu}\exp\left\{i\int d^4x\left[\mathcal{L}_M + J_{\mu}A^{\mu}\right] - \text{Tr}\ln\left[(1 + \frac{\not{A}}{i\not{\partial} + m})\right]\right\}, \tag{6.51}$$

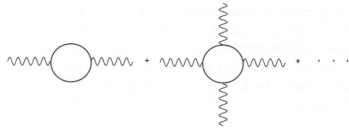

Figure 6.3 Graphs contributing to $\operatorname{Tr}\ln\left[(1+\frac{A}{i\not\partial+m})\right]$

where $\mathcal{L}_M$, the Maxwell Lagrangean, is the first term in (6.49) and we have exponentiated the determinant coming from the fermionic integrals and used the fact that $\ln\operatorname{Det}A = \operatorname{Tr}\ln A$. The trace above is given by the graphs of Fig. 6.3. These are one-loop graphs with the insertion of n A_μ fields. Only even numbers n contribute, by virtue of Furry's theorem. We see that the interaction with the electrons generates an infinity of nontrivial interaction terms among the photons.

6.6.2 Current Correlator: a Sample Calculation in QED

The electric current operator in QED is given by $j^\mu = e\bar{\psi}\gamma^\mu\psi$. Let us determine here what is the current-current correlation function. This will be very useful later on, in different applications of QFT in condensed matter systems.

The current two-point correlator is defined as

$$\langle j^\mu j^\nu\rangle = \int DA_\mu \int D\psi\, D\bar{\psi}\, \exp\left\{i\int d^4x \mathcal{L}\right\} j^\mu j^\nu. \qquad (6.52)$$

The generating functional of current correlation functions is defined in such a way that

$$\langle j^{\mu_1}\dots j^{\mu_n}\rangle = \frac{\delta^n Z[K_\mu]}{\delta K_{\mu_1}\dots\delta K_{\mu_n}}\Big|_{K_\mu=0}. \qquad (6.53)$$

A functional $Z[K_\mu]$ exhibiting this property can be written, up to a constant, as

$$Z[K_\mu] = \int DA_\mu \int D\psi\, D\bar{\psi}\, \exp\left\{i\int d^4x \left[\mathcal{L} + e\bar{\psi}\gamma^\mu\psi K_\mu\right]\right\}. \qquad (6.54)$$

It is not difficult to verify that this functional will generate (6.52), through (6.53).

The corresponding generator of connected correlation functions can be written as

$$W[K_\mu] = -i\ln Z[K_\mu].$$

Then, integrating (6.54) on the fermion fields, one obtains

$$Z[K_\mu] = \int DA_\mu \exp\left\{ i \int d^4x \mathcal{L}_M - \text{Tr} \ln\left[1 + \frac{(\slashed{A} + \slashed{K})}{i\slashed{\partial} + m} \right] \right\}$$

$$Z[K_\mu] = \int DA_\mu \exp\left\{ i \int d^4x \mathcal{L}_M + \sum_{n=0}^{\infty} \frac{1}{n!} \int d^4x_1 \ldots \int d^4x_n \right.$$

$$\left. \Pi_0^{\mu_1 \ldots \mu_n} (A_{\mu_1} + K_{\mu_1}) \ldots (A_{\mu_n} + K_{\mu_n}) \right\}, \tag{6.55}$$

where the $\Pi_0^{\mu_1 \ldots \mu_n}$, as before, are one-loop electron graphs with n field insertions. Now, using (6.53) and (6.55), we find, for the connected function

$$\langle j^\mu j^\nu \rangle_C = \int DA_\mu \exp\left\{ i \int d^4x \mathcal{L}_M \right\}$$

$$\times \left\{ \sum_{n=2}^{\infty} \frac{1}{(n-2)!} \int d^4x_1 \ldots \int d^4x_{n-2} \Pi_0^{\alpha_1 \ldots \alpha_{n-2}\, \mu\nu} A_{\alpha_1} \ldots A_{\alpha_{n-2}} \right\}$$

$$\exp\left\{ \sum_{n=0}^{\infty} \frac{1}{n!} \int d^4x_1 \ldots \int d^4x_n \Pi_0^{\mu_1 \ldots \mu_n} A_{\mu_1} \ldots A_{\mu_n} \right\}. \tag{6.56}$$

From the point of view of Feynman graphs, notice that the first term in the above expression contains a series of one-loop graphs with the insertion of an even number n of vertices. Two out of these, namely μ, ν, are loose, and the remaining $n - 2$ are connected to $A_{\alpha_1} \ldots A_{\alpha_{n-2}}$ fields. Then, the second term, when expanded, will provide terms always involving one-loop graphs with the insertion of an even number n of vertices with the corresponding fields attached to all of them. There are, consequently, no loose vertices, in this case.

The two factors in (6.56) are being integrated within the free Maxwell theory. Hence, the functional integration may be performed by the use of Wick's theorem, since it is the case of a free theory. Then, graphically, the effect of this functional integration is to close all the loose A_{α_i} lines at internal vertices with Feynman photon propagators.

Notice, however, that since all pieces contain an even number of photon lines, it turns out only the so-called proper diagrams will contribute to the resulting functional integral. Such diagrams are the ones that cannot be segmented into two disjoint sub-diagrams by cutting any of the internal lines. This property characterizes the diagrams that appear in the expression of the proper vertices $\Gamma^{(n)}$.

In the present case, the series of diagrams generated through the functional integration in (6.56) corresponds to the photon two-points proper vertex $\Pi^{\mu\nu}$, namely, the photon self-energy, also known as the vacuum polarization tensor, which is

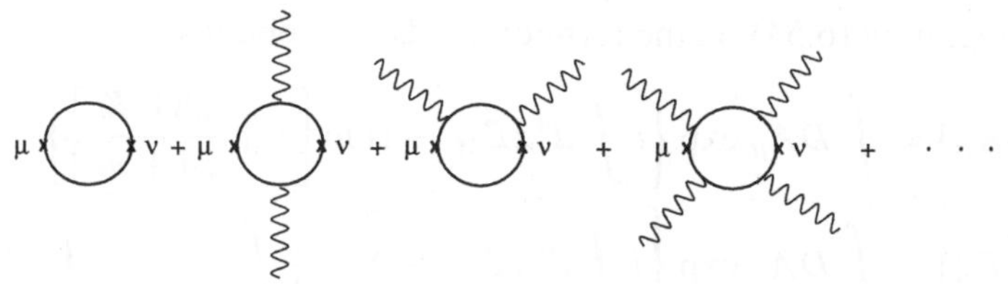

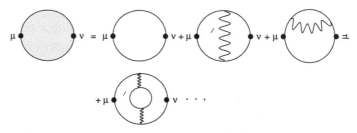

Figure 6.4 Graphs contributing to the first factor of the functional integrand in Eq. (6.56)

Figure 6.5 Graphs contributing to the vacuum polarization tensor $\Pi^{\mu\nu}$, which is also the photon self-energy and the current two-point correlation function. Notice they are obtained from the graphs of Fig. 6.4 and Fig. 6.3 by closing the photon lines.

depicted graphically in Fig. 6.5. We conclude, therefore, that the current two-point correlation function is nothing but the vacuum polarization tensor,

$$\langle j^\mu j^\nu \rangle_C = \Pi^{\mu\nu},\qquad(6.57)$$

or the photon self-energy.

6.7 Renormalization

Renormalization is one of the aspects of quantum field theory that is least understood. As a matter of fact, a lot of misconceptions exist about this procedure, perhaps the most important of them concerning the relation of renormalization with divergences. From a general point of view, the need of renormalization is imposed by the existence of interactions in the theory, irrespective of whether divergences exist or not. Renormalization, therefore, is primarily associated to interactions and not to divergences.

Let us take the example of mass renormalization. In the case of a free field theory described by action (6.19), the physical mass is the parameter m, appearing in the free Lagrangean. We know this fact because the dispersion relation appearing in the computation of the free two-point function is that of a free relativistic particle of mass m. As we saw in the previous section, this is no longer true in an interacting theory. A fundamental question, therefore, emerges: what is the physical mass of the particles in an interacting quantum field theory?

A reasonable assumption is that in some asymptotic limit, either of large or short distances, the particles associated to an interacting quantum field are effectively free. The former case, for instance, would correspond to QED and the latter to QCD.

Let us consider a quantum field theory belonging to the first case and let us examine the large-distance regime of its two-point Green function, given by (6.43). The Riemann–Lebesgue lemma implies that only the zeros of the vertex function contribute to (6.43) in the asymptotic large-distance regime.

In order to obtain these zeros, we expand the self-energy in a Taylor series about the renormalized mass m_R, namely

$$\Sigma(p) = \Sigma(p^2 = m_R^2) + (p^2 - m_R^2)\frac{\partial \Sigma}{\partial p}\Big|_{p^2=m_R^2} + \cdots \tag{6.58}$$

Inserting this expression in (6.43) and defining the renormalized mass as

$$m_R^2 = m^2 + \Sigma(p^2 = m_R^2), \tag{6.59}$$

we get

$$G_C^{(2)}(\mathbf{x}, t) = i \int \frac{d\omega}{2\pi} \int \frac{d^3k}{(2\pi)^3} \frac{e^{i[\mathbf{k}\cdot\mathbf{x}-\omega t]}}{(p^2 - m_R^2)\left[1 - \frac{\partial \Sigma}{\partial p}\Big|_{p^2=m_R^2}\right] + \cdots}. \tag{6.60}$$

Defining the renormalized field as

$$\phi_R = Z^{-1/2}\phi \qquad Z = \frac{1}{1 - \frac{\partial \Sigma}{\partial p}\Big|_{p^2=m_R^2}}, \tag{6.61}$$

we see that the renormalized two-point function becomes

$$G_C^{(2)}(\mathbf{x}, t)_R = i \int \frac{d\omega}{2\pi} \int \frac{d^3k}{(2\pi)^3} \frac{e^{i[\mathbf{k}\cdot\mathbf{x}-\omega t]}}{(p^2 - m_R^2)\left[1 + O\left(p^2 - m_R^2\right)\right]}. \tag{6.62}$$

This has poles at the renormalized mass m_R with the same residue as the free field Green function. Assuming that the asymptotic large-distance regime of the interacting theory coincides with the free field one, we are led to the conclusion that in an interacting theory the physical mass of the particles coincides with the renormalized mass m_R given by (6.59). Observe, however, that this is an implicit equation for the physical mass.

The two renormalizations above could be summarized by the conditions

$$\Gamma_R^{(2)}(p^2)\Big|_{p^2=m_R^2} = 0 \quad ; \quad \frac{\partial}{\partial p^2}\Gamma_R^{(2)}(p^2)\Big|_{p^2=m_R^2} = 1. \tag{6.63}$$

Equivalently, an effective physical coupling parameter could be defined in an analogous way. Suppose, for instance, that the interaction potential is

$$V(\varphi) = \frac{\lambda}{4!}\varphi^4. \tag{6.64}$$

Then

$$\Gamma_R^{(4)}(p_i)\Big|_{p_i^2=m_R^2} = \lambda_R. \tag{6.65}$$

Notice that no reference to divergences has been ever made in our renormalization procedure; rather, the existence of interactions is actually what imposes the necessity of renormalization. The parameter m appearing in the quadratic term of the action is no longer the physical mass, which is actually replaced by the renormalized mass m_R. Accordingly, λ is replaced by the renormalized λ_R.

Before describing the relation existing between renormalization and divergences, let us examine how these come about in quantum field theory. Consider the equal-times free two-point Green function

$$G_{0,C}^{(2)}(\mathbf{x} - \mathbf{y}, t) = \langle 0|\phi(\mathbf{x}, t)\phi(\mathbf{y}, 0)|0\rangle = i\Delta_F(\mathbf{x} - \mathbf{y}, t), \tag{6.66}$$

We show in Chapter 13 that the local field operator $\phi(\mathbf{x}, 0)$ acting on the vacuum creates the eigenstates of the position operator $\mathbf{X}$,

$$\phi(\mathbf{x}, 0)|0\rangle = |\mathbf{x}\rangle \quad ; \quad \mathbf{X}|\mathbf{x}\rangle = \mathbf{x}|\mathbf{x}\rangle, \tag{6.67}$$

widely used in quantum mechanics. Hence, the Feynman propagator satisfies the following properties:

$$\Delta_F(\mathbf{x} - \mathbf{y}, t) = \langle \mathbf{x}|e^{-iHt}|\mathbf{y}\rangle \xrightarrow{t\to 0} \langle \mathbf{x}|\mathbf{y}\rangle = \delta(\mathbf{x} - \mathbf{y}) \xrightarrow{\mathbf{x}\to\mathbf{y}} \infty. \tag{6.68}$$

The position eigenstates, which have an infinite norm, are the origin of the infinities found in quantum field theory. The price to use local fields for the description of a given system are the infinities generated by such infinite norm states. These manifest themselves as short-distance singularities in the Green functions in coordinate space and large momentum (ultraviolet (UV)) singularities in the Fourier transform of these Green functions. Should we use a smeared (non-local) field

$$\phi([f], t) = \int d^3x f(\mathbf{x})\phi(\mathbf{x}, t)$$

$$\phi([f], 0)|0\rangle = \int d^3x f(\mathbf{x})\phi(\mathbf{x}, 0)|0\rangle, \tag{6.69}$$

where $f(\mathbf{x})$ is a normalizable function of compact support, we would have no UV-divergences in our quantum field theory. From the practical, calculational point of view, however, it is much more convenient to use local fields, in spite of the price we have to pay of living with the infinities thereby generated.

Any local quantum field theory belongs to one of two broad general classes, concerning the UV infinities it may contain. The first general class contains theories where the divergences appearing in each of the terms in the expansion in (6.37) are a replication of a finite number of basic divergences appearing in the lowest order. These are the so-called primitive divergences. Theories of this class, therefore, contain a finite number of primitively divergent terms and all their divergent terms are a replication thereof. In the second general class, conversely, divergent local quantum field theories present new primitively divergent terms at each new order in the expansion (6.37) in such a way that there are infinitely many primitively divergent terms.

The key issue concerning divergences in the renormalization procedure is that in theories containing a finite number of primitively divergent terms, the process of renormaliztion of a finite number of unphysical objects into physical ones, which is imposed by the interaction itself, can be used at the same time to eliminate the divergences, order by order, by absorbing these divergences into the unphysical, unrenormalized objects, namely mass, field and coupling parameter. The quantum field theories of this type are called renormalizable, whereas the ones with an infinite number of primitively divergent terms are called unrenormalizable. For the latter, it is clearly impossible to eliminate the divergences with a finite number of renormalization of physical quantities in the expansion (6.37).

6.8 Renormalization Group

The process of obtainment of renormalized physical quantities is equivalent to imposing some renormalization conditions, namely conditions on certain functions and their derivatives at a certain point that defines an energy scale. Let us take, for instance, the proper vertices $\Gamma^{(n)}$. We may summarize the renormalization procedure by writing

$$\Gamma_R^{(n)}(p_1, \ldots, p_n, m_R, \lambda_R) = Z^{n/2}(m_R)\Gamma^{(n)}(p_1, \ldots, p_n, m, \lambda), \qquad (6.70)$$

and assuming these satisfy the renormalization conditions (6.63) and (6.65).

Performing a finite renormalization would amount to using different renormalization conditions, namely

$$\Gamma_R^{(2)}(p^2)\big|_{p^2=\mu^2} = \mu^2 - m_R^2 \;\; ; \;\; \frac{\partial}{\partial p^2}\Gamma_R^{(2)}(p^2)\big|_{p^2=\mu^2} = 1 \;\; ; \;\; \Gamma_R^{(4)}(p_i)\big|_{p_i^2=\mu^2} = \lambda_R \;,$$
$$(6.71)$$

corresponding to

$$\Gamma_R^{(n)}(p_1, \ldots, p_n, m_R(\mu), \lambda_R(\mu)) = Z^{n/2}(\mu)\Gamma^{(n)}(p_1, \ldots, p_n, m, \lambda). \qquad (6.72)$$

We may, therefore, relate the renormalized (finite) functions obtained with two different renormalization prescriptions by

$$\Gamma_R^{(n)}(p_1, \ldots, p_n, m_R(\mu), \lambda_R(\mu)) = \zeta(\mu)^{n/2} \Gamma_R^{(n)}(p_1, \ldots, p_n, m_R, \lambda_R), \quad (6.73)$$

where $\zeta(\mu) = Z(\mu)/Z(m_R)$.

The objective of the so-called renormalization group is to find equations determining how the renormalized proper vertices (or, equivalently the Green functions) change as we modify the renormalization by finite amounts, which is effectively done by choosing different renormalization prescriptions.

The first equation of this kind was obtained by Gell-Mann and Low in 1954 [5]. Subsequently, more general equations were derived by Callan and Symanzik, in 1970 [6] and by 't Hooft and Weinberg, in 1973 [7].

The most general renormalization group equations express the independence of the renormalized proper vertices on the modifications that a change in the renormalization point μ will produce in the renormalized mass m_R, coupling parameter λ_R and field operator ϕ_R. In other words, it expresses the overall invariance of the functions $\Gamma_R^{(n)}$ under the changes produced in these quantities by a finite modification of the renormalization point μ. Differentiating (6.73) with respect to μ and equating to zero, we get [6, 7]

$$\left[\mu^2 \frac{\partial}{\partial \mu^2} + \beta(\lambda_R) \frac{\partial}{\partial \lambda_R} - n\gamma + \gamma_m \, m_R^2 \frac{\partial}{\partial m_R^2} \right] \Gamma_R^{(n)}(p_i^2, m_R, \lambda_R) = 0, \quad (6.74)$$

where

$$\beta(\lambda_R) = \mu^2 \frac{\partial \lambda_R}{\partial \mu^2}, \quad (6.75)$$

$$\gamma = \frac{1}{2} \mu^2 \frac{\partial}{\partial \mu^2} \ln Z(\mu^2) \quad (6.76)$$

and

$$\gamma_m = \mu^2 \frac{\partial}{\partial \mu^2} \ln m_R^2(\mu^2). \quad (6.77)$$

The renormalization group equations have important consequences. Perhaps the most striking one is the fact that the renormalized coupling parameter and the renormalized mass, which according to the arguments presented above are the actual physical quantities of the theory, are not constants but vary as a function of the energy scale according to (6.75) and (6.77), respectively. The sign of the β-function will determine whether the coupling parameter (and therefore the effective interaction) increases or decreases as we change the energy scale.

The renormalization group has also an aspect related to the divergences found in a renormalizable local quantum field theory. Indeed, the process of renormalization that leads to a physical quantity, which also happens to be finite, implies that the sum of two divergent objects yields a finite result. When we do that, however, the finite part of this is not fixed unambiguously, since it may vary as the infinite parts are changed by adding or subtracting any finite amount.

The renormalization group equations serve precisely to unambiguously determine the finite part of renormalized quantities as a function of the energy scale and, as such, are an inseparable part of quantum field theory. The physical values of parameters such as the mass or charge (coupling parameter) in a quantum field theory must be measured at a certain energy scale μ_0, or equivalently at a certain distance scale. Thereafter, this input is used as the boundary condition for the differential equations (6.75) and (6.77) that will determine the physical mass and coupling parameter at an arbitrary scale, μ.

7

Symmetries: Explicit or Secret

Symmetry principles play a central part in many areas of physics, including particle physics, atomic and molecular physics, nuclear physics and, of course, condensed matter physics. Symmetries determine, for instance, all the properties and the detailed mechanism of the fundamental interactions of nature. They determine what are the conserved quantities of a given physical system and consequently, what are the relevant observable objects. Furthermore, symmetry principles classify the spectrum of energy eigenstates according to the multiple matrix representations of the symmetry operations. In this chapter we study symmetries, their main properties and, in particular, the consequences of the phenomenon of spontaneous symmetry breakdown.

7.1 Symmetry Principles

The properties of a physical system, both at classical and quantum levels, are determined by the action or, ultimately, by the Lagrangean or Hamiltonian of the system. These, by their turn, are functions of the basic field variables. Observable quantities, then, are, in general, expressed in terms of the latter. Symmetry transformations are operations performed on the basic variables of a system, leading to a new set of variables in such a way that the action is preserved or, in other words, is left invariant under this transformation. Examples of continuous symmetry operations involving space and/or time coordinates are rotations, space and time translations and Lorentz transformations. Conversely, examples of continuous symmetries not involving space or time are the multiplication by a phase, multiplication by a unitary 2×2 or 3×3 matrix with determinant one. Examples of discrete symmetry operations involving space and/or time are parity, space and time reversal. A discrete symmetry operation not involving space and/or time would be that of charge conjugation, which exchanges particles for anti-particles and vice versa.

Suppose a classical symmetry operation, which by definition leaves the action invariant, acts on a multicomponent classical field as

$$\varphi_i \to \varphi_i' = g_{ij}\varphi_j; \qquad (7.1)$$

g_{ij} $(i, j = 1, \ldots, M)$ can be either a discrete or a continuous operation. In the latter case we assume the matrix g_{ij} depends continuously on N real parameters $\omega^a, a = 1, \ldots, N$.

There must be a corresponding symmetry operation acting at a quantum level, where the field becomes an operator ϕ_i. A theorem due to Wigner states that, at a quantum-mechanical level, a symmetry operation is implemented by a similarity transformation associated to an operator U, which is either unitary or anti-unitary. Hence, at a quantum level

$$\phi_i \to \phi_i' = U\phi_i U^\dagger = g_{ij}\phi_j. \qquad (7.2)$$

Assuming that U is unitary, we can write, in the case of a continuous symmetry,

$$U = \exp\left\{i\omega^a G^a\right\}, \qquad (7.3)$$

where the so-called generators G^a are hermitian operators and ω^a are real parameters $(a = 1, \ldots, N)$.

Being a symmetry operation, by definition, the Hamiltonian should be left invariant by this. It follows that the Hamiltonian must commute with the symmetry operator, namely

$$UHU^\dagger = H \longrightarrow [H, U] = 0. \qquad (7.4)$$

This implies the N-generators G^a must commute with the Hamiltonian as well, namely $[G^a, H] = 0$, being therefore conserved quantities. We conclude that, when the system is invariant under a continuous symmetry operation with N independent parameters, then we will have accordingly N conserved quantities: the G^a generators of the symmetry operation. This is the quantum-mechanical version of an analogous theorem valid for classical systems, known as Noether Theorem.

The conserved quantities G_a are the volume integrals of the (density) 0th component

$$G_a = \int d^3x \rho_a \qquad (7.5)$$

of a current quadrivector $j_a^\mu = (\rho_a, \mathbf{j}_a)$ that satisfy a continuity equation $\partial_\mu j_a^\mu = 0$ or, equivalently, $\frac{\partial \rho_a}{\partial t} + \nabla \cdot \mathbf{j}_a = 0$. The continuity equation follows as a direct consequence of the dynamical evolution equation, either at a classical or quantum level. Actually, both have the same form when we formulate the quantum version in the Heisenberg picture.

The existence of a symmetry principle in a certain system has profound consequences on its physical properties. We have just seen that it determines what are the conserved quantities of such a system. Furthermore, let us see how it strongly influences the spectrum of energy eigenstates of the system. Indeed, suppose the energy spectrum is of the form

$$H|n, i\rangle = E_n|n, i\rangle \qquad i = 1, \ldots, g_n \tag{7.6}$$

such that for each energy eigenvalue E_n there are g_n linearly independent degenerate eigenvectors. Since U commutes with H, it follows that $U|n, i\rangle$ is also a degenerate eigenvector of E_n; hence, we may express it as

$$U|n, i\rangle = \sum_{j=1}^{g_a} \mathcal{R}_{ij}^{(n)} |n, j\rangle. \tag{7.7}$$

Assuming the eigenstates are orthonormal, we have

$$\mathcal{R}_{ij}^{(n)} = \langle n, j|U|n, i\rangle. \tag{7.8}$$

The matrices $\mathcal{R}_{ij}^{(n)}$ form what is called an irreducible representation of dimension g_n of the symmetry operator U, namely

$$\mathcal{R}_{ij}^{(n)} = \exp\left\{i\omega^a T^a\right\}_{ij}, \tag{7.9}$$

where the matrix T_{ij}^a is the corresponding irreducible representation of the generators G^a

$$T_{ij}^a = \langle n, j|G^a|n, i\rangle. \tag{7.10}$$

We see, therefore, that the energy eigenstates of a system are classified according to the different irreducible representations of the symmetry operator U. This is a deep and beautiful result. It is also very useful, since symmetry arguments can be invoked when we try to find the energy eigenvectors of a system. The symmetries of a system of elementary particles also play a crucial role concerning the fundamental interactions among these particles. Indeed, assuming invariance under a continuous operation, which depends on N real parameters ω^a, one can derive, in a natural way, the basic interactions of the system. This may be achieved by just imposing an extended invariance under the corresponding *local* transformations, namely, the same transformations but with local parameters: $\omega^a(x)$, where x is a space-time coordinate. As it turns out, in order to achieve local invariance one must replace the standard derivatives of the field, in the action, by the so-called covariant derivatives

$$[\partial_\mu \delta_{ij}]\varphi_j \longrightarrow [(D_\mu)_{ij}]\varphi_j = [\partial_\mu \delta_{ij} + ig A_\mu^a T_{ij}^a]\varphi_j. \tag{7.11}$$

This is expressed in terms of the gauge field A_μ^a, which is the mediator of the interaction, and g, the coupling parameter. The corresponding field intensity tensor $F_{\mu\nu}^a$ is given, in terms of $A_\mu = A_\mu^a T^a$, by

$$F_{\mu\nu} \equiv F_{\mu\nu}^a T^a = \partial_\mu A_\nu - \partial_\nu A_\mu + ig[A_\mu, A_\nu]. \tag{7.12}$$

Observe that we used the above-described procedure in (3.27) in the case of the electromagnetic interaction, and in (23.59), in the case of a non-abelian symmetry.

7.2 Symmetries: Exposed or Hidden

A fascinating aspect of a symmetry principle is that it may be either exposed or concealed. We are going to see that the condition that determines which is the case is the way the vacuum or ground state behaves under a symmetry operation. Indeed, for a symmetry to be observed in a natural system, it is not enough that the Hamiltonian, which describes its physical properties, be invariant. Consider, for instance, a magnet, described by the isotropic Heisenberg model. The Hamiltonian is invariant under arbitrary rotations. What about the ground state? When the system is in a paramagnetic phase, the ground state would contain spins pointing in arbitrary directions, in such a way that a rotation operation would not modify at all the ground state. When the system is in an ordered ferromagnetic phase, conversely, all the spins would point in the same direction. A rotation operation would clearly change the ground state in this case, producing a different ground state with the same energy. In the ordered phase of a ferromagnet, therefore, one cannot straightforwardly infer the system's rotational invariance because phenomenological results will be biased by the asymmetric nature of the ground state. The rotational invariance of the underlying Hamiltonian consequently, will be a "secret symmetry" [8], not obviously implied by the phenomenology derived from the ground state. This property of a system's ground state of not sharing the same invariance as the Hamiltonian that describes it has deep implications on the physical properties of the system, but interestingly, it only occurs in systems with an infinite number of degrees of freedom or in the thermodynamic limit. The reason is that the two ground states connected by the symmetry operations would tunnel into each other in a system with a finite number of degrees of freedom, thereby making the symmetric linear combination thereof the most stable state. In other words, the real ground state is symmetric in systems with a finite number of degrees of freedom. A good example is the ammonia molecule, NH_3, where the three hydrogen atoms form a triangle, which is the base of a tetrahedron, in the tip of which we find the nitrogen atom. There are, however two possibilities for the equilibrium position of the nitrogen atom. These are contained on a line piercing the center of the triangle and in opposite points with respect to the plane to which this triangle belongs. The system is symmetric under reflections with respect to this plane.

Both equilibrium positions would be connected by symmetry operations and therefore would be equally favored. Suppose, however, the nitrogen atom is in the tip of one given tetrahedron. It would then tunnel to the state at the tip of the opposite tetrahedron, with a probability $\gamma = e^{-\Gamma} < 1$. One can easily show, then, that,

Figure 7.1 The ammonia molecule with the two equivalent equilibrium positions of the nitrogen atom

the tunneling lifts the energy degeneracy between the two ground states. Indeed, calling $|+a\rangle$ and $|-a\rangle$ the states centered at the tip of opposite tetrahedra, which are assumed to have degenerate energy E_n in the absence of tunneling, it follows that, for a tunneling amplitude γ, the actual ground state and first excited states with the respective eigenenergies are, respectively,

$$|S\rangle = \frac{1}{\sqrt{2}} [|+a\rangle + |-a\rangle] \qquad E_S = E_n(1-\gamma)$$

$$|A\rangle = \frac{1}{\sqrt{2}} [|+a\rangle - |-a\rangle] \qquad E_A = E_n(1+\gamma). \qquad (7.13)$$

The ground state is clearly unique and given by the symmetric combination $|S\rangle$, hence it is invariant under the same reflection symmetry operation $a \rightarrow -a$, as the Hamiltonian of the system.

In this example, the symmetry is explicit: the ground state has the same symmetry as the Hamiltonian. That would also be the case for the hydrogen atom or the harmonic oscillator, for instance. This situation, however, is not the most general. Indeed, as we will see, there are systems for which the ground state has not the same symmetry of the Hamiltonian.

7.3 Spontaneous Symmetry Breaking

7.3.1 The Order Parameter

Let us start by exploring the consequences of the vacuum expectation value of the field operator, namely

$$\langle \phi_i \rangle = \langle 0 | \phi_i | 0 \rangle \qquad (7.14)$$

being different from zero or not. Assuming the system has a symmetry, which is implemented by a unitary operator U, then

$$\langle 0|\phi_i|0\rangle = \langle 0|U^\dagger U\phi_i U^\dagger U|0\rangle = g_{ij}\langle 0|U^\dagger \phi_j U|0\rangle, \tag{7.15}$$

where we used (7.2). The value of this vacuum expectation strongly depends on the behavior of the vacuum state $|0\rangle$ under the symmetry operation U. Suppose the vacuum is invariant: $U|0\rangle = |0\rangle$. Then it follows that

$$\langle 0|\phi_i|0\rangle = g_{ij}\langle 0|\phi_j|0\rangle,$$
$$[\delta_{ij} - g_{ij}]\langle 0|\phi_j|0\rangle = 0. \tag{7.16}$$

Since $\delta_{ij} \neq g_{ij}$, the above equation implies

$$\langle 0|\phi_j|0\rangle = 0. \tag{7.17}$$

We conclude that whenever the vacuum state is invariant under the symmetry operation, the vacuum expectation value of the field operator must vanish. The immediate consequence is that whenever this vacuum expectation is different from zero, the vacuum state is not invariant under the symmetry operation:

$$U|0\rangle = |0\rangle \longrightarrow \langle 0|\phi_i|0\rangle = 0$$
$$\langle 0|\phi_i|0\rangle \neq 0 \longrightarrow U|0\rangle \neq |0\rangle. \tag{7.18}$$

The result above can be used for introducing the notion of an order parameter, the value of which serves as a measure of how much the vacuum would change under a symmetry operation. The vacuum expectation value of the field operator, (7.14), serves perfectly for this purpose. The more ordered the system is, the more the vacuum state would change under a symmetry operation. Conversely, the less ordered the system is, the less its vacuum state would change under a symmetry operation. A completely disordered vacuum state would be invariant under the symmetry operation. We can, thereby quantify how ordered a given system is, by means of (7.14).

7.3.2 Vacuum Degeneracy

Since the symmetry operator commutes with the Hamiltonian, whenever the vacuum is not invariant under the symmetry operation, namely $U|0\rangle = |0'\rangle$, it produces another vacuum state, $|0'\rangle$, such that, according to (7.15),

$$H|0\rangle = E_0|0\rangle \longrightarrow HU|0\rangle = E_0 U|0\rangle \longrightarrow H|0'\rangle = E_0|0'\rangle$$
$$\langle 0'|\phi_i|0'\rangle = \langle 0|U^\dagger \phi_i U|0\rangle = g_{ij}\langle 0|\phi_j|0\rangle, \tag{7.19}$$

where E_0 is the vacuum energy, which can always be taken as $E_0 = 0$.

One could wonder whether the different vacua could tunnel among themselves as in the case of the ammonia molecule studied above. In a system with an infinite number of degrees of freedom, such as a field, or even for a system in the thermodynamic limit, such as a ferromagnetic crystal, however, the probability of a ground state to tunnel into another one would be given by $\gamma = e^{-\Gamma} < 1$ for each of the degrees of freedom. For the complete system, therefore, we would have a tunneling probability of $(e^{-\Gamma})^N \to 0$ for infinite N or even for very large N. In this situation, therefore, tunneling between symmetry-related ground states would be suppressed and we would have vacuum degeneracy. The system would have different degenerate ground states, which are connected by the symmetry operations.

7.3.3 Lost Symmetry

Whenever the vacuum state is not invariant under the symmetry operation that leaves the Hamiltonian invariant, by definition we have the situation known as spontaneous symmetry breakdown. As we have just seen above, this implies the vacuum is degenerate and the field operator possesses a nonzero vacuum expectation value, which provides a quantitative measure of how ordered the ground state is.

Consider now the action (6.33). Whenever there is spontaneous symmetry breaking and, consequently the field has a nonzero vacuum expectation value $\langle \phi_i \rangle \neq 0$, for stability reasons we must shift the classical field φ_i around this value, namely

$$\varphi_i \to \varphi'_i = \varphi_i - \langle \phi_i \rangle. \tag{7.20}$$

As a consequence, the potential changes as

$$V(\varphi_i) \to V\left(\varphi'_i + \langle \phi_i \rangle\right). \tag{7.21}$$

The original invariance under the symmetry operation $\varphi_i \to g_{ij}\varphi_j$ is lost because of the shift made in the field. For that reason, for all phenomenological purposes, spontaneous symmetry breakdown effectively hides the original symmetry of the system.

7.3.4 Goldstone Theorem

Goldstone Theorem is an important result concerning the spontaneous breakdown of a continuous symmetry [9]. In order to demonstrate it, let us suppose the symmetry is implemented by the unitary operator U, given by (7.3).

In the case where there is no spontaneous symmetry breaking and the symmetry is, therefore, explicit, we have

$$U|0\rangle = \exp\left\{i\omega^a G^a\right\}|0\rangle = |0\rangle \iff G^a|0\rangle = 0. \tag{7.22}$$

In a situation where the system presents spontaneous symmetry breaking, conversely, since the vacuum state is not invariant under the symmetry operation, we have

$$U|0\rangle = \exp\left\{i\omega^a G^a\right\}|0\rangle \neq |0\rangle \iff G^a|0\rangle = |G^a\rangle \neq 0. \tag{7.23}$$

Since $[H, G^a] = 0$, it follows that

$$H|G^a\rangle = G^a H|0\rangle = E_0|G^a\rangle, \tag{7.24}$$

implying that the N states obtained by acting with the generators G^a on the vacuum are degenerate with the vacuum. In a relativistic theory, this fact implies that such states are massless. They are known as Goldstone bosons. A mathematically rigorous demonstration of the theorem was provided in [14].

The subject acquired great importance when it was shown that by imposing the locality of the symmetry parameters ω^a in the framework of a theory with a spontaneously broken continuous symmetry, instead of the N Goldstone bosons we would have the mass generation for the N gauge fields A_μ^a, $a = 1, \dots, N$, introduced by imposing the invariance under a local symmetry. This is a mechanism of mass generation to the gauge fields, known as the Anderson–Higgs Mechanism [10, 11, 12, 13]. It involves both the spontaneous breaking of a continuous symmetry and the locality of such a symmetry.

In order to achieve the spontaneous symmetry breakdown, a multicomponent scalar field known as the Higgs field has been introduced in the standard model analogously to the complex scalar field of the Landau–Ginzburg model. What is the method, in practical terms, by which we can realize a symmetry in a spontaneously broken way?

7.4 Static × Dynamical Spontaneous Symmetry Breaking

We analyze here how to obtain spontaneous symmetry breakdown from a theory that is invariant under a certain symmetry. We have seen that this is closely related to the existence of an ordered ground state, which, therefore, is not invariant under the symmetry operation. Here we focus on two different methods of producing spontaneous symmetry breaking.

The first method consists in writing the potential of the theory in such a way that it is invariant under the symmetry operation but its minima are not. Take as an example

$$V(\varphi_i) = a(T)[\varphi_i\varphi_i] + \frac{\lambda}{2}[\varphi_i\varphi_i]^2, \tag{7.25}$$

where $a(T) = a(T - T_c)$. This is invariant under an $SO(N)$ symmetry

$$\varphi_i \longrightarrow g_{ij}\varphi_j. \tag{7.26}$$

Let us determine the minima of $V(\varphi_i)$. The condition of zero first derivative gives

$$\frac{\partial V}{\partial \varphi_i} = \{a(T) + \lambda[\varphi_i\varphi_i]\}\varphi_i = 0. \tag{7.27}$$

For $T > T_c$, the quantity between brackets is always positive, hence, we must have $\varphi_i = 0$. For $T < T_c$, conversely, there is an additional solution for which

$$[\varphi_i\varphi_i] = \frac{|a(T)|}{\lambda} \neq 0. \tag{7.28}$$

The condition of minimum requires that all eigenvalues of the Hessian matrix $\frac{\partial^2 V}{\partial\varphi_i\partial\varphi_j}$ must be positive. This condition is fulfilled for $\varphi_i^0 \neq 0$ when $T < T_c$.

According to (6.48), we may take the classical potential (7.25) as the classical contribution to the effective potential $V_{\text{eff}}(\varphi_i)$. Hence, for consistency we must have $\varphi_i = \langle\phi_i\rangle$ and $\varphi_i^0 = \langle 0|\phi_i|0\rangle$ where ϕ_i is the field operator.

When the minima of the potential occur at a nonzero value φ_i^0, it follows that the vacuum expectation value $\langle 0|\phi_i|0\rangle$ must be different of zero as well. Then, according to the results of Section 8.3, it follows that the symmetry is spontaneously broken.

We see that a practical method of producing the spontaneous breakdown of a given symmetry is to introduce a potential having nontrivial minima, namely, minima at nonzero values of the classical field, such that these minima are connected among themselves by symmetry operations. This may be called static spontaneous symmetry breaking.

Nevertheless, another method for producing the spontaneous breakdown of a certain symmetry exists. This is the dynamical symmetry breaking. In this case, the classical potential has only a trivial symmetric minimum; however, the nontrivial minima are generated by the interaction, usually with some other field.

Consider for this purpose the theory of a real scalar field coupled to a massless Dirac fermion field in 1+1D:

$$\mathcal{L} = i\overline{\psi}\partial\!\!\!/\psi + g\overline{\psi}\psi\varphi - \frac{1}{2}\varphi^2, \tag{7.29}$$

where $\gamma^0 = \sigma_x$, $\gamma^1 = i\sigma_y$, $\gamma^5 = \gamma^0\gamma^1 = -\sigma_z$ and $\overline{\psi} = \psi^\dagger\gamma^0$.

This theory possesses the discrete symmetry

$$\psi \to \gamma^5\psi$$
$$\varphi \to -\varphi. \tag{7.30}$$

The classical static and uniform minima of the potential can be found by taking derivatives with respect to the fields, namely,

$$g\overline{\psi}\psi = \varphi$$
$$\psi\varphi = 0. \tag{7.31}$$

At this level, therefore, we do not have spontaneous symmetry breaking. Now, let us determine the resulting effect of the interaction with the Dirac field on the scalar field dynamics. For this purpose, let us perform the quadratic integral over the fermion field:

$$e^{iS[\varphi]} = \frac{1}{\mathcal{N}} \int D\psi \, D\overline{\psi} \exp\left\{i \int d^2x \left[\overline{\psi}[i\slashed{\partial} + g\varphi]\psi - \frac{1}{2}\varphi^2\right]\right\}. \tag{7.32}$$

This would produce an effective action for φ, namely

$$S[\varphi] = \int d^2x \left[-\frac{1}{2}\varphi^2\right] - \mathrm{Tr}\ln\left[-\Box + g^2\varphi^2\right] - S[0]$$
$$= -\int d^2x \frac{1}{2}\varphi^2 - \frac{1}{2\pi}\int_0^\infty dk\,k\ln\left[k^2 + g^2\varphi^2\right] - S[0]. \tag{7.33}$$

This yields

$$S[\varphi] = \int d^2x \left[-\left(\frac{g^2}{4\pi} + \frac{1}{2}\right)\varphi^2 + \frac{g^2}{4\pi}\varphi^2 \ln\frac{\varphi^2}{\varphi_0^2}\right]. \tag{7.34}$$

Looking for the minima of the φ potential, we conclude they must satisfy

$$2\left[-\frac{1}{2} + \frac{g^2}{4\pi}\ln\frac{\varphi^2}{\varphi_0^2}\right]\varphi = 0. \tag{7.35}$$

Looking at the second derivative, we quickly realize that the minima occur at

$$\varphi_\pm = \pm\,\varphi_0\,\exp\left\{\frac{\pi}{g^2}\right\}. \tag{7.36}$$

These are clearly non-invariant; therefore, the symmetry is spontaneously broken. In this example, the spontaneous symmetry breaking was clearly produced by the interaction with the fermion field. Integration over these produces an infinite series of terms containing powers of the coupling g. The final result for the minima of the potential is non-analytical in g, this revealing the non-perturbative nature of this effect.

8

Classical Topological Excitations

We may subdivide mathematics roughly into four grand areas: geometry, topology, algebra and analysis. In terms of manifolds (sets), roughly speaking, we may say that geometry deals with the local properties of manifolds, topology deals with the global properties thereof, algebra deals with the structures relating the elements of a given manifold, whereas analysis deals with the relations existing between the elements belonging to different manifolds. Topological excitations are important and interesting physical objects bearing conserved physical attributes, the conservation of which does not result from any symmetries of the system, but rather, derives from the nontrivial topology of the field configurations manifold. Topology, therefore, has a strong influence upon the physical properties of a given system. In this chapter we study, from the classical point of view, different topological excitations, namely, excited states bearing some topologically conserved observable quantity. In the next chapter the quantum theory of such excitations is developed on general grounds.

8.1 Inequivalent Topological Classes

Consider the classical field configurations of a given system. A natural condition for these to be physically accessible is that their energy, which is expressed in d spatial dimensions as,

$$E = \int d^d x \mathcal{H}, \tag{8.1}$$

must be finite. For this to happen, however, the potential energy

$$E_P = \int d^d x V(\varphi) \tag{8.2}$$

must be finite as well. Then, a necessary condition for the finiteness of the potential energy is that

$$\varphi(\mathbf{x}) \xrightarrow{|\mathbf{x}| \to \infty} \varphi_0, \qquad (8.3)$$

where φ_0 is one of the minima of the potential energy density, which may be always chosen equal to zero: $V(\varphi_0) = 0$. This is so because the integral in (8.2) will contain terms such as

$$\lim_{|\mathbf{x}| \to \infty} |\mathbf{x}| V(\varphi(\mathbf{x})), \qquad (8.4)$$

which would clearly contribute an infinite amount, should we relax (8.3).

As a consequence of the condition for the finiteness of energy, imposed on the classical field configurations, it follows that these provide a mapping between the manifold defined by $|\mathbf{x}| \to \infty$ and the manifold formed by the minima of the potential energy density: $\{\varphi_0 | V(\varphi_0) \inf V(\varphi), \forall \varphi\}$. This mapping is summarized by (8.3). According to this, in $d = 1, 2, 3$ spatial dimensions, respectively, the field provides a mapping between the manifold at infinity, given respectively by the sets $\{x_{+\infty}, x_{-\infty}\}$, S_1 (a circumference) S_2 (a spherical shell) and the vacuum manifold.

Consider first the case $d = 1$, when the manifold at infinity is the discrete set $\{x_{+\infty}, x_{-\infty}\}$. Now consider the vacuum manifold. In the absence of spontaneous symmetry breaking it is just a single point, as a consequence of the fact that the order parameter vanishes and there is just a single minimum of the potential density. Now, suppose we have spontaneous breakdown of a discrete symmetry. In this case, the nontrivial minima of the potential must be connected by the discrete symmetry operation ($Z(2)$, for instance, would imply two minima at $\varphi_0 = \pm a$).

We can imagine the mapping as if it were produced by elastic strings connecting the two points of the manifold at infinity to the vacuum manifold, as we can see in Fig. 8.1. Assuming we are in a spontaneously broken phase, there are obviously

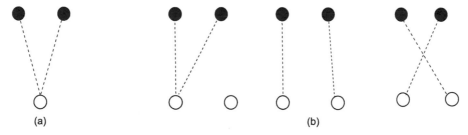

Figure 8.1 Topology of the mapping between the manifold at spatial infinity (black dots) into the vacuum manifold of a $Z(2)$ symmetric theory (white dots). (a) No spontaneous symmetry breaking. (b) With spontaneous symmetry breaking. One cannot continuously deform one situation into the other. Notice the existence of inequivalent classes of mapping in the latter.

three situations that are inequivalent: two strings, each tied at one end, respectively, to $x_{+\infty}$ and $x_{-\infty}$, and at the other end, both to $\varphi_0 = +a$; two strings, each tied at one end, respectively, to $x_{+\infty}$ and $x_{-\infty}$, and at the other end, tied, respectively, to $\varphi_0 = +a$ and $\varphi_0 = -a$; same as before but exchanging $+a \leftrightarrow -a$. These three situations are said to be topologically inequivalent, in the sense that one cannot be continuously deformed into each other without breaking a string. The configuration space, accordingly, is divided into corresponding subspaces; the elements of each may be continuously deformed among themselves but never into the elements of another subspace. We see that the configuration space, therefore, is topologically nontrivial. It subdivides into inequivalent classes that are characterized by a topological invariant, as we shall see below.

Observe that a topologically nontrivial mapping will only occur provided there is spontaneous symmetry breakdown, with the corresponding generation of a nontrivial vacuum manifold. Also, in the case of $d = 1$ (considered here), nontrivial topological mappings will only occur in the case of discrete symmetries (as in the case of $Z(2)$, examined above), where the corresponding manifold is discrete. Suppose, conversely, that the symmetry is continuous, such as $U(1)$, for instance. Since this corresponds to multiplication by a complex phase, the group elements correspond to a circumference.

Then consider the three situations just described above. Now, since the vacuum manifold is continuum (a circle), the two tying points at the vacuum manifold (formerly $\pm a$) clearly can be continuously deformed into each other in such a way that one can continuously deform the three situations among themselves. In this case, therefore, the mapping is topologically trivial.

Consider now the case of a $U(1)$ symmetry in $d = 2$. Now both the manifold at $|\mathbf{x}| \to \infty$ and the vacuum manifold are circumferences, hence the finite energy

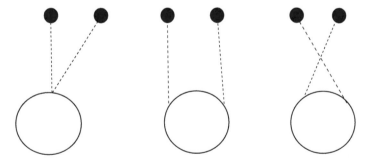

Figure 8.2 Topology of the mapping between the manifold at spatial infinity (black dots) into the vacuum manifold of a $U(1)$ symmetric theory (circle). Now one can continuously deform one situation into the other. No inequivalent classes of mappings exist and the topology is trivial.

field configurations provide a mapping between those. This mapping belongs to one of an infinite number of inequivalent topological classes characterized as follows. Let us label the points of the two circumferences, respectively, by angles: θ_1 and θ_2, both belonging to the interval $[-\pi, \pi)$. Then, each mapping is defined by $\theta_2 = n\theta_1$, $n \in \mathbb{Z}$. Because of the constraint $\theta_2 \in [-\pi, \pi)$, it follows that for a given n, the entire circumference $\theta_1 \in [-\pi, \pi)$ is mapped into the region $\theta_2 \in [-\frac{\pi}{n}, \frac{\pi}{n}]$, hence we shall map each point θ_1 of the first circumference into n different points θ_2 of the second one, in order to cover it completely. This is analogous to placing a money-rubber-string on a metal cylinder. We may simply insert it plainly, or else double- (triple-, etc.) fold it across itself before inserting it in the cylinder (see Fig. 8.3). Clearly there are infinitely many ways of doing this operation and we cannot deform the string assembled in a given situation into another without breaking it (provided we remain in the plane). The field configurations space, accordingly, in the case of theories with a $U(1)$ symmetry in $d = 2$ spatial dimensions contains an infinite number of inequivalent topological sectors.

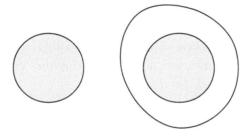

Figure 8.3 Topology of the mapping between the manifold at 2d spatial infinity (circle) into the vacuum manifold of a $U(1)$ symmetric theory (circle). Now one cannot continuously deform the mapping represented here into the one depicted in Fig. 8.4. There are infinitely many inequivalent topological classes.

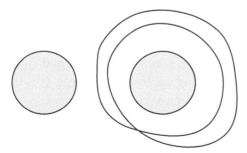

Figure 8.4 Topology of the mapping between the manifold at 2d spatial infinity (circle) into the vacuum manifold of a $U(1)$ symmetric theory (circle). Now one cannot continuously deform the mapping represented here into the one depicted in Fig. 8.3. There are infinitely many inequivalent topological classes.

There is a mathematically precise way of characterizing to which topological class a given field configuration belongs. This is accomplished by the so-called topological invariants or topological charges. The conservation of topological charge derives from the nontrivial topology of the configuration space, as is the case for instance in the situations described above for $Z(2)$ and $U(1)$ symmetries. The topological charge is conserved, just because the field configuration belonging to a given topological class cannot continuously evolve into a configuration belonging to a different class. The set of topologically inequivalent classes of the mapping of a n-sphere onto a manifold M, $n = 0, 1, 2, \ldots$ is called the nth homotopy group of M, namely, $\Pi_n(M)$. In this case, $n = 0$ corresponds to a discrete set of points, $n = 1$, a circumference, $n = 2$ a spherical shell and so on.

8.2 Topological Invariants: Identically Conserved Currents

8.2.1 $d = 1$; $Z(2)$ Group

Consider the theory of a real scalar field in one spatial dimension with a multiplicative $Z(2)$ symmetry: $\varphi \to e^{i\frac{2\pi}{2}}\varphi$ or $\varphi \to -\varphi$. As explained above, the configuration space of such class of theories has topologically inequivalent sectors. There exists a topological invariant number that classifies each field configuration according to its value. This is, for $d = 1$,

$$Q = \frac{1}{2a} \int_{-\infty}^{\infty} \partial_x \varphi(x, t) = \frac{1}{2a} [\varphi(+\infty, t) - \varphi(-\infty, t)], \tag{8.5}$$

assuming the minima of the potential are at $\varphi = \pm a$. Then, we can have $Q = 0, \pm 1$. This is equivalent to saying that $\Pi_0([0, \pm 1]) = Z(2)$, where $Z(2)$ is the additive group of integers modulo 2.

Notice that for a continuous symmetry, the points $\varphi(+\infty)$, $\varphi(-\infty)$ may be continuously deformed one into the other, since they belong to different points of the vacuum manifold, which are connected by such continuous symmetry. The topological charge Q, therefore, is zero in this case and $\Pi_0(S_1) = \emptyset$.

Observe now that the topological invariant charge Q is the spatial integral of a quantity we may call the topological charge density

$$J^0 = \frac{1}{2a}\partial_x\varphi. \tag{8.6}$$

Since Q is conserved for topological reasons as we saw, its charge density must satisfy a continuity conservation equation. This however should not follow from any symmetry of the theory via Noether theorem, or ultimately, via the field equation. The only possibility, therefore, is that the current J^μ must be *identically* conserved. Considering the above expression for the topological charge density, it follows that the topological current must be

$$J^\mu = \frac{1}{2a}\epsilon^{\mu\nu}\partial_\nu\varphi \qquad \partial_\mu J^\mu \equiv 0. \tag{8.7}$$

8.2.2 $d = 1$; $Z(N)$ *Group*

Consider a complex scalar field $\phi = \rho e^{i\theta}$ in one spatial dimension and assume the corresponding theory is invariant under the $Z(N)$ symmetry $\phi \to e^{i\frac{2\pi}{N}}\varphi$, for $N > 2$. Suppose this symmetry is spontaneously broken, such that the potential minima occur at $\phi_0 = \rho_0 e^{i\frac{2\pi n}{N}}$, for $n = 0, 1, 2, \ldots, N - 1$.

The condition of energy finiteness implies that, for $|\mathbf{x}| \to \infty$, we must have

$$\rho(\mathbf{x}) \to \rho_0$$
$$\theta(\mathbf{x}) \to \theta_{0,n} = \frac{2\pi n}{N}, \qquad n = 0, 1, \ldots, N - 1. \tag{8.8}$$

Then it becomes clear that the phase field $\theta(\mathbf{x})$ provides a mapping between the set $\{x_{+\infty}, x_{-\infty}\}$ and the discrete set $\{\theta_{0,0}, \ldots, \theta_{0,N-1}\}$. The topological invariant classifying the inequivalent topological classes of the mapping, in this case, is given by

$$Q = \frac{N}{2\pi}\int_{-\infty}^{\infty}\partial_x\theta(x, t) = \frac{N}{2\pi}[\theta(+\infty, t) - \theta(-\infty, t)]. \tag{8.9}$$

It is not difficult to see that the topological charge may assume the values $Q = 0, \pm 1, \ldots, \pm(N - 1)$, being, therefore, nontrivial. The identically conserved topological current now is given by

$$J^\mu = \frac{N}{2\pi}\epsilon^{\mu\nu}\partial_\nu\theta \qquad \partial_\mu J^\mu \equiv 0. \tag{8.10}$$

8.2.3 $d = 2$; $U(1)$ *Group*

Let us consider the theory of a complex scalar field ϕ, with a local $U(1)$ symmetry, minimally coupled to a $U(1)$ gauge field A_μ and with a self-interaction $V(|\phi|)$, such that the minima of the potential occur at $|\phi| = \rho_0$ and $V(\rho_0) = 0$. Finite energy field configurations must be such that

$$\phi \xrightarrow{|\mathbf{x}|\to\infty} \rho_0 e^{i\theta} \tag{8.11}$$
$$D_i\phi = [\partial_i - i\frac{e}{\hbar c}A_i]\varphi \xrightarrow{|\mathbf{x}|\to\infty} 0$$
$$A_i \xrightarrow{|\mathbf{x}|\to\infty} \frac{\hbar c}{e}\nabla_i\theta \qquad A_0 = 0,$$

where the third condition derives from the second.

Now, the topological invariant classifying the mapping between the circumference at spatial infinity and the circumference of the vacuum manifold is

$$\Phi = \oint_{C(\infty)} \mathbf{A} \cdot d\mathbf{l}, \tag{8.12}$$

where $C(\infty)$ is a closed curve at spatial infinity. Now, using (8.11) and polar coordinates (r, φ), we may write

$$\Phi = \frac{\hbar c}{e} \int_0^{2\pi} d\varphi \frac{d}{d\varphi}\theta(\varphi) = \frac{\hbar c}{e}[\theta(\varphi = 2\pi) - \theta(\varphi = 0)]. \tag{8.13}$$

We immediately realize that the infinite inequivalent topological classes described in the previous subsection correspond to field configurations such that $\theta(\mathbf{r}) \rightarrow n \arg(\mathbf{r}) = n\varphi$, for $n \in \mathbb{Z}$. For these the topological invariant gives

$$\Phi = n\frac{hc}{e} \tag{8.14}$$

and $\Pi_1(S_1) = \mathbb{Z}$.

Now, using Stokes' theorem, we may cast the topological invariant Φ in the form of a spatial integral sweeping the whole plane, namely

$$\Phi = \int d^2r\, B, \tag{8.15}$$

where $B = \epsilon^{ij}\partial_i A_j$; $i, j = 1, 2$ is the magnetic field associated to the vector potential $\mathbf{A}$ and, consequently, $\Phi = n\Phi_0$ is the magnetic flux throughout the entire plane, where the flux quantum is $\Phi_0 = \frac{hc}{e}$.

Observe that the topological charge Φ again is the spatial integral of a density, in this case $J^0 = \epsilon^{ij}\partial_i A_j$. As before, it must satisfy a continuity equation, hence the complete, identically conserved topological current must be

$$J^\mu = \epsilon^{\mu\nu\alpha}\partial_\nu A_\alpha \qquad \partial_\mu J^\mu \equiv 0. \tag{8.16}$$

The zeroth component of J^μ is the topological charge density.

8.2.4 $d = 2$; $SO(3)$ *Nonlinear Sigma Field*

Consider a scalar field triplet $-n^a(\mathbf{x}, t)$, $a = 1, 2, 3$, in $d = 2$ spatial dimensions – that is free except for the fact that it is subject to the constraint

$$|n|^2 = \sum_{a=1}^3 n^a n^a = 1. \tag{8.17}$$

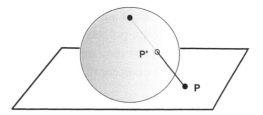

Figure 8.5 Stereographic projection of a spherical surface on the plane

As we shall see in Chapter 17, $n^a(\mathbf{x}, t)$, which is called the nonlinear sigma field, plays an important role in the physics of quantum magnets in two spatial dimensions.

From (8.17), it follows that the field manifold is a spherical shell S_2. One can then easily realize that static field configurations, $n^a(\mathbf{x}, 0)$ provide a mapping from $\mathbb{R}^2$ onto S_2.

Now, by using a stereographic projection, we can show that the $\mathbb{R}^2$ plane is equivalent to a spherical surface. This projection consists in mapping circles of $\mathbb{R}^2$ onto circles parallel to the equator in the sphere in such a way that a straight line connects the north pole, the circle at the spherical surface and the circle in the $\mathbb{R}^2$ plane. The circle at infinity, then, is mapped to the north pole, whereas the origin is mapped to the south pole.

As a consequence of the $\mathbb{R}^2 \Longleftrightarrow S^2$ equivalence, we conclude that actually a static nonlinear sigma field provides a mapping of S_2 onto S_2. Since $\Pi_n(S_n) = \mathbb{Z}$, it follows that this map contains infinitely many inequivalent topological classes.

The topological charge associated to these inequivalent classes is given by

$$Q = \frac{1}{8\pi} \int d^2x \epsilon^{ij} \epsilon^{abc} n^a \partial_i n^b \partial_j n^c \qquad (8.18)$$

and $Q \in \mathbb{Z}$. This is the spatial integral of the zeroth component of the identically conserved current

$$J^\mu = \frac{1}{8\pi} \epsilon^{\mu\nu\alpha} \epsilon^{abc} n^a \partial_\nu n^b \partial_\alpha n^c$$

$$\partial_\mu J^\mu = \frac{1}{8\pi} \epsilon^{\mu\nu\alpha} \epsilon^{abc} \partial_\mu n^a \partial_\nu n^b \partial_\alpha n^c \equiv 0. \qquad (8.19)$$

The identical conservation results from the following reason. Since $|n|^2 = n^a n^a = 1$, we have $\partial_\mu n^a \perp n^a$, hence the triple vector product in the second equation above involves three co-planar vectors and therefore vanishes.

There is another nontrivial mapping associated with the nonlinear sigma field. This is provided by the time-dependent field configurations, $n^a(\mathbf{x}, t)$, when we consider the Euclidean time τ. These produce a mapping of $\mathbb{R}^3$ onto S_2. Again, since

$\mathbb{R}^3$ is topologically equivalent to S_3, the present mapping is effectively from S_3 onto S_2, which takes closed curves in S_3 onto points in S_2. The topological invariant in this case is the Hopf invariant. It detects the number of times the mapped curves link around each other. In terms of the current (8.19), it reads

$$Q_H = \frac{1}{2\pi} \int d^3x d^3y \, \epsilon^{\mu\nu\alpha} J_\mu(x) \frac{(x-y)_\nu}{|x-y|^3} J_\alpha(y). \tag{8.20}$$

Since $\Pi_3(S_2) = \mathbb{Z}$, it follows that $Q_H \in \mathbb{Z}$.

8.2.5 $d = 3$; $SO(3)$ *Gauge Group*

Now consider an $SO(3)$ triplet scalar field Φ^a, $a = 1, 2, 3$ in 3+1D. We assume there exists a potential such that the minima of which force the condition $\Phi_0^a \Phi_0^a = \rho_0$. This would lead us to conclude that the vacuum manifold is S_2.

We then apply the condition that derives from the finiteness of energy, namely that the field configurations at spatial infinity, in this case S_2, must reduce to the vacuum manifold. We therefore immediately conclude that the field Φ^a in this case also produces a mapping $S_2 \to S_2$, similarly to the nonlinear sigma field in 2+1D. In that case, however, the target manifold was derived from the constraint. That is why we did not have to relate the finiteness of energy condition to the field behavior at infinity in that case.

The fact that the scalar fields Φ^a are not subject to the NLSM condition (just its vacuum value!) allows us to introduce the 3+1D topological current

$$J^\mu = \frac{1}{8\pi} \epsilon^{\mu\nu\alpha\beta} \epsilon^{abc} \partial_\nu \Phi^a \partial_\alpha \Phi^b \partial_\beta \Phi^c, \tag{8.21}$$

which is obviously identically conserved: $\partial_\mu J^\mu \equiv 0$.

The corresponding topological charge is

$$Q = \int d^3x J^0 = \frac{1}{8\pi} \int d^3x \, \epsilon^{ijk} \epsilon^{abc} \partial_i \Phi^a \partial_j \Phi^b \partial_k \Phi^c. \tag{8.22}$$

The derivative ∂_i can be transformed into a total derivative and then, using Gauss' theorem, we can write

$$J^0 = \nabla \cdot \mathbf{K}$$
$$K^i = \frac{1}{8\pi} \epsilon^{ijk} \epsilon^{abc} \Phi^a \partial_j \Phi^b \partial_k \Phi^c$$
$$Q = \oint_{S_2(\infty)} dS^i K^i. \tag{8.23}$$

The above expression for Q is just (8.18) with $S_2(\infty)$, the spherical shell at infinity, replacing $\mathbb{R}^2$ as the integration surface. These two manifolds, however, are mapped one into another by a stereographic projection; hence, also here $Q = \mathbb{Z}$ as it should.

8.3 Classical Solitons

The word soliton is frequently used as a generic name for solutions of the field equation, belonging to a nontrivial topological class. In this book, however, we use this name for the specific case of discrete groups in $d = 1$. In this section we examine several examples.

8.3.1 Z(2) Solitons

Consider the theory of a real scalar field with Lagrangean

$$\mathcal{L} = \frac{1}{2}\partial_\mu \varphi \partial^\mu \varphi + \frac{\mu}{2}\varphi^2 - \frac{\lambda}{4!}\varphi^4. \tag{8.24}$$

This is clearly invariant under the $Z(2)$ symmetry $\varphi \to -\varphi$. This is spontaneously broken since the field potential possesses two minima at $\varphi_0 = \pm a$, with $a = \frac{6\mu}{\lambda}$.

Time independent classical solutions must satisfy the field equation

$$\partial_x^2 \varphi = \mu\varphi \left(1 - \frac{\varphi^2}{a^2}\right), \tag{8.25}$$

which has the so-called soliton solution

$$\varphi_S = a \tanh\left(\sqrt{\frac{\mu}{2}}x\right). \tag{8.26}$$

Notice that the $Z(2)$ topological charge (8.5) calculated for this solution is $Q = +1$; hence, it belongs to a nontrivial topological sector of the configuration space.

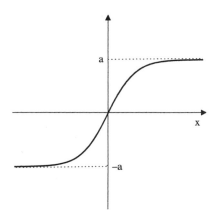

Figure 8.6 Z(2) soliton solution $\varphi_S(x)$

8.3.2 Sine-Gordon Solitons

Consider the theory of a real scalar field in one-dimensional space, described by the Lagrangean

$$\mathcal{L} = \frac{1}{2}\partial_\mu\varphi\partial^\mu\varphi + \alpha\left[\cos\beta\varphi - 1\right]. \tag{8.27}$$

This theory possesses the discrete symmetry $\varphi \to \varphi + \frac{2\pi}{\beta}$ and the potential minima occur at $\varphi_{0,n} = \frac{2\pi}{\beta}n$.

The field equation reads

$$\Box\varphi + \alpha\beta\sin\beta\varphi = 0, \tag{8.28}$$

which is the sine-Gordon (SG) equation.

The topological charge is given by a slight modification of (8.5), namely

$$Q_{SG} = \frac{\beta}{2\pi}\int_{-\infty}^{\infty}\partial_x\varphi(x,t) = \frac{\beta}{2\pi}\left[\varphi(+\infty,t) - \varphi(-\infty,t)\right]. \tag{8.29}$$

The SG equation admits static solutions

$$\varphi_{SG}(x) = \frac{4}{\beta}\arctan\left[e^{\sqrt{\alpha}\beta x}\right] \tag{8.30}$$

known as sine-Gordon solitons. Observe that $\varphi_S(+\infty) = \frac{2\pi}{\beta}$ and $\varphi_S(-\infty) = 0$, hence this soliton solution is topologically nontrivial, with $Q = +1$.

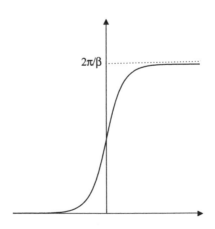

Figure 8.7 Sine-Gordon soliton solution $\varphi_{SG}(x)$

8.3.3 Z(N) (Phase) Solitons

Consider now a complex scalar field, described by the Lagrangean

$$\mathcal{L} = \partial^\mu \phi^* \partial_\mu \phi + \mu \left(\phi^N + \phi^{*N} \right) - \lambda \left(\phi^* \phi \right)^N - \frac{\mu^2}{\lambda}. \tag{8.31}$$

This is invariant under the discrete $Z(N)$ symmetry, defined as

$$\phi \to \phi \, \exp \left\{ i \frac{2\pi}{N} \right\}. \tag{8.32}$$

Using a polar representation for the complex field, namely $\phi = \frac{\rho}{\sqrt{2}} e^{i\theta}$, where ρ, θ are real fields, we can express the Lagrangean as

$$\mathcal{L} = \frac{1}{2} \partial^\mu \rho \partial_\mu \rho + \frac{\rho^2}{2} \partial^\mu \theta \partial_\mu \theta + 2\mu \left(\frac{\rho}{\sqrt{2}} \right)^N \cos N\theta - \lambda \left(\frac{\rho}{\sqrt{2}} \right)^{2N} - \frac{\mu^2}{\lambda}. \tag{8.33}$$

This possesses minima at

$$\frac{\rho_0}{\sqrt{2}} = \left(\frac{\mu}{\lambda} \right)^{\frac{1}{N}} \qquad \theta_{0,n} = \frac{2\pi}{N} n \qquad n = 0, 1, \ldots, N - 1. \tag{8.34}$$

Observe that the nontriviality of the vacuum manifold is related to the phase, rather than to the modulus of the field ϕ, since this is the one that has a nontrivial behavior in the vacuum manifold. Topologically nontrivial solutions, therefore, are expected to be related to the phase rather than to the modulus. As a consequence, the topological features of the system should not be affected by making the approximation of constant modulus $\rho = \rho_0$. Physical applications of this model, accordingly, correspond to a phase with constant ρ [17]. In this case, therefore, using (8.34), we get

$$\mathcal{L} = \frac{1}{2} \rho^2 \, \partial^\mu \theta \partial_\mu \theta + \frac{2\mu^2}{\lambda} \left[\cos N\theta - 1 \right]. \tag{8.35}$$

This is the sine-Gordon theory with $\alpha = \frac{2\mu^2}{\lambda}$ and $\beta = N$. The topological current will be given now by

$$J^\mu = \frac{N}{2\pi} \epsilon^{\mu\nu} \partial_\nu \theta \tag{8.36}$$

and the corresponding topological charge, $Q = \int J^0 dx$.

The theory will admit soliton solutions for the phase θ, given by

$$\theta_S(x) = \frac{4}{N} \arctan \left[e^{\sqrt{\frac{2\mu^2}{\lambda} \frac{N}{\rho_0} x}} \right]. \tag{8.37}$$

This will have a topological charge $Q = 1$.

8.4 Classical Vortices

We now consider the $U(1)$-symmetric abelian Higgs model in 2+1D, which is just the relativistic version of the Landau–Ginzburg theory. This is described by the Lagrangean involving a complex scalar field ϕ and a $U(1)$ gauge field A_μ, namely

$$\mathcal{L} = -\frac{1}{4}F_{\mu\nu}F^{\mu\nu} + (D_\mu\phi)^*(D^\mu\phi) + m^2\phi^*\phi - \frac{g}{2}(\phi^*\phi)^2, \tag{8.38}$$

where $D_\mu = \partial_\mu + i\frac{e}{\hbar c}A_\mu$.

This theory was shown to admit classical static solutions of the form [18]

$$\phi(\mathbf{x}) = \rho_0\left[1 - f(r)\right]\exp\{i\,\arg(\mathbf{x})\}$$
$$\mathbf{A}(\mathbf{x}) = g(r)\hat{\varphi} \qquad A_0 = 0$$
$$B = \epsilon^{ij}\partial_i A_j = h(r), \tag{8.39}$$

where $r \equiv |\mathbf{x}|$ and $\rho_0 = \frac{m^2}{g}$. Notice that this complies with the energy finiteness condition (8.12).

In the above expression, the functions $f(r)$, $g(r)$ and $h(r)$ have the following asymptotic behavior:

$$f(r) \overset{r\to\infty}{\longrightarrow} e^{-r/\xi}$$
$$g(r) \overset{r\to\infty}{\longrightarrow} \left(\frac{\hbar c}{e}\right)\frac{1}{r} - \rho_0 K_1(r/\lambda)$$
$$h(r) \overset{r\to\infty}{\longrightarrow} \frac{\rho_0}{\lambda}K_0(r/\lambda), \tag{8.40}$$

where $\xi = \frac{1}{\sqrt{2m^2}}$ is the "correlation length," $\lambda = \frac{\hbar c}{e\rho_0}$ is the "penetration depth," K_0 and K_1 are modified Bessel functions and $\hat{\varphi}$ is the angular unit polar vector in the plane.

Notice that the modulus of the Higgs field behaves as

$$|\phi(\mathbf{r})| \overset{r\to\infty}{\longrightarrow} \rho_0, \tag{8.41}$$

whereas the gauge field and corresponding magnetic field

$$\mathbf{A}(\mathbf{x}) \overset{r\to\infty}{\longrightarrow} \left(\frac{\hbar c}{e}\right)\frac{1}{r}\hat{\varphi}$$
$$B(\mathbf{r}) \overset{r\to\infty}{\longrightarrow} 0. \tag{8.42}$$

Near the origin, it can be shown that

$$|\phi(\mathbf{r})| \overset{r\to 0}{\longrightarrow} 0$$
$$B(\mathbf{r}) \overset{r\to 0}{\longrightarrow} C \neq 0. \tag{8.43}$$

We see that the Higgs field vanishes inside the vortex, whereas the magnetic field is nonzero. Considering that, for large x, the modified Bessel functions behave as

$$K_\nu(x) \to \sqrt{\frac{\pi}{2x}} e^{-x},$$
(8.44)

we see that λ sets the length scale where the magnetic field B is nonzero, hence the name penetration depth. The size of the vortex, namely, the region where the Higgs field vanishes, conversely, is determined by ξ, the correlation length.

From (8.42), it becomes clear that the topological invariant Φ given by (8.12) corresponding to the above vortex solution is

$$\Phi = \Phi_0 = \frac{hc}{e},$$
(8.45)

namely, one quantum of magnetic flux. This characterizes the vortex solution as a topological excitation.

One can immediately recognize the similarity between the vortex solution of the abelian Higgs model with the magnetic flux configurations occurring in the Landau–Ginzburg theory for a Type II superconductor, which corresponds to the situation where $\lambda > \xi$.

8.5 Classical Skyrmions

Let us consider now the $SO(3)$ invariant nonlinear sigma model in 2+1D, containing a triplet of scalar fields n^a, described by the Lagrangean

$$\mathcal{L} = \frac{\rho_0}{2} \partial_\mu n^a \partial^\mu n^a.$$
(8.46)

This is supplemented by the constraint

$$n^a n^a = 1,$$
(8.47)

which contains the interaction and makes the system nontrivial.

This system possesses the following static solution [20]:

$$\mathbf{n}_S(\mathbf{r}) = \left(\sin f(r)\hat{r}, \cos f(r) \right)$$

$$f(r) = 2 \arctan \left(\frac{\xi}{r} \right),$$
(8.48)

where $r = |\mathbf{r}|$ and $\mathbf{r} = r\hat{r}$. This was called "skyrmion" for its similarity with field configurations found by Skyrme in 3+1D [181]. Notice that n_S^a has radial symmetry, with $\mathbf{n}_S(0) = (0, -1)$, $\mathbf{n}_S(r \to \infty) = (0, 1)$ and $\mathbf{n}_S(r = \xi) = (\hat{r}, 0)$. The length scale ξ determines the skyrmion size.

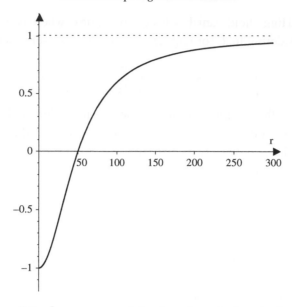

Figure 8.8 $\mathbf{n}_S^z$ component of the skyrmion solution, for $\xi = 50$

By making a stereographic projection, it is easy to see that the $\mathbf{n}_S$ solution is mapped into a configuration of unit vectors perpendicular to any point of a spherical surface. The circle with $r = \xi$, in particular, is mapped into the equator. Because of this mapping the skyrmion solution is also known as "hedgehog."

This last property suggests that the skyrmion produces a topologically nontrivial mapping $S_2 \rightarrow S_2$. Indeed, it can be easily verified that for $\mathbf{n}_S$, the topological invariant (8.18) is $Q = +1$. The classical skyrmion configuration has energy

$$E_S = \frac{\rho_0}{2} \int d^2x \nabla n^a \cdot \nabla n^a = 4\pi \rho_0. \tag{8.49}$$

We will see that skyrmion topological excitations play a central role in two-dimensional quantum magnetic systems.

8.6 Classical Magnetic Monopoles

Let us now study the non-abelian version of the Higgs model studied before, namely, the local $SO(3)$ invariant gauge theory, also known as the Georgi–Glashow model [22], given by

$$\mathcal{L} = -\frac{1}{4}F_{\mu\nu}^a F^{a\mu\nu} + (D_\mu \Phi)^a (D^\mu \Phi)^a + m^2 \Phi^a \Phi^a - \frac{\lambda}{2}(\Phi^a \Phi^a)^2, \tag{8.50}$$

where

$$F_{\mu\nu}^a = \partial_\mu A_\nu^a - \partial_\nu A_\mu^a + g\epsilon^{abc} A_\mu^b A_\nu^c \tag{8.51}$$

and

$$(D_\mu)^{ab} = \delta^{ab}\partial_\mu + g\epsilon^{abc}A_\mu^c. \tag{8.52}$$

Φ^a, $a = 1, 2, 3$ is the scalar Higgs field, transforming under the adjoint representation of the $SO(3)$ group, where the generators are given by

$$(T^a)_{bc} = i\epsilon^{abc} \tag{8.53}$$

and g is the coupling parameter.

The classical minima of the potential are such that $\Phi^a\Phi^a = |\Phi_0|^2 = m^2/\lambda$. While (classically) breaking $SO(3)$, these minima still preserve an unbroken $U(1)$ subgroup. There is, correspondingly, a $U(1)$ gauge field, the field intensity tensor of which is given by

$$F_{\mu\nu} \equiv \hat{\Phi}^a F_{\mu\nu}^a - \frac{1}{g}\epsilon^{abc}\hat{\Phi}^a[D_\mu\hat{\Phi}]^b[D_\nu\hat{\Phi}]^c, \tag{8.54}$$

where $\hat{\Phi}^a = \Phi^a/|\Phi_0|$.

The topological charge is given by

$$J^\mu = \epsilon^{\mu\nu\alpha\beta}\partial_\nu F_{\alpha\beta} \tag{8.55}$$

It follows that the magnetic field associated to the field intensity tensor above, usually defined as $B^i = \frac{1}{2}\epsilon^{ijk}F_{jk}$, is given by

$$B^i = \epsilon^{ijk}[\partial_j A_k^a]\hat{\Phi}^a - \frac{1}{g}\epsilon^{ijk}\epsilon^{abc}\hat{\Phi}^a\partial_j\hat{\Phi}^b\partial_k\hat{\Phi}^c. \tag{8.56}$$

Consequently, magnetic charge is the topological charge, namely

$$J^0 = \nabla \cdot \mathbf{B} = \epsilon^{ijk}[\partial_j A_k^a][\partial_i\hat{\Phi}^a], \tag{8.57}$$

where we used the fact that for a field $\hat{\Phi}^a$ such that $|\hat{\Phi}| = 1$, the divergence of the second term in (8.56) vanishes. Equivalently, we will have

$$Q = \int d^3x\, J^0 = g\oint_{S_2(\infty)} d\mathbf{S} \cdot \mathbf{B} = Q_M. \tag{8.58}$$

A solution for the field equations of the $SO(3)$ Georgi–Glashow model in the adjoint representation was independently obtained by 't Hooft and Polyakov [23, 24] by using the following ansatz:

$$\Phi^a(\mathbf{r}) = f(r)\frac{r^a}{r}$$

$$A_i^a(\mathbf{r}) = g(r)\epsilon^{aij}\frac{r^j}{r} \quad ; \quad A_0^a = 0, \tag{8.59}$$

where $r = |\mathbf{r}|$ and the radial functions satisfy

$$f(r) \xrightarrow{r \to \infty} |\Phi_0|$$

$$g(r) \xrightarrow{r \to \infty} \frac{1}{gr}. \tag{8.60}$$

These field configurations yield

$$[D_i \Phi]^a \xrightarrow{r \to \infty} 0$$

$$F^{ij} \xrightarrow{r \to \infty} F^{aij} \frac{\Phi^a}{\Phi_0}$$

$$\mathbf{B}(\mathbf{r}) \xrightarrow{r \to \infty} \left(\frac{1}{g}\right) \frac{\mathbf{r}}{r^3}. \tag{8.61}$$

The last expression shows that the configuration (8.59) indeed is a magnetic monopole with magnetic (topological) charge $Q_M = 4\pi/g$.

The classical magnetic monopole energy can be obtained from the above solution by integrating the energy density over all three-dimensional space. This is shown to satisfy the so-called Bogomolnyi bound [49]:

$$E_{mon} \geq \frac{4\pi M}{g^2}, \tag{8.62}$$

where M is the gauge field mass generated by the Higgs mechanism. It has actually been shown [23, 24, 50, 51] that

$$E_{mon} = f\left(\frac{\lambda}{g^2}\right) \frac{4\pi M}{g^2},$$

$$f(0) = 1 \qquad f(\infty) = 1.787, \tag{8.63}$$

hence the classical monopole energy is in the interval $[1.000, 1.787]\frac{4\pi M}{g^2}$. We will see that the quantum monopole mass also respects these bounds.

9

Quantum Topological Excitations

Given a physical system, we may in principle classify its fundamental excitations in two broad classes, namely, the Hamiltonian excitations and the topological excitations. The former are created out of the vacuum by operators, which appear themselves explicitly in the Hamiltonian and usually bear conserved quantities such as charge and spin, for instance, that are conserved by virtue of some continuous symmetry of the Hamiltonian. The latter, conversely, carry at least an observable quantity, the conservation of which derives from a nontrivial topological structure of the configuration space, as we saw in the previous chapter. There, we examined a number of classical topological excitations occurring in different dimensions. The full quantum description of such excitations is a matter of utmost importance; however, since the topological excitation degrees of freedom do not show explicitly in the Hamiltonian this is not a straightforward task. In this chapter, we develop a method, based on the concept of order-disorder duality, which allows for a systematic full quantum description of topological excitations.

9.1 Order-Disorder Duality and Quantum Topological Excitations

9.1.1 Order and Disorder Variables

Quantum topological excitations, surprisingly, are closely related to the so-called order-disorder duality structure, which is a concept playing an important role in many physical systems. We will see below how we can take advantage of this structure in order to develop a full quantum theory of topological excitations, which is widely applicable.

The stability of topological excitations, as we have seen in the past chapter, derives from a nontrivial topological structure of the configurations space. For this, a crucial condition is the existence of a nontrivial manifold of minima of the potential (vacuum manifold), which stems from the occurrence of degenerate vacuum states. This, however, invariably requires the vacuum expectation value of the basic field being different from zero, namely, a nonzero order parameter: $\langle \phi \rangle \neq 0$. We conclude the system ought to be in an ordered phase, characterized by a non-vanishing order parameter, for topological excitations to exist.

Let us consider now a full quantum-mechanical topological excitation state $|\mu\rangle$. We assume this is created out of the vacuum by the operator $\mu(\mathbf{r}, t)$, namely

$$|\mu\rangle = \mu(\mathbf{r}, t)|0\rangle. \tag{9.1}$$

In systems exhibiting classical topological excitations, we naturally expect the presence of the corresponding quantum states in the spectrum of excitations. In such a case, the topological quantum state, being a genuine excitation, must be orthogonal to the vacuum, hence $\langle 0|\mu\rangle = 0$, or equivalently, $\langle \mu \rangle = 0$, whenever a system presents classical topological excitations. We have seen, however, that a necessary condition for this is, ultimately, a non-vanishing vacuum expectation value of the basic field: $\langle \phi \rangle \neq 0$, namely, an ordered state.

Conversely, if a system does not possess classical topological excitations, the action of the operator $\mu(\mathbf{r}, t)$ on the vacuum should be trivial, producing no effect other than just the vacuum. Hence, in this case $\langle 0|\mu\rangle = 1$ or $\langle \mu \rangle = 1$. The absence of classical topological excitations, on the other hand, implies the triviality of the vacuum manifold, because this is associated with the topological triviality of the configurations space. A trivial vacuum manifold, by its turn, implies the absence of vacuum degeneracy or, in other words, a unique, symmetry invariant vacuum. Now, from (7.15), (7.16) and (7.17) we see that a unique vacuum implies $\langle \phi \rangle = 0$, namely, the absence of order.

In summary, we conclude, therefore, that

$$\langle \phi \rangle \neq 0 \longrightarrow \langle \mu \rangle = 0$$
$$\langle \mu \rangle \neq 0 \longrightarrow \langle \phi \rangle = 0. \tag{9.2}$$

We see that the vacuum expectation value of the topological excitation creation operator behaves as a disorder parameter. Indeed, $\langle \mu \rangle$ measures the amount of disorder in the system in the same way that $\langle \phi \rangle$ measures the amount of order. This order-disorder duality will be the key point for the development of a full quantum theory of topological excitations. The quantum-mechanical properties of such excitations, indeed, follow naturally from the quantization of the corresponding disorder variables. In the remainder of this chapter, we show how this method was born in the realm of statistical mechanics and thereafter generalized to quantum field theory. We also present several concrete applications of the method in the description of quantum topological excitations.

9.1.2 Dual Algebra and the Köberle–Marino–Swieca Theorem

The features deriving from the order-disorder duality that exists between the Hamiltonian field and the topological creation operator are fully captured by the so-called dual algebra, relating the two operators. In order to state what this algebraic relation is, let us assume the system is invariant under the operation of a group element g on the basic field ϕ:

$$\phi \longrightarrow U\phi U^\dagger = g \circ \phi, \tag{9.3}$$

where g belongs to some symmetry group G.

In one spatial dimension ($d = 1$) the dual algebra reads

$$\mu(x, t)\phi(y, t) = \begin{cases} [g \circ \phi](y, t)\mu(x, t) & y > x \\ \phi(y, t)\mu(x, t) & y < x. \end{cases} \tag{9.4}$$

For a $Z(N)$ group, for instance, $g \circ \phi = e^{i2\pi/N}\phi$.

For a system in d-spatial dimensions ($d > 1$), the topological excitation creation operator in principle would depend both on the point $\mathbf{r}$ and on a S_{d-1}-surface, centered on $\mathbf{r}$, namely $\mu = \mu(\mathbf{r}, t; S)$. The S_{d-1}-surface encloses the region of space $T(S)$ inside which we have the trivial vacuum. Its radius, therefore, roughly measures the extension of the topological excitation. Of course, we can always take the local limit, in which the region $T(S)$ shrinks to a point.

In the case of vortices in $d = 2$, therefore, $T(S)$ would be a circle of radius of the order of the penetration depth λ and S_1, its boundary.

In $d > 1$ dimensions, the dual algebra is given by

$$\mu(\mathbf{x}, t; T(S))\phi(\mathbf{y}, t) = \begin{cases} [g \circ \phi](\mathbf{y}, t)\mu(\mathbf{x}, t; T(S)) & \mathbf{y} \notin T(S) \\ \phi(\mathbf{y}, t)\mu(\mathbf{x}, t; T(S)) & \mathbf{y} \in T(S). \end{cases} \tag{9.5}$$

In the local limit, conversely, we would have

$$\mu(\mathbf{x}, t)\phi(\mathbf{y}, t) = [g \circ \phi](\mathbf{y}, t)\mu(\mathbf{x}, t). \tag{9.6}$$

A general theorem was rigorously demonstrated in $d = 1$ for operators satisfying the dual algebra (9.4) [317]. This is the Köberle–Marino–Swieca theorem [317, 26, 27], which contains basically three results:

$$1)\ \langle\mu\rangle\langle\phi\rangle = 0$$
$$2)\ \langle\mu(\mathbf{x}, t)\phi(\mathbf{y}, t)\rangle = 0 \tag{9.7}$$

and (3) the mass gap vanishes, whenever both $\langle\phi\rangle = 0$ and $\langle\mu\rangle = 0$.

Even though the theorem was demonstrated in $d = 1$, it should be valid in any dimension. Observe that (9.2) follows from result 1 above. Observe also that the theorem restricts the number of possible phases of the system to basically three: (a)

$\langle \phi \rangle \neq 0$ and $\langle \mu \rangle = 0$; (b) $\langle \phi \rangle = 0$ and $\langle \mu \rangle = 0$; (c) $\langle \phi \rangle = 0$ and $\langle \mu \rangle \neq 0$. A similar result was used by 't Hooft [28] in order to classify the phases of non-abelian gauge theories.

9.2 Duality in Statistical Mechanics

The concept of a duality relation connecting ordered and disordered phases of a statistical-mechanics system can be traced back to Kramers and Wannier [29]. The idea of a disorder variable satisfying a dual algebra with the Hamiltonian order variable, however, was put forward only much later by Kadanoff and Ceva [30]. These authors have also derived a method for evaluating quantum correlation functions of such disorder variables, given the Hamiltonian of the system. We have generalized their method for quantum field theory in general and applied the method to a broad variety of physical systems [31, 32, 33, 34, 35, 36, 37, 213, 38, 39, 41, 193, 42, 43]. An equivalent construction has been developed in [44].

We illustrate the method in the one-dimensional quantum Ising model, which is described by the Hamiltonian

$$H = -J \sum_n \sigma_3(n)\sigma_3(n+1), \tag{9.8}$$

where $\sigma_3(n)$ are Pauli matrices representing the S_z-operator in the $s = 1/2$ representation in the Hilbert space $\mathcal{H}_n$, such that the total Hilbert space is $\mathcal{H} = \ldots \otimes \mathcal{H}_n \otimes \mathcal{H}_{n+1} \otimes \ldots$ The Hamiltonian is invariant under the discrete symmetry $\sigma_3(n) \rightarrow -\sigma_3(n)$, for all n.

We now define a dual lattice with sites n^*, such that n^* is located right between and equidistant from the points n and $n+1$ from the original lattice. Introduce now an operator $\mu(n^*)$, through the order-disorder dual algebra [30, 46, 47],

$$\mu(n^*)\sigma_3(m) = \begin{cases} -\sigma_3(m)\mu(n^*) & m > n \\ \sigma_3(m)\mu(n^*) & m \leq n, \end{cases} \tag{9.9}$$

which explicitly uses the Hamiltonian symmetry. The algebra (9.4) actually was inspired in (9.9). An operator $\mu(n^*)$ satisfying this algebra can be written as

$$\mu(n^*) = \prod_{m>n} \sigma_1(m). \tag{9.10}$$

Now, let us determine what is the action of this operator on the ground state (vacuum) $|0\rangle$. Clearly, there are two degenerate ground states. In the representation we are using, these are expressed, respectively, as

$$\Psi_0^+ = \begin{pmatrix} 1 \\ 0 \end{pmatrix}_1 \otimes \ldots \otimes \begin{pmatrix} 1 \\ 0 \end{pmatrix}_N \tag{9.11}$$

and

$$\Psi_0^- = \begin{pmatrix} 0 \\ 1 \end{pmatrix}_1 \otimes \cdots \otimes \begin{pmatrix} 0 \\ 1 \end{pmatrix}_N. \tag{9.12}$$

The ground-state energy is $E_0 = -NJ$.

We see these are fully ordered states such that $\langle 0_\pm | \sigma_3(n) | 0_\pm \rangle = \pm 1$.

On the other hand, since

$$\sigma_1(m) \begin{pmatrix} 1 \\ 0 \end{pmatrix}_m = \begin{pmatrix} 0 \\ 1 \end{pmatrix}_m, \tag{9.13}$$

we see that, in the representation being used, $\mu(n^*)|0_+\rangle$ is given by

$$\Psi_\mu^+ = \begin{pmatrix} 1 \\ 0 \end{pmatrix}_1 \otimes \cdots \otimes \begin{pmatrix} 1 \\ 0 \end{pmatrix}_n \otimes \begin{pmatrix} 0 \\ 1 \end{pmatrix}_{n+1} \otimes \cdots \otimes \begin{pmatrix} 0 \\ 1 \end{pmatrix}_N. \tag{9.14}$$

This a Bloch wall, separating two domains with opposite spin orderings. It has precisely the characteristic features of the topological excitations we found in the field theory examples. Observe that clearly $\langle \mu \rangle_\pm = \langle 0_\pm | \mu | 0_\pm \rangle = 0$, indicating that the operator μ does indeed create states which are orthogonal to the ground state whenever this is an ordered state. Should the ground state be disordered, with (9.11) and (9.12) replaced by a random succession of up and down states, then reversing part of them would have no effect at all. Consequently, in this case we would have $\langle 0 | \mu | 0 \rangle = 1$. In a disordered phase, therefore, μ would not create a genuine excited state, as expected. This illustrates in a more concrete way the fact that topological excitation creation operators, satisfying a dual algebra, indeed act as disorder operators.

For the sake of obtaining a full description of the quantum dynamical properties of topological excitations, the knowledge of the time-dependent correlation functions $\langle \mu(\mathbf{x}, t) \mu(\mathbf{y}, t') \rangle$ would be required. In their pioneering work, Kadanoff and Ceva [30] introduced a method for evaluating a discrete (Euclidean) time version of such a correlation function. Their method was the inspiration for the derivation of a general continuum quantum field theory description of topological exciatations. We now describe this method.

Consider the classical two-dimensional system corresponding to the quantum one-dimensional Ising model, (9.8), described by the classical Hamiltonian

$$H = -J \sum_{\langle nm \rangle} \sigma(n) \sigma(m), \tag{9.15}$$

where $\sigma(n) = \pm 1$ are classical spin variables defined on the sites n of a square lattice and the sum runs over nearest neighbors. The system is clearly invariant under the global symmetry $\sigma(n) \rightarrow -\sigma(n)$.

We now introduce a dual lattice with sites n^*, located precisely at the center of each plaquette of the original square lattice. Classical disorder variables $\mu(n^*) = \pm 1$ are then introduced at the dual lattice sites. The statistical correlation function of the disorder variables, then, is introduced by the prescription [30]

$$\langle \mu(n^*)\mu(m^*) \rangle = \sum_{\{\sigma\}} e^{-H_C[\sigma]}, \tag{9.16}$$

where the sum on $\{\sigma\}$ runs over all spin configurations $\sigma(n) = \pm 1, \forall n$ and $H_C[\sigma]$ is given by $H[\sigma]$ in (9.15), with the modification that the sign of the coupling J is reversed along an arbitrary path connecting the points n^* and m^* in the dual lattice. Notice that a curve on the dual lattice crosses links connecting nearest neighbors of the original lattice.

We can write

$$H_C[\sigma] = \begin{cases} H[\sigma] & \langle ij \rangle \notin C \\ -H[\sigma] & \langle ij \rangle \in C \end{cases}, \tag{9.17}$$

where $\langle ij \rangle$ are arbitrary pairs of nearest neighbors on the square lattice. The correlation functions (9.16) are the Euclidean discrete time version of the corresponding disorder operators, which for real continuum time are given by (9.10).

A fundamental feature of this correlation function is that it is independent of the curve C, just depending on the dual lattice sites n^* and m^* where the disorder variables are located. This can be demonstrated by making a change of summation variables $\sigma(n) \rightarrow -\sigma(n)$ inside a region $\mathcal{R}$, the boundary of which is the closed curve Γ formed by C and another arbitrary curve C': $\Gamma = C - C'$. Since the sum $\sum_{\{\sigma\}}$ is clearly invariant under the change of variable above, it follows that

$$\langle \mu(n^*)\mu(m^*) \rangle_C = \langle \mu(n^*)\mu(m^*) \rangle_{C'}, \tag{9.18}$$

for arbitrary curves C and C'.

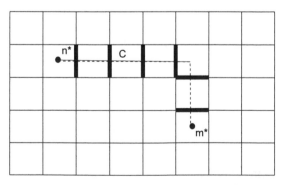

Figure 9.1 Arbitrary curve C connecting the points n^* and m^* on the dual lattice. The coupling sign is reversed for neighbors adjacent to the curve.

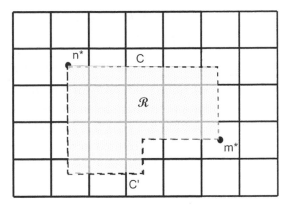

Figure 9.2 Arbitrary curves C and C' connecting the points n^* and m^* form the boundary of the region $\mathcal{R}$

The crucial feature of this construction, however, manifests itself when we evaluate mixed correlation functions, namely

$$\langle \mu(n_1^*)\sigma(m_1)\mu(n_2^*)\sigma(m_2)\rangle = \sum_{\{\sigma\}} e^{-H_C[\sigma]}\sigma(m_1)\sigma(m_2). \qquad (9.19)$$

Now, when performing the change of summation variables inside the region $\mathcal{R}$, in order to prove the path independence, we find a sign ambiguity, depending on whether $\sigma(m_1)$, $\sigma(m_2)$ or both belong to $\mathcal{R}$. This sign ambiguity, found in the classical Euclidean correlator, is precisely the manifestation of the dual algebra (9.9), satisfied by the fully quantized operator counterparts of the classical variables $\sigma(n)$ and $\mu(n^*)$. The different sign possibilities correspond to different operator orderings.

We will see that this wonderful property is shared by the Feynman functional integral formulation and we shall explore it in order to obtain a full quantum-mechanical formulation for arbitrary topological excitations in quantum field theory.

9.3 Quantum Field Theory of $Z(N)$ Solitons

Dual Algebra

We will introduce the subject for the case of a complex scalar field with a global $Z(N)$ symmetry in one-dimensional space. Later on, we will extend the method to a completely general case.

The Lagrangean could be given by (8.31), for instance, which is invariant under (8.32). We also assume the soliton creation operator $\mu(x, t)$ satisfies the dual algebra

$$\mu(x,t)\phi(y,t) = \begin{cases} e^{i2\pi/N}\,\phi(y,t)\mu(x,t) & y > x \\ \phi(y,t)\mu(x,t) & y < x. \end{cases} \tag{9.20}$$

Z(N) Quantum Soliton Correlation Functions: Derivation

Our purpose is to obtain a full quantum field theory of topological excitations, even though we do not know the appropriate Lagrangean for these degrees of freedom. In Chapter 6, we saw that for reaching this goal, it is sufficient to know all the field operator quantum correlation functions. Inspired in the construction of Kadanoff and Ceva described in the previous subsection, therefore, we are going to derive a general method for obtaining the correlation functions of the $\mu(x,t)$-operator, introduced in the theory described by a Lagrangean of the type (8.31) through the algebra (9.20), thereby fulfilling our goal.

We are going to use the fundamental features of the Kadanoff and Ceva construction as our guiding principle; namely, deformation of the Euclidean action along an arbitrary path connecting the points where the topological excitations are located, and path-independence of the corresponding correlation functions. In the case of mixed correlation functions, there must be an ambiguity, which reflects the dual order-disorder algebra in the framework of the Euclidean functional integral formalism.

Let us take the vacuum functional

$$Z_0 = \int D\phi D\phi^* \exp\left\{-\int d^2z\left[\partial^\mu\phi^*\partial_\mu\phi + V(\phi,\phi^*)\right]\right\} \tag{9.21}$$

and deform it in the following way [32, 43]:

$$\partial_\mu \longrightarrow D_\mu = \partial_\mu + i\alpha\tilde{A}_\mu(z; C(x,y)), \tag{9.22}$$

where

$$\tilde{A}^\mu(z; C(x,y)) = \int_{x,C}^{y} d\xi_\nu \epsilon^{\mu\nu}\delta^2(z-\xi) \tag{9.23}$$

and $\alpha = \frac{2\pi}{N}$. Then, the topological excitation correlation function will be

$$\langle\mu(x)\mu^\dagger(y)\rangle = Z_0^{-1}\int D\phi D\phi^* \exp\left\{-\int d^2z\left[(D^\mu\phi)^*D_\mu\phi + V(\phi,\phi^*)\right]\right\}. \tag{9.24}$$

In order to prove that the above expression does not depend on the curve C, consider a region $\mathcal{R}$ of two-dimensional Euclidean space bounded by the closed curve $\Gamma = C - C'$, where C' is an arbitrary curve also connecting x and y. We then introduce the two-dimensional Heaviside function

$$\theta(\mathcal{R}) = \begin{cases} 1 & z \in \mathcal{R} \\ 0 & z \notin \mathcal{R}. \end{cases} \tag{9.25}$$

The derivative of this is given by

$$\partial^\mu \theta(\mathcal{R}) = -\oint_{\Gamma = C - C'} \epsilon^{\mu\nu} d\xi_\nu \delta^2(z - \xi) = \tilde{A}^\mu(z; C'(x, y)) - \tilde{A}^\mu(z; C(x, y)).$$

(9.26)

Let us take (9.24) and perform the transformation

$$\phi(x) \rightarrow e^{i\alpha\theta(\mathcal{R})}\phi(x).$$ (9.27)

This would produce the following change in $\tilde{A}^\mu(z; C(x, y))$:

$$\tilde{A}^\mu(z; C(x, y)) \rightarrow \tilde{A}^\mu(z; C(x, y)) + \partial^\mu \theta(\mathcal{R}) = \tilde{A}^\mu(z; C'(x, y)),$$

(9.28)

where we used (9.26) in the last step.

Since the integration measure as well as the potential are invariant under the transformation (9.27), the only effect of such is to replace the curve C for an arbitrary curve C'. This establishes the fact that (9.24) does not depend on the curve C.

Consider now the mixed correlation function

$$\langle \mu(x_1)\phi(x_2)\mu^\dagger(y_1)\phi^\dagger(y_2) \rangle = Z_0^{-1} \int D\phi D\phi^*$$

$$\times \exp\left\{ -\int d^2z \left[(D^\mu\phi)^* D_\mu\phi + V(\phi, \phi^*) \right] \right\} \phi(x_2)\phi^*(y_2). \quad (9.29)$$

Now, when performing the transformation (9.27), in order to prove path independence, we will have extra $e^{\pm i\frac{2\pi}{N}}$-factors, in case the points x_2 or y_2 belong to the region $\mathcal{R}$. We see that the mixed correlation function above has a phase ambiguity, which is precisely a manifestation of the dual algebra (9.20) in the Euclidean functional integral, the different phase corresponding to different orderings of the operators in the left-hand side.

Z(N) Quantum Soliton Creation Operator

From (9.24), we may extract the form of the soliton creation operator by using the principle that averages are obtained through functional integration with the weight e^{-S}. Apart from a renormalization term (quadratic in $\tilde{A}_\mu$), the weighed quantity contains a linear in $\tilde{A}_\mu$ term. From this we get

$$\mu(x)\mu^\dagger(y) = \exp\left\{ i\int d^2z j_\mu \tilde{A}^\mu(z; C(x, y)) \right\}.$$ (9.30)

Inserting the external field and the complex scalar field current, and observing that

$$\tilde{A}^\mu(z; C(x, y)) = \tilde{A}^\mu(z; C(x, \infty)) - \tilde{A}^\mu(z; C(y, \infty)),$$ (9.31)

we may express the soliton operator as

$$\mu(x, t) = \exp\left\{\frac{2\pi}{N} \int_{x,C}^{\infty} dz[\phi^*\partial_t\phi - \phi\partial_t\phi^*](z, t)\right\}, \tag{9.32}$$

where we choose the curve along the spatial axis. In terms of the momenta canonically conjugated to the fields ϕ and ϕ^*, this can be written as

$$\mu(x, t) = \exp\left\{\frac{2\pi}{N} \int_{x,C}^{\infty} [\phi^*(z, t)\pi^*(z, t) - \phi(z, t)\pi(z, t)]dz\right\}. \tag{9.33}$$

From this, using the equal-times canonical commutation relations

$$[\phi(x, t), \pi(y, t)] = [\phi^*(x, t), \pi^*(y, t)] = i\delta(x - y) \tag{9.34}$$

and the Baker–Hausdorff relation

$$e^A B e^{-A} = B + [A, B] + \frac{1}{2}[A, [A, B]] + \ldots, \tag{9.35}$$

we obtain

$$\mu(x, t)\phi(y, t) = \exp\left\{i\frac{2\pi}{N}\theta(y - x)\right\}\phi(y, t)\mu(x, t), \tag{9.36}$$

which is precisely the dual algebra (9.20).

The final confirmation that we have indeed succeeded in obtaining a quantum soliton creation operator would be the demonstration that μ creates eigenstates of the topological charge operator. According to (8.36), this is given by

$$Q = \frac{N}{2\pi} \int dz\partial_z\theta. \tag{9.37}$$

Now, using the polar representation of ϕ in (9.32),

$$\mu(x, t) = \exp\left\{\frac{2\pi}{N} \int_{x,C}^{\infty} dz\pi_\theta(z, t)\right\}, \tag{9.38}$$

where $\pi_\theta = \rho^2\partial_t\theta$ is the momentum canonically conjugate to θ.

Again, using the Baker–Hausdorff formula, we can easily show that

$$Q\mu(x, t)|0\rangle = \mu(x, t)Q|0\rangle - \int_{-\infty}^{\infty} dz \int_x^{\infty} d\xi\, \partial_\xi\delta(z - \xi)\, \mu(x, t)|0\rangle$$

$$Q\mu(x, t)|0\rangle = \mu(x, t)|0\rangle, \tag{9.39}$$

where we used the fact that the vacuum is a zero charge eigenstate.

We see the operator μ creates eigenstates of the quantum topological charge operator Q, with eigenvalue equal to one. Our method of obtaining a full quantum field theory of topological excitations, based on the order-disorder duality was therefore successful. The soliton correlation functions are described by (9.24), which is nothing but the vacuum functional in the presence of the peculiar external field $\tilde{A}^\mu(z; C(x, y))$. This is very convenient from the calculational point of view, namely, standard procedures of quantum field theory allow for the calculation of these correlation functions. Arbitrary, $2n$-point functions, on the other hand, can be straightforwardly obtained by just inserting additional external field in (9.24).

The Phase Diagram

Let us consider a complex scalar field described by the Lagrangean (8.31), with $N = 4$, which is thus invariant under global $Z(4)$ transformations. This can be rewritten in terms of an auxiliary field σ that is $Z(2)$ invariant but transforms into $-\sigma$ under $Z(4)$, namely,

$$\mathcal{L} = \partial^\mu \phi^* \partial_\mu \phi - M^2 \phi^* \phi + \frac{\sigma^2}{2\mu} + \sigma \left(\phi^2 + \phi^{*2} \right) - \tilde{\lambda} \left(\phi^* \phi \right)^2 - \frac{\mu^2}{\lambda}. \qquad (9.40)$$

Integration over σ yields (8.31).

We want to determine the possible phases in which the system might exist and subsequently explore the properties of the quantum solitons in each of these. For this purpose, it is crucial to obtain the σ field properties. These are contained in the σ-dependent part of the potential in (9.40). However, in order to make the picture more faithful to reality, which is ultimately quantum-mechanical, we shall consider the full effective potential instead of just the classical one. That was introduced in Chapter 6 and allows the quantum corrections to be dealt with in the same way as we deal with the classical part of the potential. The σ-field effective potential is given by

$$V_{eff}(\sigma) = V(\sigma) + \mathcal{V}(\sigma)$$

$$V_{eff}(\sigma) = \frac{\sigma^2}{2\mu} - \sigma X - \frac{|\sigma|}{8} + \frac{M^2}{8\pi} \ln \frac{\sigma^2 + M^4}{M^4} + \frac{\sigma}{4\pi} \arctan \left(\frac{M^2}{\sigma} \right) \qquad (9.41)$$

where $X = (\phi^2 + \phi^{*2})$ and the quantum corrections are contained in $\mathcal{V}$.

The σ-field equation then reads

$$X = \frac{\sigma}{\mu} + \frac{\partial \mathcal{V}}{\partial \sigma}$$

$$X = \frac{1}{a} \left(\frac{\sigma}{M^2} \right) - \left[\frac{\pi}{2} - \arctan \left(\frac{M^2}{\sigma} \right) \right], \qquad (9.42)$$

where $a = \frac{\mu}{4\pi M^2}$.

We start looking for solutions with $\phi = 0$ or, equivalently, with $X = 0$. For $a < 1$, we can easily infer from the graphical solution in Fig. 9.3 that the only possibility in this case will be $\sigma = 0$. The two real components of the complex field, namely, $\phi = \frac{1}{\sqrt{2}}(\phi_1 + i\phi_2)$ then will have equal masses $M_i = M, i = 1, 2$.

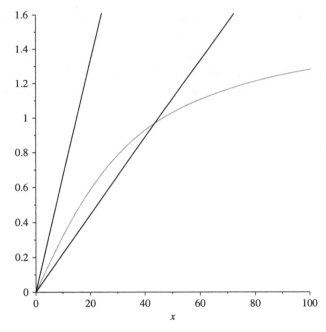

Figure 9.3 Graphical solution of Eq. (9.42) for $X = 0$: $a < 1$, left line; $a > 1$, right line

For $a > 1$ and still with $X = 0$, we start to have a solution $\sigma_0 \neq 0$. Now, shifting the field σ around its vacuum value, say $+\sigma_0$, we modify the M_i, $i = 1, 2$ masses, accordingly:

$$M_1^2 = M^2 + 2\sigma_0 \quad ; \quad M_2^2 = M^2 - 2\sigma_0. \qquad (9.43)$$

We see, therefore, that σ_0 must remain restricted within the interval $\sigma_0 \in [0, \frac{M^2}{2}]$ in order to maintain the stability of the system. For $\sigma_0 > M^2/2$, the mass of one of the components of the complex field, namely ϕ_2, becomes imaginary. According to the graphical solution, this phase will occur for $a > \kappa = [\pi - 2\arctan 2]^{-1} \simeq 1.0784$. Above this value of a, we would have necessarily $X \neq 0$, and σ_0 would remain at the value $M^2/2$. A phase with $\phi \neq 0$ consequently sets in for $a > \kappa$. In this phase, the ϕ_1-field mass will be $M_1^2 = 2M^2$. The ϕ_2-field mass, conversely, will be proportional to $|X|$.

In summary, the phase structure of the model will be: (I) an unbroken $Z(4)$ symmetry phase, with $\langle \phi \rangle = \langle \sigma \rangle = 0$, for $a < 1$; (II) a partially broken ($Z(4)$ to $Z(2)$) symmetry phase, with $\langle \phi \rangle = 0$; $\langle \sigma \rangle \neq 0$ ($< M^2/2$) for $1 < a < \kappa$; (III) a completely broken $Z(4)$ symmetry phase, with $\langle \phi \rangle \neq 0$; $\langle \sigma \rangle = M^2/2$, for $\kappa < a$.

We now determine the large distance behavior of the quantum soliton correlations function in each of these phases.

Soliton Correlation Function

From (9.24), we see that the soliton correlation function is given by

$$\langle \mu(x)\mu^\dagger(y) \rangle = \exp\left\{ \Gamma\left[\tilde{A}_\mu(x, y) \right] \right\},\tag{9.44}$$

where $\Gamma[A_\mu]$ is the generator of proper vertices, introduced in Chapter 6, and $\tilde{A}_\mu(x, y) = \tilde{A}_\mu(x) - \tilde{A}_\mu(y)$, which was defined in (9.23).

We are interested in the large distance behavior of the soliton correlation function. This tells us a lot about the physical properties of quantum solitons. Indeed, there is a theorem due to Araki, Hepp and Ruelle [48], stating that in general

$$\langle \mu(x)\mu^\dagger(y) \rangle \overset{|x-y|\to\infty}{\longrightarrow} |\langle \mu \rangle|^2 + \exp\left\{ -E_0|x - y| \right\},\tag{9.45}$$

where E_0 is the gap in the energy spectrum. The large distance behavior thus allows us to get information not only about the phase structure but also about the mass spectrum.

We have shown [39, 40] that only the two-point proper vertex contributes to the large-distance behavior of the soliton correlator (9.44), consequently

$$\Gamma\left[\tilde{A}_\mu(x, y) \right] \overset{|x-y|\to\infty}{\longrightarrow} \frac{1}{2} \int d^2z d^2z'\, \tilde{A}_\mu(z; x, y)\Pi^{\mu\nu}(z - z')\tilde{A}_\nu(z'; x, y)$$

$$= \frac{1}{2} \int \frac{d^2q}{(2\pi)^2}\, \tilde{A}_\mu(q; x, y)\left[\frac{q^2\delta^{\mu\nu} - q^\mu q^\nu}{q^2} \right]\Pi(q)\tilde{A}_\nu(-q; x, y),\tag{9.46}$$

where

$$\tilde{A}_\mu(q; x, y) = \int_{x,C}^{y} d\xi_\nu \epsilon^{\mu\nu} e^{iq\cdot\xi}\tag{9.47}$$

is the Fourier transform of the external field given by (9.23), and $\Pi^{\mu\nu}$ is the two-point proper vertex, the vacuum polarization tensor.

Inserting the external field in (9.46) and noting that

$$q^2\delta^{\mu\nu} - q^\mu q^\nu = \epsilon^{\mu\alpha}\partial_\alpha \epsilon^{\nu\beta}\partial_\beta,\tag{9.48}$$

we get, after performing the angular integration in (9.46),

$$\Gamma\left[\tilde{A}_\mu(x, y) \right] \overset{|x-y|\to\infty}{\longrightarrow} \frac{1}{2\pi} \int_0^\infty \frac{dq}{q}[1 - J_0(q|x - y|)]\Pi(q),\tag{9.49}$$

where J_0 is a modified Bessel function.

Only $\Pi(q = 0)$ contributes to the large distance regime of $\Gamma(x - y)$, because

$$\lim_{|x|\to\infty} \frac{1 - J_0(q|x|)}{q^2|x|} = \delta(q);\tag{9.50}$$

hence, we can write

$$\Gamma(|x - y|) \overset{|x-y|\to\infty}{\to} \frac{\Pi(0)}{2\pi} \int_0^\infty \frac{dq}{q} \left[1 - J_0(q|x - y|)\right]. \qquad (9.51)$$

The integral above can be evaluated with the help of an ultraviolet cutoff Λ, yielding $-\ln|x - y| - \ln\Lambda$. The cutoff can be eliminated by a multiplicative renormalization of the soliton field, namely,

$$\mu_R(x) = \mu(x)\Lambda^{\frac{\Pi(0)}{4\pi}}. \qquad (9.52)$$

The renormalized soliton correlation function is, then, given by the $|x - y|$-dependent part:

$$\langle\mu(x)\mu^\dagger(y)\rangle_R \overset{|x-y|\to\infty}{\to} \frac{1}{|x - y|^{\frac{\Pi(0)}{2\pi}}}. \qquad (9.53)$$

Soliton Correlation Function: The Unbroken and Partially Broken Phases

The scalar $\Pi(0)$ is given by the graphs of Fig. 9.5. Denoting by M_1 and M_2 the masses of the two real fields that compose the complex field ϕ, we have [33, 34],

$$\Pi(0) = \frac{1}{4\pi} \left\{\ln M_1^2 M_2^2 - 2\int_0^1 \ln\left[M_1^2 x + M_2^2(1 - x)\right]\right\}. \qquad (9.54)$$

Notice that this vanishes for $M_1 = M_2$. For $M_1 \neq M_2$ we have

$$\Pi(0) = \frac{1}{2\pi} \left[1 - \frac{1}{2}\frac{M_1^2 + M_2^2}{M_1^2 - M_2^2} \ln\frac{M_1^2}{M_2^2}\right]. \qquad (9.55)$$

In the partially broken phase, where $\langle\sigma\rangle \neq 0$, we have $M_1 \neq M_2$, according to (9.43). Then it follows that $\Pi_{II}(0) \neq 0$, as can be inferred from the expression above.

Conversely, in the unbroken phase, where $\langle\sigma\rangle = 0$, we see from (9.43) that $M_1 \to M_2$. Then, we see from (9.54) that $\Pi(0) \to 0$ in this limit. In the unbroken

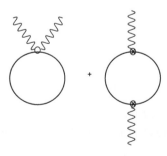

Figure 9.4 Diagrams contributing to the large distance behavior of the soliton correlation function. Phase I

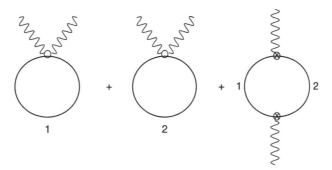

Figure 9.5 Diagrams contributing to the large distance behavior of the soliton correlation function. Phase II

phase, we have, therefore, $\Pi_I(0) = 0$. These properties of $\Pi(0)$ have far-reaching consequences for the quantum soliton physics. We immediately conclude from (554a) that in Phase I, we have

$$\langle \mu(x)\mu^\dagger(y)\rangle^I_R \stackrel{|x-y|\to\infty}{\longrightarrow} 1, \tag{9.56}$$

implying $\langle \mu \rangle_I = 1$. This result signifies that the soliton operator does not create genuine excitations in this phase.

In Phase II, conversely,

$$\langle \mu(x)\mu^\dagger(y)\rangle^{II}_R \stackrel{|x-y|\to\infty}{\longrightarrow} \frac{1}{|x-y|^{\frac{\Pi_{II}(0)}{2\pi}}}, \tag{9.57}$$

with $\Pi_{II}(0)$ given by (9.54). This now implies $\langle \mu \rangle_{II} = 0$, which means the soliton operator now creates true physical excitations. The power-law behavior in this case tells us these quantum solitons are massless.

Soliton Correlation Function: the Completely Broken Phase

We now turn to the completely broken phase, where $\langle \phi \rangle = \rho_0 \neq 0$. In this phase we must shift the field around its vacuum expectation value, thereby generating the new vertices depicted in Fig. 9.5. The proper vertex function is modified accordingly,

$$\Pi^{\mu\nu}_{III} = \Pi'^{\mu\nu}_{II} + \delta^{\mu\nu}\frac{M^2\rho_0^2}{q^2 + M^2}, \tag{9.58}$$

where $\Pi'^{\mu\nu}_{II}$ is the same as in Phase II, but with the modification

$$\Pi(q^2) \to \Pi(q^2) + \rho_0^2\frac{q^2}{q^2 + M^2}. \tag{9.59}$$

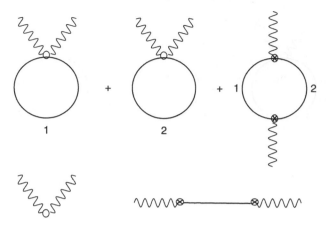

Figure 9.6 Diagrams contributing to the large distance behavior of the soliton correlation function. Phase III

The added piece above vanishes at $q = 0$, hence it will not change the contribution to the large distance behavior of the soliton correlator. The last one in (9.58), however, will. Inserting it in (9.46), we get

$$\Gamma_{III}(x - y) \overset{|x-y|\to\infty}{\Longrightarrow} \Gamma_{II}(x - y)$$

$$-\frac{M^2\rho_0^2}{2}\int_{x,C}^{y} d\xi_v \int_{x,C}^{y} d\eta_v \int \frac{d^2q}{(2\pi)^2}\frac{e^{iq\cdot(\xi-\eta)}}{q^2 + M^2}. \tag{9.60}$$

The q-integral above yields a modified Bessel function $K_0(M|\xi - \eta|)$; therefore, changing the integration variables to $r = \xi - \eta$, $R = \frac{1}{2}(\xi + \eta)$, we get

$$\Gamma_{III}(x - y) \overset{|x-y|\to\infty}{\Longrightarrow} \Gamma_{II}(x - y) - \frac{M^2\rho_0^2}{4\pi}|x - y|\int_0^{|x-y|} dr\, K_0(Mr). \tag{9.61}$$

The large distance behavior of the quantum soliton correlator in Phase III, therefore, will be

$$\langle \mu(x)\mu^\dagger(y)\rangle_R^{III} \overset{|x-y|\to\infty}{\Longrightarrow} \frac{e^{-\mathcal{M}|x-y|}}{|x - y|^{\frac{\Pi_{II}(0)}{2\pi}}}, \tag{9.62}$$

where $\mathcal{M} = \frac{M\rho_0^2}{8}$ is the quantum soliton mass.

9.4 Quantum Field Theory of Vortices

Quantum Vortex Correlation Functions: Derivation

We consider now local $U(1)$-symmetric gauge field theories in $d = 2$ spatial dimensions, with Lagrangean of the type

$$\mathcal{L} = -\frac{1}{4} F_{\mu\nu} F^{\mu\nu} + (D_\mu \phi)^* (D^\mu \phi) - V(\phi^* \phi), \tag{9.63}$$

where $D_\mu = \partial_\mu + i \frac{e}{\hbar c} A_\mu$, such as (8.38). Our purpose is to obtain a functional integral expression for the quantum vortex correlation functions, such that the (local) vortex creation operator satisfies the dual algebra (9.6) and the states it creates out of the vacuum are eigenstates of the topological charge operator (8.12) and (8.15), which is the magnetic flux piercing the plane. For this, we will make use of the same principles contained in the Kadanoff–Ceva construction.

Maxwell Theory

To begin with, let us consider just pure Maxwell theory. Even though the vortex states shall be trivial in this case, we may still introduce the vortex operator and evaluate its correlation functions. Following [38, 35], we write

$$\langle \mu(x) \mu^\dagger(y) \rangle_M = Z_0^{-1} \int DA_\mu \exp\left\{ -\int d^3z \frac{1}{4} \left[F^{\mu\nu} + \tilde{B}^{\mu\nu} \right] \left[F_{\mu\nu} + \tilde{B}_{\mu\nu} \right] \right\}, \tag{9.64}$$

where the external field is given by

$$\tilde{B}^{\mu\nu}(x, y; L) = 2\pi \frac{\hbar c}{e} \int_{x,L}^y \epsilon^{\mu\nu\alpha} \delta^3(z - \xi) d\xi_\alpha. \tag{9.65}$$

In this expression, L is an arbitrary curve connecting the points x and y in Euclidean space.

In order to show that the above correlation function is independent of the curve L, let us note that for the case where $L = L_0$ is a straight line going from $(\mathbf{x}, x^3)$ to infinity and returning to $(\mathbf{y}, y^3)$ through another straight line, then the external field $\tilde{B}^{\mu\nu}(x, y; L)$ will have just the spatial components $i, j = 1, 2$. We can write, therefore, in this case,

$$\tilde{B}^{ij}(x, y; L_0) = \frac{\hbar c}{e} \left(\partial_i \partial_j - \partial_j \partial_i \right) \alpha_{L_0}(z; x, y). \tag{9.66}$$

Here,

$$\alpha_{L_0}(z; x, y) = \theta(z^3 - x^3) \arg(\mathbf{z} - \mathbf{x}) - \theta(z^3 - y^3) \arg(\mathbf{z} - \mathbf{y}). \tag{9.67}$$

The previous result follows from the fact that the function $\arg(\mathbf{z})$, being the imaginary part of the analytic complex function $\ln z = \ln |\mathbf{z}| + i \arg(\mathbf{z})$, must satisfy the Cauchy–Riemann equation

$$\epsilon^{ij} \partial_j \arg(\mathbf{z}) = -\partial^i \ln |\mathbf{z}|$$
$$\left(\partial_i \partial_j - \partial_j \partial_i \right) \arg(\mathbf{z}) = -\epsilon^{ij} \partial^2 \ln |\mathbf{z}| = 2\pi \epsilon^{ij} \delta(\mathbf{z}). \tag{9.68}$$

It is clear that the δ-singularity of $\alpha_{L_0}(z; x, y)$ is located precisely on the curve L_0. It is not difficult to generalize $\alpha_{L_0}(z; x, y) \rightarrow \alpha_L(z; x, y)$ in such a way that that singularity lies at an arbitrary curve L, connecting x and y in 3d Euclidean space.

Then we can express (9.65) in terns of $\alpha_L(z; x, y)$ as follows,

$$
\tilde{B}^{\mu\nu}(x, y; L) = 2\pi \frac{\hbar c}{e} \int_{x,L}^{y} \epsilon^{\mu\nu\alpha} \delta^3(z - \xi) d\xi_\alpha = \frac{\hbar c}{e} \left(\partial_\mu \partial_\nu - \partial_\nu \partial_\mu \right) \alpha_L(z; x, y).
$$

(9.69)

Path independence of (9.64) follows immediately. Indeed, making a change of functional integration variable

$$
A_\mu \longrightarrow A_\mu + \frac{\hbar c}{e} \partial_\mu \omega(L', L),
$$

(9.70)

where $\omega(L', L) = \alpha_{L'}(z; x, y) - \alpha_L(z; x, y)$, we conclude, using (9.69), that under the above change of variable $F^{\mu\nu}$ transforms as

$$
F^{\mu\nu} \longrightarrow F^{\mu\nu} + \tilde{B}^{\mu\nu}(x, y; L') - \tilde{B}^{\mu\nu}(x, y; L),
$$

(9.71)

thus establishing the path independence of (9.64).

Abelian Higgs Model

We have found the method for obtaining quantum vortex correlation functions in 2+1D Maxwell theory. This can easily be generalized for theories of the type (9.63). Following the above procedure in order to obtain a path-independent vortex correlation function, we now face a new situation when considering the mixed correlator: $\langle \mu(x_1)\mu^\dagger(y_1)\phi(x_2)\phi^\dagger(y_2) \rangle_{AHM}$. Now, the transformation (9.70) must be accompanied by the corresponding transformation in the complex scalar field, namely

$$
\phi(x) \longrightarrow \exp\left\{ \frac{i}{\hbar} \omega(L', L) \right\} \phi(x).
$$

(9.72)

According to (9.67), this would imply the mixed correlation function

$$
\langle \mu(x_1)\mu^\dagger(y_1)\phi(x_2)\phi^\dagger(y_2) \rangle_{AHM}
$$

is defined by a functional integral similar to (9.64), up to a phase

$$
\exp\left\{ \frac{i}{\hbar} \left[\arg(x_1 - x_2) - \arg(x_1 - y_2) + \arg(y_1 - y_2) - \arg(y_1 - x_2) \right] \right\}
$$

depending on whether or not x_2 and y_2 belong to the region corresponding to $\omega(L', L)$. This ambiguity is the expression, within the functional integral framework, of the nontrivial commutation relations that exist between the vortex operator $\mu(x)$ and $\phi(x)$, which in the present case must be

$$\mu(\mathbf{x}, t)\phi(\mathbf{y}, t) = \exp\left\{\frac{i}{\hbar}\arg(\mathbf{y} - \mathbf{x})\right\}\phi(\mathbf{y}, t)\mu(\mathbf{x}, t). \tag{9.73}$$

We conclude that the vortex operator, defined by means of the above correlation function, in a similar way as in the Kadanoff–Ceva construction, satisfies a dual algebra with the Higgs scalar field operator ϕ.

Abelian Higgs Model: Gauge-Invariant Formulation

As it turns out, it will be much more convenient to work with a Lagrangean that is in an explicitly gauge-invariant form, as the Maxwell Lagrangean is. Hence, before applying the method in a theory such as the abelian Higgs model, let us recast (9.63) in a form that is explicitly gauge-invariant.

For this purpose, let us use the polar representation of the complex scalar field, $\phi = \frac{\rho}{\sqrt{2}}e^{i\theta}$. Inserting this in (9.63), we get

$$\mathcal{L} = -\frac{1}{4}F_{\mu\nu}F^{\mu\nu} + \frac{1}{2}\tilde{e}^2\rho^2(A^\mu + \frac{1}{\tilde{e}}\partial^\mu\theta)(A_\mu + \frac{1}{\tilde{e}}\partial_\mu\theta) + \frac{1}{2}\partial^\mu\rho\partial_\mu\rho - V(\rho), \tag{9.74}$$

where $\tilde{e} = \frac{e}{\hbar c}$. This is clearly invariant under the gauge transformation

$$A^\mu \to A^\mu + \frac{1}{\tilde{e}}\partial^\mu\Lambda$$
$$\theta \to \theta - \Lambda. \tag{9.75}$$

We now introduce the new field

$$W^\mu = A^\mu + \frac{1}{\tilde{e}}\partial^\mu\theta$$
$$\partial_\mu W^\mu = 0, \tag{9.76}$$

where, in the last step, we used the θ-field equation: $\Box\theta = -\tilde{e}\partial_\mu A^\mu$.

Using (9.76), we can rewrite (9.74) as

$$\mathcal{L} = -\frac{1}{4}W_{\mu\nu}W^{\mu\nu} + \frac{1}{2}\tilde{e}^2\rho^2 W_\mu\left[\frac{-\Box\delta^{\mu\nu} + \partial^\mu\partial^\nu}{-\Box}\right]W_\nu + \frac{1}{2}\partial^\mu\rho\partial_\mu\rho - V(\rho), \tag{9.77}$$

or, equivalently,

$$\mathcal{L} = -\frac{1}{4}W_{\mu\nu}\left[1 + \frac{\tilde{e}^2\rho^2}{-\Box}\right]W^{\mu\nu} + \frac{1}{2}\partial^\mu\rho\partial_\mu\rho - V(\rho), \tag{9.78}$$

which is explicitly gauge invariant.

Now we are ready to apply the method developed for obtaining the vortex correlation functions in the Maxwell theory. Using the same procedure employed there, we write the quantum vortex correlation functions in the abelian Higgs model as

$$
\langle \mu(x)\mu^\dagger(y)\rangle_{AHM} = Z_0^{-1} \int DW_\mu D\rho \exp \left\{ - \int d^3z \right.
$$
$$
\times \left[\frac{1}{4}\left[W^{\mu\nu} + \tilde{B}^{\mu\nu}\right]\left[1 + \frac{\tilde{e}^2\rho^2}{-\Box}\right]\left[W_{\mu\nu} + \tilde{B}_{\mu\nu}\right] + \frac{1}{2}\partial^\mu\rho\partial_\mu\rho + V(\rho)\right]\right\},
$$

$$(9.79)$$

where the external field $\tilde{B}^{\mu\nu}$ is given by (9.65).

Performing (9.70) and (9.71) to W_μ, we can promptly show the path independence of (9.79). In what follows, after extracting from this expression the form of the vortex creation operator, we evaluate its large distance behavior.

Vortex Creation Operators and Their Commutation Relations

From the above expression we may extract the explicit form of the vortex creation operator, namely

$$
\mu(x) = \exp\left\{ i\int d^3z \frac{1}{2}\tilde{B}^{\mu\nu}(x, \infty; L)\left[1 + \frac{\tilde{e}^2\rho^2}{-\Box}\right]W_{\mu\nu}\right\}. \tag{9.80}
$$

The corresponding operator in Maxwell theory would be obtained by just making $\rho \to 0$ in the expression above. Inserting the external field $\tilde{B}^{\mu\nu}(x, \infty; L)$, we get both in the abelian Higgs model and in Maxwell theory

$$
\mu(\mathbf{x}, t) = \exp\left\{ i2\pi\frac{\hbar c}{e}\int_{\mathbf{x},L}^\infty d\mathbf{z}^i \epsilon^{ij}\Pi^j(\mathbf{z}, t)\right\}, \tag{9.81}
$$

where $\Pi^i = \partial\mathcal{L}/\partial\dot{W}^i$ is the momentum canonically conjugate to W^i, which satisfies

$$
[W^i(\mathbf{x}, t), \Pi^j(\mathbf{y}, t)] = i\delta^{ij}\delta^2(\mathbf{x} - \mathbf{y}). \tag{9.82}
$$

Now, let us prove that indeed the vortex operator creates eigenstates of the topological charge operator

$$
Q = \int d^2z\epsilon^{ij}\partial^i W^j(\mathbf{z}, t). \tag{9.83}
$$

Using the relation

$$
Be^A = e^A B + [B, A]e^A, \tag{9.84}
$$

derived from (9.35) when $[A, [A, B]] = [B, [A, B]] = 0$, we immediately get

$$
[Q, \mu(\mathbf{x}, t)] = 2\pi\frac{\hbar c}{e}\mu(\mathbf{x}, t). \tag{9.85}
$$

Considering that for the vacuum state $Q|0\rangle = 0$, the above relation implies that the vortex quantum state $|\mu\rangle = \mu(\mathbf{x}, t)|0\rangle$ is an eigenstate of the topological charge operator with eigenvalue equal to one vorticity unit, namely

$$Q|\mu\rangle = \frac{hc}{e}|\mu\rangle. \tag{9.86}$$

It is very instructive to evaluate the commutator of the vortex operator with the W^i-field. Using (9.82) and (9.84), we readily obtain

$$[\mu(\mathbf{x}, t), W^i(\mathbf{y}, t)] = \mu(\mathbf{x}, t) 2\pi \frac{hc}{e} \int_{\mathbf{x}, L}^{\infty} dz^j \epsilon^{ij} \delta^2(\mathbf{z} - \mathbf{y}). \tag{9.87}$$

The integral above can be written as

$$I = 2\pi \frac{hc}{e} \int_{\mathbf{x}, L}^{\infty} dz^j \epsilon^{ij} \delta^2(\mathbf{z} - \mathbf{y}) = \frac{hc}{e} \oint_{C_x} dz^j \epsilon^{ij} \arg(\mathbf{z} - \mathbf{x}) \delta^2(\mathbf{z} - \mathbf{y}), \tag{9.88}$$

where we chose the contour going around the cut of the $\arg(\mathbf{z} - \mathbf{x})$-function and closing at infinity. Then, using Stokes' theorem, we have

$$I = \frac{hc}{e} \int_{T(C_x)} d^2 z \partial^i_{(z)} \left[\arg(\mathbf{z} - \mathbf{x}) \delta^2(\mathbf{z} - \mathbf{y}) \right]$$

$$I = \frac{hc}{e} \partial^i_{(y)} \arg(\mathbf{y} - \mathbf{x}) + \frac{hc}{e} \arg(\mathbf{y} - \mathbf{x}) \int_{T(C_x)} d^2 z \partial^i_{(z)} \delta^2(\mathbf{z} - \mathbf{y}). \tag{9.89}$$

The last integral vanishes when we remove the cutoffs; therefore we obtain, by inserting (9.89) in (9.87),

$$\mu(\mathbf{x}, t) W^i(\mathbf{y}, t) = \left[W^i(\mathbf{y}, t) + \frac{hc}{e} \partial^i_{(y)} \arg(\mathbf{y} - \mathbf{x}) \right] \mu(\mathbf{x}, t). \tag{9.90}$$

Using (9.76), we also infer that

$$\mu(\mathbf{x}, t) \theta(\mathbf{y}, t) = \left[\theta(\mathbf{y}, t) + \arg(\mathbf{y} - \mathbf{x}) \right] \mu(\mathbf{x}, t), \tag{9.91}$$

or, equivalently, we derive Eq. (9.73).

The commutators of the μ-operator with the fields W^i, θ and ϕ, along with the fact that it creates topological charge eigenstates, inequivocally characterize it as the quantum vortex creation operator. Notice that (9.73) is an example of (9.6), thus also exhibiting the fact that the μ-operator satisfies a dual algebra with ϕ.

Quantum Vortex Correlation Functions: Evaluation

We now conclude our study of quantum vortices by explicitly evaluating their two-point correlation functions in both phases of the theory.

In the Higgs phase, ρ will develop a vacuum expectation value ρ_0, consequently generating a mass $M = \tilde{e}\rho_0$ to the gauge field W_μ. We may, therefore, expand ρ around ρ_0 in expression (9.79), thereby obtaining in the leading order

$$\langle \mu(x)\mu^\dagger(y)\rangle_{AHM} = Z_0^{-1} \int DW_\mu \exp\left\{-\int d^3z\right.$$

$$\left.\times \frac{1}{4}\left[W^{\mu\nu} + \tilde{B}^{\mu\nu}\right]\left[1 + \frac{M^2}{-\Box}\right]\left[W_{\mu\nu} + \tilde{B}_{\mu\nu}\right]\right\},$$

$$= \exp\left\{\Lambda(x, y; L) - \int d^3z \frac{1}{4}\tilde{B}^{\mu\nu}\left[1 + \frac{M^2}{-\Box}\right]\tilde{B}_{\mu\nu}\right\},$$

$$(9.92)$$

where $\Lambda(x, y; L)$ results from the quadratic functional integral over W_μ, namely

$$\Lambda(x, y; L) = \frac{1}{8}\int d^3z d^3z' \, \tilde{B}^{\mu\nu}(z)\tilde{B}^{\alpha\beta}(z')P_\lambda^{\mu\nu}P_\rho^{\alpha\beta}$$

$$\times \left[1 + \frac{M^2}{-\Box}\right]D^{\lambda\rho}(z - z')\left[1 + \frac{M^2}{-\Box}\right],\qquad (9.93)$$

where $P_\lambda^{\mu\nu} = \partial^\mu\delta_\lambda^\nu - \partial^\nu\delta_\lambda^\mu$ and $D^{\lambda\rho}(z - z')$ is the massive gauge field propagator, given by

$$D^{\lambda\rho} = \frac{\delta^{\lambda\rho}}{-\Box + M^2} + \text{gauge terms}. \qquad (9.94)$$

Now, inserting (9.65) in (9.93), considering that

$$\left[1 + \frac{M^2}{-\Box}\right]\left[\frac{1}{-\Box + M^2}\right]\left[1 + \frac{M^2}{-\Box}\right] = \frac{1}{-\Box}\left[1 + \frac{M^2}{-\Box}\right] \qquad (9.95)$$

and introducing

$$F(z - z') = \left(2\pi\frac{hc}{e}\right)^2\left[\frac{1}{-\Box} + \frac{M^2}{(-\Box)^2}\right], \qquad (9.96)$$

we get four identical terms given by

$$\int_{x^a}^\infty d\xi^i \int_{x^b}^\infty d\eta^j\left[-\Box\delta^{ij} - \partial_{(\xi)}^i\partial_{(\eta)}^j\right]F(\xi - \eta). \qquad (9.97)$$

In view of (9.95), we see that the first term above exactly cancels the quadratic $\tilde{B}_{\mu\nu}$-term in (9.92). We are left, therefore, with

$$\langle \mu(x)\mu^\dagger(y)\rangle_{AHM} = \exp\left\{-\frac{1}{2}\sum_{ij=1}^2 \lambda_i\lambda_j F(x_i - x_j)\right\} = \exp\left\{F(x - y) - F(\epsilon)\right\},$$

$$(9.98)$$

where ϵ is a short-distance (ultraviolet) regulator.

Now, we use the inverse Fourier transforms

$$\mathcal{F}^{-1}\left[\frac{1}{-\Box}\right] = \frac{1}{4\pi|x|}$$

$$\mathcal{F}^{-1}\left[\frac{M^2}{(-\Box)^2}\right] = \lim_{m\to 0}\frac{M^2}{4\pi}\left[\frac{1}{m} - \frac{|x|}{2}\right], \qquad (9.99)$$

where m is an infrared regulator introduced in order to make the inverse Fourier transform of $(-\Box)^{-2}$ meaningful.

Using the above result in (9.96) and inserting in (9.98), we immediately see that the m-regulator cancels out and completely disappears from the vortex correlation function. This is so because this correlator preserves the vortex topological number selection rule. If we should calculate a vortex number violating correlation function such as $\langle\mu\mu\rangle$ or $\langle\mu^\dagger\mu^\dagger\rangle$, we would have, instead of (9.98) ($\lambda_1 = \lambda_2$),

$$\langle\mu(x)\mu(y)\rangle_{AHM} = \langle\mu^\dagger(x)\mu^\dagger(y)\rangle_{AHM} = \exp\{-F(x-y) - F(\epsilon)\}. \quad (9.100)$$

Then, the regulator m would no longer cancel. Rather, it would provide an overall term that would force these correlators to vanish:

$$\langle\mu(x)\mu(y)\rangle_{AHM} = \langle\mu^\dagger(x)\mu^\dagger(y)\rangle_{AHM} \propto \exp\left\{-\frac{M^2}{\tilde{e}^2 m}\right\} \xrightarrow{m\to 0} 0. \quad (9.101)$$

Notice, however, that this only happens in the phase where $M \neq 0$ and, consequently $\rho_0 \neq 0$ and $\langle\phi\rangle \neq 0$.

We still have left the ultraviolet regulator, ϵ. This may be eliminated by renormalizing the vortex field as

$$\mu_R(x) = \exp\left\{\frac{F(\epsilon)}{2}\right\}\mu(x). \qquad (9.102)$$

Then, we finally get the renormalized vortex correlation function

$$\langle\mu(x)\mu^\dagger(y)\rangle_{AHM}^R = \exp\left\{-\mathcal{M}|x-y| + \frac{1}{2\tilde{e}^2|x-y|}\right\}, \qquad (9.103)$$

where $\mathcal{M} = \frac{\pi}{2}(\frac{M}{\tilde{e}})^2 = \frac{\pi\rho_0^2}{2}$.

Considering the fact that, at large distances, according to Araki's theorem [48] (9.104),

$$\langle\mu(x)\mu^\dagger(y)\rangle_{AHM}^R \xrightarrow{|x-y|\to\infty} |\langle\mu\rangle|^2, \qquad (9.104)$$

it follows from (9.103) that, whenever $\langle\phi\rangle \neq 0$ (and consequently $M \neq 0$), then $\langle\mu\rangle = 0$. Conversely, if $\langle\mu\rangle \neq 0$, the same theorem [48] then implies that $\langle\phi\rangle = 0$ (and consequently $M = 0$). We can see here the duality relation existing between the vortex field operator and the Higgs field, manifested in a very clear way though the form of the quantum vortex correlation functions. The existence of genuine quantum vortex excitations requires a phase in which the Higgs field possesses a nonzero vacuum expectation value. In such a phase, the quantum vortex state $|\mu\rangle$ will be orthogonal to the vacuum and there will be a mass gap separating the minimum vortex energy from the vacuum energy, namely, the vortex mass, given by

$$\mathcal{M} = \frac{\pi \rho_0^2}{2}. \tag{9.105}$$

9.5 Quantum Field Theory of Magnetic Monopoles

Quantum Magnetic Monopoles Correlation Functions: Derivation

Let us establish now the quantum field theory of magnetic monopoles. We take as a basic theory the non-abelian version of the Higgs model, given by (8.50), (8.51) and (8.52), having a local SO(3) symmetry, also known as the Georgi–Glashow model. It is convenient to define

$$A_\mu = A_\mu^a T^a \qquad D_\mu = \partial_\mu \mathbb{I} + i g A_\mu$$
$$F_{\mu\nu} = \partial_\mu A_\nu - \partial_\nu A_\mu + i g [A_\mu, A_\nu] = F_{\mu\nu}^a T^a, \tag{9.106}$$

in terms of the group generators T^a. Under a gauge transformation $\mathcal{G} = e^{i\omega^a T^a}$, we have

$$A_\mu \longrightarrow \mathcal{G}^{-1} A_\mu \mathcal{G} + \frac{i}{g} \mathcal{G}^{-1} \partial_\mu \mathcal{G}$$
$$F_{\mu\nu} \longrightarrow \mathcal{G}^{-1} F_{\mu\nu} \mathcal{G}. \tag{9.107}$$

Our main purpose is to derive a general expression for the magnetic monopole quantum correlation functions. For this, we are going to follow again the guidelines inspired on the Kadanoff–Ceva construction for disorder variable correlation functions [30], as we did for vortices in the abelian case. First of all, let us recast the second term in (8.50) in terms of the field intensity tensor, as we did in that case. We have

$$(D_\mu \Phi)^a (D^\mu \Phi)^a = \mathrm{tr}(D_\mu \Phi)^T (D^\mu \Phi). \tag{9.108}$$

We can express the Higgs field in "polar" form, in terms of the scalar fields ρ^a, θ^a:

$$\Phi^a = \frac{1}{\sqrt{2}} \mathcal{H}^{ab} \rho^b \qquad \mathcal{H}^{bc} = \left[e^{i\theta^a T^a} \right]^{bc}. \tag{9.109}$$

Inserting this in (9.108), we may write

$$\text{tr}(D_\mu \Phi)^T (D^\mu \Phi) = \text{tr}\rho^T \mathcal{H}^{-1} (\partial_\mu \mathbb{I} + ig A_\mu) \mathcal{H} \mathcal{H}^{-1} (\partial_\mu \mathbb{I} + ig A_\mu) \mathcal{H}\rho + \frac{1}{2}\partial_\mu \rho^a \partial^\mu \rho^a$$

$$= \frac{1}{2} g^2 \rho^2 \text{tr} \left(\frac{-i}{g} \mathcal{H}^{-1} \partial_\mu \mathcal{H} + \mathcal{H}^{-1} A_\mu \mathcal{H} \right)$$

$$\times \left(\frac{-i}{g} \mathcal{H}^{-1} \partial_\mu \mathcal{H} + \mathcal{H}^{-1} A_\mu \mathcal{H} \right) + \frac{1}{2}\partial_\mu \rho^a \partial^\mu \rho^a,$$

$$\equiv \frac{1}{2} g^2 \rho^2 \text{tr} W^\mu W_\mu + \frac{1}{2}\partial_\mu \rho^a \partial^\mu \rho^a, \tag{9.110}$$

where $\rho^2 = \rho^a \rho^a$, and we introduced

$$W_\mu = \frac{-i}{g} \mathcal{H}^{-1} \partial_\mu \mathcal{H} + \mathcal{H}^{-1} A_\mu \mathcal{H}.$$

Now, using the fact that $\partial_\mu W^\mu = 0$, we can re-write

$$\frac{1}{2} g^2 \rho^2 \text{tr} W^\mu W_\mu = \frac{1}{2} g^2 \rho^2 \text{tr} W_\mu \left[\frac{-\Box \delta^{\mu\nu} + \partial^\mu \partial^\nu}{-\Box} \right] W_\nu. \tag{9.111}$$

Adding and subtracting a term

$$\frac{1}{2} g^2 \rho^2 \text{tr} W_\mu \left[\frac{\delta^{\mu\nu} (W^\alpha W^\alpha - \partial^\alpha W^\alpha)}{-\Box} \right] W_\nu, \tag{9.112}$$

we finally get

$$\frac{1}{2} g^2 \rho^2 \text{tr} W^\mu W_\mu = -\frac{1}{4} \text{tr} W^{\mu\nu} \left[\frac{g^2 \rho^2}{\Box} \right] W_{\mu\nu}, \tag{9.113}$$

where

$$W_{\mu\nu} = \partial_\mu W_\nu - \partial_\nu W_\mu + ig[W_\mu, W_\nu] \equiv W^a_{\mu\nu} T^a.$$

Consequently, we may write the Lagrangean (8.50) in the convenient form

$$\mathcal{L} = -\frac{1}{4} W^a_{\mu\nu} \left[1 + \frac{g^2 \rho^2}{\Box} \right] W^{a\mu\nu} + \frac{1}{2}\partial_\mu \rho^a \partial^\mu \rho^a + V(\rho^2). \tag{9.114}$$

We are now ready for obtaining a general expression for the magnetic monopole correlation functions. Following the same procedure adopted before, we have

$$\langle \mu(x) \mu^\dagger(y) \rangle_{NAHM} =$$

$$Z_0^{-1} \int DW^a_\mu \exp \left\{ -\int d^4z \left[\frac{1}{4} \left[W^{\mu\nu a} + \tilde{B}^{\mu\nu a} \right] \left[1 + \frac{g^2 \rho^2}{-\Box} \right] \left[W_{\mu\nu a} + \tilde{B}_{\mu\nu a} \right] \right. \right.$$

$$\left. \left. + \frac{1}{2}\partial^\mu \rho^a \partial_\mu \rho^a + V(\rho^2) \right] \right\},$$

$$\tag{9.115}$$

where the external field is now given by

$$\tilde{B}^{\mu\nu a}(x, y; L) = \frac{4\pi}{M} \int_{x,L}^{y} d\xi_\alpha \epsilon^{\alpha\mu\nu a} \delta^4(z - \xi),$$ (9.116)

where $M = g\rho_0$ is the gauge field mass. If we choose the integration path along a straight line going from $-\infty$ to $(\mathbf{x}, x^4)$ and returning to $-\infty$ from $(\mathbf{y}, y^4)$ through another straight line, the external field $\tilde{B}^{\mu\nu}(x, y; L)$ will have just the spatial components $i, j = 1, 2, 3$.

We show below that we can write, in this case,

$$\tilde{B}^{ija}(x, y; L_0) = \frac{1}{M} \left(\partial_i \partial_j - \partial_j \partial_i \right) \omega_{L_0}^a(z; x, y),$$ (9.117)

where

$$\omega_{L_0}^a(z; x, y) = \theta(x^4 - z^4)\omega_{L_0}^a(z; -\infty, x) - \theta(y^4 - z^4)\omega_{L_0}^a(z; -\infty, y)$$ (9.118)

with

$$\omega_{L_0}^a(z; -\infty, x) = \frac{1 - \cos\theta}{r \sin\theta} \hat{\varphi}^a.$$ (9.119)

In the above expression, θ and φ are the angles of the spherical coordinate system centered at $\mathbf{x}$.

The vector function ω^a exhibits the property

$$\epsilon^{ijk} \partial_j \omega^k(\mathbf{r}) = \partial^i \left(\frac{1}{|\mathbf{r}|} \right)$$

$$\epsilon^{ijk} \partial_i \partial_j \omega^k(\mathbf{r}) = -\nabla^2 \left(\frac{1}{|\mathbf{r}|} \right) = 4\pi \delta^3(\mathbf{r}),$$ (9.120)

being therefore the classic magnetic monopole vector potential.

In terms of the line integral along the singularity at $\theta = \pi$, we have

$$\epsilon^{ijk} \partial_i \partial_j \omega^k(\mathbf{r}) = 4\pi \int_{-\infty, L_0}^{(x^4, \mathbf{r})} dx^4 \delta^4(z - r).$$ (9.121)

Then (9.117) immediately follows. Thereafter, we can generalize $\omega_{L_0}^a(z; x, y) \rightarrow \omega_L^a(z; x, y)$, in such a way that that singularity lies at an arbitrary curve L, connecting x and y in Euclidean space; thereby we obtain (9.116).

We may establish the path independence of the monopole correlation function (9.115) by making the functional integral change of variable

$$W_\mu^a \longrightarrow W_\mu^a + \frac{1}{g\rho_0} \partial_\mu \omega^a(L', L),$$

$$W_\mu^a T^a \longrightarrow \mathcal{G}^{-1} \left[W_\mu^a + \frac{1}{g\rho_0} \partial_\mu \omega^a(L', L) \right] T^a \mathcal{G}, \tag{9.122}$$

where $\omega^a(L', L) = \omega_{L'}^a(z; x, y) - \omega_L^a(z; x, y)$ and $\mathcal{G} = e^{i\omega^a T^a}$. Under this transformation, we have

$$F^{\mu\nu a} T^a \longrightarrow \mathcal{G}^{-1} \left[F^{\mu\nu a} + \tilde{B}^{\mu\nu a}(x, y; L') - \tilde{B}^{\mu\nu a}(x, y; L) \right] T^a \mathcal{G}, \tag{9.123}$$

thus establishing the path independence of (9.115).

Magnetic Monopoles Creation Operator and Its Commutation Relations

We now obtain the explicit expression for the quantum magnetic monopole creation operator. From (9.115), we conclude that this is given by

$$\mu(x) = \exp \left\{ i \int d^4 z \frac{1}{2} \tilde{B}^{\mu\nu a}(x, \infty; L) \left[1 + \frac{g^2 \rho^2}{-\Box} \right] W_{\mu\nu a} \right\}. \tag{9.124}$$

Inserting the external field $\tilde{B}^{\mu\nu a}(x, \infty; L)$ given by (9.116), we can write the monopole operator as

$$\mu(\mathbf{x}, t) = \exp \left\{ i \frac{4\pi}{e\rho_0} \int_{\mathbf{x}, L}^{\infty} d\mathbf{z}^k \epsilon^{kia} \Pi_i^a(\mathbf{z}, t) \right\}, \tag{9.125}$$

where $\Pi_i^a = \partial \mathcal{L} / \partial \dot{W}^{ia}$ is the momentum canonically conjugate to W^{ia}, satisfying the canonical commutation relation

$$[W_i^a(\mathbf{x}, t), \Pi_j^b(\mathbf{y}, t)] = i\delta^{ij} \delta^{ab} \delta^3(\mathbf{x} - \mathbf{y}). \tag{9.126}$$

We must now demonstrate that indeed the μ-operator creates, out of the vacuum, eigenstates of the magnetic charge operator given by (8.56) and (8.57). Using (9.84) and writing $\mu = e^A$, we have

$$[\nabla^i B^i, A] = \frac{4\pi}{g} \int_{-\infty}^{\mathbf{x}} d\mathbf{z}^l \epsilon^{ijk} \partial_j \epsilon^{lrb} [W_k^a(\mathbf{y}, t), \Pi_r^b(\mathbf{z}, t)] \nabla^i \frac{\phi^a}{\rho_0}. \tag{9.127}$$

Then, using (9.126) and the fact that the magnetic charge operator is

$$Q_M = \int d^3 y \nabla^i B^i(\mathbf{y}), \tag{9.128}$$

we get

$$[Q_M, A(\mathbf{y}, t)] = \frac{4\pi}{g}, \tag{9.129}$$

where we used the fact that $\nabla^i \hat{\phi}^i = 1$, for $\hat{\phi} = \hat{\mathbf{r}}$.

It follows immediately that

$$[Q_M, \mu(\mathbf{y}, t)] = \frac{4\pi}{g} \mu(\mathbf{y}, t), \qquad (9.130)$$

hence

$$Q_M |\mu\rangle = \frac{4\pi}{g} |\mu\rangle, \qquad (9.131)$$

as it should.

Quantum Magnetic Monopole Correlation Functions: Evaluation

Let us now evaluate the quantum magnetic monopole correlation function. Starting from (9.115) and just retaining the external-field-dependent terms as well as the ones that are quadratic in the gauge field, we can write

$$\langle \mu(x)\mu^{\dagger}(y) \rangle_{NAHM} = Z_0^{-1} \int DW_\mu^a \exp \left\{ -\int d^4z \left[\frac{1}{2} W_\mu^a \left[(-\Box + M^2)\delta^{\mu\nu} + \partial^\mu\partial^\nu \right] \right. \right.$$

$$\times W_\nu^a \left[\frac{\Box + M^2}{\Box} \right] P_\lambda^{\mu\nu} W_\lambda^a \tilde{B}_{\mu\nu a} + g\epsilon^{abc} \left[\frac{\Box + M^2}{\Box} \right] W_\mu^b W_\nu^c \tilde{B}_{\mu\nu a}$$

$$\left. \left. + \frac{1}{4} \tilde{B}_{\mu\nu a} \left[\frac{\Box + M^2}{\Box} \right] \tilde{B}_{\mu\nu a} + \dots \right] \right\}, \qquad (9.132)$$

where $P_\lambda^{\mu\nu} = \partial^\mu \delta_\lambda^\nu - \partial^\nu \delta_\lambda^\mu$.

Expanding the exponential of the second and third terms above, we shall get products of the two-point functions $\langle W_\mu^a W_\nu^b \rangle$, according to the Wick theorem. We call the last term above T_3. The second one yields

$$T_1 = 6\alpha^2 \int_x^y d\xi_\mu \int_x^y d\eta_\nu \left[-\Box\delta^{\mu\nu} + \partial^\mu\partial^\nu \right] \left[\frac{1}{\Box} + \frac{M^2}{\Box^2} \right], \qquad (9.133)$$

whereas the third one gives

$$T_2 = \frac{18}{\pi^2} g^2 \alpha^2 \rho_0^4 \int_x^y d\xi_\mu \int_x^y d\eta_\nu \delta^{\mu\nu} \frac{1}{\Box^2}, \qquad (9.134)$$

where $\alpha = \frac{4\pi}{M}$.

The first term in T_1 exactly cancels the last term in (9.132), namely T_3 and the remaining one gives

$$T_1 + T_3 = 12\alpha^2 [F(x - y) - F(0)]$$

$$F(x) = \mathcal{F}^{-1} \left[\frac{1}{\Box} + \frac{M^2}{\Box^2} \right] = \frac{1}{4\pi^2 |x - y|^2} - \frac{M^2}{8\pi^2} \ln|x - y|. \quad (9.135)$$

In T_2, making the change of integration variables $(\xi, \eta) \rightarrow (R, r)$, where $r = \xi - \eta$ and $R = \frac{1}{2}(\xi + \eta)$, and using the Fourier representation of $\frac{1}{\Box^2}$, we get, after integration over R and r:

$$T_2 = -\left(\frac{36}{\pi^3}\right)\left(\frac{4\pi}{g^2\rho_0^2}\right)\frac{g^2\rho_0^4}{M}|x - y|. \tag{9.136}$$

Collecting all the terms back in the exponential and using the fact that $M = g\rho_0$, we finally obtain for the renormalized quantum monopole correlation function,

$$\langle\mu(x)\mu^\dagger(y)\rangle^R_{NAHM} = \exp\left\{-\frac{36}{\pi^3}\left(\frac{4\pi M}{g^2}\right)|x - y|\right.$$
$$\left. -24\ln|x - y| + \frac{48}{M^2|x - y|^2}\right\}. \tag{9.137}$$

From the exponential decay, we may infer that the quantum monopole mass is given by

$$\mathcal{M}_{mon} = \frac{36}{\pi^3}\left(\frac{4\pi M}{g^2}\right) = 1.161\left(\frac{4\pi M}{g^2}\right). \tag{9.138}$$

Notice that this satisfies the bounds (8.63) imposed on the classical monopole energy.

Recalling (9.45), observe that for $\langle\phi\rangle \neq 0(M \neq 0)$, it follows that $\langle\mu\rangle = 0$. Conversely, if $\langle\mu\rangle \neq 0$, then, we would have $M - 0(\langle\psi\rangle = 0)$. The phase with both $\langle\phi\rangle = 0$ and $\langle\mu\rangle = 0$ apparently is not realized in this case.

10

Duality, Bosonization and Generalized Statistics

We have seen in the previous chapter how the existence of an order-disorder duality structure allows the obtainment of a full quantum theory of topological excitations. In the present chapter, conversely, we will see how the same structure is at the very roots of a method by which one can generate, out of bosonic fields, new composite fields with different statistics, either fermionic or generalized. In the first case, the method is usually called bosonization and allows a full description of fermions within the bosonic theory, whereas in the second, the method provides a complete description of anyons, as the particles with generalized statistics have been called [59], also in the framework of the bosonic theory. From their inception, both bosonization and the construction of fields with generalized statistics are deeply related to the order-disorder duality structure, since both fermion and anyon fields can be expressed as products of order and disorder operators, which respectively carry charge and topological charge [31, 32, 45]. The statistics of the resulting composite field, then, is proportional to the product of the charge and topological charge borne by this field. In this chapter we will expose the basic features of bosonization, as well as its relation with generalized statistics, and apply it to different quantum field theory systems in D = 2, 3 and 4, thereby verifying it is an extremely powerful tool for solving interacting field theories. Indeed, in some cases, this method can lead to the exact solution of highly nonlinear systems.

10.1 The Symmetrization Postulate and Its Violation: Bosons, Fermions and Anyons

Physical systems most frequently possess identical objects. Such is the case, for instance, of electrons, photons, phonons, etc. Also billiard balls, perhaps. The description of identical objects, however, changes dramatically according to whether we use a classical or quantum-mechanical approach. In the first case, one can always distinguish identical objects because we can follow their trajectories, whereas in the second case, this is no longer possible. Indeed, within a quantum-mechanical framework the wave-functions of different identical particles may overlap, thus causing us at the end of a physical process, to completely loose track of which particle corresponds to each of the initial ones.

Stating it more precisely, suppose a quantum-mechanical system possesses one-particle states $|\varphi_n\rangle$, $n = 0, 1, 2, \ldots$ An N-particle state would belong to the direct product Hilbert space,

$$|\varphi_{n_1}(1)\varphi_{n_2}(2)\ldots\varphi_{n_N}(N)\rangle = |\varphi_{n_1}(1)\rangle \otimes |\varphi_{n_2}(2)\rangle \otimes \ldots \otimes |\varphi_{n_N}(N)\rangle. \quad (10.1)$$

Now, the new states obtained by making arbitrary permutations of the N identical particles are, physically, precisely the same in spite of the fact that they are mathematically completely different. Indeed, there are infinitely many inequivalent state vectors corresponding to exactly the same physical situation. These are given by

$$\sum_{\alpha=1}^{N!} C_\alpha P_\alpha\{1, 2, \ldots, N\}|\varphi_{n_1}(1)\varphi_{n_2}(2)\ldots\varphi_{n_N}(N)\rangle, \quad (10.2)$$

where C_α are arbitrary complex coefficients and $P_\alpha\{1, 2, \ldots, N\}$, permutations in the set $\{1, 2, \ldots, N\}$.

In order to select the physical states out of these infinitely many possibilities, the foundations of quantum mechanics were supplemented by the symmetrization postulate. According to this, only the completely symmetric and anti-symmetric states correspond to physical situations in nature. For these states we have, respectively, $C_\alpha = \frac{1}{\sqrt{N!}}, \alpha = 1, 2, , N!$ or $C_\alpha = (-1)^{\pi(\alpha)}\frac{1}{\sqrt{N!}}, \alpha = 1, 2, , N!$, where $\pi(\alpha)$, the parity of the permutation, is $2n$ or $2n+1$, according to whether the number of transpositions performed by the permutation $P_\alpha\{1, 2, \ldots, N\}$ is even or odd. The only two possible N-particle physical states, according to the symmetrization postulate are thus

$$|\varphi_{n_1}(1)\varphi_{n_2}(2)\ldots\varphi_{n_N}(N)\rangle_S = \frac{1}{\sqrt{N!}} \sum_{\alpha=1}^{N!} P_\alpha\{1, 2, \ldots, N\}|$$
$$\times \varphi_{n_1}(1)\varphi_{n_2}(2)\ldots\varphi_{n_N}(N)\rangle \quad (10.3)$$

or

$$|\varphi_0(1)\varphi_1(2)\ldots\varphi_k(N)\rangle_A$$
$$= \frac{1}{\sqrt{N!}} \sum_{\alpha=1}^{N!} (-1)^{\pi(\alpha)} P_\alpha\{1, 2, \ldots, N\}|\varphi_{n_1}(1)\varphi_{n_2}(2)\ldots\varphi_{n_N}(N)\rangle. \quad (10.4)$$

It turns out that the state $|\varphi_{n_1}(1)\varphi_{n_2}(2)\ldots\varphi_{n_N}(N)\rangle_S$ is symmetric under the exchange of any two particles, whereas the state $|\varphi_{n_1}(1)\varphi_{n_2}(2)\ldots\varphi_{n_N}(N)\rangle_A$ is anti-symmetric. The occupation number of each particle state of the latter is, consequently, $n_i = 0, 1$, while of the former is $n_i = 0, 1, 2, 3, \ldots$ This implies the particles satisfy, respectively, Fermi–Dirac and Bose–Einstein statistics, being in each case, respectively, fermions or bosons.

Admitting that the $|\varphi_n\rangle$ particle states are obtained by the action of a creation operator on the vacuum state, namely,

$$|\varphi_{n_1}(1)\varphi_{n_2}(2)\ldots\varphi_{n_N}(N)\rangle_{A,S} = a_{n_1}^\dagger a_{n_2}^\dagger \ldots a_{n_N}^\dagger |0\rangle, \tag{10.5}$$

it follows immediately that the creation operators must either commute or anti-commute, respectively, for bosons and fermions, or

$$a_{\varphi_i}^\dagger a_{\varphi_j}^\dagger = (-1)^{2s} a_{\varphi_j}^\dagger a_{\varphi_i}^\dagger, \tag{10.6}$$

where the statistics s satisfies $2s \in \mathbb{N}$, being an integer for bosons and a half-integer for fermions. We have already seen examples of such operators in Chapters 2, 3 and 4, for electrons and phonons. A remarkable and unexpected result, demonstrated by Pauli on very general hypotheses, is the spin-statistics theorem. According to this, particles with an integer spin quantum number j ($S^2 = j(j+1)\hbar^2$) correspond to a symmetric state vector and commuting creation operators, being therefore bosons, while particles with a half-integer spin correspond to an anti-symmetric vector state and anticommuting operators, being therefore fermions. In other words, the theorem states that $s = j$.

The fact that the spin quantum number can only be an integer or half-integer is a consequence of the algebra of generators of the rotation group in 3D. This implies we can reach all states of a multiplet spanning an irreducible representation of this group, namely $|jm\rangle$, $m = -j, \ldots, +j$ by an integer number of steps, hence $2j$ is an integer and, consequently, the spin quantum number j (and also the statistics s) is an integer or half-integer. Consequently, we will only have either bosons or fermions in $d = 3$.

Surprisingly, when the number of spatial dimensions is less than three, the symmetrization postulate is violated. A first indication of that is the fact that in $d = 2$, the rotation group is abelian, hence there is no algebra of generators and the spin can be any real number: $j \in \mathbb{R}$. In $d = 1$, conversely, there is no rotation subgroup of the Lorentz group, but still, there is a real number j, called "Lorentz spin," that determines how the states transform under Lorentz transformations. We now show that the statistics s of particle states corresponding to the spin j, accordingly, can be any real number in $d = 1, 2$, thus violating the symmetrization postulate. In order to see this, suppose the operators in (10.5) are fields $\psi^\dagger(\mathbf{r})$ creating particles in position eigenstates. Then we can cast (10.6) in the form

$$\psi^\dagger(\mathbf{x})\psi^\dagger(\mathbf{y}) = e^{i2s\pi}\psi^\dagger(\mathbf{y})\psi^\dagger(\mathbf{x}). \tag{10.7}$$

Let us consider first the case when $d = 1$. If we commute once more the operators on the right-hand side, we arrive at the same expression on the left-hand side, multiplied by $e^{i2s \cdot 2\pi}$. Thus, for consistency, we must have $e^{i2s \cdot 2\pi} = 1$, hence an operator algebra such as (10.7) would only admit $2s \in \mathbb{N}$.

It follows that, if we want to preserve the $s = j$ relation by allowing arbitrary real values for $2s$, we must modify the commutation relation above in such a way that the inconsistency is removed. This can be achieved by changing the phase sign whenever we exchange $x \leftrightarrow y$, namely (in $d = 1$)

$$\psi^\dagger(x)\psi^\dagger(y) = e^{i2s\pi \, \epsilon(y-x)} \psi^\dagger(y)\psi^\dagger(x), \tag{10.8}$$

where $\epsilon(y - x)$ is the sign function. Now the statistics s can be any real number as well.

Now we switch to the $d = 2$ case. We must find a generalization of (10.8) appropriate for this dimension of space. The guideline for this is that the sign of the phase must change when we exchange $\mathbf{x} \leftrightarrow \mathbf{y}$. Using the fact that

$$\arg(\mathbf{y} - \mathbf{x}) = \arg(\mathbf{x} - \mathbf{y}) + \pi\epsilon(\pi - \arg(\mathbf{x} - \mathbf{y})), \tag{10.9}$$

we obtain the following generalization of (10.8), valid in $d = 2$:

$$\psi^\dagger(\mathbf{x})\psi^\dagger(\mathbf{y}) = \exp\{i2s\pi \, \epsilon(\pi - \arg(\mathbf{x} - \mathbf{y}))\} \psi^\dagger(\mathbf{y})\psi^\dagger(\mathbf{x}). \tag{10.10}$$

The statistics factor s, again, can be any real number.

Let us consider now the case $d = 3$. For this, the function generalizing $\arg(\mathbf{x} - \mathbf{y})$ is the solid angle $\Omega(C_\mathbf{y}; \mathbf{x})$ comprising the point $\mathbf{x}$ and a curve $C_\mathbf{y}$, which we take as a circle centered at $\mathbf{y}$ and belonging to a plane perpendicular to the z-axis in a coordinate system centered at $\mathbf{x}$. Indeed, the relation that generalizes (10.9) is

$$\Omega(C_\mathbf{y}; \mathbf{x}) = \Omega(C_\mathbf{x}; \mathbf{y}) + 2\pi\epsilon(2\pi - \Omega(C_\mathbf{y}; \mathbf{x})). \tag{10.11}$$

The corresponding commutation rule (in $d = 3$) would be

$$\psi^\dagger(C_\mathbf{x})\psi^\dagger(C_\mathbf{y}) = \exp\{i2s\pi \, \epsilon(2\pi - \Omega(C_\mathbf{y}; \mathbf{x}))\} \psi^\dagger(C_\mathbf{y})\psi^\dagger(C_\mathbf{x}). \tag{10.12}$$

The results above have far-reaching implications. They imply that in three-dimesional space, it would be possible to define an arbitrary statistics $s \in \mathbb{R}$, but only for nonlocal fields $\psi^\dagger(C_\mathbf{x})$, creating strings along the curve $C_\mathbf{x}$ as described above. Notice, then, that a local limit of the field ($\psi(C_\mathbf{x}) \to \psi(\mathbf{x})$) can be obtained by shrinking the curve to its central point $C_\mathbf{x} \to \mathbf{x}$. In this limit, however, the solid angles collapse to zero, the phase factor in (10.12) becomes a constant and we encounter again the same inconsistency found in (10.7), unless $2s \in \mathbb{N}$. This shows unequivocally that, for local objects, only fermion or boson statistics are allowed in $d = 3$. Nonlocal objects, however, can have arbitrary statistics.

In $d = 1$ and $d = 2$, nevertheless, we can have local fields with arbitrary statistics, thereby violating the symmetrization postulate. In Chapter 30, we will see that this postulate is violated in an even stronger way by the occurrence of non-abelian statistics, also in $d < 3$.

The process of obtaining fields with arbitrary statistics out of boson fields involves two main steps: (a) the structural construction of the former; (b) the determination of the specific boson field Lagrangean corresponding to a given fermion or generalized statistics theory. In the next section we mostly focus on the former, while in the subsequent section we concentrate on the latter.

10.2 Order × Disorder Fields

10.2.1 The Charge-Topological Charge Duality

We have seen that topological excitations bear a conserved quantity, which is associated to an identically conserved current. This is the topological charge. Conversely, charge is in general a quantity that is conserved by force of the dynamical field equation. We are going to explore the duality existing between the quantum excitations carrying these two kinds of conserved quantities in $d = 1, 2, 3$. We shall see that this basic duality is the foundation both of the bosonization method and of the construction of quantum excitations with a generalized statistics.

Topologically Charged Excitations

The form of the topological current depends crucially on the space-time dimension D. Also, it is naturally expressed in terms fields, which are tensors of different ranks. For $D = d + 1 = 2, 3$ and 4, respectively, the topological current is expressed as

$$J_2^\mu = \epsilon^{\mu\nu}\partial_\nu\phi$$
$$J_3^\mu = \epsilon^{\mu\nu\alpha}\partial_\nu B_\alpha$$
$$J_4^\mu = \epsilon^{\mu\nu\alpha\beta}\partial_\nu B_{\alpha\beta}, \tag{10.13}$$

where ϕ, B_α, $B_{\alpha\beta}$ are bosonic scalar, vector and 2-tensor fields in the respective dimension. Their kinetic Lagrangean density is, respectively, assumed to be

$$\mathcal{L}_2 = \frac{1}{2}H_\mu H^\mu$$
$$\mathcal{L}_3 = -\frac{1}{4}H_{\mu\nu}H^{\mu\nu}$$
$$\mathcal{L}_4 = \frac{1}{12}H_{\mu\nu\alpha}H^{\mu\nu\alpha}, \tag{10.14}$$

where

$$H_\mu = \partial_\mu \phi$$
$$H_{\mu\nu} = \partial_\mu B_\nu - \partial_\nu B_\mu$$
$$H_{\mu\nu\alpha} = \partial_\mu B_{\nu\alpha} + \partial_\nu B_{\alpha\mu} + \partial_\alpha B_{\mu\nu} \tag{10.15}$$

are, respectively, the field intensity tensors of the fields ϕ, B_μ and $B_{\mu\nu}$.

The topological charge corresponding to the currents above is given, in general, by

$$Q_T = \int d^d x \, J_D^0. \tag{10.16}$$

Now, within a quantum-mechanical framework, one needs to describe the topological charge eigenstates. We show below that, in any dimension, the operator

$$\mu(\mathbf{x}, t) = \exp\left\{ ia \int_{-\infty}^{\mathbf{x}} dz^\mu J_D^\mu(\mathbf{z}, t) \right\} \tag{10.17}$$

creates eigenstates of the topological charge with eigenvalue a.

Inserting (10.13) in (10.17) and using (10.14), we have

$$\mu(\mathbf{x}, t) = \exp\left\{ ia \int_{-\infty}^{\mathbf{x}} d\xi \, \Pi(\xi, t) \right\}$$
$$\mu(\mathbf{x}, t) = \exp\left\{ ia \int_{-\infty}^{\mathbf{x}} d\xi^i \epsilon^{ij} \Pi^j(\xi, t) \right\}$$
$$\mu(\mathbf{x}, t) = \exp\left\{ ia \int_{-\infty}^{\mathbf{x}} d\xi^i \epsilon^{ijk} \Pi^{jk}(\xi, t) \right\}, \tag{10.18}$$

where

$$\Pi(\mathbf{x}, t) = \partial_0 \phi(\mathbf{x}, t)$$
$$\Pi^j(\mathbf{x}, t) = H^{0j}(\mathbf{x}, t)$$
$$\Pi^{jk}(\mathbf{x}, t) = H^{0jk}(\mathbf{x}, t) \tag{10.19}$$

are the momenta canonically conjugate to the fields ϕ, B_α, $B_{\alpha\beta}$. These satisfy the commutation rules

$$\left[\phi(\mathbf{x}, t), \Pi(\mathbf{y}, t)\right] = i\delta(\mathbf{x} - \mathbf{y})$$
$$\left[B^i(\mathbf{x}, t), \Pi^j(\mathbf{y}, t)\right] = i\delta^{ij}\delta^2(\mathbf{x} - \mathbf{y})$$
$$\left[B^{ij}(\mathbf{x}, t), \Pi^{kl}(\mathbf{y}, t)\right] = i\{\delta^{ik}\delta^{jl} - \delta^{il}\delta^{jk}\}\delta^3(\mathbf{x} - \mathbf{y}). \tag{10.20}$$

One can easily show, using (9.35), (10.18) and (10.20) that

$$Q_T\left(\mu(\mathbf{x}, t)|0\rangle\right) = a\left(\mu(\mathbf{x}, t)|0\rangle\right), \tag{10.21}$$

thereby implying that the operators in (10.18) indeed create, out of the vacuum, topological charge eigenstates with eigenvalue a.

Quantum Charged Excitations

We now construct operators that will prove to be dual to the topological excitation creation operator $\mu(\mathbf{x}, t)$ introduced above. We will see that these operators create quantum states bearing a generalized charge, while the former created quantum topologically charged states.

To begin with, consider the following current densities, which correspond, respectively, to particle, string and membrane (2-brane) densities, in D spacetime dimensions:

$$j^\mu(\mathbf{z}, \mathbf{x}) = \int_{L(\mathbf{x})} d\xi^\mu \delta^D(\mathbf{z} - \xi)$$

$$j^{\mu\nu}(\mathbf{z}, L) = \int_{S(L)} d^2\xi^{\mu\nu} \delta^D(\mathbf{z} - \xi)$$

$$j^{\mu\nu\alpha}(\mathbf{z}, S) = \int_{V(S)} d^3\xi^{\mu\nu\alpha} \delta^D(\mathbf{z} - \xi). \tag{10.22}$$

Now, the dual operators are constructed by coupling the above currents to the field intensity tensors, extracted from (10.14), namely

$$\sigma = \exp\left\{ib \int d^D z j^{\mu_1 \cdots \mu_{D-1}} H_{\mu_1 \cdots \mu_{D-1}}\right\}, \tag{10.23}$$

where $H_{\mu_1 \cdots \mu_{(D-1}}$ are the field intensity tensors given by (10.15). The resulting operators obtained after integrating over one of the ξ components are, respectively, functions of $\mathbf{x}$, $L(\mathbf{x})$ and $S(C)$:

$$\sigma(\mathbf{x}, t) = \exp\{-ib\phi(\mathbf{x}, t)\}$$

$$\sigma(L(\mathbf{x}), t) = \exp\left\{-ib \int_{-\infty}^{\mathbf{x}} d\xi^i B_i(\xi, t)\right\}$$

$$\sigma(S(C), t) = \exp\left\{-ib \int_{S(C)} d^2\xi^{ij} B_{ij}(\xi, t)\right\}. \tag{10.24}$$

The generalized charge density operators may be expressed in terms of each of the fields in (10.24) as the familiar operator appearing in the Gauss' Law constraint,

$$\rho(\mathbf{x}, t) = H^0(\mathbf{x}, t) = \Pi(\mathbf{x}, t)$$

$$\rho(\mathbf{x}, t) = \partial_i H^{0i}(\mathbf{x}, t) = \partial_i \Pi^i(\mathbf{x}, t)$$

$$\rho^i(\mathbf{x}, t) = \partial_j H^{0ij}(\mathbf{x}, t) = \partial_j \Pi^{ij}(\mathbf{x}, t). \tag{10.25}$$

Notice the vector character of the last generalized charge density.

The charge operators corresponding to the densities above are given by

$$Q = \int d^d z \rho(\mathbf{z}, t)$$

$$Q^i = \int d^d z \rho^i(\mathbf{z}, t), \tag{10.26}$$

where the last expression corresponds to the third one in (10.25).

With the help of the Baker–Hausdorff formula, we may determine the commutation relation $[Q, \sigma]$. Using (10.24) and (10.20), we readily find $[Q, \sigma] = b\sigma$, for ϕ and B_μ, thus implying, in these cases

$$Q\sigma(\mathbf{x}, t)|0\rangle = b\sigma(\mathbf{x}, t)|0\rangle. \tag{10.27}$$

For the case of the field $B_{\mu\nu}$, conversely, we get

$$\rho^i(\mathbf{x}, t)\,[\sigma(S(C), t)]\,|0\rangle = \int_C d\xi^i \delta^3(\xi - \mathbf{x})\,[\sigma(S(C), t)|0\rangle]$$

$$Q^i\,[\sigma(S(C), t)|0\rangle] = b \int_C d\xi^i\,[\sigma(S(C), t)|0\rangle]. \tag{10.28}$$

Introducing the flux of Q^i along the surface R as

$$\Phi_R = \int_R d^2\xi^i \rho^i(\xi, t), \tag{10.29}$$

we get

$$\Phi_R\,[\sigma(S(C), t)|0\rangle] = b \int_C d\xi^i \int_R d^2\eta^i \delta^3(\xi - \eta)\,[\sigma(S(C), t)|0\rangle]$$

$$\Phi_R\,[\sigma(S(C), t)|0\rangle] = nb\,[\sigma(S(C), t)|0\rangle]\ ;\quad n \in \mathbb{Z}. \tag{10.30}$$

We see that the operator $\sigma(S(C), t)$ creates states carrying a unit of magnitude b of the charge Q^i along the closed curve C. Accordingly, it will create as many units of the flux Φ_R as the number of times, here denoted by n, the curve C pierces the surface R.

10.3 Arbitrary Statistics Out of Bosons in D = 2, 3, 4

We now demonstrate that the operators σ and μ satisfy algebraic relations identical to the dual algebras introduced in the previous chapter. For this purpose, we use the relation

$$e^A e^B = e^{A+B} e^{\frac{1}{2}[A,B]}$$

$$e^A e^B = e^B e^A e^{[A,B]}, \tag{10.31}$$

valid when $[A, [A, B]] = [B, [A, B]] = 0$, in $D = 2, 3, 4$, in order to derive the proper duality relation for each case.

We also show that the composite field, obtained from the product $\sigma \mu$ possesses statistics given by $s = \frac{ab}{2\pi}$, where a and b are, respectively, the units of topological charge and (generalized) charge carried, respectively, by μ and σ. For local composite fields, arbitrary values of s can be obtained in $D = 2, 3$, whereas only $s = 1/2$ is allowed in $D = 4$. Nonlocal, string-like operators, however, may have an arbitrary statistics s in $D = 4$.

These results are a concrete realization of the commutation rules introduced in the first section of this chapter and make evident the fact that order-disorder duality is the foundation both of bosonization and generalized statistics, which are two aspects of the same process.

Duality and Arbitrary Statistics in D=2

In this case, let us take

$$\sigma(\mathbf{x}, t) = \exp\left\{-ib\phi(\mathbf{x}, t)\right\} \quad ; \quad \mu(\mathbf{y}, t) = \exp\left\{ia \int_{-\infty}^{\mathbf{y}} d\xi \, \Pi(\xi, t)\right\}. \quad (10.32)$$

Using (10.31), we obtain

$$\sigma(\mathbf{x}, t)\mu(\mathbf{y}, t) = e^{iab\theta(\mathbf{y}-\mathbf{x})}\sigma(\mathbf{x}, t)\mu(\mathbf{y}, t), \quad (10.33)$$

where $\theta(\mathbf{y} - \mathbf{x})$ is the Heaviside function. We immediately recognize the above expression as the order-disorder dual algebra (9.4). It follows that the composite order$\times$disorder operator $\psi(x) = \sigma(x)\mu(x)$ will have the commutation rule

$$\psi(\mathbf{x}, t) = \sigma(\mathbf{x}, t)\mu(\mathbf{x}, t)$$
$$\psi(\mathbf{x}, t)\psi(\mathbf{y}, t) = e^{iab\epsilon(\mathbf{y}-\mathbf{x})}\psi(\mathbf{y}, t)\psi(\mathbf{x}, t), \quad (10.34)$$

where $\epsilon(\mathbf{y} - \mathbf{x})$ is the sign function. The above relation coincides with (10.8) and inequivocally characterizes the mixed order-disorder product $\psi(x)$ as a field with arbitrary statistics $s = \frac{ab}{2\pi}$, despite the fact it is completely expressed in terms of the bosonic field $\phi(x)$. For the particular case $ab = \pi$, the composite field would be a fermion.

Duality and Arbitrary Statistics in D=3

Consider now

$$\sigma(L(\mathbf{x}), t) = \exp\left\{-ib \int_{-\infty}^{\mathbf{x}} d\xi^i \, B_i(\xi, t)\right\}$$
$$\mu(\mathbf{x}, t) = \exp\left\{ia \int_{-\infty}^{\mathbf{x}} d\xi^i \, \epsilon^{ij} \, \Pi^j(\xi, t)\right\}. \quad (10.35)$$

Using (10.31), we obtain

$$\sigma(L(\mathbf{x},t)\mu(\mathbf{y},t) = \exp\left\{iab\int_{-\infty,L}^{\mathbf{x}} d\xi^i \int_{-\infty,L'}^{\mathbf{y}} d\eta^k \epsilon^{ki}\delta^2(\xi-\eta)\right\}$$
$$\times \mu(\mathbf{y},t)\sigma(L(\mathbf{x},t). \tag{10.36}$$

We now use the following analytic properties of the $\arg(\mathbf{x})$ function,

$$\int_{-\infty,L'}^{\mathbf{y}} d\eta^k \epsilon^{ki}\delta(\xi-\eta) = \partial^i \arg(\xi-\mathbf{y}) + \epsilon^{ki}\partial_k \ln|\xi-\mathbf{y}|, \tag{10.37}$$

in (10.36). Then, integrating on ξ^i and choosing L in such a way that it goes through $\mathbf{y}$ before reaching $\mathbf{x}$, it follows that the last term in (10.37) does not contribute to this integral. The result is

$$\sigma(L(\mathbf{x},t)\mu(\mathbf{y},t) = \exp\left\{i\frac{ab}{\pi}\left[\arg(\mathbf{x}-\mathbf{y})\right]\right\}\mu(\mathbf{y},t)\sigma(L(\mathbf{x},t). \tag{10.38}$$

This algebraic relation coincides with the one found before in the context of quantum vortices, namely (9.73). Again, we identify here the general order-disorder duality structure. The composite order×disorder operator $\psi(\mathbf{x},t)$ will now have the commutation rule

$$\psi(\mathbf{x},t) = \sigma(\mathbf{x},t)\mu(\mathbf{x},t)$$
$$\psi(\mathbf{x},t)\psi(\mathbf{y},t) = e^{i\frac{ab}{\pi}[\arg(\mathbf{x}-\mathbf{y})-\arg(\mathbf{y}-\mathbf{x})]}\psi(\mathbf{y},t)\psi(\mathbf{x},t)$$
$$= e^{iabe(\pi-\arg(\mathbf{x}-\mathbf{y}))}\psi(\mathbf{y},t)\psi(\mathbf{x},t), \tag{10.39}$$

where we used (10.9). This coincides with (10.10).

Again we see, in this case, that the composite order-disorder field $\psi(x)$ will have arbitrary statistics $s = \frac{ab}{2\pi}$, being now completely expressed in terms of the bosonic vector field B_i. For the particular case $ab = \pi$, the composite field would be again a fermion.

We see that the existence of an order-disorder duality structure, mathematically reflected in a dual algebra, allows one to express fields with arbitrary statistics, in particular fermionic, in terms of bosonic ones. In the two previous examples, in $D = 2, 3$, the former fields were local.

Duality and Arbitrary Statistics in D=4

Let us take now the operators

$$\sigma(S(C),t) = \exp\left\{-i\frac{b}{2}\int_{S(C)} d^2\xi^{ij} B_{ij}(\xi,t)\right\}$$
$$\mu(\mathbf{x},t) = \exp\left\{ia\int_{-\infty}^{\mathbf{x}} d\xi^i \epsilon^{ijk}\Pi^{jk}(\xi,t)\right\}. \tag{10.40}$$

Notice that here $\sigma(S(C), t)$ is defined on a closed curve C, being therefore nonlocal.

Again using (10.31) and (10.20), we obtain

$$\sigma(S(C), t)\mu(\mathbf{y}, t) = \exp\left\{iab\int_{S(C)} d^2\xi^{ij}\int_{-\infty,L}^{\mathbf{y}} d\eta^k \epsilon^{kij}\delta^3(\xi - \eta))\right\}$$
$$\times \mu(\mathbf{y}, t)\sigma(S(C), t). \tag{10.41}$$

Now, consider the following property of the magnetic monopole vector potential. Starting from (9.120) and integrating in η^k along L, we get

$$\int_{-\infty,L}^{\mathbf{y}} d\eta^k\delta^3(\xi - \eta) = \frac{1}{4\pi}\epsilon^{ijk}\partial^i\omega^j(\xi - \mathbf{y}) = \partial^k\left[\frac{1}{4\pi|\xi - \mathbf{y}|}\right]. \tag{10.42}$$

Introducing the vector surface integration element $d^2\xi^k = \frac{1}{2}\epsilon^{ijk}d^2\xi^{ij}$ and using (9.120) once more we can write the exponent of the phase factor in (10.41) as

$$i\frac{ab}{2\pi}\int_{S(C)} d^2\xi^i\frac{[\xi - \mathbf{y}]^i}{|\xi - \mathbf{y}|^3} = i\frac{ab}{2\pi}\Omega(\mathbf{y}; C), \tag{10.43}$$

where $\Omega(\mathbf{y}; C)$ is the solid angle comprising the curve C with respect to the point $\mathbf{y}$.

We have, therefore,

$$\sigma(S(C), t)\mu(\mathbf{y}, t) = \exp\left\{i\frac{ab}{2\pi}\Omega(\mathbf{y}; C)\right\}\mu(\mathbf{y}, t)\sigma(S(C), t). \tag{10.44}$$

This is the dual algebra satisfied by the order and disorder fields (10.40) in $D = 4$. Again we can form the composite operator

$$\psi(\mathbf{x}; C_{\mathbf{x}}, t) = \sigma(S(C_{\mathbf{x}}), t)\mu(\mathbf{x}, t).$$

According to (10.44), it will have the commutation

$$\psi(\mathbf{x}; C_{\mathbf{x}}, t)\psi(\mathbf{y}; C_{\mathbf{y}}, t) = \exp\left\{i\frac{ab}{2\pi}\left[\Omega(\mathbf{x}; C_{\mathbf{y}}) - \Omega(\mathbf{y}; C_{\mathbf{x}})\right]\right\}$$
$$\times \psi(\mathbf{y}; C_{\mathbf{y}}, t)\psi(\mathbf{x}; C_{\mathbf{x}}, t)$$
$$\psi(\mathbf{x}; C_{\mathbf{x}}, t)\psi(\mathbf{y}; C_{\mathbf{y}}, t) = \exp\left\{iab\epsilon(2\pi - \Omega(\mathbf{x}; C_{\mathbf{y}}))\right\}\psi(\mathbf{y}; C_{\mathbf{y}}, t)\psi(\mathbf{x}; C_{\mathbf{x}}, t), \tag{10.45}$$

where we used (10.11). This coincides with (10.12).

The composite field possesses arbitrary statistics $s = \frac{ab}{2\pi}$, however, it is nonlocal, depending on a closed curve C. This is the only possibility for the occurrence of generalized statistics, namely, neither fermionic nor bosonic in $D=4$, as we saw before. Indeed, if we take the local limit, where the curve C shrinks to a point, then the solid angle $\Omega(\mathbf{x}; C_{\mathbf{y}}) \to 0$. The phase factor in (10.45) then becomes a constant e^{iab}. In this case, as we have seen, by commuting the composite ψ field twice, we would conclude that necessarily $e^{i2ab} = 1$. Thus, in the local limit, we must have $ab = \pi$ or $s = 1/2$, implying that only fermion local fields may be constructed by the bosonization process in $3 + 1$ dimensions. Fields with generalized statistics may be obtained by this method, however they are necessarily nonlocal.

10.4 The Bosonic Fields Associated to a Dirac Field in D-Spacetime Dimensions

10.4.1 The Current Correlator

We are going to obtain here the bosonic Lagrangeans associated to a free massless Dirac field in D=2, 3, 4, as well as the respective current bosonization formulas.

We start with the generating functional of the fermionic current correlation functions, for arbitrary dimension D:

$$Z[J] = Z_0^{-1} \int D\psi \, D\bar{\psi} \exp\left\{ -\int d^D z \left[i\bar{\psi} \, \partial\!\!\!/ \psi - \bar{\psi} \gamma^\mu \psi J_\mu \right] \right\}$$
$$= Z_2[J] Z_{N>2}[J]$$
$$Z_2[J] = \exp\left\{ \frac{1}{2} \int d^D z J_\mu \Pi_1^{\mu\nu} J_\nu \right\}, \tag{10.46}$$

where $\Pi_1^{\mu\nu}$ is the one-loop vacuum polarization tensor. It follows that the fermion current $j^\mu = \bar{\psi} \gamma^\mu \psi$ two-point correlation function is

$$\langle j^\mu(x) j^\nu(y) \rangle = \frac{\delta^2}{\delta J(x) \delta J(y)} Z[J]\Big|_{J=0} = \Pi_1^{\mu\nu}(x, y). \tag{10.47}$$

In the following subsections, we show how to reproduce (10.47) in the framework of bosonic theories in D=2, 3, 4.

10.4.2 D = 2, 3, 4

D = 2

We have shown in the previous section that the bosonic field is a scalar in D = 2; hence, assuming the fermionic current $j^\mu = \bar{\psi} \gamma^\mu \psi$ is expressed as $j^\mu = K^\mu \phi$, in terms of the bosonic field, ϕ, we write

$$Z[J] = Z_0^{-1} \int D\phi \exp\left\{ -\int d^2 z \left[\frac{1}{2}\phi A\phi + J_\mu K^\mu \phi \right] \right\}$$
$$Z[J] = \exp\left\{ \frac{1}{2} \int d^2 z J_\mu [K^\mu A^{-1} K^\nu] J_\nu \right\}, \tag{10.48}$$

where the operators A and K^μ are to be determined.

Comparing (10.48) with (10.46), we see that $[K^\mu A^{-1} K^\nu] = \Pi_1^{\mu\nu}$. Considering that the vacuum polarization tensor in D=2 is

$$\Pi_1^{\mu\nu} = [k^2 \delta^{\mu\nu} - k^\mu k^\nu] \frac{1}{\pi k^2} = \frac{\epsilon^{\mu\alpha} k_\alpha \epsilon^{\nu\beta} k_\beta}{\pi k^2}, \tag{10.49}$$

we then conclude that

$$A = k^2 \qquad K^\mu = \frac{1}{\sqrt{\pi}} \epsilon^{\mu\alpha} k_\alpha \tag{10.50}$$

or, in Minkowski coordinate space

$$A = -\Box \qquad K^\mu = \frac{1}{\sqrt{\pi}} \epsilon^{\mu\alpha} \partial_\alpha. \tag{10.51}$$

We arrive, therefore, at the bosonization formulas

$$i\bar{\psi}\,\slashed{\partial}\psi = \frac{1}{2}\phi[-\Box]\phi \qquad \bar{\psi}\gamma^\mu\psi = \frac{1}{\sqrt{\pi}} \epsilon^{\mu\alpha} \partial_\alpha\phi. \tag{10.52}$$

D=3

Now we turn to the case of D=3, where the natural bosonic field is a vector B_μ. Assuming the fermionic current $j^\mu = \bar{\psi}\gamma^\mu\psi$ is expressed as $j^\mu = K^{\mu\nu}B_\nu$, in terms of the bosonic field, B_ν, we write

$$Z[J] = Z_0^{-1} \int DB_\mu \exp\left\{-\int d^3z\left[\frac{1}{2}B_\mu A^{\mu\nu} B_\nu + J_\mu K^{\mu\nu} B_\nu\right]\right\}$$

$$Z[J] = \exp\left\{\frac{1}{2}\int d^3z J_\mu[K^{\mu\alpha}(A^{-1})^{\alpha\beta} K^{\beta\nu}]J_\nu\right\}, \tag{10.53}$$

where the operators $A^{\mu\nu}$ and $K^{\mu\nu}$ are to be determined. Comparing (10.53) with (10.46), we have that

$$[K^{\mu\alpha}(A^{-1})^{\alpha\beta} K^{\beta\nu}] = \Pi_1^{\mu\nu}.$$

This is easily solved by

$$A^{\mu\nu} = K^{\mu\nu} = \Pi_1^{\mu\nu}. \tag{10.54}$$

We, therefore obtain

$$i\bar{\psi}\,\slashed{\partial}\psi = \frac{1}{2}B_\mu[\Pi_1^{\mu\nu}]B_\nu \qquad \bar{\psi}\gamma^\mu\psi = \Pi_1^{\mu\nu} B_\nu. \tag{10.55}$$

Then, considering that in D=3, the vacuum polarization tensor is given by

$$\Pi_1^{\mu\nu}(p) = \frac{1}{16\sqrt{p^2}}[p^2\delta^{\mu\nu} - p^\mu p^\nu] + \theta\epsilon^{\mu\nu\alpha} p_\alpha, \tag{10.56}$$

we derive the bosonization formulas

$$i\bar{\psi}\ \partial\!\!\!/\psi = -\frac{1}{4}B_{\mu\nu}[\frac{1}{16\Box^{1/2}}]B^{\mu\nu} + \theta\epsilon^{\mu\nu\alpha}B_{\mu}\partial_{\nu}B_{\alpha}$$

$$\bar{\psi}\gamma^{\mu}\psi = \theta\epsilon^{\mu\nu\alpha}\partial_{\nu}B_{\alpha} + \frac{1}{32\Box^{1/2}}\partial_{\nu}B^{\mu\nu}, \tag{10.57}$$

where $B^{\mu\nu} = \partial^{\mu}B^{\nu} - \partial^{\nu}B^{\mu}$.

$$D=4$$

Consider now the case D=4, where the natural bosonic field is a 2-tensor $B_{\mu\nu}$. Assuming the fermionic current $j^{\mu} = \bar{\psi}\gamma^{\mu}\psi$ is expressed as $j^{\mu} = K^{\mu\alpha\beta}B_{\alpha\beta}$, in terms of the bosonic field, $B_{\alpha\beta}$, we may write

$$Z[J] = Z_0^{-1}\int DB_{\alpha\beta}\exp\left\{-\int d^4z\left[\frac{1}{2}B_{\mu\nu}A^{\mu\nu\alpha\beta}B_{\alpha\beta} + J_{\mu}K^{\mu\alpha\beta}B_{\alpha\beta}\right]\right\}$$

$$Z[J] = \exp\left\{\frac{1}{2}\int d^4z J_{\mu}[K^{\mu\alpha\beta}(A^{-1})^{\alpha\beta\lambda\rho}K^{\lambda\rho\nu}]J_{\nu}\right\}, \tag{10.58}$$

where the operators $A^{\mu\nu\alpha\beta}$ and $K^{\mu\alpha\beta}$ are to be determined.

Comparing (10.58) with (10.46), we have that

$$[K^{\mu\alpha\beta}(A^{-1})^{\alpha\beta\lambda\rho}K^{\lambda\rho\nu}] = \Pi_1^{\mu\nu} - C[k^2\delta^{\mu\nu} - k^{\mu}k^{\nu}], \tag{10.59}$$

where $C = \frac{1}{24\pi^2}$.

We now make the following ansätze

$$(A^{-1})^{\alpha\beta\lambda\rho} = f(k)\Delta^{\alpha\beta\lambda\rho} \quad ; \quad K^{\mu\alpha\beta} = \sqrt{C}g(k)\epsilon^{\mu\nu\alpha\beta}k_{\nu}, \tag{10.60}$$

where

$$\Delta^{\mu\nu\alpha\beta} = k^2[\delta^{\mu\alpha}\delta^{\nu\beta} - \delta^{\mu\beta}\delta^{\nu\alpha}] - \delta^{\mu\alpha}k^{\nu}k^{\beta} - \delta^{\mu\beta}k^{\nu}k^{\alpha} = \epsilon^{\gamma\rho\mu\nu}k_{\rho}\epsilon^{\gamma\lambda\alpha\beta}k_{\lambda} \tag{10.61}$$

Then, since

$$K^{\lambda\alpha\beta}K^{\lambda\mu\nu} = Cg^2(k)\Delta^{\alpha\beta\mu\nu}$$
$$\Delta^{\alpha\beta\lambda\rho}\Delta^{\lambda\rho\mu\nu} = 2k^2\Delta^{\alpha\beta\mu\nu}, \tag{10.62}$$

it follows that $f(k) = \frac{1}{4k^2g^2(k)}$. Imposing the canonical dimension, equal to one in mass units, to the 2-tensor field $B_{\mu\nu}$, we are led to the choice $g(k) = k$ and $f(k) = \frac{1}{4k^4}$. We, then, obtain

$$(A^{-1})^{\alpha\beta\mu\nu} = \frac{1}{4k^4}\Delta^{\alpha\beta\mu\nu}$$
$$A^{\alpha\beta\mu\nu} = \Delta^{\alpha\beta\mu\nu}. \tag{10.63}$$

Now, using (10.61), we derive the bosonization formulas

$$i\bar{\psi}\,\partial\!\!\!/\psi = \frac{1}{2}B_{\mu\nu}[A^{\mu\nu\alpha\beta}]B_{\alpha\beta} \qquad \bar{\psi}\gamma^\mu\psi = \sqrt{\frac{k^2}{24\pi^2}}\epsilon^{\mu\nu\alpha\beta}k_\nu B_{\alpha\beta} \qquad (10.64)$$

or, equivalently,

$$i\bar{\psi}\,\partial\!\!\!/\psi = \frac{1}{12}H_{\mu\nu\alpha}H^{\mu\nu\alpha} \qquad \bar{\psi}\gamma^\mu\psi = \sqrt{\frac{\Box}{24\pi^2}}\epsilon^{\mu\nu\alpha\beta}\partial_\nu B_{\alpha\beta}. \qquad (10.65)$$

In the remainder of this chapter, we combine the results of the two last sections in order to obtain a complete bosonization of fermion fields in D= 2, 3, 4.

10.5 Bosonization in One Spatial Dimension

10.5.1 *The Massless Free Fermion Field*

Field Bosonization

Let us consider the massless free Dirac fermion field in D = 2, which corresponds to the Lagrangean density

$$\mathcal{L} = i\bar{\psi}\gamma^\mu\partial_\mu\psi, \qquad (10.66)$$

where $\bar{\psi} = \psi^\dagger\gamma^0$ and the Dirac matrices are

$$\gamma^0 = \begin{pmatrix} 0 & 1 \\ 1 & 0 \end{pmatrix} \quad \gamma^1 = \begin{pmatrix} 0 & -1 \\ 1 & 0 \end{pmatrix} \quad \gamma^5 = \gamma^0\gamma^1 = \begin{pmatrix} 1 & 0 \\ 0 & -1 \end{pmatrix}. \qquad (10.67)$$

We may write

$$\mathcal{L} = i\psi^\dagger\left[\mathbb{I}\partial_0 + \gamma^5\partial_1\right]\psi$$
$$\mathcal{H} = i\psi^\dagger\gamma^5\partial_1\psi. \qquad (10.68)$$

The Hamiltonian density tells us that the two components of the Dirac field ψ_1 and ψ_2 are, respectively, right-movers and left-movers.

The system is invariant under the continuous symmetries

$$\psi \to e^{i\theta}\psi \qquad \psi \to e^{i\theta\gamma^5}\psi \qquad (10.69)$$

to which correspond, respectively, the conserved currents

$$j^\mu = \bar{\psi}\gamma^\mu\psi \qquad j^{\mu5} = \bar{\psi}\gamma^\mu\gamma^5\psi. \qquad (10.70)$$

The correlation function, according to (6.30) and (10.68) is

$$\langle\psi\psi^\dagger\rangle = -i\left[\mathbb{I}\partial_0 + \gamma^5\partial_1\right]^{-1}$$
$$= -i\begin{pmatrix} \frac{1}{\partial_0+\partial_1} & 0 \\ 0 & \frac{1}{\partial_0-\partial_1} \end{pmatrix} = -i\begin{pmatrix} \partial_0 - \partial_1 & 0 \\ 0 & \partial_0 + \partial_1 \end{pmatrix}\frac{1}{\Box} \longrightarrow$$

$$-i \begin{pmatrix} -\partial_1 + i\partial_2 & 0 \\ 0 & \partial_1 + i\partial_2 \end{pmatrix} \frac{1}{-\Box_E} = \frac{i}{4\pi} \begin{pmatrix} -\partial_{z^*} & 0 \\ 0 & \partial_z \end{pmatrix} \times [\ln z + \ln z^*].$$

$$(10.71)$$

Where in the last step, we went to Euclidean space ($x^2 = ix^0$, $z = x^1 + ix^2$) and used the fact that

$$\frac{1}{-\Box_E} = \lim_{\mu \to 0} \int \frac{d^2k}{(2\pi)^2} \frac{e^{ik \cdot x}}{k^2 + \mu^2} = -\frac{1}{4\pi} \ln \mu^2 z^* z \qquad (10.72)$$

where $z^* z = |\mathbf{x}|^2$. We obtain, therefore,

$$\langle \psi \psi^\dagger \rangle = \frac{i}{4\pi} \begin{pmatrix} -\frac{1}{z^*} & 0 \\ 0 & \frac{1}{z} \end{pmatrix} \qquad (10.73)$$

or, equivalently,

$$\langle \psi_1(\mathbf{r}) \psi_1^\dagger(\mathbf{0}) \rangle = -\frac{i}{4\pi} \frac{e^{i \arg(\mathbf{r})}}{|\mathbf{r}|} = \frac{-i}{4\pi z^*} \to \frac{-i}{4\pi u}$$

$$\langle \psi_2(\mathbf{r}) \psi_2^\dagger(\mathbf{0}) \rangle = \frac{i}{4\pi} \frac{e^{-i \arg(\mathbf{r})}}{|\mathbf{r}|} = \frac{i}{4\pi z} \to \frac{-i}{4\pi v}$$

$$\langle \psi_1(\mathbf{r}) \psi_2^\dagger(\mathbf{0}) \rangle = \langle \psi_2(\mathbf{r}) \psi_1^\dagger(\mathbf{0}) \rangle = 0, \qquad (10.74)$$

where u and v are the light-cone variables: $u = z^0 + z^1$, $v = z^0 - z^1$.

Let us now take the dual field operators, which are expressed by (10.32) in terms of a scalar bosonic field ϕ. On one side we have seen that in D=2, according to (10.52), the massless Dirac field is bosonized in terms of ϕ. On the other side, according to (10.34), the composite order-disorder field, in general, will possess arbitrary statistics, $s = ab/2\pi$. Hence it is natural to take the dual operators in (10.32) as the basic building blocks we shall use in order to bosonize the Dirac field. For this purpose, we now evaluate the mixed four-point correlation function in the framework of the free massless scalar field theory, which we have seen to correspond to the massless Dirac field,

$$\langle \sigma(x_1)\mu(x_2)\mu^\dagger(y_2)\sigma^\dagger(y_1) \rangle = \int D\phi \exp\left\{ -\int d^2z \left[-\frac{1}{2}\phi\Box\phi + (\alpha(z; x_1, y_1) \right.\right.$$

$$\left.\left. + \beta(z; x_2, y_2)) \phi \right] \right\} \qquad (10.75)$$

or

$$\langle \sigma(x_1)\mu(x_2)\mu^\dagger(y_2)\sigma^\dagger(y_1) \rangle = \exp\left\{ -\frac{1}{4\pi} \int d^2z d^2z' \left(\alpha(z; x_1, y_1) + \beta(z; x_2, y_2) \right) \right.$$

$$\left. \times [\ln \mu|z - z'|] \left(\alpha(z'; x_1, y_1) + \beta(z'; x_2, y_2) \right) \right\},$$

$$(10.76)$$

where, according to (10.32), in Euclidean space,

$$\alpha(z; x_1, y_1) = ib[\delta^2(z - x_1) - \delta^2(z - y_1)]$$

$$\beta(z; x_2, y_2) = a \int_{x_2}^{y_2} d\xi^\mu \epsilon^{\mu\nu} \partial_\nu \delta^2(z - \xi). \tag{10.77}$$

Observe that the exponent in (10.76) is the electrostatic interaction energy of a two-dimensional system with charge density given by (10.77). We have three terms, namely,

$$T_{\alpha\alpha} = -\frac{b^2}{2\pi}[\ln\mu|x_1 - y_1| - \ln\mu|\epsilon|]$$

$$T_{\beta\beta} = -\frac{a^2}{4\pi} \int_{x_2}^{y_2} d\xi^\mu \int_{x_2}^{y_2} d\eta^\nu \epsilon^{\mu\alpha} \epsilon^{\nu\beta} \partial_\alpha^{(\xi)} \partial_\beta^{(\eta)} \ln\mu|\xi - \eta|$$

$$= -\frac{a^2}{2\pi}[\ln\mu|x_2 - y_2| - \ln\mu|\epsilon|] + T_{\beta\beta}(L, \epsilon)$$

$$T_{\alpha\beta} = -i\frac{ab}{2\pi} \int_{x_2}^{y_2} d\xi^\mu \epsilon^{\mu\alpha} \partial_\alpha^{(\xi)}[\ln\mu|\xi - x_1| - \ln\mu|\xi - y_1|]$$

$$= i\frac{ab}{2\pi}[\arg(x_2 - y_1) + \arg(y_2 - x_1) - \arg(x_2 - x_1) - \arg(y_2 - y_1)],$$

$$\tag{10.78}$$

where we used the Cauchy–Riemann equation (9.68). Notice that the infrared regulator μ completely cancels and can be ignored in correlation functions, such as (10.75), that respect the conservation of quantities, which are carried by the fields σ and μ. Should we have, for instance, the correlation function $\langle\sigma\mu\mu^\dagger\sigma\rangle$ and the $T_{\alpha\alpha}$ term would produce an overall $\ln\mu$-factor, that would make the correlation function vanish, thus enforcing the selection rule. The infrared regulator, therefore, provides an interesting and efficient mechanism of enforcing the selection rules provenient from fermionic conservation laws in the framework of the bosonized theory.

The unphysical, ϵ-dependent, self-interaction terms appearing in the $\alpha\alpha$ and $\beta\beta$ terms may be removed by renormalizing, respectively, the operators σ and μ in the correlation functions.

We have seen in 10.1.3 that the product $\sigma\mu$ is a fermion for the choice $ab = \pi$. Starting with (10.78) and taking the limit $x_1 \to x_2 = x$, $y_1 \to y_2 = y$, we infer that

$$\langle[\sigma\mu](x)[\sigma\mu]^\dagger(y)\rangle = \frac{e^{i\frac{ab}{\pi}\arg(x-y)}}{|x - y|^{\frac{a^2+b^2}{2\pi}}}. \tag{10.79}$$

We now conclude that, with the choice $a = b = \sqrt{\pi}$ and $\psi_1 = \sigma\mu$, we reproduce the correlator of the first component of the Dirac field in (10.74). By just changing $b \to -b$, we reproduce the second. We therefore establish the bosonization formulas

$$\psi_1(x) = \sqrt{\frac{-i}{4\pi}}\sigma(x)\mu(x) = \sqrt{\frac{-i}{4\pi}}\exp\left\{-ib\phi(\mathbf{x}, t) + ia\int_{-\infty}^{\mathbf{x}} d\xi\,\Pi(\xi, t)\right\}$$

$$\psi_2(x) = \sqrt{\frac{-i}{4\pi}}\sigma^\dagger(x)\mu(x) = \sqrt{\frac{-i}{4\pi}}\exp\left\{ib\phi(\mathbf{x}, t) + ia\int_{-\infty}^{\mathbf{x}} d\xi\,\Pi(\xi, t)\right\}$$

$$(10.80)$$

for σ and μ given by (10.32) with $a = b = \sqrt{\pi}$.

Current Bosonization

Let us consider now the bosonization of the fermionic current, given by (10.70). We take firstly the 0-component of the current correlator, namely

$$\langle j^0(\mathbf{x})j^0(\mathbf{y})\rangle = \langle\psi_1(\mathbf{x})\psi_1^\dagger(\mathbf{y})\rangle\langle\psi_1^\dagger(\mathbf{x})\psi_1(\mathbf{y})\rangle + 1 \leftrightarrow 2, \qquad (10.81)$$

where we used Wick's theorem. Now, considering the result (10.74), we get

$$\langle j^0(\mathbf{x})j^0(\mathbf{y})\rangle = \frac{1}{(4\pi)^2}\left(\frac{1}{z^{*2}} + \frac{1}{z^2}\right) \longrightarrow \frac{1}{(4\pi)^2}\left(\frac{1}{u^2} + \frac{1}{v^2}\right). \qquad (10.82)$$

We now evaluate the bosonic correlator

$$\begin{aligned}
\langle\partial_1^{(x)}\phi(x)\partial_1^{(y)}\phi(y)\rangle &= \partial_1^{(x)}\partial_1^{(y)}\langle\phi(x)\phi(y)\rangle \\
&= -\frac{\pi}{(4\pi)^2}(\partial_u - \partial_v)_x(\partial_u - \partial_v)_y[\ln u + \ln v] \\
&= \frac{\pi}{(4\pi)^2}\left(\frac{1}{u^2} + \frac{1}{v^2}\right).
\end{aligned} \qquad (10.83)$$

We conclude that the 2-point correlators of j^0 and of $\frac{1}{\sqrt{\pi}}\partial_1\phi$ are identical. It is straightforward to show that those of j^1 and of $-\frac{1}{\sqrt{\pi}}\partial_0\phi$ also are. Furthermore, not only are the two-point functions of these operators identical but, because Wick's theorem reduces n-point functions to products of 2-point functions, actually all correlators are identical. Hence we can establish the following operator identities:

$$j^\mu = \frac{1}{\sqrt{\pi}}\epsilon^{\mu\nu}\partial_\nu\phi \qquad j^{\mu 5} = \frac{1}{\sqrt{\pi}}\partial^\mu\phi, \qquad (10.84)$$

the first of which was already derived in (10.52). We see that under bosonization, the fermionic charge becomes the topological charge and the chirality becomes the generalized charge of the ϕ-field given by (10.25) and (10.26). The field bosonization formulas (10.77) reflects the fact that μ and σ, respectively, each carries one of them.

Lagrangean Bosonization

We now turn to the bosonization of the Lagrangean density (10.66). For this purpose, consider the correlator

$$\langle i\bar{\psi}(x)\gamma^{\mu}\partial_{\mu}\psi(y)\rangle = i\partial_{u_y}\langle\psi_1^{\dagger}(x)\psi_1(y)\rangle + i\partial_{v_y}\langle\psi_2^{\dagger}(x)\psi_2(y)\rangle$$
$$= \frac{1}{4\pi}\left(\frac{1}{u^2} + \frac{1}{v^2}\right), \tag{10.85}$$

where we used (10.74). This we compare with

$$\frac{1}{2}\langle\partial_{\mu}^{(x)}\phi(x)\partial^{\mu(y)}\phi(y)\rangle = \partial_{\mu}^{(x)}\partial^{\mu(y)}\langle\phi(x)\phi(y)\rangle$$
$$= -\frac{1}{4\pi}[\partial_0^x\partial_0^y - \partial_1^x\partial_1^y]\ln\left[(x_0 - y_0)^2 - (x_1 - y_1)^2\right]$$
$$= \frac{1}{4\pi}\frac{u^2 + v^2}{(uv)^2} = \frac{1}{4\pi}\left(\frac{1}{u^2} + \frac{1}{v^2}\right). \tag{10.86}$$

We now see that the operators $i\bar{\psi}(x)\gamma^{\mu}\partial_{\mu}\psi(y)$ and $\frac{1}{2}\partial_{\mu}^{(x)}\phi(x)\partial^{\mu(y)}\phi(y)$ are identical, for the same reasons that led to the current bosonization formulas. This identification also holds in the limit $x \to y$, provided we add some c-number renormalization factor to both operators. Hence we obtain the bosonization formula

$$i\bar{\psi}\gamma^{\mu}\partial_{\mu}\psi = \frac{1}{2}\partial_{\mu}\phi\partial^{\mu}\phi \tag{10.87}$$

that corresponds to (10.52).

10.5.2 The Massless Thirring Model

In the previous subsection, we saw how the duality structure can be used to map the free massless Dirac field into the free massless scalar field. We consider now the massless Thirring model [55] described by the Lagrangean density

$$\mathcal{L} = i\bar{\psi}\gamma^{\mu}\partial_{\mu}\psi - \frac{g}{2}(\bar{\psi}\gamma^{\mu}\psi)(\bar{\psi}\gamma_{\mu}\psi). \tag{10.88}$$

Using the bosonization formulas (10.87) and (10.84), we may cast this in the form

$$\mathcal{L} = \frac{1}{2}\left(1 - \frac{g}{\pi}\right)\partial_{\mu}\phi\partial^{\mu}\phi = \frac{v\lambda}{2}\left[\frac{\dot{\phi}^2}{v^2} - (\partial_x\phi)^2\right], \tag{10.89}$$

where $\lambda = 1 - \frac{g}{\pi}$. The momentum canonically conjugate to ϕ, then, is given by $\Pi = \frac{\lambda}{v}\dot{\phi}$. The Hamiltonian density, accordingly, will be

$$\mathcal{H} = \frac{v}{2}\left(\frac{\Pi^2}{\lambda} + \lambda(\partial_x\phi)^2\right). \qquad (10.90)$$

Now, performing the canonical transformation

$$\tilde{\Pi} = \frac{\Pi}{\sqrt{\lambda}} \qquad \tilde{\phi} = \sqrt{\lambda}\phi, \qquad (10.91)$$

we reobtain the free Hamiltonian density for the new field operators. However, when we express the bosonization formulas (10.77) and (10.32) in terms of the new free field operators, the a and b parameters must be modified in the following way: instead of $a = b = \sqrt{\pi}$, we now have

$$a = \sqrt{\pi\lambda} \qquad b = \sqrt{\frac{\pi}{\lambda}}. \qquad (10.92)$$

From (10.75), we obtain the following field correlators for the massless Thirring model in Euclidean space:

$$\langle\psi_1(\mathbf{r})\psi_1^\dagger(0)\rangle = -\frac{i}{4\pi}\frac{e^{i\,\text{arg}\,(\mathbf{r})}}{|\mathbf{r}|^{\nu/2}}$$

$$\langle\psi_2(\mathbf{r})\psi_2^\dagger(0)\rangle = \frac{i}{4\pi}\frac{e^{-i\,\text{arg}\,(\mathbf{r})}}{|\mathbf{r}|^{\nu/2}}$$

$$\langle\psi_1(\mathbf{r})\psi_2^\dagger(0)\rangle = \langle\psi_2(\mathbf{r})\psi_1^\dagger(0)\rangle = 0, \qquad (10.93)$$

where

$$\nu = \lambda + \frac{1}{\lambda} \quad ; \quad \lambda = 1 - \frac{g}{\pi} \quad ; \quad g = \pi\left(1 - \frac{a}{b}\right). \qquad (10.94)$$

Notice that the statistics $s = \frac{ab}{2\pi}$, according to (10.92), is still $s = 1/2$.

The current bosonization formulas now, become

$$j^\mu = \frac{1}{\sqrt{\pi\lambda}}\epsilon^{\mu\nu}\partial_\nu\phi \qquad j^{\mu 5} = \frac{1}{\sqrt{\pi\lambda}}\partial^\mu\phi. \qquad (10.95)$$

Observe also that the coupling constant has a limiting upper value, where the system becomes unstable and a phase transition occurs, namely $g < \pi$. We have

$$0 \leq g < \pi \quad \Leftrightarrow \quad \lambda \leq 1 \quad \Leftrightarrow \quad a \leq b$$
$$g < 0 \quad \Leftrightarrow \quad \lambda > 1 \quad \Leftrightarrow \quad a > b. \qquad (10.96)$$

The above result coincides with the famous operator solution for the massless Thirring model obtained by Klaiber in 1967 [52]. It shows the remarkable power of the bosonization method, by which a complicated nonlinear system is mapped into a free one.

10.5.3 The Massive Thirring/Sine-Gordon System

We now consider the effect of adding a mass term to the Thirring Lagrangean (10.88), namely

$$\mathcal{L} = i\bar{\psi}\gamma^{\mu}\partial_{\mu}\psi - M\bar{\psi}\psi - \frac{g}{2}(\bar{\psi}\gamma^{\mu}\psi)(\bar{\psi}\gamma_{\mu}\psi), \tag{10.97}$$

the massive Thirring model.

Using the bosonization formulas (10.77), (10.84) and (10.87), we readily obtain

$$\mathcal{L} = \frac{1}{2}\left(1 - \frac{g}{\pi}\right)\partial_{\mu}\phi\partial^{\mu}\phi - \frac{M}{2\pi}\cos\left[2\sqrt{\pi}\phi\right]. \tag{10.98}$$

Going through the same procedure as in the massless case, we obtain the Hamiltonian density

$$\mathcal{H} = \frac{1}{2}\left(\tilde{\Pi}^2 + (\partial_x\tilde{\phi})^2\right) + \frac{M}{2\pi}\cos\left[2\sqrt{\frac{\pi}{\lambda}}\tilde{\phi}\right]. \tag{10.99}$$

Then, we may write the Lagrangean density corresponding to the Hamiltonian above as

$$\mathcal{L} = \frac{1}{2}\partial_{\mu}\phi\partial^{\mu}\phi - \alpha\cos\beta\phi, \tag{10.100}$$

where we have dropped the tilde in order to simplify the notation, and introduced the parameters

$$\alpha = \frac{M}{2\pi} \quad ; \quad \beta = 2\sqrt{\frac{\pi}{\lambda}}. \tag{10.101}$$

We immediately recognize in (10.100), the sine-Gordon Lagrangean.

By taking (10.92) into account, we then have the fermionic massive Thirring field corresponding to (10.97), expressed by means of the bosonization formulas (10.77), with parameters

$$b = \frac{\beta}{2} \quad ; \quad a = \frac{2\pi}{\beta} \tag{10.102}$$

in terms of the sine-Gordon field (10.99).

Notice that still $ab = \pi$, hence $s = 1/2$ and

$$g = \pi\left(1 - \frac{4\pi}{\beta^2}\right). \tag{10.103}$$

10.6 Quantum Sine–Gordon Solitons

Observe that since the topological current of the bosonic theory is identified with the fermionic current of the associated theory, we conclude that the dynamics of sine-Gordon solitons is described by the massive Thirring model, and the quantum sine-Gordon soliton creation operator is just the Dirac massive Thirring field. The coupling parameter determining the magnitude of the solitons' interaction is given by g, whereas the one of the bosons, by β. Both are related by (10.103). From this, we can arrive at two remarkable conclusions. Firstly, for $\beta^2 = 4\pi$, the coupling g vanishes and the sine-Gordon solitons become free massive fermions! Secondly, for $\beta^2 < 4\pi$, when the soliton coupling is attractive, the more we decrease β, the more we increase the magnitude of the soliton coupling, namely $\beta \to 0 \Leftrightarrow g \to -\infty$. Conversely, for $\beta^2 > 4\pi$, we have $\beta \to \infty \Leftrightarrow g \to \pi$. We see that through bosonization we map the strong into the weak coupling regimes of the corresponding bosonic and fermionic theories.

In the attractive soliton regime, $\beta^2 < 4\pi$, the use of advanced mathematical methods has enabled the obtainment of the exact energy spectrum of soliton bound states, namely [53, 54]

$$
M_n = 2M \sin \left(n \frac{\pi}{2} \xi \right) \quad n = 1, 2, \ldots < \frac{1}{\xi} \; ; \; \xi = \frac{\beta^2}{8\pi - \beta^2}, \tag{10.104}
$$

where M is the one-soliton mass.

10.6.1 QED$_2$: the Schwinger Model

The Model

The Schwinger model [56] is the quantum electrodynamics of a massless Dirac field in D = 2, which is described by the Lagrangean density

$$
\mathcal{L} = i \bar{\psi} \gamma^\mu \partial_\mu \psi - \frac{1}{4} F_{\mu\nu} F^{\mu\nu} - e \bar{\psi} \gamma^\mu \psi A_\mu. \tag{10.105}
$$

The theory possesses the same global symmetries (10.69), hence charge and chirality, associated to the currents (10.70), are conserved quantities at the classical level.

This is, in many ways, a remarkable system. Besides of admitting an exact solution, it shares many features of QCD, in D = 4. It presents, for instance, the same vacuum structure as QCD, reflecting the nontrivial topological mapping produced by the gauge field at infinity. Also, similarly to this, the original Lagrangean degrees of freedom do not show in the spectrum of physical excitations, in the same way that quarks and gluons do not appear in the hadronic spectrum. Also all charge and chirality will disappear from the physical spectrum, in the same way color does not appear in the spectrum of QCD.

Bosonization

Using the bosonization formulas (10.84) and (10.87), we can express the Lagrangean as

$$\mathcal{L} = \frac{1}{2}\partial_\mu\phi\partial^\mu\phi - \frac{1}{4}F_{\mu\nu}F^{\mu\nu} - \frac{e}{\sqrt{\pi}}\epsilon^{\mu\nu}\partial_\nu\phi A_\mu. \tag{10.106}$$

The field equations corresponding to this are

$$\partial_\nu F^{\mu\nu} = \frac{e}{\sqrt{\pi}}\epsilon^{\mu\nu}\partial_\nu\phi$$

$$\Box\phi = \frac{e}{\sqrt{\pi}}\epsilon^{\mu\nu}\partial_\nu A_\mu \tag{10.107}$$

and they are solved by the operator identity

$$eA^\mu = \sqrt{\pi}\epsilon^{\mu\nu}\partial_\nu\phi, \tag{10.108}$$

provided the consistency between (10.107) and (10.108),

$$\Box\phi = \frac{e^2}{\pi}\phi, \tag{10.109}$$

is satisfied.

Now, integrating over the gauge field A_μ,

$$Z = \int D\phi DA_\mu \exp\left\{\frac{1}{2}\partial_\mu\phi\partial^\mu\phi + \frac{1}{2}A_\mu[-\Box\delta\mu\nu + \partial_\mu\partial_\nu]A_\nu - \frac{e}{\sqrt{\pi}}\epsilon^{\mu\nu}\partial_\nu\phi A_\mu\right\}$$

$$= \int D\phi \exp\left\{\frac{1}{2}\partial_\mu\phi\partial^\mu\phi + \frac{e^2}{2\pi}\epsilon^{\mu\alpha}\partial_\alpha\phi\left[\frac{-\Box\delta^{\mu\nu} + \partial_\mu\partial_\nu}{\Box^2}\right]\epsilon^{\nu\beta}\partial_\beta\phi\right\}$$

$$= \int D\phi \exp\left\{\frac{1}{2}\partial_\mu\phi\partial^\mu\phi + \frac{e^2}{2\pi}\phi^2\right\}. \tag{10.110}$$

We conclude that the resulting effective theory is a free scalar field with mass $m = \frac{e}{\sqrt{\pi}}$, thus satisfying the Klein–Gordon field equation, which is nothing but (10.109).

This is a remarkable result. The physical content of the QED_2 spectrum is a neutral massive scalar field. The charge and chirality have completely disappeared from the spectrum of physical excitations. This is analogous to what happens in QCD_4, where conserved quantities carried by quarks and gluons, such as color and chirality, also do not show in the physical spectrum.

Effective Higgs Mechanism and Spontaneous Symmetry Breaking

The Schwinger model, furthermore, provides an example where an effective Higgs mechanism occurs without the presence of any Higgs field. Indeed, from (10.108) and the bosonization formula (10.84), we conclude that

$$j^\mu = \frac{e}{\pi} A_\mu. \tag{10.111}$$

Inserting this in (10.105), we can see that the gauge field A_μ acquires a mass $M = \frac{e}{\sqrt{\pi}}$.

Considering the field equation (10.109) and the current bosonization formulas (10.84), we see that an anomaly occurs in the axial current, preventing its conservation at a quantum level

$$\partial_\mu j^{\mu 5} = \frac{e^2}{\pi} \phi. \tag{10.112}$$

Charge, in a similar way, also cannot exist in this theory. Already at a classical level we see that this quantity cannot be defined. Indeed, charge is expressed as

$$Q = \int_{-\infty}^{\infty} dx j^0(x) = \frac{e}{\sqrt{\pi}} [\phi(x = +\infty) - \phi(x = -\infty)] = 0. \tag{10.113}$$

This must be zero because the Klein–Gordon equation does not admit finite energy solutions, which are non-vanishing at infinity. We see that charge and chirality just disappear from the physical spectrum of excitations.

It is quite instructive to evaluate the two-point correlation function of the gauge-invariant Dirac field operator

$$\hat\psi(x) = \psi(x) \exp\left\{ -ie \int_{-\infty}^{x} A_\mu d\xi^\mu \right\}. \tag{10.114}$$

Using the bosonization formula (10.80) and the operator identity (10.108), we have

$$\hat\psi_1(x) = \sqrt{\frac{-i}{4\pi}} \sigma(x) \qquad \hat\psi_2(x) = \sqrt{\frac{-i}{4\pi}} \sigma^\dagger(x). \tag{10.115}$$

The gauge invariant correlator, therefore, is

$$\langle \hat\psi_i(x) \hat\psi_j^\dagger(y) \rangle = \frac{1}{4\pi} \int D\phi DA_\mu \exp\left\{ -\int d^2x \left[\frac{1}{2} \partial_\mu \phi \partial^\mu \phi \right.\right.$$
$$\left.\left. + \frac{e^2}{2\pi} \phi^2 + \alpha_{ij}(x, y)\phi \right] \right\}, \tag{10.116}$$

where

$$\alpha_{ij}(x, y) = -i\sqrt{\pi}[\lambda_i \delta^2(z - x) + \lambda_j \delta^2(z - y)] \quad \lambda_1 = 1 \; ; \; \lambda_2 = -1 \tag{10.117}$$

Performing the ϕ-integration using (5.49), we readily get

$$\langle \hat{\psi}_i(x) \hat{\psi}_j^\dagger(y) \rangle = \frac{1}{4\pi} \exp\left\{ -2\pi \left[-\lambda_i \lambda_j \Delta(x-y) + \Delta(0) \right] \right\},$$

$$\langle \hat{\psi}_i(x) \hat{\psi}_j(y) \rangle = \frac{1}{4\pi} \exp\left\{ -2\pi \left[\lambda_i \lambda_j \Delta(x-y) + \Delta(0) \right] \right\}, \qquad (10.118)$$

where $\Delta(x)$ is the Euclidean Green function of the Klein–Gordon operator in $D=2$, which is given by the modified Bessel function $\Delta(x) = K_0(M|\mathbf{x}|)$, $M = e/\sqrt{\pi}$.

Now, we no longer have the infrared regulator $\mu \to 0$ enforcing the charge and the chirality selection rules for the different components of the above correlators. Absorbing the last term in (10.118) and (10.154) in a renormalization of the field $\hat{\psi}$, we get

$$\langle \hat{\psi}_i(x) \hat{\psi}_j^\dagger(y) \rangle_R = \frac{1}{4\pi} \exp\left\{ 2\pi \lambda_i \lambda_j \Delta(x-y) \right\}$$

$$\langle \hat{\psi}_i(x) \hat{\psi}_j(y) \rangle_R = \frac{i}{4\pi} \exp\left\{ -2\pi \lambda_i \lambda_j \Delta(x-y) \right\}. \qquad (10.119)$$

The large-distance behavior of the above correlators, obtained using the fact that $\Delta(x) \to 0$ in this limit, implies that $\langle 0|\hat{\psi}_i|0\rangle \neq 0$, thus violating both the charge and chirality selection rules. We therefore have the spontaneous breakdown of both the U(1) and chiral U(1) global symmetries of the theory.

Topological Vacua

The condition of action finiteness requires the abelian U(1) gauge field A_μ to be a pure gauge in the asymptotic manifold at $r_E \to \infty$ in a two-dimensional Euclidean space. This produces a mapping between such manifold, a circumference S_1 and the U(1) group manifold, also S_1. Since this mapping presents infinite inequivalent topological classes, reflecting the fact that $\Pi_1(S_1) = n \in \mathbb{Z}$, it follows that the A_μ gauge field itself will belong to one of the infinitely many inequivalent topological classes characterized by the Chern number

$$q = \frac{e}{2\pi} \int d^2x \epsilon^{\mu\nu} \partial_\mu A_\nu$$

$$q = \frac{e}{2\pi} \oint_{C(\infty)} d\xi^\mu A_\mu = \frac{1}{2\pi}[\Lambda(2\pi) - \Lambda(0)], \qquad (10.120)$$

where we used the condition that A_μ must be a pure gauge at infinity:

$$A_\mu(x) \overset{|x|\to\infty}{\longrightarrow} \frac{1}{e}\partial_\mu \Lambda. \qquad (10.121)$$

Choosing $\Lambda = n\varphi$, where φ is the polar angle, we have $q = n$. Since classical pure gauge configurations must correspond to quantum vacuum states in the Hilbert space, for each of these classical inequivalent topological classes we must have different vacuum states $|n\rangle$.

Using a temporal gauge, $A_0 = 0$, we have $q = n(\tau = +\infty) - m(\tau = -\infty)$, so, according to the Feynman functional integral expression (5.63), when we integrate over a topological class q we are calculating the amplitude between different vacua, namely

$$\langle n|\mathcal{O}|m\rangle = Z_0^{-1} \int DA_\mu^{(n-m)} \mathcal{O} e^{-S[A_\mu]}, \tag{10.122}$$

where $DA_\mu^{(q)}$ means we are integrating over fields belonging to the topological class characterized by the Chern number q.

Now, defining the θ-vacua as

$$|\theta\rangle = \frac{1}{2\pi} \sum_n e^{in\theta} |n\rangle \tag{10.123}$$

$(0 \le \theta < 2\pi)$, we readily obtain from (10.122)

$$\langle \theta|\mathcal{O}|\theta\rangle = Z_0^{-1} \sum_{q\in\mathbb{Z}} \int DA_\mu^{(q)} \mathcal{O} e^{-S[A_\mu]-i\theta q[A_\mu]}. \tag{10.124}$$

The θ-vacua correlation functions, therefore, are obtained by integrating over all topological classes of the gauge field, each one weighed by the Chern number term $e^{iq\theta}$. Observe that our previous calculation of the gauge invariant correlator was actually carried on for $\theta = 0$. It is straightforward to obtain the corresponding results for nonzero θ [31]. The vacuum expectation value of the gauge invariant Dirac field, in particular, will be

$$\langle \theta|\hat{\psi}_i|\theta\rangle = \sqrt{\frac{-i}{4\pi}} \exp\left\{\frac{i}{2}\lambda_i\theta\right\}. \tag{10.125}$$

The θ-vacuum structure found in QED$_2$ is equal to the one of QCD$_4$ [57, 58]. There, θ is an input parameter that must be determined experimentally. It was shown that a nonzero θ in QCD implies the neutron must have a nonzero electric dipole moment. This has not been observed experimentally, thus implying that for some reason θ must be zero in nature.

10.6.2 *Features of Bosonization in* $d > 1$ *Spatial Dimension*

Bosonization has produced such spectacular results in $d = 1$ that, consequently, its generalization to higher dimensions was soon explored [60, 61, 62, 63]. Several important results, then, clearly indicated that it would be possible to generalize the method of bosonization to higher dimensions [64, 65, 66, 67, 68, 69]. Generalizations, however, were invariably partial, in the sense that bosonized expressions were found for the current or for the Lagrangean but not for the fermion field itself. The first complete bosonization in $d > 1$, which therefore included a bosonic expression for the fermion field, was obtained in [71], for the free massless Dirac fermion field in $D = 2 + 1$ dimensions. Recently, the complete bosonization of free Weyl fermions has been achieved in $D = 3 + 1$ [72]. This may be useful in the description of the recently observed Weyl semimetals in condensed matter systems such as $TaAs$, TaP and $NbAs$ [73, 74, 75].

It soon became clear that the method of bosonization indeed could be generalized for $d > 1$ spatial dimensions, even though it would only be an exact mapping in the case of free theories. This fact, however, does not remove the interest or the usefulness of bosonization in higher dimensions, because it will always provide invaluable new insights and unanticipated points of view about the system under consideration. It can also certainly be used to devise new approximation procedures for complex interacting systems.

In the next two sections, we describe the method of complete bosonization for the free Dirac fermion in $D = 2 + 1$ dimensions and for free Weyl fermions in $D = 3 + 1$ dimensions, respectively, according to the results contained in [71] and [72].

10.7 Bosonization in Two Spatial Dimensions

We now consider the bosonization of the massless, two-component Dirac field in $D = 2 + 1$, following the procedure contained in [71]. The Lagrangean is given by (10.66), with the γ-matrices being now

$$\gamma^0 = \begin{pmatrix} 1 & 0 \\ 0 & -1 \end{pmatrix} \quad \gamma^1 = \begin{pmatrix} 0 & -i \\ -i & 0 \end{pmatrix} \quad \gamma^2 = \begin{pmatrix} 0 & -1 \\ 1 & 0 \end{pmatrix}. \tag{10.126}$$

The massless Dirac Lagrangean can be written, in momentum space, as

$$\mathcal{L} = \psi^\dagger \gamma^0 \gamma^\mu k_\mu \psi. \tag{10.127}$$

Now, introducing the rapidity variable $\chi \in [0, \infty)$, such that

$$k^0 = k \cosh \chi \quad ; \quad |\mathbf{k}| = k \sinh \chi \tag{10.128}$$

where $k = \sqrt{k_\mu k^\mu}$ in the positive energy, time-like region of Minkowski space, we may write

$$\gamma^0 \gamma^\mu k_\mu = k \left[\mathbb{I} \cosh \chi + \hat{\varphi} \cdot \sigma \sinh \chi \right]. \tag{10.129}$$

Furthermore, using

$$T = \left[\mathbb{I} \cosh \frac{\chi}{2} + \hat{\varphi} \cdot \sigma \sinh \frac{\chi}{2} \right] \tag{10.130}$$

we can express the Dirac Lagrangean as

$$\mathcal{L} = \psi^\dagger \gamma^0 \gamma^\mu k_\mu \psi = \psi^\dagger T^\dagger T \psi. \tag{10.131}$$

Then, introducing the new fermion field $\Psi \equiv T\psi$, we have

$$\mathcal{L} = \psi^\dagger \sigma^\mu k_\mu \psi = \Psi^\dagger \begin{pmatrix} k & 0 \\ 0 & k \end{pmatrix} \Psi. \tag{10.132}$$

The two-point correlation function of the Dirac field may be expressed, in momentum space, in terms of the correlation functions of the Ψ field, namely,

$$\langle \Psi \Psi^\dagger \rangle(k) = \begin{pmatrix} 1 & 0 \\ 0 & 1 \end{pmatrix} \frac{1}{k}. \tag{10.133}$$

Performing the Wick rotation to Euclidean space and considering the inverse Fourier transform $\mathcal{F}[\frac{1}{k}] = \frac{1}{2\pi^2|\mathbf{r}|^2}$, we have

$$\langle \Psi(x) \Psi^\dagger(y) \rangle = \begin{pmatrix} 1 & 0 \\ 0 & 1 \end{pmatrix} \frac{1}{2\pi^2|x-y|^2}. \tag{10.134}$$

The bosonization of the Lagrangean and current were established in (10.57). In order to achieve the field bosonization, we introduce the dual operators

$$\mu(x) = \exp\left\{ ia \int_{-\infty}^{\mathbf{x}} d\xi^\mu \Box^{-1/2} \epsilon^{\mu\alpha\beta} \partial_\alpha B_\beta(\xi) \right\} \tag{10.135}$$

and

$$\sigma(y) = \exp\left\{ -ib \int_{-\infty}^{\mathbf{y}} d\eta^\mu B_\mu(\xi) \right\}. \tag{10.136}$$

Notice that a and b are dimensionless, as they should be.

In order to reproduce the correlation functions (10.134) in the framework of the bosonic theory given by (10.57), we now evaluate the four-points order-disorder correlation function

$$\langle \sigma(x_1)\mu(x_2)\mu^\dagger(y_2)\sigma^\dagger(y) \rangle = \int DB_{\mu\nu} \exp\left\{ -\int d^4z \left[\frac{1}{2} B_\mu A^{\mu\nu} B_\nu \right. \right.$$
$$\left. \left. + \left(\alpha_\mu(z; x_1, y_1) + \beta_\mu(z; x_2, y_2) \right) B_\mu \right] \right\}$$
$$= \exp\left\{ \frac{1}{2} \int d^3z d^3z' \left(\alpha_\mu(z; x_1, y_1) + \beta_\mu(z; x_2, y_2) \right) \right.$$
$$\left. (A^{-1})^{\mu\nu} \left(\alpha_\mu(z'; x_1, y_1) + \beta_\mu(z'; x_2, y_2) \right) \right\}. \tag{10.137}$$

In the above expression, $A^{\mu\nu}$ is given by (10.56) and, according to (10.135) and (10.136),

$$\alpha_\mu(z; x_1, y_1) = -ib \int_{-\infty}^{y} d\eta^\mu \delta^3(z - \xi)$$

$$\beta_{\mu\nu}(z; x_2, y_2) = ia \int_{-\infty}^{x} d\xi^\mu \Box^{-1/2} \epsilon^{\mu\alpha\beta} \partial_\alpha \delta^3(z - \xi). \tag{10.138}$$

The quadratic functional integral over the bosonic B_μ field produces the following terms:

$$T_{\alpha\alpha} = b^2 \frac{A}{\pi^2}[-\ln\mu|x_1 - y_1| + \ln\mu|\epsilon|]$$

$$T_{\beta\beta} = a^2 \frac{A}{\pi^2}[-\ln\mu|x_2 - y_2| + \ln\mu|\epsilon|]$$

$$T_{\alpha\beta} = ab \frac{B}{\pi^2}[-\ln\mu|x_1 - y_2| - \ln\mu|x_2 - y_1| + 2\ln\mu|\epsilon|] \tag{10.139}$$

and

$$\tilde{T}_{\alpha\alpha} = ib^2 \frac{B}{4\pi}[\varphi(x_1 - y_1)]$$

$$\tilde{T}_{\beta\beta} = ia^2 \frac{B}{4\pi}[\varphi(x_2 - y_2)]$$

$$\tilde{T}_{\alpha\beta} = -iab \frac{A}{4\pi}[\varphi(x_1 - y_2) + \varphi(x_2 - y_1)], \tag{10.140}$$

where we used the fact that, in D = 3 Euclidean space, we have

$$\mathcal{F}^{-1}\left[\frac{1}{k^3}\right] = \lim_{\mu \to 0} -\frac{1}{2\pi^2} \ln \mu^2[|\mathbf{r}|^2 + |\epsilon|^2].$$

In the above expressions

$$A = \frac{16}{1 + (16\theta)^2} \quad ; \quad B = 16\theta A \tag{10.141}$$

and

$$\varphi(x - y) = \int_x^y d\xi^\mu \int_x^y d\eta^\nu \epsilon^{\mu\nu\alpha} \partial_\alpha \frac{1}{4\pi|\xi - \eta|}. \tag{10.142}$$

Choosing the parameters a and b in such a way that

$$(a^2 + b^2)A + 2abB = \pi^2 \quad ; \quad (a^2 + b^2)B - 2abA = 0$$

$$a^2 + b^2 = \frac{\pi^2}{16} \quad ; \quad ab = \theta \frac{\pi^2}{2} \tag{10.143}$$

and expressing the field Ψ in the bosonized form

$$\Psi = \begin{pmatrix} \sigma\mu \\ \sigma^\dagger\mu \end{pmatrix}, \tag{10.144}$$

we can reproduce the correlation functions (10.134) completely within the bosonic theory, as we did in $D=2$. Since it is a free theory, all many-points functions will be reproduced as well.

Notice that the infrared regulator μ, as happened in $D=2$, completely cancels in the correlation function (10.137). As before, it will not vanish in selection rule, violating correlation functions, thereby making these vanish.

10.8 Bosonization in Three Spatial Dimensions

10.8.1 Lagrangean and Current Bosonization

In this section, we will first describe the bosonization of free Weyl fermions and, subsequently, the complete bosonization of free Dirac fermions in $D=3+1$, following the method contained in [72]. We start by considering the free massless Dirac field, which is described by the Lagrangean

$$\mathcal{L} = i\bar\psi\gamma^\mu\partial_\mu\psi = \psi_L^\dagger\sigma^\mu\partial_\mu\psi_L + \psi_R^\dagger\bar\sigma^\mu\partial_\mu\psi_R, \tag{10.145}$$

where we use the Weyl representation of the Dirac matrices, namely

$$\gamma^0 = \begin{pmatrix} 0 & \mathbb{I} \\ \mathbb{I} & 0 \end{pmatrix} \quad \gamma^i = \begin{pmatrix} 0 & \sigma^i \\ -\sigma^i & 0 \end{pmatrix} \quad \gamma^5 = \begin{pmatrix} \mathbb{I} & 0 \\ 0 & -\mathbb{I} \end{pmatrix}, \tag{10.146}$$

where each block is 2×2. Then the Dirac Fermion decomposes in two-components Weyl fermions, such that

$$\psi = \begin{pmatrix} \psi_L \\ \psi_R \end{pmatrix} \tag{10.147}$$

and

$$\sigma^\mu = (\mathbb{I}, \sigma^i) \quad ; \quad \bar\sigma^\mu = (\mathbb{I}, -\sigma^i). \tag{10.148}$$

The current and axial current are

$$\bar\psi\gamma^\mu\psi = \psi_L^\dagger\sigma^\mu\psi_L + \psi_R^\dagger\bar\sigma^\mu\psi_R$$
$$\bar\psi\gamma^5\gamma^\mu\psi = \psi_L^\dagger\sigma^\mu\psi_L - \psi_R^\dagger\bar\sigma^\mu\psi_R. \tag{10.149}$$

Let us take the Weyl component ψ_L. Defining the rapidity variable χ as

$$k^0 = k\cosh\chi \quad ; \quad |\mathbf{k}| = k\sinh\chi, \tag{10.150}$$

where $k = \sqrt{k_\mu k^\mu}$ in the positive energy, time-like region of Minkowski space, we have

$$\sigma^\mu k_\mu = k \begin{pmatrix} \cosh\chi + \sinh\chi\cos\theta & \sinh\chi\sin\theta e^{-i\varphi} \\ \sinh\chi\sin\theta e^{i\varphi} & \cosh\chi - \sinh\chi\cos\theta \end{pmatrix}$$

$$= k\left[\mathbb{I}\cosh\chi + \hat{r}\cdot\sigma\sinh\chi\right], \tag{10.151}$$

where $\hat{r}$ is the radial unit vector of the spherical coordinate system.

Before bosonizing the Weyl fermion field ψ_L, we introduce the new spinor field

$$\Psi_L = T_L\psi_L$$
$$T_L = \left[\mathbb{I}\cosh\frac{\chi}{2} + \hat{\varphi}\cdot\sigma\sinh\frac{\chi}{2}\right]\left[\mathbb{I} - i\hat{\theta}\cdot\sigma\right] \tag{10.152}$$

such that

$$\psi_L^\dagger\sigma^\mu k_\mu\psi_L = \Psi_L^\dagger\begin{pmatrix} k & 0 \\ 0 & k \end{pmatrix}\Psi_L. \tag{10.153}$$

The canonical transformation T_L renders the Lagrangean diagonal. It is similar to the Foldy–Wouthuysen transformation but is not unitary. We can obtain a similar transformation for the R Weyl spinor, namely

$$\Psi_R = T_R\psi_R$$
$$T_R = \left[\mathbb{I}\cosh\frac{\chi}{2} - \hat{\varphi}\cdot\sigma\sinh\frac{\chi}{2}\right]\left[\mathbb{I} - i\hat{\theta}\cdot\sigma\right] \tag{10.154}$$

such that

$$\psi_R^\dagger\bar{\sigma}^\mu k_\mu\psi_R = \Psi_R^\dagger\begin{pmatrix} k & 0 \\ 0 & k \end{pmatrix}\Psi_R. \tag{10.155}$$

The Ψ-field Euclidean correlation functions are

$$\langle\Psi_L(x)\Psi_L^\dagger(y)\rangle = \langle\Psi_R(x)\Psi_R^\dagger(y)\rangle = \begin{pmatrix} 1 & 0 \\ 0 & 1 \end{pmatrix}\frac{1}{2\pi^3|x-y|^3}$$

$$\langle\Psi_L(x)\Psi_R^\dagger(y)\rangle = \langle\Psi_R(x)\Psi_L^\dagger(y)\rangle = 0. \tag{10.156}$$

The Lagrangean and current bosonization formulas appropriate to the chiral Weyl fields are the following:

$$\psi_L^\dagger\sigma^\mu\partial_\mu\psi_L = \frac{1}{24}H_{\mu\nu\alpha}^L H_L^{\mu\nu\alpha}$$

$$\psi_R^\dagger\bar{\sigma}^\mu\partial_\mu\psi_R = \frac{1}{24}H_{\mu\nu\alpha}^R H_R^{\mu\nu\alpha}, \tag{10.157}$$

where $H_{\mu\nu\alpha} = \partial_\mu B_{\nu\alpha} + \partial_\nu B_{\alpha\mu} + \partial_\alpha B_{\mu\nu}$.

For the axial current, we have, in the absence of an electromagnetic field,

$$\bar{\psi}\gamma^\mu\gamma^5\psi = \frac{1}{2}\sqrt{\frac{\Box}{24\pi^2}}\epsilon^{\mu\nu\alpha\beta}\partial_\nu\left[B_{\alpha\beta}^L - B_{\alpha\beta}^R\right]. \tag{10.158}$$

If there is an applied EM field A_μ, however, the axial current will acquire a topological term [77]. For the chiral, L, R Weyl currents, according to (10.149), we must have, consequently,

$$j_L^\mu = \psi_L^\dagger \sigma^\mu \psi_L = \frac{1}{2}\sqrt{\frac{\Box}{24\pi^2}} \epsilon^{\mu\nu\alpha\beta} \partial_\nu B_{\alpha\beta}^L + \frac{1}{2} I^\mu$$

$$j_R^\mu = \psi_R^\dagger \bar\sigma^\mu \psi_R = \frac{1}{2}\sqrt{\frac{\Box}{24\pi^2}} \epsilon^{\mu\nu\alpha\beta} \partial_\nu B_{\alpha\beta}^R - \frac{1}{2} I^\mu. \tag{10.159}$$

The Dirac chiral current, j_5^μ, hence, is given by

$$\bar\psi \gamma^\mu \gamma^5 \psi = \frac{1}{2}\sqrt{\frac{\Box}{24\pi^2}} \epsilon^{\mu\nu\alpha\beta} \partial_\nu \left[B_{\alpha\beta}^L - B_{\alpha\beta}^R \right] + I^\mu. \tag{10.160}$$

In the absence of an external EM field, the I^μ topological term just vanishes. When there is an EM background field A_μ, then [77]

$$I^\mu = \frac{1}{4\pi^2} \epsilon^{\mu\nu\alpha\beta} A_\nu \partial_\alpha A_\beta, \tag{10.161}$$

implying that

$$\partial_\mu j_5^\mu = \partial_\mu I^\mu = -\frac{1}{16\pi^2} F^{\mu\nu} \tilde F^{\mu\nu}, \tag{10.162}$$

where the last term is the Chern–Pontryagin topological charge density of the EM field, namely, the axial anomaly [76].

10.8.2 Field Bosonization

We now turn to the bosonization of the fields $\Psi_{L,R}$. We have seen that the natural bosonic field in $D = 3+1$ is the antisymmetric tensor gauge-field $B_{\mu\nu}$, also known as Kalb–Ramond field. The relevant dual operators introduced before, namely, $\sigma(C)$ and $\mu(x)$, are expressed in terms of this field by (10.40). A few adjustments, however, are required before we use these operators in the bosonization of the Weyl fermion.

Firstly, the $\mu(x)$ operator. We have seen that this bears the topological charge associated with the current J_4^μ, given by (10.13). When bosonizing the fermionic current in $D = 3+1$, however, we had to identify it with this topological current multiplied by the pseudo-differential operator $\Box^{1/2}$ as we can see in (10.157). Consequently, the operator we must use for bosonizing the fermion field carrying the charge associated to this current must contain the Green function of such operator. We shall use, then,

$$\mu(\mathbf{x}, t) = \exp\left\{ ia \int_{-\infty}^{\mathbf{x}} d\xi^\mu \Box^{-1/2} \epsilon^{\mu\nu\alpha\beta} \partial_\nu B_{\alpha\beta}(\xi) \right\}. \tag{10.163}$$

Notice that a is dimensionless, as it should be.

Then, there is the $\sigma(C)$ operator. This is nonlocal, in the sense it is defined on a curve, rather than on a point. In order to keep the canonical dimension of the Kalb–Ramond field, we have also added the pseudo-differential operator $\Box^{1/2}$ in the exponent, namely

$$\sigma(S(C), t) = \exp\left\{-ib \int_{S(C)} d^2\xi^{\mu\nu}\Box^{1/2}B_{\mu\nu}(\xi)\right\}, \qquad (10.164)$$

where C is assumed to be a circle of infinitesimal radius ρ. Notice again that b is dimensionless, as it should be.

Let us evaluate now the four-points order-disorder correlation function in the framework of the bosonic theory associated to the Weyl fermions in $D = 4$ Euclidean space, given by (10.157), namely

$$\langle\sigma(C_{x_1})\mu(x_2)\mu^\dagger(y_2)\sigma^\dagger(C_{y_1})\rangle = \int DB_{\mu\nu}\exp\left\{-\int d^4z\left[\frac{1}{2}B_{\mu\nu}A^{\mu\nu\alpha\beta}B_{\alpha\beta}\right.\right.$$

$$\left.\left. + \left(\alpha_{\mu\nu}(z; C_{x_1}, C_{y_1}) + \beta_{\mu\nu}(z; x_2, y_2)\right)B_{\mu\nu}\right]\right\}$$

$$= \exp\left\{-\frac{1}{4\pi^3}\int d^4z d^4z'\left(\alpha_{\mu\nu}(z; C_{x_1}, C_{y_1}) + \beta_{\mu\nu}(z; x_2, y_2)\right)\right.$$

$$\left. [\ln\mu|z - z'|]\Delta^{\mu\nu\alpha\beta}\left(\alpha_{\alpha\beta}(z'; C_{x_1}, C_{y_1}) + \beta_{\alpha\beta}(z'; x_2, y_2)\right)\right\}. \qquad (10.165)$$

In the above expression, $A^{\mu\nu\alpha\beta}$ is given by (10.63) and, according to (10.163) and (10.164),

$$\alpha_{\mu\nu}(z; C_{x_1}, C_{y_1}) = i\frac{b}{2\pi\rho}\int_{S(C_{x_1})-S(C_{y_1})} d^2\xi^{\mu\nu}\Box^{1/2}\delta^4(z - \xi)$$

$$\beta_{\mu\nu}(z; x_2, y_2) = a\int_{x_2}^{y_2} d\xi_\lambda\Box^{-1/2}\epsilon^{\lambda\alpha\mu\nu}\partial_\alpha\delta^4(z - \xi). \qquad (10.166)$$

We have also used the fact that in $D = 4$ Euclidean space,

$$\mathcal{F}^{-1}\left[\frac{1}{k^4}\right] = \lim_{\mu\to0} -\frac{1}{4\pi^3}\ln\mu^2[|\mathbf{r}|^2 + |\epsilon|^2].$$

Now, using (10.61), we get the three following terms:

$$T_{\alpha\alpha} = -\frac{b^2}{2\pi^3}[\ln\mu|x_1 - y_1| - \ln\mu|\epsilon|]$$

$$T_{\beta\beta} = -\frac{a^2}{4\pi^3}\int_{x_2}^{y_2} d\xi^\mu \int_{x_2}^{y_2} d\eta^\nu\left[-\Box\delta^{\mu\nu} + \partial_\mu^{(\xi)}\partial_\nu^{(\eta)}\right]\ln\mu|\xi - \eta|$$

$$= -\frac{a^2}{2\pi^3}[\ln\mu|x_2 - y_2| - \ln\mu|\epsilon|] + T_{\beta\beta}(L, \epsilon)$$

$$T_{\alpha\beta} = -i\frac{ab}{6\pi^2}\int_{S(C_{x_1})-S(C_{y_1})}d^2\xi^{\mu\nu}\int_{x_2}^{y_2}d\eta^\lambda\epsilon^{\lambda\mu\nu\alpha}\partial_\alpha[\frac{1}{|\xi - \eta|}]\overset{\rho\to 0}{\longrightarrow} 0.$$

$$(10.167)$$

Notice that the infrared regulator μ completely cancels in this correlation function. Conversely, for the correlator $\langle\sigma\mu\mu^\dagger\sigma\rangle$, for instance, the $T_{\alpha\alpha}$ term, would produce an overall $\ln\mu$-factor that would force it to vanish. The ultraviolet regulator appearing in the $\alpha\alpha$ and $\beta\beta$ terms may be removed by a multiplicative field renormalization, respectively, of the operators σ and μ in the correlation functions. The renormalized fields acquire the correct dimension, which corresponds to the respective correlators.

We have, consequently,

$$\langle\sigma(C_{x_1})\mu(x_2)\mu^\dagger(y_2)\sigma^\dagger(C_{y_1})\rangle = \exp\left\{T_{\alpha\alpha} + T_{\beta\beta} + 2T_{\alpha\beta}\right\}$$

$$\overset{x_1\to x_2=x; y_1\to y_2=y}{\longrightarrow}\exp\left\{-\frac{a^2 + b^2}{2\pi^3}\ln|x - y|\right\}.\qquad(10.168)$$

The natural choice for the bosonization of the Weyl fermions, therefore, is

$$\psi_L = \begin{pmatrix}\sigma\mu \\ \sigma\mu^\dagger\end{pmatrix}_L \qquad \psi_R = \begin{pmatrix}\sigma^\dagger\mu \\ \sigma^\dagger\mu^\dagger\end{pmatrix}_R.\qquad(10.169)$$

As before, the unphysical, ϵ-dependent, self-interaction terms appearing in the $\alpha\alpha$ and $\beta\beta$ terms may be removed by renormalizing, respectively, the operators σ and μ in the correlation functions. With the choice

$$\frac{a^2 + b^2}{2\pi^3} = 3,\qquad(10.170)$$

we reproduce the correlation functions (10.134) completely within the framework of the bosonic $B_{\mu\nu}$ field theory.

Observe that the $T_{\alpha\beta}$ vanishes in the local limit where the radius of the circle defining the operator $\sigma(C_x)$ is taken to zero. This fact has important consequences. If it would not vanish it would allow for the construction of local fields with generalized statistics in $D = 3+1$, which is not possible. Hence, the fact that the crossed $\alpha\beta$ term is zero is a clear manifestation of the fact that local fields in $D = 3 + 1$ can only be bosons or fermions.

The dual operators used in the bosonization of the Weyl fields satisfy the dual algebra (10.45). As we saw, in the local limit only fermion statistics is allowed, thus implying

$$ab = \pi. \tag{10.171}$$

This and (10.170) define the value of the parameters a and b.

10.8.3 Application: Left-Right Imbalance in Weyl Semimetals

An interesting application of the bosonization of Weyl fermions consists in the determination of the imbalance between the electronic population of the two chiralities, L and R. This follows directly from the bosonized form of the chiral currents (10.159), which is also related to the axial anomaly [76]. Indeed, from (10.159), it follows that

$$j_L^\mu - j_R^\mu = J^\mu + \frac{1}{4\pi^2}\epsilon^{\mu\nu\alpha\beta}A_\nu\partial_\alpha A_\beta, \tag{10.172}$$

where $\partial_\mu J^\mu = 0$. From this, we get

$$\partial_\mu\left[j_L^\mu - j_R^\mu\right] = \frac{1}{4\pi^2}\partial_\mu\epsilon^{\mu\nu\alpha\beta}A_\nu\partial_\alpha A_\beta = -\frac{1}{16\pi^2}F_{\mu\nu}\tilde{F}^{\mu\nu}, \tag{10.173}$$

which is precisely the axial anomaly.

Let $\Delta\rho = \rho_R - \rho_L$ be the particle density difference between the two chiralities. Then, the continuity equation tells us that

$$\frac{\partial\Delta\rho}{\partial t} = \nabla\cdot\left(\mathbf{j}_L - \mathbf{j}_R\right). \tag{10.174}$$

In the presence of a constant electric field $\mathbf{E}$ and a constant magnetic field $\mathbf{B}(\mathbf{A} = \frac{1}{2}\mathbf{r}\times\mathbf{B})$, we have

$$\nabla\cdot\left(\mathbf{j}_L - \mathbf{j}_R\right) = \frac{1}{4\pi^2}\nabla\cdot[\mathbf{r}(\mathbf{E}\cdot\mathbf{B}) - \mathbf{B}(\mathbf{E}\cdot\mathbf{r})] = \frac{1}{2\pi^2}\mathbf{E}\cdot\mathbf{B}, \tag{10.175}$$

and therefore we find the equation determining the time evolution of the difference of Weyl particle chiralities:

$$\frac{d\Delta\rho}{dt} = \frac{1}{2\pi^2}\mathbf{E}\cdot\mathbf{B}. \tag{10.176}$$

This can be improved by including the chirality backscattering. Assuming this process occurs at a time rate $1/\tau$, we have

$$\frac{d\Delta\rho}{dt} = \frac{1}{2\pi^2}\mathbf{E}\cdot\mathbf{B} - \frac{\Delta\rho}{\tau}. \tag{10.177}$$

This is solved by

$$\Delta\rho(t) = \frac{\tau}{2\pi^2}\mathbf{E}\cdot\mathbf{B}\left(1 - e^{-\frac{t}{\tau}}\right). \tag{10.178}$$

For large times, the imbalance between the two chiralities of Weyl fermions will stabilize at

$$\Delta\overline{\rho} = \frac{\tau}{2\pi^2}\mathbf{E}\cdot\mathbf{B}. \tag{10.179}$$

From this expression, different calculations have shown one can derive the magneto-conductance behavior observed in Weyl semimetals transport experiments [78, 164, 80, 73, 74].

10.8.4 Bosonization of Electrons

Once we have derived the bosonization of Weyl fermions, it is straightforward to obtain the bosonization of Dirac fermions. According to (10.147) and (10.169), we can write the Dirac field in the following bosonized form:

$$\Psi = \begin{pmatrix} \begin{pmatrix} \sigma\mu \\ \sigma\mu^\dagger \end{pmatrix}_L \\ \begin{pmatrix} \sigma^\dagger\mu \\ \sigma^\dagger\mu^\dagger \end{pmatrix}_R \end{pmatrix}, \tag{10.180}$$

where the operators in the first two rows are expressed in terms of $B_{\alpha\beta}^L$, while the ones in the last two, in terms of $B_{\alpha\beta}^R$.

The Dirac field kinetic term, then, according to (10.157), is bosonized as

$$i\overline{\Psi}\slashed{\partial}\Psi = \frac{1}{24}H_{\mu\nu\alpha}^R H_R^{\mu\nu\alpha} + \frac{1}{24}H_{\mu\nu\alpha}^L H_L^{\mu\nu\alpha}, \tag{10.181}$$

while the Dirac current is bosonized as

$$\overline{\Psi}\gamma^\mu\Psi = \frac{1}{2}\sqrt{\frac{\Box}{24\pi^2}}\epsilon^{\mu\nu\alpha\beta}\partial_\nu\left[B_{\alpha\beta}^L + B_{\alpha\beta}^R\right]. \tag{10.182}$$

The electron is associated to a massive Dirac field. A Dirac mass term is given by (10.183), in terms of the fields $\Psi_{L,R}$. Using the bosonization formula (10.169) and the expression for the dual operators (10.40), we obtain a bosonic form for the mass term that resembles the sine-Gordon Lagrangean, namely

$$\mathcal{L}_M = M\overline{\Psi}\Psi = M\cos\{F_L + F_R\}\cos\{G_L - G_R\}, \tag{10.183}$$

where F and G are the functionals of the $B_{\mu\nu}$ field, appearing, respectively, in the exponents of the σ and μ operators, given by (10.40).

It is very instructive to investigate how the bosonized field behaves under a gauge transformation of the bosonic gauge field, namely $B_{\mu\nu} \to B_{\mu\nu} + \partial_\mu\Lambda_\nu - \partial_\nu\Lambda_\mu$. We immediately see that the operator μ is gauge invariant, whereas

$$\sigma(S(C_{\mathbf{y}}), t) \to \sigma(S(C_{\mathbf{y}}), t)\exp\left\{-i\frac{b}{2\pi\rho}\oint_{C_{\mathbf{y}}}d\xi^i\Lambda_i(\xi, t)\right\}$$

$$\equiv \sigma(S(C_{\mathbf{y}}), t)e^{-i\varphi(y)}$$

$$\varphi(y) = \frac{b}{2\pi\rho} \oint_{C_{\mathbf{y}}} d\xi^i \Lambda_i(\xi, t). \tag{10.184}$$

We see that a gauge transformation of the bosonic tensor field emerges as a U(1) gauge transformation of the Weyl fermions Ψ_L and Ψ_R. The U(1) gauge transformation of a Dirac field such as the electron field for instance, according to (10.169) and (10.147), would be obtained by simultaneous gauge transformations of the chiral bosonic tensor fields $B^L_{\mu\nu}$ and $B^L_{\mu\nu}$ with opposite gauge parameters $\Lambda^R_\mu = -\Lambda^L_\mu \equiv \Lambda_\mu$. Then, the U(1) transformation of a Dirac field is such that

$$\psi \to e^{i\varphi(x)}\psi$$

$$\varphi(x) = b \oint_{C_{\mathbf{x}}} d\xi^i \Lambda_i(\xi, t) = b\Phi_\Lambda(x). \tag{10.185}$$

The phase of the U(1) transformation on the Dirac field is the flux of $\mathcal{B} = \nabla \times \Lambda$ through the surface delimited by the curve $C_{\mathbf{x}}$.

11

Statistical Transmutation

We have seen that fields with arbitrary statistics may exist in D = 2, 3 and 4 spacetime dimensions. In D = 4, however, this is only possible for nonlocal fields, namely, field operators creating extended objects, such as strings and n-branes. In D = 2, 3, conversely, this is allowed for local fields, which are associated to point particles. Through the order-disorder duality procedure, we have been able to build such fields with arbitrary statistics out of bosonic fields. This method, known as bosonization in the case of fermion fields, consists in combining operator pairs each one of them carrying, respectively, the charge and the topological charge of the bosonic field) into a composite field that will replace the fermionic or anyonic field. The statistics of the latter is determined by the product of charge × topological charge of the bosonic field. In this chapter, we explore a general method, related to bosonization and known as statistical transmutation, by means of which we may continuously change the statistics of a field, and consequently of the objects it creates, in D = 3, 4. This is achieved by a mechanism, working at the Lagrangean level, through which a certain amount of topological charge is imparted on charged fields and on the objects they create. The resulting objects, consequently, change their amount of the product charge × topological charge, thereby modifying their statistics. We shall see that both bosonization and statistical transmutation have many interesting applications in condensed matter physics.

11.1 Generalized BF Theories

In the previous chapter, we introduced the topological current, in D = 2, 3, 4, expressed in terms of a generalized field $B_{ij\ldots}$ (10.13), which was, respectively, a scalar, a vector and a rank-2 tensor (10.14) and (10.15). There, we have shown that $B_{ij\ldots}$ is the bosonic field used both in the bosonization process and also for constructing fields with generalized statistics.

We now introduce a field theory involving the $B_{ij\ldots}$ field, coupled to a vector U(1) abelian gauge field A_μ, where the topological charge of the former is the source of the latter. The magnetic field of the latter, conversely, is the "charge," acting as the source of the former. This kind of theory is known as the BF-theory.

The generalized BF-theories are obtained by coupling, in each case, the proper topological current to an abelian vector gauge field A_μ, namely

$$\mathcal{L}_{2BF} = \frac{1}{2} H_\mu H^\mu - J_2^\mu A_\mu - \frac{1}{4} F^{\mu\nu} F_{\mu\nu}$$

$$\mathcal{L}_{3BF} = -\frac{1}{4} H_{\mu\nu} H^{\mu\nu} - J_3^\mu A_\mu - \frac{1}{4} F^{\mu\nu} F_{\mu\nu}$$

$$\mathcal{L}_{4BF} = \frac{1}{12} H_{\mu\nu\alpha} H^{\mu\nu\alpha} - J_4^\mu A_\mu - \frac{1}{4} F^{\mu\nu} F_{\mu\nu}, \qquad (11.1)$$

where J_D^μ is given by (10.13) and $F_{\mu\nu}$ is the field intensity tensor of A_μ. It follows that $\partial_\nu F^{\mu\nu} = J_D^\mu$, hence the topological charge is the source of the A_μ vector gauge field.

The generalized $B_{ij\ldots}$ field equations are

$$\partial_\mu H^\mu = \Box \phi = \epsilon^{\mu\nu} \partial_\mu A_\nu$$

$$\partial_\nu H^{\mu\nu} = \epsilon^{\mu\alpha\beta} \partial_\alpha A_\beta$$

$$\partial_\alpha H^{\mu\nu\alpha} = \epsilon^{\mu\nu\alpha\beta} \partial_\alpha A_\beta \qquad (11.2)$$

Taking the zeroth component of these field equations in D = 3, 4, we obtain, respectively,

$$\partial_j H^{0j} = \epsilon^{ij} \partial_i A_j = \mathcal{B}$$

$$\partial_j H^{0ij} = \epsilon^{ijk} \partial_j A_k = \mathcal{B}^i. \qquad (11.3)$$

Here, $\mathcal{B}$ is the magnetic field in D = 3, or magnetic flux density, on the plane. Observe that it is a scalar. $\mathcal{B}^i$, by its turn, is the magnetic field in D = 4, which is a vector. Notice that $\mathcal{B}$ and $\mathcal{B}^i$ are, respectively, the generalized "charge" densities that effectively act as the sources of the B_μ and $B_{\mu\nu}$ fields in each case.

11.2 The Chern–Simons Theory

11.2.1 Generalized Chern–Simons Theories in $D \geq 4$

Let us introduce now a peculiar class of gauge field theories involving the U(1) abelian vector gauge field A_μ and the generalized $B_{\mu\nu\ldots}$ tensor field. The Lagrangean is such that when we couple the current associated to charged point particles to the U(1) vector field, a certain amount of topological charge is attached to these particles. Conversely, when we couple to the $B_{\mu\nu\ldots}$ tensor field the current corresponding to the appropriate extended object (string in D = 4, …), the magnetic field of the abelian vector field is attached to this extended object.

The Lagrangean density of a theory presenting this property in a spacetime of $D \geq 4$ dimensions is given, in general, by

$$\mathcal{L}_{D,CS} = \theta J_D^\mu A_\mu - j^\mu A_\mu - j^{\mu\nu\cdots} B_{\mu\nu\ldots}, \tag{11.4}$$

where θ is an arbitrary real parameter, J_D^μ is the topological current in $D \geq 4$ dimensions, given by (10.13), and j^μ and $j^{\mu\nu\cdots}$ are the current densities associated, respectively, to point particles and extended objects such as strings, given by (10.22).

We have seen in Section 10.2, from the operator point of view, in D = 4, Eqs. (10.40)–(10.45), that the composite object consisting of a closed string carrying a magnetic flux and a point particle carrying the topological charge will possess generalized statistics in D = 4. As we will see, there exists a Lagrangean mechanism corresponding to this, which we now describe.

For D = 4, we would have

$$\mathcal{L}_{4,CS} = \theta \epsilon^{\mu\nu\alpha\beta} A_\mu \partial_\nu B_{\alpha\beta} - j^\mu A_\mu - j^{\mu\nu} B_{\mu\nu}. \tag{11.5}$$

By varying with respect to A_μ, we can see that the particle matter current is identified with the topological current J_D^μ. Conversely, varying with respect to $B_{\mu\nu}$, we infer that

$$j^{\mu\nu} = \theta \epsilon^{\mu\nu\alpha\beta} \partial_\alpha A_\beta. \tag{11.6}$$

The string density in the i-direction is obtained by taking $\mu = 0$ component, namely

$$j^{0i} = \theta \epsilon^{ijk} \partial_j A_k = \mathcal{B}^i, \tag{11.7}$$

and we see that a magnetic field $\mathcal{B}^i$ is attached to the string along its length.

The generalized Chern–Simons theory appropriate for a spacetime with D = 4 dimensions is given by (11.5). It provides a mechanism leading to the same effect already studied before employing fully quantized operators.

Indeed, from Eqs. (11.5)–(11.7), we see that a magnetic field is attached along the closed string, while a topological charge is attached to the particle associated to the current density j_μ. The composite object formed by the point particle and closed string, consequently, will have arbitrary statistics proportional to the product of topological charge and magnetic flux [81, 85, 86, 87].

11.2.2 Chern–Simons Theory: The D = 3 Case

The case of a spacetime with D = 3 is very special because the B-field is a vector and we can make it equal to A_μ, thereby obtaining the so-called Chern–Simons theory [6], namely

$$\mathcal{L}_{3,CS} = \frac{\theta}{2}\epsilon^{\mu\alpha\beta}A_\mu\partial_\alpha A_\beta - qj^\mu A_\mu. \tag{11.8}$$

The field equation deriving from this Lagrangean is

$$qj^\mu = \theta\epsilon^{\mu\alpha\beta}\partial_\alpha A_\beta, \tag{11.9}$$

hence for a static point particle, for which $j^0 = \delta(\mathbf{r})$, the field equation appends a point magnetic field

$$\mathcal{B} = \epsilon^{ij}\partial_i A_j = \frac{1}{\theta}\delta(\mathbf{r}) \tag{11.10}$$

to the charged particles interacting with the Chern–Simons field A_μ. A magnetic flux $\Phi = \frac{q}{\theta}$ is, therefore, attached to the charged point particle.

We have shown, in the process of building composite fields out of dual bosonic ones, that the statistics of the resulting field is $s = \frac{ab}{2\pi}$ where a is the topological charge and b the generalized "charge" carried by this bosonic field. Hence, we may expect that in the case of Chern–Simons theory the charged particles coupled to the gauge field will undergo a change of spin-statistics (statistical transmutation),

$$\Delta s = \frac{Qq}{2\pi\theta},$$

where Q is the total charge. From (11.9) we see that the total current coupled to A_μ is

$$j^\mu_{\text{TOT}} = qj^\mu - \frac{\theta}{2}\epsilon^{\mu\alpha\beta}\partial_\alpha A_\beta,$$

hence, using (11.9), we have that

$$j^\mu_{\text{TOT}} = \frac{q}{2}j^\mu, \tag{11.11}$$

implying that the total charge will be

$$Q = \int d^2x\, j^\mu_{\text{TOT}} = \frac{q}{2}. \tag{11.12}$$

It follows that the change in spin/statistics will be

$$\Delta s = \frac{q^2}{4\pi\theta}. \tag{11.13}$$

We will see in the next section that this is indeed the case.

11.3 Statistical Transmutation in D = 3

We examine in this section, from a general point of view, how the spin/statistics of charged particles in two-dimensional space, is affected by attaching to them a point magnetic flux Φ.

Consider the initial state-vector of a two-particles system $|\Psi(t_0)\rangle$. It evolves, at subsequent times, as

$$|\Psi(t)\rangle = \exp\left\{-\frac{i}{\hbar}H(t - t_0)\right\}|\Psi(t_0)\rangle, \tag{11.14}$$

where H is the Hamiltonian of the system. Choosing the coordinate representation, the wave function for $t > t_0$ factorizes as $\Psi_{CM}(\mathbf{R})\Psi(\mathbf{r}, t)$, where $\mathbf{R}$ and $\mathbf{r}$ are, respectively, the center-of-mass and relative coordinates. The latter evolves in time as a one-particle wave-function, namely

$$\Psi(\mathbf{r}, t) = \int d^2r'\, G(\mathbf{r}, t; \mathbf{r}', t_0)\Psi(\mathbf{r}', t_0). \tag{11.15}$$

The Green function, which is the coordinate representation of the unitary time-evolution operator, is given in the Feynman formulation by the one-particle quantum-mechanics version of (5.63), namely

$$G(\mathbf{r}, t; \mathbf{r}', t_0) = \int D\mathbf{z}\, \exp\left\{\frac{i}{\hbar}S[\mathbf{z}]\right\}\Big|_{\mathbf{z}(t_0)=\mathbf{r}'}^{\mathbf{z}(t)=\mathbf{r}}, \tag{11.16}$$

where

$$S[\mathbf{z}] = \int_{t_0}^t L(\mathbf{z}, \dot{\mathbf{z}}). \tag{11.17}$$

Now, attaching a point magnetic flux Φ to the particle described by the Lagrangean L amounts to adding to this a piece

$$L_\Phi(\mathbf{z}, \dot{\mathbf{z}}) = \int d^2r\, \mathbf{j} \cdot \mathbf{A}_\Phi, \tag{11.18}$$

where

$$\mathbf{j} = q\dot{\mathbf{r}}\delta(\mathbf{r} - \mathbf{z}) \qquad \dot{\mathbf{r}} = \dot{r}\hat{r} + r\dot{\varphi}\hat{\varphi}$$
$$\mathbf{A}_\Phi = \frac{\Phi}{2\pi}\nabla \arg(\mathbf{r}) \qquad \nabla \arg(\mathbf{r}) = \frac{1}{r}\hat{\varphi}. \tag{11.19}$$

Inserting (11.19) in (11.18), we get

$$L_\Phi(\mathbf{z}, \dot{\mathbf{z}}) = \left(\frac{q\Phi}{2\pi}\right)\frac{d}{dt}\arg(\mathbf{z}) \tag{11.20}$$

Now, adding L_Φ to L in (11.17) and the resulting action in (11.16), we conclude that by attaching a point magnetic flux to a charged particle, the corresponding Green function is modified as

$$G(\mathbf{r}, t; \mathbf{r}', t_0) \longrightarrow \exp\left\{is\arg(\mathbf{r}) - is\arg(\mathbf{r}')\right\}G(\mathbf{r}, t; \mathbf{r}', t_0), \tag{11.21}$$

where $s = \frac{q\Phi}{2\pi}$.

Using the modified Green function in (11.15), we immediately see that under a reflection $\mathbf{r} \to -\mathbf{r}$, which corresponds to the exchange of identical particles, the new wave-function acquires a phase $e^{is\pi}$, because of (10.9). The corresponding change in spin/statistics is, therefore, $\frac{q\Phi}{4\pi}$, as anticipated.

We see that the Chern–Simons theory produces, through a Lagrangean mechanism, the composite state of charge and magnetic flux, which exhibits arbitrary spin/statistics. The corresponding operator mechanism was described in the sequence (10.35)–(10.39) of Section 10.2.

11.4 Topological Aspects of the Chern–Simons Theory in D = 3

The Chern–Simons Lagrangean in D = 3, as well as its $D \geq 4$ generalizations and the BF coupling in (11.1), possess a peculiar common feature, namely, they do not depend on the metric tensor of the manifold on which the theory is defined. This remarkable property has far-reaching consequences. The physical properties of the theory are insensitive to the geometric details of the manifold; rather, they depend on the topology thereof. For this reason, such theories are known as topological theories.

A striking property of this class of theories is that the energy-momentum tensor, which can be defined as

$$T^{\mu\nu} = \frac{\delta}{\delta g_{\mu\nu}} \int d^D x \mathcal{L},$$

(11.22)

identically vanishes because of the metric independence of the action. It follows that the Hamiltonian of a topological theory vanishes, namely

$$H = \int d^{D-1} x T^{00} = 0.$$

(11.23)

The zero-energy ground states are degenerate and the degree of degeneracy turns out to be determined by the topology of the manifold where the CS theory is defined. For the Chern–Simons theory in D = 3, it can be shown that for a manifold of genus g, the ground-state degeneracy is k^g, where k, known as the level of the theory, is related to the parameter θ as

$$\theta = \frac{k}{2\pi}.$$

(11.24)

Notice that for a level k, the point particle coupled to the Chern–Simons theory will undergo a change in statistics given by $\Delta s = \frac{1}{k}$.

We can show that, as a consequence of gauge invariance, the level of a Chern–Simons theory is an integer. For this purpose, choose the spatial components of the gauge field as the ones producing one unit of magnetic flux along the manifold M_2, such that the CS action is

$$S_{CS} = \frac{k}{4\pi} \int_M d^3x \epsilon^{\mu\alpha\beta} A_\mu \partial_\alpha A_\beta \qquad (11.25)$$

with $M = M_2 \otimes S_1$. The variation produced by a gauge transformation in this is

$$\Delta S_{CS} = \frac{k}{2\pi} \int_M d^3x \epsilon^{\mu\alpha\beta} \delta A_\mu \partial_\alpha A_\beta, \qquad (11.26)$$

where $\delta A_\mu = \partial_\mu \Lambda$. Then we get

$$\Delta S_{CS} = \frac{k}{2\pi} \int_M d^3x \partial_\mu \left[\Lambda J^\mu \right], \qquad (11.27)$$

where $J^\mu = \epsilon^{\mu\alpha\beta} \partial_\alpha A_\beta$. From this we obtain $\Delta S_{CS} = \frac{k}{2\pi} 2\pi [\Lambda(2\pi) - \Lambda(0)]$, where we used the fact that in the units we are using the magnetic flux quantum is $\phi_0 = 2\pi$. Now, we must have $[\Lambda(2\pi) - \Lambda(0)] = 2\pi m$, $m \in \mathbb{Z}$, for the corresponding gauge transformation of a matter field coupled to the CS field ought to be univalent. This yields $\Delta S_{CS} = 2\pi m k$. Finally, gauge invariance of the functional integral defining the quantum CS theory requires $\Delta S_{CS} = 2\pi n$, $n \in \mathbb{Z}$, thus implying the level k is an integer. This result is valid in general.

12

Pseudo Quantum Electrodynamics

All evidences indicate that we live in a three-dimensional space. Should there be extra dimensions, they must be curled in tiny coils in such a way that we normally would not perceive their existence. Consider, for instance, a one-meter-long tube of micrometric radius. This would resemble a one-dimensional structure. By the same token, there are systems in condensed matter that for all purposes behave as if they had a dimension lower than three, because some degrees of freedom are just not available. Systems in this category include polymers such as polyacetylene and one-atom-wide materials such as graphene that are, respectively one- and two-dimensional. For most planar materials, however, the electromagnetic field by which the corresponding particles interact is not confined to the plane. Consequently, such interaction is ruled by three-dimensional, rather than by planar Maxwell theory, despite the fact that kinematics is two-dimensional. Nevertheless, both for practical and esthetical reasons, it would be highly desirable to have available a full two-dimensional theory that, yet, would describe the genuine three-dimensional electromagnetic interaction of the particles confined to a plane. Such a theory, called pseudo quantum electrodynamics (PQED), is the subject of this chapter.

12.1 Electrodynamics of Particles Confined to a Plane

Consider a general system of charged particles described by the QED Lagrangean, which can be written as $\mathcal{L}_{QED}[A_\mu, j^\mu_{3+1}] + \mathcal{L}_M$, where

$$\mathcal{L}_{QED}[A_\mu, j^\mu_{3+1}] = -\frac{1}{4}F^{\mu\nu}F_{\mu\nu} - ej^\mu_{3+1}A_\mu, \qquad (12.1)$$

$\mathcal{L}_M$ is the matter Lagrangean and j^μ_{3+1} is the matter current density in three-dimensional space.

Performing the quadratic functional integral over the electromagnetic field, with the help of an appropriate gauge-fixing term [89], we obtain the effective functional of the matter current, namely,

$$
\begin{aligned}
Z[j^\mu_{3+1}] &= Z_0^{-1} \int DA_\mu \exp\left\{-\int d^4x \mathcal{L}_{QED}[A_\mu, j^\mu_{3+1}]\right\} \\
&= \exp\left\{\frac{e^2}{2}\int d^4x d^4y\, j^\mu_{3+1}(x)\frac{1}{-\Box}j^\mu_{3+1}(y)\right\} \\
&= \exp\left\{S_{eff}[j^\mu_{3+1}]\right\},
\end{aligned}
\tag{12.2}
$$

where we used the Green function of the A_μ field, namely

$$
G^{\mu\nu}(k) = \frac{\delta^{\mu\nu}}{k^2} + \text{g.t.},
\tag{12.3}
$$

"g.t." standing for "gauge dependent terms," which do not contribute.

Let us assume now the matter particles are constrained to move on a plane. In this case the current density has the form

$$
j^\mu_{3+1} = \begin{cases} j^\mu(x^0, \mathbf{r})\,\delta(z) & ; \quad \mu = 0, 1, 2 \\ 0 & ; \quad \mu = 3 \end{cases},
\tag{12.4}
$$

where $\mathbf{r} = (x, y)$ is the position vector on the plane.

Inserting (12.4) in (12.2) and integrating over the z-coordinates, we get

$$
\begin{aligned}
S_{eff}[j^\mu_{3+1}] = S_{eff}[j^\mu] = \frac{e^2}{2}\int d^2r d^2r' d\tau d\tau' \\
\times j^\mu(\mathbf{r}, \tau)\left[\int \frac{d\omega d^2k dk_3}{(2\pi)^4}\frac{e^{i[\mathbf{k}\cdot(\mathbf{r}-\mathbf{r}')-\omega(\tau-\tau')]}}{\omega^2 + \mathbf{k}^2 + k_3^2}\right] j^\mu(\mathbf{r}', \tau').
\end{aligned}
\tag{12.5}
$$

Now, integrating over k_3, we obtain

$$
\begin{aligned}
S_{eff}[j^\mu] = \frac{e^2}{2}\int d^2r d^2r' d\tau d\tau' \\
\times \frac{1}{2}j^\mu(\mathbf{r}, \tau)\left[\int \frac{d\omega d^2k}{(2\pi)^3}\frac{e^{i[\mathbf{k}\cdot(\mathbf{r}-\mathbf{r}')-\omega(\tau-\tau')]}}{[\omega^2 + \mathbf{k}^2]^{1/2}}\right] j^\mu(\mathbf{r}', \tau'),
\end{aligned}
\tag{12.6}
$$

which is completely in D = 2+1.

It is not difficult to realize that the above effective action, which was obtained by integrating over the photon momenta out of the plane, can be derived directly from the full 2+1-dimensional gauge theory [89]

$$
\mathcal{L}_{PQED} = -\frac{1}{4}F^{\mu\nu}\left[\frac{2}{\sqrt{\Box}}\right]F_{\mu\nu} - ej^\mu A_\mu + \mathcal{L}_M,
\tag{12.7}
$$

which we call pseudo quantum electrodynamics.

In the next sections, we explore some of the beautiful properties of PQED, in particular the ones that make it so useful in the realm of planar condensed matter systems.

12.2 Coulomb Potential

Let us determine what is the interaction energy, according to PQED, for a pair of static point charges located at $\mathbf{x}$ and $\mathbf{y}$ in $D = 2+1$. For this purpose, consider the corresponding current density

$$j^\mu(\mathbf{r}, t) = \begin{cases} \delta(\mathbf{r} - \mathbf{x}) + \delta(\mathbf{r} - \mathbf{y}) & ; \quad \mu = 0 \\ 0 & ; \quad \mu = 1, 2 \end{cases}. \tag{12.8}$$

The interaction energy of a charged matter distribution associated with the current density $j^\mu(\mathbf{r}, t)$, and interacting through a gauge field having a propagator $G^{\mu\nu}(\omega, \mathbf{k})$ is given, for a static configuration such as (12.8), by

$$E = \frac{e}{2} \int d^2r j^\mu(\mathbf{r}) A_\mu(\mathbf{r}), \tag{12.9}$$

where A_μ is the field created by the matter distribution $j^\mu(\mathbf{r}, t)$.

Expressing the field in terms of the sources, by means of the propagator, we may write

$$E = \frac{e^2}{2} \int d^2r d^2r' dt'$$
$$\times j^\mu(\mathbf{r}) \left[\int \frac{d^2k}{(2\pi)^2} \int \frac{d\omega}{2\pi} e^{i[\mathbf{k}\cdot(\mathbf{r}-\mathbf{r}')-\omega(t-t')]} G^{\mu\nu}(\omega, \mathbf{k}) \right] j^\nu(\mathbf{r}'). \tag{12.10}$$

Integrating over t' and ω, we obtain

$$E = \frac{e^2}{2} \int d^2r d^2r' j^\mu(\mathbf{r}) \left[\int \frac{d^2k}{(2\pi)^2} e^{i\mathbf{k}\cdot(\mathbf{r}-\mathbf{r}')} G^{\mu\nu}(\omega = 0, \mathbf{k}) \right] j^\mu(\mathbf{r}'). \tag{12.11}$$

The potential interaction energy between static point particles then, is given by

$$E = \frac{e^2}{2} \int d^2r d^2r' \rho(\mathbf{r}) \left[\int \frac{d^2k}{(2\pi)^2} e^{i\mathbf{k}\cdot(\mathbf{r}-\mathbf{r}')} G^{00}(\omega = 0, \mathbf{k}) \right] \rho(\mathbf{r}'). \tag{12.12}$$

The free PQED propagator, in momentum space, is given in a transverse gauge by

$$G^{\mu\nu}(\omega, \mathbf{k}) = \frac{P^{\mu\nu}}{2\left[-\omega^2 + \mathbf{k}^2\right]^{1/2}}, \tag{12.13}$$

where

$$P^{\mu\nu} = \delta^{\mu\nu} - \frac{k^\mu k^\nu}{k^2} \tag{12.14}$$

is the transverse projector.

Then, inserting (12.8) in (12.12), neglecting the unphysical self-interaction terms and using (12.13), we obtain

$$E = \frac{e^2}{2} \int \frac{d^2k}{(2\pi)^2} \frac{e^{i\mathbf{k}\cdot(\mathbf{x}-\mathbf{y})}}{\sqrt{\mathbf{k}^2}}$$

$$E = \frac{e^2}{4\pi |\mathbf{x} - \mathbf{y}|}, \tag{12.15}$$

which is the familiar $1/r$ Coulomb potential of QED$_4$. Should we perform the same calculation with QED$_3$, we would have instead

$$E = -\frac{e^2}{2\pi} \ln |\mathbf{x} - \mathbf{y}|. \tag{12.16}$$

12.3 Green Functions

Let us consider here the different Green functions we obtain for PQED in coordinate space. We start from the Green function in energy-momentum space corresponding to (12.7). Choosing a transverse gauge, this is given by (12.13).

The corresponding Green function in coordinate space can be written as

$$G^{\mu\nu}(t, \mathbf{r}) = \frac{1}{2} P^{\mu\nu} D(t, \mathbf{r}), \tag{12.17}$$

where $P^{\mu\nu}$ here is meant to be in coordinate space, where $k^\mu \to i\partial^\mu$.

Now, the function $D(x)$ will strongly depend on the prescription used in order to handle the singularities occurring in (12, 13). The advanced and retarded functions, for instance, are given by

$$D_\pm(t, \mathbf{r}) = \int \frac{d^3k}{(2\pi)^3} \frac{e^{ik\cdot x}}{[(k^0 \pm i\epsilon)^2 - \mathbf{k}^2]^{1/2}}, \tag{12.18}$$

D_+ being the retarded function and D_-, the advanced one. Evaluating the integrals [90], we get

$$D_\pm(t, \mathbf{r}) = \frac{1}{2\pi^2} \theta(\pm t) \left[\frac{1}{[t^2 - \mathbf{r}^2 + i\epsilon]} - \frac{1}{[t^2 - \mathbf{r}^2 - i\epsilon]} \right]$$

$$= D_\pm(t, \mathbf{r}) = -\frac{1}{\pi^2} \theta(\pm t) \frac{i\epsilon}{(t^2 - \mathbf{r}^2)^2 + \epsilon^2}. \tag{12.19}$$

Let us consider now the Feynman Green function. This is defined by the prescription

$$D_F(t, \mathbf{r}) = \int \frac{d^3k}{(2\pi)^3} \frac{e^{ik\cdot x}}{[k_0^2 - \mathbf{k}^2 + i\epsilon]^{1/2}}. \tag{12.20}$$

The integrals have been evaluated in [90], yielding the result

$$D_F(t, \mathbf{r}) = \frac{1}{2\pi^2} \frac{1}{[t^2 - \mathbf{r}^2 + i\epsilon]}. \tag{12.21}$$

Notice the interesting property that the coordinate space form of the Feynman Green function of PQED coincides with the energy-momentum space form of the Feynman Green function of QED_3 and vice versa.

12.4 Scale Invariance

PQED has no dimensionful parameters, differently from QED_3 and similarly to QED_4. This property has interesting consequences when we couple massless fermions to PQED. Due the absence of any parameter with dimension of length, it follows that the corrected gauge field propagator is given, at a certain order, by

$$D_F^{\mu\nu}(k_0, \mathbf{k}) = \frac{AP^{\mu\nu} + B\epsilon^{\mu\nu\alpha}\frac{k_\alpha}{\sqrt{k^2}}}{[k_0^2 - \mathbf{k}^2 + i\epsilon]^{1/2}}, \tag{12.22}$$

where A and B are constants. Corrections due to the interaction at any order will just modify the constants A and B, with no further modifications in the propagator.

12.5 Causality: Huygens Principle

The Lagrangean of PQED is nonlocal in space and time, as it usually happens in systems where we integrate out part of the degrees of freedom, thereby trading their influence by some effective Lagrangean. A well-known example is the Caldeira–Leggett quantum-mechanical dissipative system [271].

As a consequence of its nonlocality, one may wonder as to whether PQED respects causality or not. A necessary and sufficient condition for this is that the relevant Green functions should vanish outside the light-cones. A related (but more stringent) issue is whether the system complies with the Huygens principle. The condition in this case is that the Green functions should vanish both outside and inside the light-cones, thus having support on their surface.

In order to address these problems, let us examine the retarded and advanced functions, $D_\pm(t, \mathbf{r})$, given by (12.19). If we take the limit $\epsilon \to 0$, we obtain

$$D_\pm(t, \mathbf{r}) = -\frac{1}{\pi^2}\theta(\pm t)\frac{i\epsilon}{(t^2 - \mathbf{r}^2)^2 + \epsilon^2} \xrightarrow{\epsilon \to 0} -\frac{i}{\pi}\theta(\pm t)\delta(t^2 - \mathbf{r}^2). \tag{12.23}$$

This result shows that both the retarded and advanced Green functions of PQED have support on the light-cone surfaces, respectively, at $t > 0$ and $t < 0$. It becomes, therefore, evident that the theory does respect causality.

The delta function in (12.23), furthermore, implies that PQED satisfies the Huygens principle, similarly to QED$_4$. Conversely, we would like to remark that the corresponding Green functions of QED$_3$ have support inside the whole light-cones, which means that, despite respecting causality, QED$_3$ does not satisfy the Huygens principle.

12.6 Unitarity

Unitarity is a basic consistency condition of any quantum theory. This assertion derives from the fact that a unitary time-evolution operator preserves the norm of the state vector, which, by its turn, coincides with the sum of the probabilities for the possible outcomes of measurements made, if any observable quantity. Any sensible probabilistic interpretation of a quantum theory requires this to be equal to one at any time.

Unitarity of the time-evolution operator implies the unitarity of the S-matrix, which has its elements defined by

$$S_{\alpha\beta} = \lim_{t_0 \to -\infty} \lim_{t \to +\infty} \langle \alpha | U(t, t_0) | \beta \rangle, \qquad (12.24)$$

where $U(t, t_0)$ is the time-evolution operator and $|\alpha\rangle, |\beta\rangle \in \{|\alpha\rangle\}$, a complete set of free asymptotic states. Knowledge of the S-matrix conveys the probability amplitude for a given initially free state $|\beta\rangle$ to be scattered, a long time later, and after undergoing interaction, into another, possibly different free state $|\alpha\rangle$. It is, therefore, a crucial element in the determination of the scattering cross-section in a scattering process.

Needless to say, scattering experiments have been playing a central role in physics since the discovery of the atomic nucleus by Rutherford, to the experimental observation of quarks in the SLAC, and going through the determination of crystalline structures in experiments involving x-ray scattering by solids.

Determination of unitarity of a quantum theory is, therefore, an issue of utmost importance and it would be highly desirable to have available a practical and efficient method to test it. This is what we derive below.

For this purpose, let us write the S-matrix in the form $\mathbb{S} = \mathbb{I} + i\mathbb{T}$, which follows from the form of the $U(t, t_0)$ operator itself. Unitarity of $\mathbb{S}$, namely, $\mathbb{S}^\dagger \mathbb{S} = \mathbb{I}$, then implies

$$i[\mathbb{T}^\dagger - \mathbb{T}] = \mathbb{T}^\dagger \mathbb{T}. \qquad (12.25)$$

Taking the $\mathbb{T}_{\alpha\alpha} = \langle \alpha | T | \alpha \rangle$ element of the above equation, we have

$$2 \operatorname{Im} \mathbb{T}_{\alpha\alpha} = \sum_\beta \mathbb{T}_{\alpha\beta}^\dagger \mathbb{T}_{\beta\alpha} = \sum_\beta \mathbb{T}_{\beta\alpha}^* \mathbb{T}_{\beta\alpha}. \qquad (12.26)$$

This relation, known as the Optical Theorem, is a necessary and sufficient condition for the theory to be unitary. It is, therefore, a practical and efficient method to test unitarity of a given system.

We now choose the states $|\alpha\rangle$ as energy-momentum eigenstates and introduce the $\mathcal{M}$ matrix through

$$\mathbb{T}_{\alpha\beta} \equiv (2\pi)^3 \, \delta(p_\alpha^\mu - p_\beta^\mu) \, \mathcal{M}_{\alpha\beta}. \tag{12.27}$$

It follows from the LSZ formula [92], which relates the S-matrix elements to the corresponding quantum field correlation functions, that $\mathcal{M}_{\alpha\beta}(p_\alpha^\mu; p_\beta^\nu)$ is given by the corresponding Feynman graphs in energy-momentum space, with all external legs removed and calculated at the external energy-momenta that correspond to the asymptotic energy-momenta p_α^μ of the associated S-matrix element. The element $\mathbb{T}_{\alpha\alpha}$, for instance, apart from a delta function, becomes the propagator

$$\mathbb{T}_{\alpha\alpha} \equiv (2\pi)^3 \, \delta(p_\alpha^\mu - p_\alpha^\mu) \, G^{\mu\nu}(p_\alpha^0, \mathbf{p}_\alpha). \tag{12.28}$$

Then, using (12, 13) and neglecting an overall delta function factor, we can write (12.26) as [93]

$$2 \, \mathrm{Im} \, \frac{1}{\sqrt{p_0^2 - \mathbf{p}^2 + i\epsilon}}$$
$$= \frac{1}{4} \int d\Phi \int d^3k \, \frac{i}{\sqrt{k_0^2 - \mathbf{k}^2 + i\epsilon}} \, \frac{i}{\sqrt{(p_0 - k_0)^2 - (\mathbf{p} - \mathbf{k})^2 + i\epsilon}}, \tag{12.29}$$

where we used the fact that the transverse projector satisfies $P^2 = P$. The integral over Φ is a phase space factor needed to ensure that the completeness of the intermediate states has been taken properly.

We now use the fact that the Fourier transform of a product is a convolution in order to transform the above expression back to coordinate space. Then, using (12.20) and (12.21), we get

$$\mathrm{Im} \, \frac{1}{t^2 - \mathbf{r}^2 + i\epsilon} = -\frac{\mathcal{T}^2}{8} \, \frac{1}{[t^2 - \mathbf{r}^2 - i\epsilon]} \, \frac{1}{[t^2 - \mathbf{r}^2 + i\epsilon]}$$
$$\frac{\epsilon}{(t^2 - \mathbf{r}^2)^2 + \epsilon^2} = \frac{\mathcal{T}^2}{8} \, \frac{1}{(t^2 - \mathbf{r}^2)^2 + \epsilon^2}, \tag{12.30}$$

where $\mathcal{T}$ is the resulting phase space factor. Notice that it must have the dimension of time in order to guarantee that the above equation is dimensionally correct. We immediately see that with the choice $\mathcal{T}^2 = 8\epsilon$, PQED satisfies the Optical Theorem.

The above demonstration was made using the free propagator. In the case of interaction with massless Dirac fermions, we will have, according to (12.22),

$$P^{\mu\nu} \longrightarrow A P^{\mu\nu} + B \epsilon^{\mu\nu\alpha} \frac{k_\alpha}{\sqrt{k^2}}$$

$$(P^2)^{\mu\nu} \longrightarrow (A^2 - B^2) P^{\mu\nu} + 2AB \epsilon^{\mu\nu\alpha} \frac{k_\alpha}{\sqrt{k^2}}. \tag{12.31}$$

It is not difficult to see that now the corrected propagator will satisfy the Optical Theorem, provided we choose the regulators as

$$\mathcal{T}^2 = 8 \frac{A}{A^2 - B^2} \epsilon \tag{12.32}$$

for the A-term of the propagator and

$$\mathcal{T}^2 = 8 \frac{1}{2A} \epsilon' \tag{12.33}$$

for the B-term.

For PQED with massless Dirac fermions, therefore, we have an exact demonstration of unitarity, achieved with the help of the Optical Theorem.

12.7 Screening

12.7.1 Coupling to a Higgs Field

It is instructive to inquire about the behavior of the static potential in PQED, when we couple the gauge field A_μ to a complex scalar field possessing a nonzero vacuum expectation value. In QED, this mechanism, known as the Anderson–Higgs mechanism, leads to a mass term for the gauge field, which has the consequence of modifying the large-distance behavior of the static potential from $1/r$ to an exponential decay.

Let us choose the matter Lagrangean $\mathcal{L}_M$ in such a way that

$$\mathcal{L}_{PQED} = -\frac{1}{4} F^{\mu\nu} \left[\frac{2}{\sqrt{\Box}} \right] F_{\mu\nu} - e j^\mu A_\mu + (D^\mu \phi)^* D_\mu \phi + \mu |\phi|^2 - \lambda |\phi|^4, \tag{12.34}$$

where $D_\mu = \partial_\mu + i A_\mu$. The potential has nontrivial minima at $|\phi_0| = \frac{\mu}{2\lambda}$. Shifting the field around the minimum generates a "mass term" for A_μ, with $M = 2|\phi_0|$.

The A_μ-field propagator now becomes

$$G^{\mu\nu}(\omega, \mathbf{k}) = \frac{P^{\mu\nu}}{2 \left[[-\omega^2 + \mathbf{k}^2]^{1/2} + M \right]}. \tag{12.35}$$

The interaction energy of static charges then becomes

$$E = \frac{e^2}{2} \int d^2r \, d^2r' \, j^0(\mathbf{r}) \left[\int \frac{d^2k}{(2\pi)^2} \frac{e^{i\mathbf{k}\cdot(\mathbf{x}-\mathbf{y})}}{|\mathbf{k}| + M} \right] j^0(\mathbf{r}'). \tag{12.36}$$

Then, neglecting the unphysical self-interaction terms, we have

$$E = \frac{e^2}{2} \int \frac{d^2k}{(2\pi)^2} \frac{e^{i\mathbf{k}\cdot(\mathbf{x}-\mathbf{y})}}{\sqrt{\mathbf{k}^2} + M}$$

$$E(r) = e^2 \left\{ \frac{1}{4\pi r} - \frac{M}{16} \left[\mathbf{H}_0 \left(\frac{Mr}{2} \right) - Y_0 \left(\frac{Mr}{2} \right) \right] \right\}, \tag{12.37}$$

where $r = |\mathbf{x} - \mathbf{y}|$, $\mathbf{H}_0$ is a Struve function and Y_0 is a Neumann function.

At large distances, the interaction potential behaves as [82]

$$E(r) \sim \frac{1}{M^2 r^3}.$$

We conclude that, contrary to QED, the Anderson–Higgs mechanism in PQED has the effect of modifying the electrostatic potential from $1/r$ to $1/r^3$ at large distances, whereas in QED it becomes an exponential decaying potential.

12.7.2 Coupling to a Charged Dirac Field

Massive Dirac Field

Let us consider PQED coupled to a charged Dirac field in 2+1 D. As we know, the inverse free propagator is additively corrected by the vacuum polarization tensor to yield the inverse exact propagator, namely

$$G^{-1}_{E,\mu\nu} = G^{-1}_{\mu\nu} - \Pi_{\mu\nu}. \tag{12.38}$$

Hence, using the exact propagator in (12.12), we obtain instead of (12.15),

$$E = \frac{e^2}{2} \int \frac{d^2k}{(2\pi)^2} \frac{e^{i\mathbf{k}\cdot(\mathbf{x}-\mathbf{y})}}{\sqrt{\mathbf{k}^2 + \Pi^{00}(\omega = 0, \mathbf{k})}}. \tag{12.39}$$

In the large fermion mass, (M) regime, we have [84]

$$\Pi^{00}(\omega = 0, \mathbf{k}) \simeq -\frac{\alpha}{3M} |\mathbf{k}|^2,$$

where α is the fine-structure constant.

We see that expression (12.41) is the Fourier transform of

$$\frac{\mathcal{M}}{|\mathbf{k}| \, [|\mathbf{k}| + \mathcal{M}]} = \frac{1}{|\mathbf{k}|} - \frac{1}{|\mathbf{k}| + \mathcal{M}},$$

where $\mathcal{M} = 3M/\alpha$.

Using the relation above and (12.37), we conclude that in the present case the screened potential energy will be

$$E(r) = e^2 \frac{\mathcal{M}}{16} \left[\mathbf{H}_0 \left(\frac{\mathcal{M}r}{2} \right) - Y_0 \left(\frac{\mathcal{M}r}{2} \right) \right], \tag{12.40}$$

where $r = |\mathbf{x} - \mathbf{y}|$. From (12.37), we can see that it behaves as $\sim 1/r$ at large distances.

This interaction potential is sometimes called the Keldysh potential and was derived in order to describe the effective Coulomb interaction in thin films [83]. The fact that we can re-obtain it from pseudo quantum electrodynamics is a great success of this theory.

Massless Dirac Field

An interesting result occurs when PQED is coupled to massless Dirac fermions. As a consequence of scale invariance, it follows that

$$\Pi^{00}(\omega = 0, \mathbf{k}) \simeq F(\alpha)|\mathbf{k}|,$$

hence,

$$E(\mathbf{x} - \mathbf{y}) = \frac{e^2 A(\alpha)}{2} \int \frac{d^2 k}{(2\pi)^2} \frac{e^{i\mathbf{k}\cdot(\mathbf{x}-\mathbf{y})}}{\sqrt{\mathbf{k}^2}} = e^2 A(\alpha) \frac{1}{4\pi |\mathbf{x} - \mathbf{y}|}, \tag{12.41}$$

where $A(\alpha)$ is the constant in (12.22).

We conclude here the second part of this book, which is concerned with an introduction to QFT and the study of the main features thereof that may be relevant in CMP. In the third part, starting with the next chapter, we present a QFT approach to several CMP systems.

Part III

Quantum Field Theory Approach to Condensed Matter Systems

Part III

Quantum Field Theory Approach to Condensed Matter Systems

13

Quantum Field Theory Methods in Condensed Matter

Condensed matter physics invariably exhibits many-particle systems, which must be treated according to the laws of quantum-mechanics. Such particles may be electrons, holes, phonons, magnons, polarons, Cooper pairs and so on. A quantum field theory, conversely, describes the dynamics of fields according to the same laws. It turns out that the energy eigenstates of a quantum field are precisely quantum many-particle states: photons, in the case of the quantized electromagnetic field; phonons, in the case of the elastic vibrating field of a crystal; magnons, in the case of the oscillating magnetization vector of magnetic materials; electrons and holes, in the case of Schrödinger or Dirac matter fields. Because of this fact, quantum field theory has become a powerful instrument in the realm of condensed matter systems, in the same way as it used to be in particle physics.

In this chapter we describe the contact point between a quantum field theory and a quantum many-particle system. This may be summarized by the fact that a particle position eigenstate, which forms the base for its full quantum-mechanical description, is obtained by acting on the vacuum state with a local quantum field operator, which is the basic piece of a quantum field theory. After introducing this point, we derive several results that will be relevant for applications of the latter in condensed matter systems.

13.1 Quantum Fields and Many-Particles

We have seen in Chapter 3 that a particle in a state of definite momentum $\mathbf{p}$, such that

$$\mathbf{P}|\mathbf{p}\rangle = \mathbf{p}|\mathbf{p}\rangle \tag{13.1}$$

has the state-vector given by

$$|\mathbf{p}\rangle = c^{\dagger}(\mathbf{p})|0\rangle. \tag{13.2}$$

Here $c(\mathbf{p})$ is a commuting or anti-commuting operator, according to whether the particle is a boson or a fermion. For the sake of simplifying the notation, we are neglecting any spin index.

The same particle in a state of definite position $\mathbf{x}$, namely $|\mathbf{x}\rangle$, obeys

$$\mathbf{X}|\mathbf{x}\rangle = \mathbf{x}|\mathbf{x}\rangle. \tag{13.3}$$

Then, from the completeness relation

$$\int d^3p |\mathbf{p}\rangle\langle\mathbf{p}| = \mathbb{I} \tag{13.4}$$

it follows that

$$
\begin{aligned}
|\mathbf{x}\rangle &= \int d^3p |\mathbf{p}\rangle\langle\mathbf{p}|\mathbf{x}\rangle \\
&= \int \frac{d^3p}{(2\pi\hbar)^3} e^{-\frac{i}{\hbar}\mathbf{p}\cdot\mathbf{x}} c^\dagger(\mathbf{p})|0\rangle,
\end{aligned}
\tag{13.5}
$$

where we used (13.2) and the fact that the position representation of momentum eigenfunctions is

$$\langle\mathbf{x}|\mathbf{p}\rangle = \frac{1}{(2\pi\hbar)^3} e^{\frac{i}{\hbar}\mathbf{p}\cdot\mathbf{x}}. \tag{13.6}$$

We clearly see, then, that the Schrödinger picture field operator

$$\psi^\dagger(\mathbf{x}) = \int \frac{d^3p}{(2\pi\hbar)^3} e^{-\frac{i}{\hbar}\mathbf{p}\cdot\mathbf{x}} c^\dagger(\mathbf{p}) \tag{13.7}$$

creates a one-particle position eigenstate, namely,

$$
\begin{aligned}
\mathbf{X}\left[\psi^\dagger(\mathbf{x})|0\rangle\right] &= \mathbf{x}\left[\psi^\dagger(\mathbf{x})|0\rangle\right] \\
\psi^\dagger(\mathbf{x}, 0)|0\rangle &= |\mathbf{x}\rangle.
\end{aligned}
\tag{13.8}
$$

Going to the Heisenberg picture, the operator $\psi^\dagger(\mathbf{x})_S$ becomes a dynamical field $\psi^\dagger(\mathbf{x}, t)_H$ with Hamiltonian density (dropping the H subscript) given by

$$\mathcal{H}_0 = \psi^\dagger\left(-\frac{\hbar^2}{2m}\nabla^2\right)\psi \quad ; \quad H_0 = \int d^3x \mathcal{H}_0 \tag{13.9}$$

in the case of non-interacting particles. The corresponding energy eigenvalues are shown in (3.21).

The associated Lagrangean density is given by

$$\mathcal{L}_0 = i\hbar\psi^\dagger\partial_t\psi - \psi^\dagger\left(-\frac{\hbar^2}{2m}\nabla^2\right)\psi, \tag{13.10}$$

out of which we obtain the following operator equation:

$$i\hbar \frac{\partial}{\partial t}\psi(\mathbf{r}, t) = -\frac{\hbar^2}{2m}\nabla^2\psi(\mathbf{r}, t). \tag{13.11}$$

This has the form of the Schrödinger equation for the wave-function of a one-particle quantum-mechanical system. Since this has become itself an operator, the present formalism is sometimes called "second quantization." The solutions of this have the form

$$\psi(\mathbf{r}, t) = \int \frac{d\omega}{(2\pi)^{1/2}} \int \frac{d^3k}{(2\pi)^{3/2}} \delta\left(\omega - \frac{\hbar k^2}{2m}\right) \exp\{i(\mathbf{k}\cdot\mathbf{r} - \omega t)\} c(\mathbf{k}) \tag{13.12}$$

or, integrating in ω,

$$\psi(\mathbf{r}, t) = \int \frac{d^3k}{(2\pi)^{3/2}} \exp\{i(\mathbf{k}\cdot\mathbf{r} - \omega(\mathbf{k})t)\} c(\mathbf{k}), \tag{13.13}$$

where the dispersion relation is $\omega(\mathbf{k}) = \frac{\hbar k^2}{2m}$.

We have considered here a non-relativistic system. The relativistic generalization thereof is straightforward. Perhaps the most significant difference is that we will have in (13.12) a frequency delta of the form

$$\delta\left(\omega^2 - \omega^2(\mathbf{k})\right) = \frac{1}{2\omega(\mathbf{k})}\left[\delta(\omega - \omega(\mathbf{k})) + \delta(\omega + \omega(\mathbf{k}))\right], \tag{13.14}$$

where now $\omega(\mathbf{k}) = \sqrt{|\mathbf{k}|^2 + m^2}$. The immediate consequence is the presence of positive and negative energies corresponding, respectively, to the two terms in (13.14). This leads to the existence of particles and anti-particles in relativistic systems, the latter being associated to the absence of a negative energy particle.

The above free-particle system is modified by interactions as follows. In the case of interactions of each particle with an external potential V, we would have

$$\mathcal{L} = i\hbar\psi^\dagger\partial_t\psi - \psi^\dagger\left(-\frac{\hbar^2}{2m}\nabla^2\right)\psi - \psi^\dagger\psi V. \tag{13.15}$$

In this case, the particles still do not interact among themselves. The interaction among particles, then, would be described by the Lagrangean density

$$\mathcal{L} = i\hbar\psi^\dagger\partial_t\psi - \psi^\dagger\left(-\frac{\hbar^2}{2m}\nabla^2\right)\psi - V[\psi, \psi^\dagger], \tag{13.16}$$

where $V[\psi, \psi^\dagger]$ is a higher-than-quadratic function of the fields. In the following chapters we will see examples of both types of interaction in several condensed matter systems.

13.2 The Time-Evolution Operator and the Green Operator

Consider a system, such that the state-vector at a time t_0 is $|\Psi(t_0)\rangle$. If let by itself, this system will evolve, at a later time t, to a state $|\Psi(t)\rangle$, which is obtained from the former by the action of the forward time-evolution operator,

$$|\Psi(t)\rangle = U_+(t, t_0)|\Psi(t_0)\rangle, \tag{13.17}$$

which is given by

$$U_+(t, t_0) = \theta(t - t_0)e^{-\frac{i}{\hbar}H(t - t_0)}, \tag{13.18}$$

where H is the Hamiltonian of the system. The operator $U_+(t, t_0)$ satisfies the differential equation

$$\left(i\hbar\frac{d}{dt} - H\right)U_+(t, t_0) = i\hbar\delta(t - t_0) \tag{13.19}$$

with the initial condition $U_+(t_0, t_0) = \mathbb{I}$.

Fourier transforming $U_+(t, 0)$ to energy space, we get

$$\int_{-\infty}^{\infty} dt e^{\frac{i}{\hbar}(E+i\epsilon)t} U_+(t, 0) = \int_0^{\infty} dt e^{\frac{i}{\hbar}(E-H+i\epsilon)t} = i\hbar G(E)$$

$$G(E) = \frac{1}{E - H + i\epsilon}. \tag{13.20}$$

$G(E)$ is called the Green operator or resolvent. This could be also obtained by Fourier transforming Eq. (13.19). Suppose now the Hamiltonian spectrum is such that

$$H|n\rangle = E_n|n\rangle \quad ; \quad \sum_n |n\rangle\langle n| = \mathbb{I}. \tag{13.21}$$

The spectral decomposition of $G(E)$ and $U(t, 0)$, then, are given respectively by

$$G(E) = \sum_n \frac{|n\rangle\langle n|}{E - E_n + i\epsilon} \tag{13.22}$$

and

$$U_+(t, 0) = \theta(t) \sum_n |n\rangle\langle n|e^{-\frac{i}{\hbar}E_n t}. \tag{13.23}$$

13.3 The Spectral Operator and the Spectral Weight

Let us now introduce the spectral operator

$$A(\omega) = \sum_n |n\rangle\langle n|\,\delta(\omega - \omega_n), \tag{13.24}$$

where $E_n = \hbar\omega_n$.

From the spectral operator we define the spectral weight as

$$N(\omega) \equiv \mathrm{Tr}A(\omega) = \sum_n \delta(\omega - \omega_n). \tag{13.25}$$

It expresses the contribution of frequency ω to the sum in (13.24).

According to (13.23), we may express the time-evolution operator in terms of the spectral operator as

$$U_+(t, 0) = \int_{-\infty}^{\infty} d\omega e^{-i\omega t} A(\omega). \tag{13.26}$$

Now, using

$$\mathrm{Im}\frac{1}{\omega - \omega_0 + i\epsilon} = \mathrm{Im}\frac{\omega - \omega_0 - i\epsilon}{(\omega - \omega_0)^2 + \epsilon^2} \xrightarrow{\epsilon \to 0} -\pi\delta(\omega - \omega_0), \tag{13.27}$$

we obtain the relation

$$A(\omega) = -\frac{1}{\pi} \mathrm{Im}\, G(\omega). \tag{13.28}$$

Inserting this in (13.26), we get

$$U_+(t, 0) = -\frac{1}{\pi} \theta(t) \int_{-\infty}^{\infty} d\omega e^{-i\omega t} \mathrm{Im}G(\omega). \tag{13.29}$$

Using (13.22), we can further write the spectral operator as

$$A(\omega) = -\frac{1}{\pi} \mathrm{Im} \sum_n \frac{|n\rangle\langle n|}{\hbar\omega - E_n + i\epsilon}. \tag{13.30}$$

The time-evolution operator, accordingly, can be written as

$$U_+(t, 0) = -\frac{1}{\pi} \theta(t) \sum_n \int_{-\infty}^{\infty} d\omega e^{-i\omega t} \mathrm{Im} \frac{|n\rangle\langle n|}{\hbar\omega - E_n + i\epsilon}. \tag{13.31}$$

From this expression and (12.24) we may understand, for instance, why the S-matrix elements, and consequently the scattering cross-section, will have poles (resonances) at energies corresponding to the Hamiltonian energy eigenvalues.

13.4 The Green Function

The Green function is the coordinate representation of the forward time-evolution operator, namely

$$\mathcal{G}_+(\mathbf{y}, t; \mathbf{x}, 0) = \langle \mathbf{y}|U_+(t, 0)|\mathbf{x}\rangle = \theta(t) \sum_n \langle \mathbf{y}|n\rangle\langle n|\mathbf{x}\rangle e^{-i\omega_n t}. \tag{13.32}$$

Here, $|\mathbf{x}\rangle$ and $|\mathbf{y}\rangle$ are position eigenstates.

Then, using (13.8), the Green function can be expressed, respectively, in the Schrödinger and Heisenberg pictures as

$$\mathcal{G}_+(\mathbf{y}, t; \mathbf{x}, 0) = \langle 0|\psi(\mathbf{y})e^{-\frac{i}{\hbar}Ht}\psi^\dagger(\mathbf{x})|0\rangle_S = \langle 0|\psi(\mathbf{y}, t)\psi^\dagger(\mathbf{x}, 0)|0\rangle_H. \quad (13.33)$$

From (13.17) and (13.33), we have

$$\Psi(\mathbf{x}, t) = \int d^3y \, \mathcal{G}_+(\mathbf{x}, t; \mathbf{y}, 0)\Psi(\mathbf{y}, 0). \quad (13.34)$$

13.5 The Spectral Function

The Discrete Energy Representation

Out of the spectral operator, we can obtain the spectral function

$$A(\omega; \mathbf{x}, \mathbf{y}) = \langle \mathbf{y}|A(\omega)|\mathbf{x}\rangle = \sum_n \langle \mathbf{y}|n\rangle\langle n|\mathbf{x}\rangle\delta(\omega - \omega_n), \quad (13.35)$$

which is the coordinate representation of $A(\omega)$.

From the previous equation, we derive the expression for the spectral weight, namely,

$$N(\omega) = \int d^3x \, A(\omega; \mathbf{x}, \mathbf{x}) = \sum_n \int d^3x \, \langle n|\mathbf{x}\rangle\langle \mathbf{x}|n\rangle\delta(\omega - \omega_n) = \sum_n \delta(\omega - \omega_n). \quad (13.36)$$

We can, alternatively, write the spectral weight, as

$$N(\omega) = -\frac{1}{\pi}\text{Im}\sum_n \frac{1}{\hbar\omega - E_n + i\epsilon}. \quad (13.37)$$

We may express the Green function in terms of the spectral function as

$$\mathcal{G}_+(\mathbf{y}, t; \mathbf{x}, 0) = \theta(t)\int_{-\infty}^{\infty} d\omega e^{-i\omega t} A(\omega; \mathbf{x}, \mathbf{y}) \quad (13.38)$$

and using (13.28)

$$\mathcal{G}_+(\mathbf{y}, t; \mathbf{x}, 0) = -\frac{1}{\pi}\theta(t)\int_{-\infty}^{\infty} d\omega e^{-i\omega t}\text{Im}\, G(\omega; \mathbf{x}, \mathbf{y}), \quad (13.39)$$

where

$$G(\omega; \mathbf{x}, \mathbf{y}) = \sum_n \frac{\langle \mathbf{y}|n\rangle\langle n|\mathbf{x}\rangle}{E - E_n + i\epsilon}. \quad (13.40)$$

We immediately see that the energy eigenvalues appear as poles in the Fourier transform of the Green function.

The Momentum Representation

The spectral function is very useful for determining important properties of quantum many-particle systems. We have so far considered it with a discrete base. Let us illustrate these properties when we use a continuum base, namely of momentum:

$$\sum_n \langle y|n\rangle \langle n|x\rangle = \frac{1}{V} \sum_{\mathbf{k}_n} e^{i\mathbf{k}_n \cdot (\mathbf{x}-\mathbf{y})} = \int \frac{d^3k}{(2\pi)^3} e^{i\mathbf{k} \cdot (\mathbf{x}-\mathbf{y})}. \qquad (13.41)$$

The spectral function is given by

$$A(\omega; \mathbf{x}, \mathbf{y}) = \frac{1}{V} \sum_{\mathbf{k}_n} e^{i\mathbf{k}_n \cdot (\mathbf{x}-\mathbf{y})} \delta(\omega - \omega(\mathbf{k}_n)). \qquad (13.42)$$

Then the average number of particles is given by

$$\int d^3x \int_{-\infty}^{\infty} d\omega A(\omega; \mathbf{x}, \mathbf{x}) = \sum_{\mathbf{k}_n} = N, \qquad (13.43)$$

the average density, by

$$\int_{-\infty}^{\infty} d\omega A(\omega; \mathbf{x}, \mathbf{x}) = \frac{1}{V} \sum_{\mathbf{k}_n} = \frac{N}{V} \qquad (13.44)$$

and the average spectral weight by

$$\int d^3x A(\omega; \mathbf{x}, \mathbf{x}) = \sum_{\mathbf{k}_n} \delta(\omega - \omega(\mathbf{k}_n)) = N(\omega). \qquad (13.45)$$

In the full continuum limit, we have, assuming a certain dispersion relation $\epsilon(\mathbf{k})$,

$$\sum_{\mathbf{k}_n} \Longrightarrow \int d^3k = \int_0^{\infty} dk\, 4\pi k^2 = \int_0^{\infty} d\epsilon\, g(\epsilon), \qquad (13.46)$$

where $g(\epsilon)$, the density of states with energy ϵ is given by

$$g(\epsilon) = 4\pi k^2 \frac{dk}{d\epsilon} = \frac{4\pi k^2}{\frac{d\epsilon}{dk}}. \qquad (13.47)$$

Inserting (13.46) in (13.45), we find

$$N(\omega) = \int d^3k \delta(\omega - \epsilon(\mathbf{k})) = \int_0^{\infty} d\epsilon g(\epsilon)\delta(\omega - \epsilon)) = g(\omega) \qquad (13.48)$$

and we see that the average spectral weight coincides with the density of states.

In two spatial dimensions, we would have, instead of (13.46),

$$\sum_{\mathbf{k}_n} \Longrightarrow \int d^2k = \int_0^{\infty} dk\, 2\pi k = \int_0^{\infty} d\epsilon\, g(\epsilon), \qquad (13.49)$$

where $g(\epsilon)$, the density of states with energy ϵ is now given by

$$g(\epsilon) = 2\pi k \frac{dk}{d\epsilon} = \frac{2\pi k}{\frac{d\epsilon}{dk}}. \tag{13.50}$$

The probability of the field $\psi^\dagger(\mathbf{x})$ creating one-particle states out of the vacuum is another quantity, which is instructive to calculate. It is given by

$$\mathcal{P}_1 = \frac{1}{N} \int d^3x \int_{-\infty}^{\infty} d\omega A(\omega; \mathbf{x}, \mathbf{x}) = \frac{1}{N} \sum_{\mathbf{k}_n} = \frac{1}{N} \int_{-\infty}^{\infty} d\omega N(\omega), \tag{13.51}$$

which is equal to one for the free theory.

We may write the Fourier expansion of the spectral function as

$$A(\omega; \mathbf{x}, \mathbf{y}) = \int \frac{d^3k}{(2\pi)^3} e^{i\mathbf{k}\cdot(\mathbf{x}-\mathbf{y})} A(\omega, \mathbf{k}). \tag{13.52}$$

Then, it follows that

$$A(\omega, \mathbf{k}) = \frac{1}{\pi} \operatorname{Im} G(\omega - i\epsilon, \mathbf{k}) = \mathcal{G}(\omega - i\epsilon, \mathbf{k}), \tag{13.53}$$

where $\mathcal{G}(\omega, \mathbf{k})$ is the space and time Fourier transform of the Green function (13.32).

According to (13.45), as a consequence of the previous expression, we can write the spectral weight as

$$N(\omega) = \int d^3x \int \frac{d^3k}{(2\pi)^3} A(\omega, \mathbf{k}) = V \int \frac{d^3k}{(2\pi)^3} A(\omega, \mathbf{k}) \rightarrow$$
$$N(\omega) = \sum_{\mathbf{k}_n} A(\omega, \mathbf{k}_n)$$
$$N(\omega) = \sum_{\mathbf{k}_n} \mathcal{G}(\omega - i\epsilon, \mathbf{k}_n). \tag{13.54}$$

This is a useful expression for determining the spectral density, given the knowledge of the Green function.

In the next chapter, we will see a few applications of this formalism to relevant systems in condensed matter.

14

Metals, Fermi Liquids, Mott and Anderson Insulators

The energy eigenstates of non-interacting electrons moving on a perfect lattice, according to Bloch's Theorem, form continuous bands. In a situation where a completely filled band precisely coincides with the first Brillouin zone (see Subsection 1.3.3), an energy gap forms and no electron states in the valence band are adjacent to empty available states in the conduction band. The system is an insulator when the energy gap is much larger than $k_B T$, or a semiconductor when it is of the order of $k_B T$ or less. Conversely, when the occupied electron states form a Fermi surface, such that all states belonging to it have adjacent empty states in the conduction band, then the system is a metal. When only part of the states on the Fermi surface exhibit this property, the system is a semi-metal. This scenario, which was derived for free electrons on an ideal lattice, nevertheless is profoundly modified by the introduction either of electronic interactions or of disorder. A substance that would be a metal, according to this scenario, for instance, may become either a Fermi liquid or an insulator when the new features are taken into account. In this chapter, we present a few selected examples where the application of quantum field theory methods in condensed matter systems leads to clarifying results. We first consider metals and describe their non-interacting quasi-particles. We then introduce weak electronic interactions that lead to the so-called Fermi liquid and present a QFT approach to this, studying in particular the mass renormalization and finite lifetime of the quasi-particles. We also calculate the conductivity of such systems in the presence of electronic scattering by impurities. We conclude the chapter by presenting two special types of insulators that would be otherwise metals but, in the presence of interactions and disorder, become special insulators: the Mott and Anderson insulators, respectively.

14.1 Metals

14.1.1 The Quasi-Particle Spectrum of Metals

Let us first consider a metal, a system supposed to have non-interacting electrons, with Hamiltonian given by

$$H_0 = \int d^3r \psi_\sigma^\dagger(\mathbf{r}) \left(-\frac{\hbar^2}{2m} \nabla^2 - \epsilon_F \right) \psi_\sigma(\mathbf{r}), \tag{14.1}$$

where the energy is referred to the Fermi level ϵ_F. The corresponding Lagrangean is

$$L_0 = \int d^3r \psi_\sigma^\dagger(\mathbf{r}) \left[i\hbar \frac{\partial}{\partial t} + \frac{\hbar^2}{2m} \nabla^2 + \epsilon_F \right] \psi_\sigma(\mathbf{r}). \tag{14.2}$$

The two-point Green function is given by

$$G^{(2)}(\mathbf{r}, t; \mathbf{0}, 0) = \langle 0|T\psi(\mathbf{r}, t)\psi^\dagger(\mathbf{0}, 0)|0\rangle \tag{14.3}$$

and satisfies

$$\left[i\hbar \frac{\partial}{\partial t} + \frac{\hbar^2}{2m} \nabla^2 - \epsilon_F \right] G^{(2)}(\mathbf{r}; t) = i\hbar \delta(\mathbf{r})\delta(t). \tag{14.4}$$

The two-point vertex function in momentum space reads

$$\Gamma_0^{(2)}(\omega, \mathbf{p}) = \left[\hbar\omega - \frac{\hbar^2 \mathbf{p}^2}{2m} + \epsilon_F \right], \tag{14.5}$$

hence, according to Eq. (6.18), we have

$$G_0^{(2)}(\omega, \mathbf{p}) = i\hbar \left[\hbar\omega - \frac{\hbar^2 \mathbf{p}^2}{2m} + \epsilon_F \right]^{-1},$$

$$G^{(2)}(\mathbf{x}, t; \mathbf{y}, 0) = \int_{-\infty}^{\infty} d\omega e^{-i\omega t} \int \frac{d^3p}{(2\pi)^3} G_0^{(2)}(\omega, \mathbf{p}) e^{i\mathbf{p}\cdot(\mathbf{x}-\mathbf{y})}. \tag{14.6}$$

From this and (13.38), we have the spectral function in this case given by

$$A(\omega; \mathbf{x}, \mathbf{y}) = \int \frac{d^3p}{(2\pi)^3} G_0^{(2)}(\omega, \mathbf{p}) e^{i\mathbf{p}\cdot(\mathbf{x}-\mathbf{y})}. \tag{14.7}$$

Even though the electrons in a metal are taken as non-interacting, the constraints imposed by the existence of a Fermi surface have a strong influence on the nature of the elementary excitations one would produce, when adding energy to the system. These, indeed, may be rather different from simply free electrons, as we show below.

A crucial step in the determination of the corresponding Green function is the prescription for approaching the poles that occur at $\frac{\hbar^2 \mathbf{p}^2}{2m} - \epsilon_F$. These will correspond to energies larger or smaller than the Fermi energy ϵ_F, according to whether the modulus of the momentum $|\mathbf{p}|$ is larger or smaller than the Fermi momentum $|\mathbf{p}_F|$. Then, we have

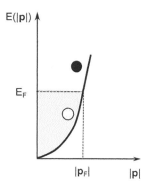

Figure 14.1 The quasi-electrons and quasi-holes in a metal

$$\frac{\hbar^2 \mathbf{p}^2}{2m} - \epsilon_F = \xi_+(\mathbf{p}) > 0 \qquad |\mathbf{p}| > |\mathbf{p}_F|$$

$$-\frac{\hbar^2 \mathbf{p}^2}{2m} + \epsilon_F = \xi_-(\mathbf{p}) > 0 \qquad |\mathbf{p}| < |\mathbf{p}_F|. \tag{14.8}$$

As we did in the relativistic case, we shift the poles in such a way that we can perform the Wick rotation in an analytic way. This leads to

$$G_0^{(2)}(\omega, \mathbf{p}) = \frac{i\theta(|\mathbf{p}| - |\mathbf{p}_F|)}{\hbar\omega - \left[\xi_+(\mathbf{p}) - i\epsilon\right]} + \frac{i\theta(|\mathbf{p}_F| - |\mathbf{p}|)}{\hbar\omega - \left[\xi_-(\mathbf{p}) + i\epsilon\right]}. \tag{14.9}$$

Notice that with the prescription choice made above, the pole at $\hbar\omega_+ = \xi_+(\mathbf{p})$ corresponds to electrons above the Fermi surface and $\hbar\omega_- = \xi_-(\mathbf{p})$ to holes below it. These excitations are called, respectively, quasi-electrons and quasi-holes.

The corresponding spectral weight, according to what we saw in the previous section will be

$$N(\omega) = N_e(\omega) + N_h(\omega)$$

$$= \int \frac{d^3 p}{(2\pi)^3} \left\{ \theta(|\mathbf{p}| - |\mathbf{p}_F|)\delta\left(\omega - \omega_+(\mathbf{p})\right) - \theta(|\mathbf{p}_F| - |\mathbf{p}|)\delta\left(\omega - \omega_-(\mathbf{p})\right) \right\},$$

$$\tag{14.10}$$

thus containing contributions coming both from outside and inside the Fermi surface. These are the physical excitations in the metal. Interestingly, notice that the Fermi surface profoundly influences the properties of the electrons in a metal, even in the absence of interactions. The form of the elementary excitations, namely, quasi-electrons and quasi-holes, is a manifestation of that.

Using (13.47), we obtain for the spectral weight,

$$N(\omega_+, \omega_-) = \frac{m^{3/2}}{2^{1/2}\pi^2} \left[\theta(\omega_+)\sqrt{\omega_+ + \epsilon_F} + \theta(\omega_-)\sqrt{-\omega_- + \epsilon_F} \right] \tag{14.11}$$

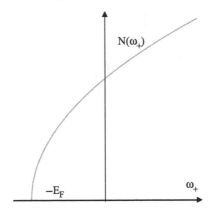

Figure 14.2 The density of states of the quasi-electrons in a metal. The physical region is $\omega_+ > 0$.

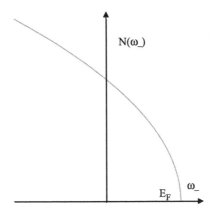

Figure 14.3 The density of states of the quasi-holes in a metal. The physical region is $\omega_- > 0$.

The Fermi energy is located at the origin. The physical region corresponding to $E > \epsilon_F$ corresponds to $\omega_+ > 0$; that corresponding to $E < \epsilon_F$ corresponds to $\omega_- > 0$.

In the rest of this chapter, we will consider how the effects of interactions and disorder will modify the present picture.

14.2 The Fermi Liquid: a Quantum Field Theory Approach

The independent electron description of condensed matter systems works surprisingly well, and for that reason has been a fundamental paradigm for a long time. The first attempt to include interactions in the picture is Landau's Fermi liquid theory. The underlying ideas behind this theory allow us to understand why the independent electron approximation is so good. When the electromagnetic interaction is taken into account, for instance, in the case of an electron away from the Fermi surface, it would excite an arbitrary number of electron-hole pairs from the Fermi sea, thereby decaying into lesser energy states. Now, the closer the electron is to the Fermi surface, because of the Pauli exclusion principle, the lesser the number of possible final states would be available for such electron to decay into. Ultimately, an electron on the Fermi surface, despite the interactions, would not decay for the absence of available possible final states.

Landau's Fermi liquid theory is based on the concept of "quasi-particle." This would have an infinite lifetime when belonging to the Fermi surface and a finite lifetime when away from it. When these quasi-particles are not very far from the Fermi surface, however, they will behave very much like free electrons, with a relatively long lifetime, because of the restriction on the phase space imposed by the Pauli principle. According to this theory, therefore, the only effect of the interaction is basically to redefine the parameters of the ideal Fermi gas, leaving the physical properties essentially unchanged. This fact explains why the non-interacting description of the electron gas has been so successful.

Here, we present a quantum field theory approach to the Fermi liquid theory that, besides of fostering a deeper understanding of its foundations, frames its contents within a general context.

The Fermi liquid theory describes a situation where interactions are taken into account. Let us assume, therefore, that the Hamiltonian, in addition to H_0, contains an interacting term H_I. A primary effect of the interactions is to modify the proper vertex two-point function by adding to it the electron self-energy, namely,

$$\Gamma^{(2)}(\omega, \mathbf{p}) = \left[\hbar\omega - \frac{\hbar^2 \mathbf{p}^2}{2m} + \Sigma(\omega, \mathbf{p}) \right], \tag{14.12}$$

to which there corresponds the two-point Green function in coordinate space,

$$G^{(2)}(t, \mathbf{x}) = i \int \frac{d\omega}{2\pi} \int \frac{d^3 k}{(2\pi)^3} \frac{e^{i[\mathbf{k} \cdot \mathbf{x} - \omega t]}}{\hbar\omega - \frac{\hbar^2 \mathbf{p}^2}{2m} + \Sigma(\omega, \mathbf{p})}. \tag{14.13}$$

The physical excitations correspond to the poles of the integrand, which occur at

$$\hbar\omega_R(\mathbf{p}) = \frac{\hbar^2 \mathbf{p}^2}{2m} - \mathrm{Re}\,\Sigma(\omega_R, \mathbf{p}). \tag{14.14}$$

We now expand the real part of the electron self-energy around $\omega_R(\mathbf{p})$, for a fixed $\mathbf{p}$:

$$\mathrm{Re}\,\Sigma(\omega, \mathbf{p}) = \mathrm{Re}\,\Sigma(\omega_R, \mathbf{p}) + (\hbar\omega - \hbar\omega_R)\frac{\partial \mathrm{Re}\,\Sigma(\omega_R, \mathbf{p})}{\partial \omega_R} + \dots \tag{14.15}$$

Inserting this in (14.13), we get

$$G^{(2)}(\omega_R, \mathbf{p}) = \frac{iZ}{[(\hbar\omega - \hbar\omega_R) + i\epsilon][1 + ZO(\hbar\omega - \hbar\omega_R)]}, \tag{14.16}$$

where

$$Z = \frac{1}{1 + \frac{\partial \mathrm{Re}\Sigma(\omega_R, \mathbf{p})}{\partial \omega_R}}. \tag{14.17}$$

We see that the two-point function possesses a pole at ω_R. We again use the prescription $i\epsilon$ in order to define the integral.

Now we make the assumption, corresponding to the Fermi liquid theory, that for sufficiently weak interactions, the physical dispersion relation on the Fermi surface can be written as

$$\hbar\omega_R(\mathbf{p}_F) = \frac{\hbar^2 \mathbf{p}_F^2}{2m} - \mathrm{Re}\Sigma(\omega_R, \mathbf{p}_F) = \frac{\hbar^2 \mathbf{p}_F^2}{2m^*}, \tag{14.18}$$

where m^* is the so-called effective mass. Notice that we have the same dispersion relation as the one in the free theory, the only effect of the interaction being that the electron mass m is replaced by the effective mass m^*.

Since $\omega_R = \omega_0 \frac{m}{m^*}$, it follows that the ratio between the masses is given by

$$\frac{m}{m^*} = \frac{\partial \omega_R}{\partial \omega_0}$$

$$\frac{m}{m^*} = 1 - \frac{\partial \mathrm{Re}\Sigma(\omega_R, \mathbf{p}_F)}{\partial \omega_R} \frac{\partial \omega_R}{\partial \omega_0}. \tag{14.19}$$

Using the first part of the equation above, we get

$$\frac{m}{m^*} = \frac{1}{1 + \frac{\partial \mathrm{Re}\Sigma(\omega_R, \mathbf{p}_F)}{\partial \omega_R}}$$

$$m^* = m \, Z^{-1}, \tag{14.20}$$

where Z is given by (14.17).

The Green function of the interacting system will be

$$G^{(2)}(\omega_R, \mathbf{p}) = \frac{iZ}{\left(\hbar\omega - \frac{\hbar^2 \mathbf{p}^2}{2m^*} + i\epsilon\right)[1 + O(\hbar\omega - \hbar\omega_R)]}, \tag{14.21}$$

where the momentum is assumed to be close to the Fermi surface.

The quantum field theory approach to the Fermi liquid system shows that, when the interactions can be described in the Fermi liquid regime, the fermionic quasi-particles behave as free particles with an effective mass m^*. If we write the interacting Green function as

$$\langle 0|T\psi_{qp}(\mathbf{r}, t)\psi_{qp}^{\dagger}(\mathbf{0}, 0)|0\rangle, \tag{14.22}$$

then (14.21) implies the quasi-particle and the electron fields are related as

$$\psi_{qp}(\mathbf{r}, t) = Z^{1/2}\psi(\mathbf{r}, t). \tag{14.23}$$

Knowing that the local electron field creates eigenstates of the position operator, according to (6.67), it follows that the quasiparticle operator will have a probability of less than one, given by (14.17), of creating local quasi-particle states. The spectral weight, accordingly, is given by

$$N(\omega) = Z\delta(\omega - \omega_R(\mathbf{p})), \tag{14.24}$$

and the probability of the basic field creating one-particle states is now, according to (13.51),

$$\mathcal{P}_1 = Z, \tag{14.25}$$

where $Z < 1$ is given by (14.17). Because of the interaction, the field operator is no longer a one-particle creation operator.

As an example of a physical quantity that is well described by the Fermi liquid theory, we take the electronic specific heat of metals. This is given, respectively, by

$$c_V = \kappa m T$$
$$c_V = \kappa m^* T \tag{14.26}$$

in the free and Fermi liquid regimes, the electron mass being replaced by the effective mass in the latter. Many systems present a behavior compatible with the predictions of the Fermi liquid theory and are, consequently, well-described by this theory.

14.3 Quasi-Particles and Their Lifetime

When the quanta associated to a certain quantized field are elementary particles such as electrons or quarks, these have an infinite lifetime or, in other words, are stable. Conversely, when these quanta are quasi-particles, as in the previous example of the Fermi liquid system, they are unstable when away from the Fermi surface, therefore presenting a finite lifetime. As we will see below, the quasi-particle lifetime corresponds to the inverse of the imaginary part of the self-energy.

We have seen that the energy eigenvalues appear as poles of the Green function in momentum space. Suppose the self-energy possesses an imaginary part

$$\Sigma(\omega_R, \mathbf{p}) = \Delta\omega_R + i\frac{\hbar}{\tau}. \tag{14.27}$$

This would imply, according to (14.14), that the eigenfrequencies $\hbar\omega_R$ should possess an imaginary part as well,

$$\omega_R = \omega_0 - \Delta\omega_R - i\frac{1}{\tau}. \tag{14.28}$$

The stability condition for the quasi-particles is

$$\frac{\hbar}{\tau} \ll \hbar\omega_R \quad ; \quad \tau \gg \frac{1}{\omega_R}.$$

Now, consider the time evolution of the one-quasi-particle state $|\mathbf{r}, 0\rangle$,

$$|\mathbf{r}, t\rangle = \exp\{-\frac{i}{\hbar} Ht\}|\mathbf{r}, 0\rangle.$$

Using the spectral representation of the time-evolution operator, namely

$$U(t) = \sum_n |n\rangle e^{-i\omega_n t} \langle n|, \tag{14.29}$$

we clearly see that if the eigenfrequencies are real, then

$$|| \, |\mathbf{r}, t\rangle ||^2 = \sum_n \langle \mathbf{r}, 0|n\rangle \langle n|\mathbf{r}, 0\rangle = || \, |\mathbf{r}, 0\rangle ||^2 \tag{14.30}$$

and the norm of the quasi-particle states is preserved. Conversely, if the frequencies possess an imaginary part $1/\tau$, then according to (14.29), we would have

$$|| \, |\mathbf{r}, t\rangle ||^2 = \sum_n \langle \mathbf{r}, 0|n\rangle e^{-2t/\tau} \langle n|\mathbf{r}, 0\rangle = e^{-2t/\tau} || \, |\mathbf{r}, 0\rangle ||^2. \tag{14.31}$$

We see that the quasi-particle state vanishes with a decaying rate $e^{-2t/\tau}$, having therefore a lifetime $\Delta t \sim \tau/2$.

It is instructive to compare the quasi-particle Green function to the electron-hole Green function, when the self-energy possesses an imaginary part. In this case (14.21) will have the form

$$G_\tau^{(2)}(\omega_R, \mathbf{p}) = \frac{iZ}{\left(\hbar\omega - \frac{\hbar^2 \mathbf{p}^2}{2m^*} - i\frac{\hbar}{\tau}\right)[1 + O(\hbar\omega - \hbar\omega_R)]}. \tag{14.32}$$

Now, there is a natural imaginary part in $G(\omega)$ and the spectral weight is now

$$N(\omega) = Z\frac{\frac{1}{\tau}}{(\omega - \omega_R)^2 + \frac{1}{\tau^2}}. \tag{14.33}$$

The probability for the field to create a quasi-particle of energy in the interval $\hbar\omega$ and $\hbar\omega + d\omega$ has now a Lorentzian shape. This should be compared to the stable particle case, given by (14.10), namely

$$Z\delta(\omega - \omega_R). \tag{14.34}$$

For stable particles, the lifetime is infinite ($\tau \to \infty$) and the Lorentzian in (14.33) becomes a delta. For a finite lifetime, conversely, the energy eigenvalues will always have an uncertainty of the order of $\hbar/\tau$ and the corresponding transition lines accordingly will have a width of the same order.

14.4 Quantum Field Theory Model for Conductivity

In this section, we consider a free electron system, which is subject to scattering from localized centers, occurring at an average time interval τ. We are going to determine the conductivity using the Kubo formula, Eq. (4.61), and for this purpose we need the current correlation function. In order to obtain this, we use quantum field theory methods, which prove to be very convenient and powerful. In spite of the electrons' being free, the fact that they suffer scattering from localized centers makes us expect a finite conductivity, depending crucially on τ.

Let us assume the Hamiltonian of the system is

$$H = \int d^3r \; \psi_\sigma^\dagger(\mathbf{r}) \left[-\frac{\hbar^2}{2m}\nabla^2 + \frac{\hbar}{\tau} \right] \psi_\sigma(\mathbf{r}), \tag{14.35}$$

where the second term represents the interaction of electrons with a scattering center. τ is a characteristic time interval between these scatterings.

Using the Lagrangean corresponding to (4.52) and (14.35) and considering (5.66), we can express the partition functional as

$$Z[\mathbf{A}_{\text{ext}}] = \frac{1}{\mathcal{N}} \int D\psi_\sigma D\psi_\sigma^\dagger \exp\left\{ -\int_0^{\hbar\beta} dt \int d^3x\, \psi_\sigma^\dagger \left[i\hbar\frac{\partial}{\partial t} + \frac{\hbar^2}{2m}\nabla^2 - \frac{\hbar}{\tau} \right. \right.$$
$$\left. \left. + \left[-ie\frac{\hbar}{mc}\overleftrightarrow{\nabla}^i + \frac{e^2}{mc}\mathbf{A}_{\text{ext}}^i \right] \mathbf{A}_{\text{ext}}^i \right] \psi_\sigma \right\}, \tag{14.36}$$

where $\mathcal{N} = Z[0]$. The last term takes into account (4.49) and, it is assumed that the external field starts to act at $t > 0$.

Performing the functional integration over the fermion fields, we get

$$Z[\mathbf{A}_{\text{ext}}] = \frac{1}{\mathcal{N}} \exp\left\{ -\text{Tr}\ln\left[-i\hbar\frac{\partial}{\partial t} - \frac{\hbar^2}{2m}\nabla^2 + \frac{i\hbar}{\tau} \right. \right.$$
$$\left. \left. + \left[-ie\hbar\overleftrightarrow{\nabla}^i + \frac{ne^2}{mc}\mathbf{A}_{\text{ext}}^i \right] \mathbf{A}_{\text{ext}}^i \right] \right\}. \tag{14.37}$$

where we take into account the fact that the electron density is considered with respect to as its average value n. The free energy corresponding to (14.37) is conveniently expressed as a trace in frequency-momentum space, namely

$$F[\mathbf{A}_{\text{ext}}] = \frac{1}{\beta}\text{Tr}\ln\left[1 + \frac{\left[ek^i + \frac{ne^2}{mc}\mathbf{A}_{\text{ext}}^i \right] \mathbf{A}_{\text{ext}}^i}{\hbar\omega - \frac{\hbar^2k^2}{2m} + \frac{i\hbar}{\tau}} \right]. \tag{14.38}$$

Now, the average current density is given by (4.53). Hence, applying it to the latter expression for the free energy and using (4.58), we readily obtain, for $T \to 0$

$$\langle J^i \rangle = \int \frac{d\omega}{2\pi} \langle J^i J^j \rangle (\omega, \mathbf{k} = 0) \Big|_{A_{\text{ext}}=0} \frac{A_{\text{ext}}^j}{\omega}$$

or, equivalently,

$$\langle J^i \rangle = \int_{-\infty}^{\infty} dt \, \langle J^i J^j \rangle (t) \theta(t) \, A_{\text{ext}}^j.$$

From the two previous, equations and applying (4.53) in (14.38), we obtain:

$$\langle J^i J^j \rangle (\omega, \mathbf{k} = 0) \Big|_{A_{\text{ext}}=0} = \left(\frac{ne^2}{mc} \right) \frac{\omega}{\omega + i/\tau} \delta^{ij}, \tag{14.39}$$

where we also used the facts: (a) that the external fields start to act at $t = 0$, and, (b) that the Fourier transform of the Heaviside function is i/ω.

Now, considering that

$$\langle J^i J^j \rangle (\omega, \mathbf{k}) = \langle j^i j^j \rangle (\omega, \mathbf{k}) + \delta^{ij} \frac{ne^2}{mc}, \tag{14.40}$$

it follows, by using (14.39), that

$$\langle j^i j^j \rangle (\omega, \mathbf{k} = 0) \Big|_{A_{\text{ext}}=0} = \left(\frac{ne^2}{mc} \right) \frac{-i/\tau}{\omega + i/\tau} \delta^{ij}. \tag{14.41}$$

Then, inserting (14.39) in the Kubo formula, we have the conductivity given by the famous Drude formula:

$$\sigma^{ij}(\omega) = \delta^{ij} \frac{\sigma_0}{1 - i\omega\tau}$$

$$\sigma_0 = \frac{ne^2 \tau}{m}. \tag{14.42}$$

This frequency-dependent conductivity is usually called "optical conductivity." The corresponding expression as a function of time is given by the inverse Fourier transform of (14.42), namely

$$\sigma^{ij}(t - t') = \delta^{ij} \frac{ne^2}{m} e^{-(t-t')/\tau} \theta(t - t'). \tag{14.43}$$

Hence, according to (4.58), we write

$$\langle \mathbf{J}^i \rangle (t) = \int_{-\infty}^{t} dt' \, \sigma^{ij}(t - t') \mathbf{E}^j(t'). \tag{14.44}$$

Let us now probe the response of the system in two different situations: firstly, in the presence of a uniform and static applied electric field that starts to act at $t = 0$: $\mathbf{E} = \mathbf{E}_0 \, \theta(t)$; and secondly, under the application of a uniform electric pulse at $t = 0$, namely, an electric field given by $\mathbf{E} = \mathbf{A}_0\,\delta(t)$, or equivalently, a vector potential $\mathbf{A}(t) = \mathbf{A}_0\,\theta(t)$.

Inserting (14.43) in (14.44), we obtain in the first case,

$$\langle \mathbf{J}\rangle(t) = \frac{ne^2}{m}\mathbf{E}_0 \int_{-\infty}^{t} dt'\, e^{-(t-t')/\tau}\theta(t'), \tag{14.45}$$

from which we immediately get

$$\langle \mathbf{J}\rangle(t) = \sigma_0 \mathbf{E}_0 \left[1 - e^{-t/\tau}\right], \tag{14.46}$$

where σ_0 is given by (14.42). Notice that, after a transient regime, the current produced by the static external electric field stabilizes at a constant value corresponding to Drude's DC-conductivity σ_0.

In the case of the applied electric pulse, conversely, we obtain

$$\langle \mathbf{J}\rangle(t) = \frac{ne^2}{m}\mathbf{A}_0\, e^{-t/\tau} \tag{14.47}$$

and we see that the current decays exponentially on a time-scale τ, after the pulse is applied. We may write the above equation as

$$\langle \mathbf{J}\rangle(t) = \frac{ne^2}{m}\mathbf{A}_0 - \frac{ne^2}{m}\mathbf{A}_0\left[1 - e^{-t/\tau}\right]$$

$$\langle \mathbf{J}\rangle(t) = \frac{ne^2}{m}\mathbf{A}_0 + \langle \mathbf{j}\rangle(t). \tag{14.48}$$

Comparing with (3.49), we conclude that, after the pulse is applied, the current eventually vanishes in a metal because the j-term of the current builds up so as to cancel the other term. Observe that in the case of a superconductor, Eq. (14.48) is replaced with (4.75) and no longer vanishes.

It is also instructive to examine the behavior of the model in the two different regimes: $\tau \to \infty$ and $\tau \to 0$, corresponding, respectively, to a very low and very high density of scatterers. Let us consider now just the case of an applied uniform electric field. Then, from (14.46), we get

$$\langle \mathbf{J}\rangle(t) \xrightarrow{\tau \to \infty} \frac{ne^2\,t}{m}\mathbf{E}_0$$

$$\langle \mathbf{J}\rangle(t) \xrightarrow{\tau \to 0} 0. \tag{14.49}$$

In the large τ regime (low density of scatterers), the current increases linearly with time, whereas in the small τ regime (high density of scatterers), it tends to zero. The latter would correspond to the situation occurring in disordered systems, where the conductivity vanishes, due to localization, whereas the former reflects the situation found in a perfect ideal lattice, where the electron would move without being scattered. In this case, application of an electric field would produce an ever-growing current that increases linearly with time.

14.5 The Mott Insulator: Interaction-Induced Gap

Let us consider an electronic system described by the Hamiltonian

$$H = \int d^3r \left[\psi_\sigma^\dagger(\mathbf{r}) \left(-\frac{\hbar^2}{2m}\nabla^2 - \epsilon_F \right) \psi_\sigma(\mathbf{r}) \right]$$
$$+ \frac{e^2}{2} \int d^3r \int d^3r' \left[(\psi_\sigma^\dagger(\mathbf{r})\psi_\sigma(\mathbf{r})) \frac{1}{4\pi|\mathbf{r}-\mathbf{r}'|} (\psi_{\sigma'}^\dagger(\mathbf{r}')\psi_{\sigma'}(\mathbf{r}')) \right]. \quad (14.50)$$

The first term is the free Hamiltonian (14.1) and the second one describes the electrostatic interaction among the electrons. The corresponding Lagrangean is

$$L = \int d^3r \; \psi_\sigma^\dagger(\mathbf{r}) i\hbar \frac{\partial}{\partial t} \psi_\sigma(\mathbf{r}) - H. \quad (14.51)$$

Considering that

$$-\nabla^2 \left[\frac{1}{4\pi|\mathbf{r}-\mathbf{r}'|} \right] = \delta(\mathbf{r}-\mathbf{r}'), \quad (14.52)$$

we can write

$$\int D\psi_\sigma D\psi_\sigma^\dagger e^{i \int d^4x \mathcal{L}[\psi,\psi^\dagger]} = \int D\psi_\sigma D\psi_\sigma^\dagger D\phi e^{i \int d^4x \mathcal{L}[\psi,\psi^\dagger,\phi]}, \quad (14.53)$$

where $\mathcal{L}[\psi, \psi^\dagger]$ is the Lagrangean corresponding to (14.50) and

$$\mathcal{L}[\psi, \psi^\dagger, \phi] = \psi_\sigma^\dagger(\mathbf{r}) \left[i\hbar \frac{\partial}{\partial t} - \frac{\hbar^2}{2m}\nabla^2 - \epsilon_F + \phi \right] \psi_\sigma(\mathbf{r}) + \frac{1}{2e^2}\nabla\phi \cdot \nabla\phi.$$
$$(14.54)$$

Since the Lagrangean is quadratic in the fermion field, we may integrate on this, thus deriving an effective action for the field ϕ. With the help of the Nambu fermion field $\Phi^\dagger = (\psi_\downarrow^\dagger \; \psi_\uparrow)$, we can rewrite (14.54) as

$$\mathcal{L}[\Psi, \phi] = \frac{1}{2e^2}\nabla\phi \cdot \nabla\phi + \Phi^\dagger \mathcal{A}\Phi, \quad (14.55)$$

where the matrix $\mathcal{A}$ is given, in momentum space, by

$$\mathcal{A} = \begin{pmatrix} \xi(\mathbf{k}) + \phi + \hbar\omega & 0 \\ 0 & \xi(\mathbf{k}) + \phi - \hbar\omega \end{pmatrix}, \quad (14.56)$$

with $\xi(\mathbf{k}) = \frac{\hbar^2\mathbf{k}^2}{2m} - \epsilon_F$.

Integrating over the fermion fields, we obtain

$$\mathcal{Z} = \frac{1}{\mathcal{Z}_{0,\phi}} \int D\phi \; e^{iS_{eff}[\phi]}, \quad (14.57)$$

where

$$S_{eff}[\phi] = \int d^4x \left(\frac{1}{2e^2}\nabla\phi \cdot \nabla\phi \right) - i \ln \text{Det} \left[\frac{\mathcal{A}[\phi]}{\mathcal{A}[0]} \right]. \quad (14.58)$$

The determinant of the matrix $\mathcal{A}$ is

$$\det \mathcal{A}[\phi] = [\xi + \phi]^2 - (\hbar \omega)^2, \tag{14.59}$$

hence the effective potential corresponding to the above expression becomes, in Euclidean space,

$$
\begin{aligned}
V_{eff}[\phi] &= \frac{1}{2e^2} \nabla \phi \cdot \nabla \phi - \int_{-\infty}^{\infty} \frac{d\omega}{2\pi} \int_{-\Lambda}^{\Lambda} d\xi N(\xi) \ln \left[\frac{\omega^2 + (\xi + \phi)^2}{\omega^2 + \xi^2} \right] \\
&= \frac{1}{2e^2} \nabla \phi \cdot \nabla \phi - \int_{-\Lambda}^{\Lambda} d\xi 2N(\xi) \phi, \tag{14.60}
\end{aligned}
$$

where $N(\xi)$ is the density of states at the energy ξ and Λ is a momentum cutoff. Since our calculation is made at a zero temperature, it is natural to expect that most of the contributions for the ξ-integral will come from the Fermi surface; hence, we assume

$$N(\xi) \simeq N(E_F).$$

Then, the effective potential is given by

$$V_{eff}[\phi] = \frac{1}{2e^2} \nabla \phi \cdot \nabla \phi - 4N(E_F) \Lambda \phi.$$

Then, the field equation for ϕ, derived from $V'_{eff}[\phi] = 0$, reads

$$-\nabla^2 \phi = 4N(E_F) e^2 \Lambda. \tag{14.61}$$

This presents the following solution in momentum space:

$$\phi_0(\mathbf{p}) = \frac{4N(E_F) e^2 \Lambda}{\mathbf{p}^2}. \tag{14.62}$$

Inserting this solution in (14.56), we see that the dispersion relation of the electronic system is modified by the interaction as

$$\xi(\mathbf{p}) \to \bar{\xi} = \xi(\mathbf{p}) + \frac{4N(E_F) e^2 \Lambda}{\mathbf{p}^2}. \tag{14.63}$$

As a consequence, the poles of the Green function shift in such a way that now

$$G_0^{(2)}(\omega, \mathbf{p}) = \frac{i\theta(|\mathbf{p}| - |\mathbf{p}_F|)}{\hbar \omega - \left[\bar{\xi}_+(\mathbf{p}) - i\epsilon \right]} + \frac{i\theta(|\mathbf{p}_F| - |\mathbf{p}|)}{\hbar \omega - \left[\bar{\xi}_-(\mathbf{p}) + i\epsilon \right]}, \tag{14.64}$$

where

$$
\begin{aligned}
\bar{\xi}_+(\mathbf{p}) &= \frac{\hbar^2 \mathbf{p}^2}{2m} + \phi_0(\mathbf{p}) - \epsilon_F \\
\bar{\xi}_-(\mathbf{p}) &= -\left[\frac{\hbar^2 \mathbf{p}^2}{2m} + \phi_0(\mathbf{p}) \right] + \epsilon_F. \tag{14.65}
\end{aligned}
$$

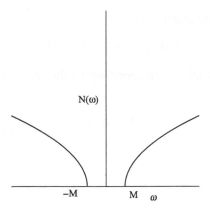

Figure 14.4 The density of states of the Mott insulator

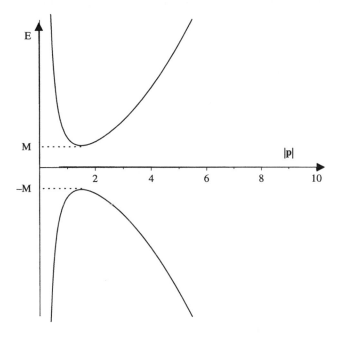

Figure 14.5 The interaction-induced gap in a Mott insulator

Observe that the interaction shifts the poles to $\hbar\omega_\pm = \bar{\xi}_\pm(\mathbf{p})$, thereby opening a gap in the spectrum (see Fig. 14.5). This implies that the system is actually an insulator.

Expanding around the point where the bands are the closest, we obtain the following expressions (assuming $\epsilon_F = 0$):

$$\overline{\xi}_+(\mathbf{p}) = \frac{\hbar^2 \mathbf{p}^2}{2m^*} + M + O(\mathbf{p}^3)$$

$$\overline{\xi}_-(\mathbf{p}) = -\frac{\hbar^2 \mathbf{p}^2}{2m^*} - M + O(\mathbf{p}^3). \tag{14.66}$$

These insulators, which have a gap produced by the Coulomb interaction but which otherwise would be metals, are called Mott insulators [96]. It is instructive to evaluate the density of states. From the previous expression and (13.48), we derive the following expression, at low momentum

$$N(\omega) == \frac{m^{3/2}}{2^{1/2}\pi^2} \left[\theta(\omega - M)\sqrt{\omega - M} + \theta(-\omega + M)\sqrt{-\omega - M} \right], \tag{14.67}$$

where $M = \phi_0(\mathbf{p}_0) \xrightarrow{\mathbf{p}_0 \to 0} 2\sqrt{2N(E_F)e^2\Lambda}$ provided the coupling $N(E_F)e^2 \neq 0$.

We have seen in this section that the Coulomb interaction alone is capable of turning a metal into an insulator. This is just an example of how the presence of interactions may profoundly modify the physical properties of a system. In the next section, we will see how the occurrence of disorder may cause deep changes as well.

14.6 Anderson Localization: Disorder-Induced Insulators

Another interesting class of insulators is revealed when we include the effects of disorder in a given electronic system. Consider the Hamiltonian given by

$$H = \int d^3r \, \psi_\sigma^\dagger(\mathbf{r}) \left[-\frac{\hbar^2}{2m}\nabla^2 + \eta(\mathbf{r}) \right] \psi_\sigma(\mathbf{r}), \tag{14.68}$$

describing free electrons in the presence of a field η. This is a generalization of (14.35), where the spatially dependent external field $\eta(\mathbf{r}) \equiv \frac{\hbar}{\tau(\mathbf{r})}$ represents the scattering of electrons from randomly distributed scattering centers, each one with a characteristic scattering time $\tau(\mathbf{r})$.

We assume the dynamics of the electrons is much faster than that of the scatterers, in such a way that the disorder is treated in the so-called quenched approach. In this, we first evaluate the free energy for a given external configuration $\eta(\mathbf{r})$ and subsequently average over this with a Gaussian distribution of width Δ:

$$\overline{F} = \int D\eta \exp\left\{ -\frac{1}{2\Delta} \int d^3x \, \eta^2 \right\} F[\eta]$$

$$F[\eta] = -k_B T \ln Z[\eta]. \tag{14.69}$$

In order to perform the above average, we use the replica method [94], writing the logarithm of the partition function as

$$\ln Z[\eta] = \lim_{n \to 0} \frac{Z^n[\eta] - 1}{n}, \tag{14.70}$$

where $Z^n[\eta]$ is the replicated partition function, given by

$$Z^n[\eta] = \frac{1}{Z^n[0]} \int \prod_{\alpha=1}^{n} D\psi_\sigma^\alpha D\psi_\sigma^{\alpha\dagger}$$

$$\times \exp\left\{ -\int_0^\beta dt \int d^d x \psi_\sigma^{\alpha\dagger} \left[-\frac{\hbar^2}{2m} \nabla^2 - \eta \right] \psi_\sigma^\alpha \right\}. \tag{14.71}$$

Inserting this in (14.70) and (14.69), we get the averaged replicated partition function

$$\overline{Z}^n = \frac{1}{Z^n[0]} \int \prod_{\alpha=1}^{n} D\psi_\sigma^\alpha D\psi_\sigma^{\alpha\dagger} D\eta$$

$$\times \exp\left\{ -\int d^3 x \frac{\eta^2}{2\Delta} \right\} \cdot \exp\left\{ -\int_0^\beta dt \int d^d x \psi_\sigma^{\alpha\dagger} \left[-\frac{\hbar^2}{2m} \nabla^2 - \eta \right] \psi_\sigma^\alpha \right\}, \tag{14.72}$$

which becomes, upon integration on η,

$$\overline{Z}^n = \frac{1}{Z^n[0]} \int \prod_{\alpha=1}^{n} D\psi_\sigma^\alpha D\psi_\sigma^{\alpha\dagger} \times \exp\left\{ -\int_0^\beta dt \int d^d x \psi_\sigma^{\alpha\dagger} \left(-\frac{\hbar^2}{2m} \nabla^2 \right) \psi_\sigma^\alpha \right.$$

$$\left. + \frac{\Delta}{2} \int_0^\beta dt \int_0^\beta dt' \psi_\sigma^{\alpha\dagger}(t) \psi_{\sigma'}^\beta(t') \psi_\sigma^\alpha(t) \psi_{\sigma'}^{\beta\dagger}(t') \right\}. \tag{14.73}$$

Now, introducing the Hubbard–Stratonovitch field $Q_{\sigma\sigma'}^{\alpha\beta}(t, t')$, we can cast the above expression in the form

$$\overline{Z}^n = \frac{1}{Z^n[0]} \int \prod_{\alpha=1}^{n} D\psi_\sigma^\alpha D\psi_\sigma^{\alpha\dagger} D Q_{\sigma\sigma'}^{\alpha\beta} \exp\left\{ -\int_0^\beta dt \int d^d x \psi_\sigma^{\alpha\dagger} \left(-\frac{\hbar^2}{2m} \nabla^2 \right) \psi_\sigma^\alpha \right.$$

$$\left. + \Delta \int_0^\beta dt \int_0^\beta dt' \left[\frac{Q_{\sigma\sigma'}^{\alpha\beta} Q_{\sigma\sigma'}^{\alpha\beta}}{2} + \psi_\sigma^{\alpha\dagger}(t) Q_{\sigma\sigma'}^{\alpha\beta} \psi_{\sigma'}^\beta(t') \right] \right\}. \tag{14.74}$$

The field equation for the Hubbard–Stratonovitch field is

$$Q_{\sigma\sigma'}^{\alpha\beta} = \langle \psi_\sigma^{\alpha\dagger}(t) \psi_{\sigma'}^\beta(t') \rangle. \tag{14.75}$$

From this, we may infer that

$$Q_{\sigma\sigma'}^{\alpha\beta} = \delta^{\alpha\beta} \delta^{\sigma\sigma'} \chi(t, t'). \tag{14.76}$$

We now determine the effective action associated to (14.74). This will be given by

$$\exp\{-S\} = \frac{1}{Z[0]} \int \prod_{\alpha=1}^{n} D\psi_\sigma^\alpha D\psi_\sigma^{\alpha\dagger} D\chi$$

$$\times \exp\left\{-\int dt d^d x \left[n\Delta \frac{\chi^2}{2} + \psi_\sigma^{\alpha\dagger}\left(i\hbar\frac{\partial}{\partial t} - \frac{\hbar^2}{2m}\nabla^2 - \Delta\chi\right)\psi_\sigma^\alpha\right]\right\}.$$

(14.77)

Notice there is a product of n fermionic integrals. Hence, there is an overall n factor in the exponent that will cancel when we remove the replicas, leading to a finite free energy. The corresponding effective potential will be obtained by integration over the fermion fields. Following the procedure we used in the previous subsection, we get

$$V_{eff}[\chi] = -\frac{1}{2}\Delta\chi^2 + \int \frac{d\omega}{2\pi}\int d\xi N(\xi)\ln\left[\frac{\omega^2 + (\xi - \Delta\chi)^2}{\omega^2 + \xi^2}\right].$$

(14.78)

For $d = 3$, we obtain, following the same procedure as in (14.60),

$$V_{eff}[\chi] = -\frac{1}{2}\Delta\chi^2 + 2\int_0^{\Delta\chi} d\xi N(\xi)\left[(\Delta\chi - 2\xi)\right] - 2\int_{\Delta\chi}^{\Lambda} d\xi N(\xi)\left[(\Delta\chi)\right],$$

(14.79)

whereupon the effective potential is, up to a constant,

$$V_{eff}[\chi] = -\frac{1}{2}\Delta\chi^2 + \alpha\Delta^2\chi^2 - \gamma\Delta\chi,$$

(14.80)

where

$$\alpha = N(E_F) \qquad \gamma = \Lambda N(E_F)$$

(14.81)

$$\frac{dV_{eff}}{d\chi} = [-1 + 2\alpha\Delta]\chi - \gamma, = 0$$

(14.82)

$$\frac{d^2V_{eff}}{d\chi^2} = [-1 + 2\alpha\Delta] > 0.$$

(14.83)

We see the effective potential derivative vanishes at

$$\chi_0 = \frac{\gamma}{2\Delta\alpha - 1}.$$

(14.84)

This will be a minimum, provided $\Delta > \frac{1}{2\alpha}$.

For an amount of disorder above ant the threshold $\Delta \in [\frac{1}{2\alpha}, +\infty)$, we have $M = -\chi_0 < 0$, and the spectral weight corresponding to (14.67) will be

$$N(\omega) = \frac{m^{3/2}}{2^{1/2}\pi^2}\left[\theta(\omega + |M|)\sqrt{\omega + |M|} + \theta(-\omega - |M|)\sqrt{-\omega - M}\right].$$

(14.85)

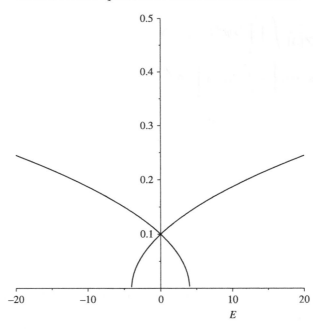

Figure 14.6 The density of states of the Anderson insulator. The states outside the central region of the band are localized and therefore non-conducting.

Observe that, contrary to the Mott insulator – which had allowed states for $\{\omega | \omega > |M|\} \cup \{\omega | \omega < -|M|\}$, therefore exhibiting a gap for $\{\omega | \omega \in [-|M|, +|M|]\}$ – the Anderson insulator possesses allowed states for the set $\{\omega | \omega > -|M|\} \cup \{\omega < +|M|\} = \{\omega | \omega \in (-\infty, +\infty)\}$, namely, for the whole band. It happens, however, that only for the states with $\{\omega | \omega \in [-|M|, +|M|]\}$ will the spectral weight will be real. This means that only for this interval in the center of the band, the system will present propagating, extended states. For the rest of the band, namely, on the band edges, the states will be localized and therefore non-conducting. This phenomenon is known as Anderson localization [95]. The line separating conducting from non-conducting states is called "mobility edge."

Notice that the onset of the localized phase occurs at the critical amount of disorder $\Delta_c = \Delta > \frac{1}{2\alpha}$. For $\Delta < \frac{1}{2\alpha}$, the system is a normal metal. Observe that $|M(\Delta_c)| \to \infty$, implying the width of the conducting region is infinite at the transition point. As we increase the amount of disorder, namely Δ, the position of the mobility edge, $|M(\Delta)|$, decreases, thus making the conducting region ever narrower and eventually vanishing for $\Delta \to \infty$.

While in a Mott insulator the interaction opens a gap in the middle of a metallic band, in an Anderson insulator, disorder generates a mobility edge in the band, separating a region of extended, conducting states in the middle from localized, non-conducting states in its edges.

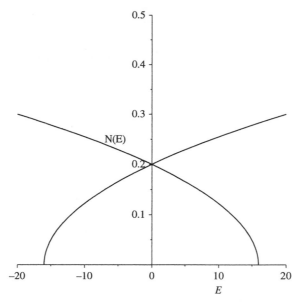

Figure 14.7 The density of states of the Anderson insulator. The amount of disorder is less than that in the previous figure, thus producing a larger conducting region. Below the disorder threshold, the width of such region becomes infinite and the system is a metal.

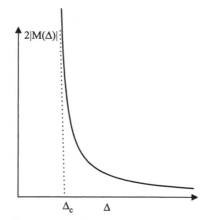

Figure 14.8 The width of the mobility region of the band of a 3D Anderson insulator as a function of the amount of disorder Δ. Notice that below a critical amount of disorder, Δ_c, the mobility region covers the whole band.

It is a characteristic feature of the Anderson transition that in d = 3 there exists a finite threshold for the amount of disorder, below which the system is a normal metal. In d = 1, 2, on the other hand, it can be shown that there is no such a threshold, the effect occurring with any amount of disorder.

15

The Dynamics of Polarons

The motion of electrons in a crystal lattice may be described at different levels of increasing profundity. We start with electrons moving on a static, ideal Bravais lattice. Next, we include the lattice vibrations, which manifest as phonons at a quantum level. Through the detailed study of the electron-phonon interaction we can understand, for instance, the phenomenon of superconductivity. There is an interesting effect, which arises when the full interplay between the electron and the crystal background is taken into account. This is the backreaction an electron suffers after creating, by its very presence, a lattice distortion. Indeed, since the lattice sites are usually occupied by charged species, one should expect that the presence of an electron will create a local distortion in such a lattice. This generates a polarization charge density and current that backreacts on the electron, thus affecting its properties. As the electron moves through the lattice, it will be unavoidably yoked to this polarization cloud, which will be dragged along and will ever accompany it. The effect will be particularly strong in the case of ionic crystals. The resulting effective excitation consisting of the combination of the electron and the attached lattice distortion is called a polaron. In this chapter, we address the full dynamics of the many-polaron problem in one spatial dimension, thus deriving an effective field theory, which describes their resulting interaction. This, thereafter, can be exactly solved through the bosonization method [97].

15.1 The Polaron Hamiltonian

Consider a one-dimensional lattice with spacing a. The tight-binding electron Hamiltonian, after diagonalization in the k-space, reads

$$H_{el} = \sum_k \epsilon(k) C^\dagger(k) C(k), \tag{15.1}$$

where $C(k)$ is the annihilation operator for an electron of momentum $\hbar k$, and $\epsilon(k)$ is the tight-binding energy eigenvalues given by $\epsilon(k) = -2t \cos ka$ (1.52). Admitting the system has one electron per site, the band will be half-filled, with momenta ranging from $-\pi/2a \le k \le +\pi/2a$. Expanding the energy around the two Fermi points, we get two sets of electrons, respectively with dispersion relations $\epsilon_{R,L} = \pm v_F k$, namely right-movers and left-movers. The Hamiltonian, expressed in terms of the corresponding creation and annihilation operators, will be

$$H_{el} = \sum_k v_F k \left[c_R^\dagger(k)c_R(k) - c_L^\dagger(k)c_L(k) \right], \tag{15.2}$$

where v_F is the Fermi velocity.

We assume the phonons are acoustic with a dispersion relation $\tilde{\epsilon}(k) = s|k|$, where s is the speed of sound. Hence the phonon Hamiltonian will be

$$H_{ph} = \sum_k s|k|a^\dagger(k)a(k), \tag{15.3}$$

where $a(k)$ is the annihilation operator of a longitudinal phonon with momentum k. Notice that we have linearized the phonon energy around the origin in (2.9).

Now the electron-phonon interaction. This is a special case of (3.55) and (3.56):

$$H_{e-ph} = \sum_{k,q} \sum_{i=R,L} W(q)c_i^\dagger(k+q)c_i(k)\left[a(q) + a^\dagger(-q)\right], \tag{15.4}$$

where

$$W(q) = i\lambda \frac{q}{\sqrt{|q|}}. \tag{15.5}$$

The total Hamiltonian of the polaron problem is

$$H_{pol} = H_{el} + H_{ph} + H_{el-ph}. \tag{15.6}$$

This problem was first addressed by Landau and Pekar [98] and subsequently by Fröhlich [99] and Feynman [100].

In the next section, we derive a quantum field theory that describes acoustic polarons in one spatial dimension and eventually obtain an exact solution for it [97].

15.2 Field Theory Description

We start by introducing the electron and phonon quantum fields in the Schrödinger picture. The first one is given by

$$\psi(x) = \int \frac{dk}{2\pi} \begin{pmatrix} c_R(k) \\ c_L(k) \end{pmatrix} e^{ikx}. \tag{15.7}$$

The phonon field, conversely, is given by

$$\varphi(x) = \int \frac{dk}{2\pi} \frac{1}{\sqrt{s|p|}} \left[a(p)e^{ipx} + a^\dagger(p)e^{-ipx}\right] \tag{15.8}$$

and

$$\dot{\varphi}(x) = i \int \frac{dk}{2\pi} \frac{\sqrt{s|p|}}{2} \left[-a(p)e^{ipx} + a^\dagger(p)e^{-ipx} \right]. \tag{15.9}$$

We may easily re-express the above Hamiltonians in terms of the electron and phonon fields. Indeed, for the electron Hamiltonian we get

$$H_{el} = -i\hbar \int dx \psi^\dagger(x)\sigma^z \partial_x \psi(x), \tag{15.10}$$

whereas for the phonon Hamiltonian,

$$H_{ph} = \int dx \left[\dot{\varphi}^2(x) + (s\partial_x\varphi(x))^2 \right]. \tag{15.11}$$

Now, for the interaction Hamiltonian. Consider the two terms

$$H_I^{(1)} = \frac{1}{2} \int dx \psi^\dagger(x)\psi(x)(s\partial_x\varphi(x)) \tag{15.12}$$

and

$$H_I^{(2)} = \frac{1}{2} \int dx \psi^\dagger(x)\sigma^z\psi(x)\dot{\varphi}(x). \tag{15.13}$$

Inserting (15.7), (15.8) and (15.9) in the two expressions above, we get

$$H_I^{(1)} = i\frac{1}{2} \int_0^\infty dq \int dp \sqrt{\frac{s|q|}{2}} \left\{ \left[c_R^\dagger(k+q)c_R(k) + c_L^\dagger(k+q)c_L(k) \right] \right.$$
$$\left. \times \left[a(q) + a^\dagger(-q) \right] - \left[c_R^\dagger(k-q)c_R(k) + c_L^\dagger(k-q)c_L(k) \right] \left[a(-q) + a^\dagger(q) \right] \right\} \tag{15.14}$$

and

$$H_I^{(2)} = i\frac{1}{2} \int_0^\infty dq \int dp \sqrt{\frac{s|q|}{2}} \left\{ \left[c_R^\dagger(k+q)c_R(k) - c_L^\dagger(k+q)c_L(k) \right] \right.$$
$$\left. \times \left[-a(q) + a^\dagger(-q) \right] + \left[c_R^\dagger(k-q)c_R(k) - c_L^\dagger(k-q)c_L(k) \right] \left[-a(-q) + a^\dagger(q) \right] \right\}. \tag{15.15}$$

Combining $H_I^{(1)}$ and $H_I^{(2)}$, we obtain

$$H_I^{(1)} + H_I^{(2)} = i \int_0^\infty dq \int dp \sqrt{\frac{s}{2|q|}} \left\{ q \left[c_R^\dagger(k+q)c_R(k) + c_L^\dagger(k+q)c_L(k) \right] \right.$$
$$\left. \times \left[a(q) + a^\dagger(-q) \right] - q \left[c_R^\dagger(k-q)c_R(k) + c_L^\dagger(k-q)c_L(k) \right] \left[a(-q) + a^\dagger(q) \right] \right\}. \tag{15.16}$$

Changing the integration variable, $q \leftrightarrow -q$, in the last term, we get

$$H_I^{(1)} + H_I^{(2)} = i\sqrt{2s} \int_{-\infty}^{\infty} dq \int dp \, \frac{q}{\sqrt{|q|}} \left\{ \left[c_R^\dagger(k+q)c_R(k) + c_L^\dagger(k+q)c_L(k) \right] \right.$$
$$\left. \times \left[a(q) + a^\dagger(-q) \right] \right\}. \tag{15.17}$$

This is precisely the continuum version of the electron-phonon interaction Hamiltonian (15.4) and (15.5). We conclude that the dynamical polaron problem in one spatial dimension may be described by a quantum field theory with Hamiltonian density given by

$$\mathcal{H}_{Pol} = -i\hbar v_F \psi^\dagger \sigma^z \partial_x \psi + \frac{1}{2} \left[\pi^2 + (s\partial_x \varphi)^2 \right]$$
$$+ \frac{\lambda}{2\sqrt{2s}} \left[\psi^\dagger \psi (s\partial_x \varphi) + \psi^\dagger \sigma^z \psi \dot{\varphi} \right], \tag{15.18}$$

where $\pi = \dot{\varphi}$.

This Hamiltonian describes the kinematics of both electrons and phonons as well as the electron-phonon interaction originated by the local charge density associated to the lattice distortion produced by the electron itself. Here we will appreciate how powerful the quantum field theory formulation of the problem actually is. Indeed, it will allow in a very simple way, the obtainment of the effective polaron dynamics, which describes the whole "dressing" of the electrons by the phonons generated by the lattice distortion.

In order to achieve such an effective description, consider the partition functional of the fermions, namely

$$Z[\psi, \psi^\dagger] = \int D\pi \, D\varphi \exp\left\{ \frac{i}{\hbar} \int dx dt \left[\pi\dot{\varphi} - \mathcal{H}_{Pol}[\pi, \partial_x\varphi, \psi] \right] \right\}$$
$$= \int D\pi \, D\varphi \exp\left\{ \frac{i}{\hbar} \int dx dt \left[-i\hbar v_F \psi^\dagger \sigma^z \partial_x \psi + \frac{1}{2} \left[\pi^2 + (s\partial_x\varphi)^2 \right] \right. \right.$$
$$\left. \left. + \frac{\lambda}{2\sqrt{2s}} \left[\psi^\dagger \psi (s\partial_x\varphi) + \psi^\dagger \sigma^z \psi \pi \right] \right] \right\}. \tag{15.19}$$

Notice that Z factors out in two functional integrals over π and φ. Integrating out the phonon field, we obtain

$$Z = \int D\psi \, D\psi^\dagger \exp\left\{ \frac{i}{\hbar} \int dx dt \left[-i\hbar\psi^\dagger \partial_t \psi - i\hbar v_F \psi^\dagger \sigma^z \partial_x \psi \right. \right.$$
$$\left. \left. + \frac{\lambda^2}{16s} \left[(\psi^\dagger \psi)^2 - (\psi^\dagger \sigma^z \psi)^2 \right] \right] \right\}. \tag{15.20}$$

The effective polaron Hamiltonian can be read from the expression above. Rescaling the fermion field as

$$\sqrt{v_F}\psi \rightarrow \psi, \tag{15.21}$$

we obtain

$$H_{Pol} = \int dx \left[-i\hbar\psi^\dagger \sigma^z \partial_x \psi - \frac{\lambda^2}{4v_F^2 s}\psi_R^\dagger \psi_R \psi_L^\dagger \psi_L \right]. \tag{15.22}$$

The interaction term can be written in terms of the current as (see (14.45))

$$-\frac{\lambda^2}{4v_F^2 s}\psi_R^\dagger \psi_R \psi_L^\dagger \psi_L = \frac{g}{2}\left[(j^0)^2 - (j^1)^2\right] = \frac{g}{2}\, j^\mu j_\mu, \tag{15.23}$$

where

$$j^\mu = (j^0, j^1) = \left(\psi^\dagger \psi, \psi^\dagger \sigma^z \psi\right) \tag{15.24}$$

is the fermion current and

$$g = -\frac{\lambda^2}{2v_F^2 s}. \tag{15.25}$$

We see that the effective polaron dynamics is governed by the massless Thirring model, which we saw in Chapter 10. The coupling g is strictly negative and corresponds, therefore, to the attractive regime.

15.3 Exact Solution

15.3.1 Bosonization

We saw in Chapter 10 that an exact solution to the massless Thirring model is provided by the method of bosonization. According to this, the polaron field correlation functions are given by (10.93), with ν such that

$$\frac{\nu}{2} = 1 + \frac{g^2}{2\pi(\pi + |g|)} \equiv 1 + \alpha. \tag{15.26}$$

The current is expressed, in terms of the bosonic scalar field, by (10.84). This, by its turn, according to (10.90) has its dynamical properties determined by the Lagrangean

$$\mathcal{L} = \frac{1}{2}\left(\frac{\pi}{\pi + |g|}\right)\dot{\phi}^2 - \frac{1}{2}\left(\frac{\pi + |g|}{\pi}\right)(s\partial_x\phi)^2. \tag{15.27}$$

From this, we may extract the bosonic field correlation function

$$\langle \phi(x, t)\phi(0, 0)\rangle = \int \frac{d\omega}{2\pi}\int \frac{dk}{2\pi}\, \frac{e^{i[kx - \omega t]}}{\left(\frac{\pi}{\pi+|g|}\right)\omega^2 - \left(\frac{\pi+|g|}{\pi}\right)s^2 k^2}. \tag{15.28}$$

The current-current correlator may be obtained by using (10.84) and (15.28), and yields

$$\langle jj \rangle(\omega, k) = \frac{1}{\pi} \frac{\omega^2}{\left(\frac{\pi}{\pi+|g|}\right)\omega^2 - \left(\frac{\pi+|g|}{\pi}\right)s^2 k^2}. \tag{15.29}$$

15.4 Optical Conductivity

Let us derive now the optical conductivity for the polaron system under consideration. Using the Kubo formula, (4.61) and (15.29), we immediately obtain

$$\sigma(\omega, k) = \frac{1}{\pi\omega} \frac{(\omega + i\epsilon)^2}{\left[\left(\frac{\pi}{\pi+|g|}\right)(\omega + i\epsilon)^2 - \left(\frac{\pi+|g|}{\pi}\right)s^2 k^2\right]}, \tag{15.30}$$

where, following the rule, we chose the prescription leading to the retarded Green function.

$\mathrm{Re}\,\sigma(\omega, k)$ is the so-called optical conductivity. From the expression above, we get

$$\sigma(\omega, k) = \frac{1}{\pi\omega} \frac{(\omega + i\epsilon)^2}{\left(\frac{\pi}{\pi+|g|}\right)(\omega + i\epsilon)^2 - \left(\frac{\pi+|g|}{\pi}\right)s^2 k^2}. \tag{15.31}$$

The optical conductivity is, then, given by

$$\sigma(\omega, k) = \frac{\left(\frac{\pi}{\pi+|g|}\right)\omega^2 - \left(\frac{\pi+|g|}{\pi}\right)s^2 k^2}{\pi\omega\left[\left(\frac{\left(\frac{\pi}{\pi+|g|}\right)\omega^2 - \left(\frac{\pi+|g|}{\pi}\right)s^2 k^2}{\omega}\right)^2 + 4\epsilon^2\right]}. \tag{15.32}$$

This is plotted in Fig. 15.1. Notice the existence of a cut in the optical absorption spectrum for frequencies less than the threshold frequency

$$\omega_c = \left(\frac{\pi + |g|}{\pi}\right)\omega_0(k), \tag{15.33}$$

where $\omega_0(k) = sk$ is the phonon frequency. There is an excellent agreement with experimental data for different one-dimensional systems, as one can see in [101].

15.5 Bipolarons

The fact that the effective polaron interaction is always attractive makes us wonder whether bound states could possibly form. These bipolarons have been suggested to be, upon condensation in the ground state, the carriers of the supercurrent, thereby being the agents of a new mechanism of superconductivity [102].

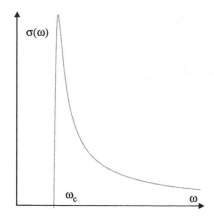

Figure 15.1 The polaronic optical conductivity showing a threshold frequency cut

From (10.80), we can write the bipolaron operator in bosonized form as

$$\Psi(x) = \lim_{x \to y} \psi_R(x)\psi_L(y) = \sqrt{\frac{-i}{4\pi}} \exp\left\{2ia \int_{-\infty}^{x} d\xi\, \Pi(\xi, t)\right\}, \qquad (15.34)$$

where $a = \sqrt{\pi + |g|}$.

From this and (10.170) we can readily obtain the bipolaron correlation function, namely,

$$\langle \Psi(x)\Psi^\dagger(y)\rangle = \frac{1}{4\pi |x - y|^{2a^2/\pi}}. \qquad (15.35)$$

From this we may extract the bipolaron operator vacuum expectation value

$$\langle \Psi(x)\Psi^\dagger(y)\rangle \xrightarrow{|x-y|\to\infty} |\langle\Psi\rangle| = 0. \qquad (15.36)$$

We see that for any value of the coupling g whatsoever, the vacuum expectation value of the bipolaron operators vanishes, as we may infer from the large distance behavior of the two-point correlation function (15.44). This means a polaronic superconducting phase there will never happen in this system.

15.6 Polaronic Excitons

One can also form neutral composite states of a polaron and an anti-polaron

$$\Phi(x) = \lim_{x \to y} \psi_R^\dagger(x)\psi_L(y) = \sqrt{\frac{-i}{4\pi}} \exp\left\{2ib\phi(\mathbf{x}, t)\right\}, \qquad (15.37)$$

where $b = \frac{\pi}{\sqrt{\pi+|g|}}$.

16

Polyacetylene

Carbon has four electrons in its outer electronic shell, which contains one s and three p orbitals. This, however, is the picture for an isolated atom. When the carbon atom is part of a material with a given crystal structure, almost invariably, for energetic reasons imposed by geometric constraints associated with the crystal structure of the material, two or more of these orbitals combine in a hybrid form that is a linear combination of the former. This interesting phenomenon is known as hybridization. When the four orbitals combine in the so-called sp^3 hybridization, the hybrid orbitals assemble in the form of a tetrahedron, which is the basic building block of diamond. Conversely, when three orbitals, namely, one s and two p, combine to create hybrid sp^2 orbitals, these are now co-planar, pointing to directions that make an angle of 120° among themselves. An unexpectedly vast and extremely interesting amount of physical phenomena emerge in materials formed by this form of carbon. Among these we find polyacetylene and graphene, respectively, possessing one- and two-dimensional structures.

Polyacetylene is a polymer presenting a sequence of CH radicals, formed by carbon atoms, each one with three sp^2 hybridized orbitals having covalent bonds with two adjacent carbon atoms, thus forming a zig-zag chain (trans-polyacetylene). The third hybridized orbital of each carbon atom is covalently bonded to a hydrogen atom. There remains a p-orbital, which does not hybridize, which is occupied by a single electron. Since this orbital admits up to two electrons with opposite spins, the carbon p-electrons in polyacetylene can move all over the chain, being therefore responsible for most of the interesting physics of this polymer. Especially interesting effects derive from the interplay of the carbon p-electrons with the lattice, imparticular with deformations thereof possessing nontrivial topological properties and, for this reason, called topological solitons.

Polyacetylene exhibits remarkable effects, such as the Peierls mechanism, induced by the electron-lattice interaction, by which a gap is generated where otherwise there would be a Fermi surface. This mechanism is completely analogous to the Yukawa mechanism of the Standard Model of the fundamental interactions, by which the whole mass of the elementary particles composing matter, such as quarks and leptons, is generated by a coupling to the Higgs field. It is amazing that essentially the same mechanism governs systems with typical energies separated by twelve orders of magnitude.

16.1 The Su–Schrieffer–Heeger Model

We now describe the standard model for polyacetylene, known as the Su–Schrieffer–Heeger Model [103].

Consider a one-dimensional Bravais lattice with spacing a, having a CH radical occupying each site. We shall describe the kinematics of the carbon p-electrons, which occur in the number of one per site, by means of a tight-binding Hamiltonian, which reads

$$H_e = -t \sum_n \sum_{\sigma=\uparrow,\downarrow} \left[c_\sigma^\dagger(n+1)c_\sigma(n) + c_\sigma^\dagger(n)c_\sigma(n+1) \right], \qquad (16.1)$$

where t is the hopping parameter and $c_\sigma^\dagger(n)$ is the creation operator of a carbon p-electron with spin $\sigma = \uparrow, \downarrow$ at the site n.

Diagonalizing the tight-binding Hamiltonian, we get

$$H_e = \sum_k \sum_{\sigma=\uparrow,\downarrow} \epsilon(k)c_\sigma^\dagger(k)c_\sigma(k), \qquad (16.2)$$

where $\epsilon(k) = -2t\cos ka$, according to (1.52). Since each state accommodates up to two electrons with opposite spins, the p-electrons will fill half the first Brillouin zone, namely, the states with $|k| \in [0, \pi/2a]$. Within this non-interacting electron approach, therefore, the system will be a metal with two Fermi points at $k = \pm\pi/2a$. For trans-polyacetylene, $t \simeq 2.5 \, eV$ and $a \simeq 0.122 \, nm$. Polyacetylene and other one-dimensional systems with one electron per site provide a first concrete example of the Dirac sea, a concept put forward by Dirac in order to manage the negative energy solutions of his equation. Further examples can be found in two-dimensional systems such as graphene, for instance. An electron-hole pair is created whenever an electron from the valence band is promoted to the conduction band, leaving there a vacancy.

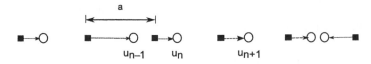

Figure 16.1 The lattice of polyacetylene: $u_n = x_n - na$

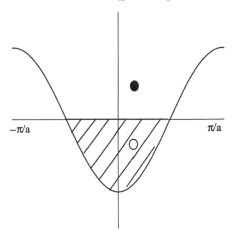

Figure 16.2 The realization of the Dirac sea in polyacetylene

Now consider the lattice Hamiltonian. Call u_n the displacement of the CH radical located nearest to the Bravais lattice site n, with respect to this site. This is given by $u_n = x_n - na$, where x_n is the position of the CH radical. Within the harmonic approximation, the lattice elastic energy is described by

$$H_L = \sum_n \frac{p_n^2}{2M} + \frac{1}{2} K \sum_n (u_{n+1} - u_n)^2 , \qquad (16.3)$$

where M is the CH radical mass, $p_n = M\dot{u}_n$, its momentum and K, the lattice elastic constant, being $K \simeq 21 \frac{eV}{A^2}$, for trans-polyacetylene.

Let us now turn to the electron-lattice interaction. Hamiltonian (16.1) describes the motion of electrons on a perfect Bravais lattice. Then, according to Bloch's Theorem, considering there is one electron per site, we would have a metallic system. This is confirmed by the tight-binding solution (16.2).

In the actual system, however, the real CH-radical positions differ from the Bravais lattice sites by u_n. It is intuitive, therefore, to change $t_{n,n+1}$, the hopping parameter between sites n and $n + 1$ in (16.1), by a term that is linear in the actual distance between adjacent radicals and reduces to zero when such distance equals the original Bravais lattice spacing a. Then, following [103], we write

$$t_{n,n+1} = t + \alpha \left(u_n - u_{n+1} \right) , \qquad (16.4)$$

where t is the original hopping parameter in (16.1) and α is an electron-lattice coupling constant, being $\alpha \simeq 4.1 \frac{eV}{A}$, for trans-polyacetylene. Introducing now the so-called dimerization parameter, which is defined as

$$y_n = (-1)^n u_n, \qquad (16.5)$$

we have

$$t_{n,n+1} = t + \alpha(-1)^n (y_n + y_{n+1}).$$ (16.6)

For a constant $y_n = y_0$, we see that the hopping is alternately enhanced or attenuated, as the CH radical positions are alternately shifted to the left or to the right of the Bravais lattice sites.

The SSH model is described by the Hamiltonian [103]

$$H_{SSH} = H_e + H_L,$$ (16.7)

where the hopping parameter in H_e is given by (16.4).

16.2 The Takayama–Lin-Liu–Maki Model

For excitations with low momentum, k, compared with the Fermi momentum, which has a modulus $|k_F| = \pi/2a$, we may linearize the tight-binding dispersion relation about the two Fermi points, $k_F = \pm\pi/2a$, namely,

$$\epsilon(k) \simeq \epsilon_F \pm 2tak = \pm v_F k$$ (16.8)

for $\epsilon_F = 0$, $v_F = 2ta$ and $-\Lambda \le k \le \Lambda$, where Λ is a momentum cutoff such that $\Lambda \ll |k_F|$.

The corresponding electronic excitations have momentum, respectively, given by

$$k_R = +\pi/2a + k \qquad -\Lambda \le k \le \Lambda$$
$$k_L = -\pi/2a - k \; ; \quad -\Lambda \le k \le \Lambda.$$ (16.9)

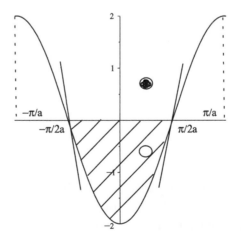

Figure 16.3 The linearization around the Fermi points of the Dirac sea

Since k_R and k_L are, respectively, positive and negative, the corresponding electronic excitations are called right-movers and left-movers. The corresponding excitations will have energies, with respect to the Fermi level,

$$\epsilon_R(k) = +v_F k \qquad -\Lambda \leq k \leq \Lambda$$
$$\epsilon_L(k) = -v_F k \qquad -\Lambda \leq k \leq \Lambda. \tag{16.10}$$

The ground state will present each of the negative energy states occupied with two electrons with opposite spins. It consists, therefore, in a concrete realization of the Dirac sea, the concept created by Dirac to solve the problem of the negative solutions of his equation. This concept, even though it did not correspond to the reality, led Dirac to predict the existence of antimatter, which was soon confirmed experimentally. As it turns out, however, several condensed matter systems exhibit the ground state proposed by Dirac. In connection to this, the state obtained by removing from it an electron with a negative energy behaves precisely like an electron with positive energy and an opposite charge. Such a "hole" plays the same role in condensed matter as the positron plays in Dirac's theory.

Within a momentum cutoff Λ, such that $|k| \leq \Lambda \ll k_F$, we can write the electronic Hamiltonian as

$$H_e = \sum_k \sum_{\sigma=\uparrow,\downarrow} v_F k \left[c_{R,\sigma}^\dagger(k) c_{R,\sigma}(k) - c_{L,\sigma}^\dagger(k) c_{L,\sigma}(k) \right], \tag{16.11}$$

where $c_{R,\sigma}^\dagger(k)$ creates an electron with wave-vector $+\pi/2 + k$ and energy $\epsilon(k) = v_F k$, whereas $c_{L,\sigma}^\dagger(k)$ creates an electron with wave-vector $-\pi/2 + k$ and energy $\epsilon(k) = -v_F k$, both for $|k| \leq \Lambda \ll k_F$.

Introducing the electron field as

$$\psi_\sigma(x) = \begin{pmatrix} \psi_{L,\sigma}(x) \\ \psi_{R,\sigma}(x) \end{pmatrix} = \int \frac{d^2k}{(2\pi)^2} \begin{pmatrix} c_{L,\sigma}(k) \\ c_{R,\sigma}(k) \end{pmatrix} e^{ik\cdot x}, \tag{16.12}$$

we may express the continuum electron Hamiltonian as

$$H_e = -iv_F \int dx\, \psi_\sigma^\dagger \partial_x \sigma^z \psi. \tag{16.13}$$

Consider now the electron-lattice interaction Hamiltonian

$$H_{eL} = \alpha \sum_n \sum_{\sigma=\uparrow,\downarrow} (-1)^n (y_n + y_{n+1}) \left[c_\sigma^\dagger(n+1) c_\sigma(n) + c_\sigma^\dagger(n) c_\sigma(n+1) \right]. \tag{16.14}$$

Going to momentum space, and writing

$$F_n = (-1)^n (y_n + y_{n+1}) = \sum_q F(q) e^{iqna}, \tag{16.15}$$

one finds, using the fact the Fourier transform of a product is a convolution,

$$H_{eL} = \alpha \sum_{k,q} \sum_{\sigma=\uparrow,\downarrow} F(q) \left[c_\sigma^\dagger(k+q)c_\sigma(k) + c_\sigma^\dagger(k)c_\sigma(k+q) \right]. \tag{16.16}$$

We now assume $y_n + y_{n+1}$ is slowly varying, which implies $F_n = -F_{n+1}$. Then, it follows from (16.15) that $e^{iqa} = -1$ and, consequently $q = \frac{\pi}{a}$. Since $q = 2p_F$, we see from (16.16) that the lattice interaction backscatters L-electrons into R-electrons and vice versa.

Returning to coordinate space, and introducing the canonical transformation

$$c_{L,\sigma}(n) \longrightarrow i^n c_{L,\sigma}(n)$$
$$c_{R,\sigma}(n) \longrightarrow (-i)^n c_{R,\sigma}(n), \tag{16.17}$$

we get

$$H_{eL} = 2\alpha \sum_n \sum_{\sigma=\uparrow,\downarrow} y_n \left[c_{R,\sigma}^\dagger(n)c_{L,\sigma}(n) + c_{L,\sigma}^\dagger(n)c_{R,\sigma}(n) \right]. \tag{16.18}$$

The lattice and electron-lattice continuum Hamiltonians can be obtained by taking the continuum counterpart of y_n, after a rescaling,

$$y_n \longrightarrow \sqrt{\frac{a}{4K}} \Delta(x). \tag{16.19}$$

Using (16.12) and (16.18), we find

$$H_{eL} = \frac{\alpha}{\sqrt{4Ka}} \int dx \Delta \psi_\sigma^\dagger \sigma^x \psi_\sigma. \tag{16.20}$$

The lattice Hamiltonian corresponding to the elastic potential energy is given by

$$H_L = \frac{1}{2} \int dx \Delta^2. \tag{16.21}$$

The lattice kinetic energy may be neglected in a situation where the lattice characteristic time is much larger than the corresponding electronic time. Admitting such is the situation in polyacetylene, we obtain the continuum Lagrangean density describing this polymer [104]

$$\mathcal{L} = -i\psi_\sigma^\dagger \partial_t \psi_\sigma - iv_F \psi_\sigma^\dagger \partial_x \sigma^z \psi_\sigma - \frac{\alpha}{\sqrt{4Ka}} \Delta \psi_\sigma^\dagger \sigma^x \psi_\sigma - \frac{1}{2}\Delta^2. \tag{16.22}$$

16.3 The Gross–Neveu Model

16.3.1 The Model

The Gross–Neveu (GN) model describes the quartic self-interaction of a many-flavored fermion field [243]. It has the Lagrangean density given by

$$\mathcal{L}_{GN} = i\overline{\psi}_a\gamma^\mu\partial_\mu\psi_a - \frac{g^2}{2}\left(\overline{\psi}_a\psi_a\right)^2, \tag{16.23}$$

where $a = 1, \ldots, N$ is a flavor index and $\overline{\psi} = \psi^\dagger\gamma^0$, $\gamma^0 = \sigma^x$ and $\gamma^1 = \gamma^0\sigma^x$.

Interestingly, the GN model is closely related to the Lagrangean that describes the physics of the carbon p-electrons in polyacetylene [106], given by (16.22). In order to see that, let us transform the quartic interaction into a trilinear one by means of a Hubbard–Stratonovitch transformation as follows:

$$\exp\left\{-i\int d^2x\,\frac{g^2}{2}\left(\overline{\psi}_a\psi_a\right)^2\right\} = \int D\sigma\,\exp\left\{i\int d^2x\left[\frac{\sigma^2}{2} - g\sigma\overline{\psi}_a\psi_a\right]\right\}. \tag{16.24}$$

Hence, in terms of the σ field, we may express the Gross–Neveu Lagrangean as

$$\mathcal{L}_{GN} = i\overline{\psi}_a\gamma^\mu\partial_\mu\psi_a - \frac{1}{2}\sigma^2 - g\sigma\overline{\psi}_a\psi_a. \tag{16.25}$$

Now, choosing the Dirac matrices as $\gamma^0 = \sigma^x$, $\gamma^0\gamma^1 = \sigma^z$ and the coupling $g = \frac{\alpha}{\sqrt{4Ka}}$, we can write the polyacetylene Lagrangean in the static lattice regime, (16.22), as the Gross–Neveu Lagrangean. The σ-field corresponds to Δ and the fermion flavors correspond to the two spin components, $\sigma = \uparrow, \downarrow$.

Notice the system presents a discrete symmetry

$$\psi \to \gamma^5\psi \quad ; \quad \overline{\psi}\psi \to -\overline{\psi}\psi$$
$$\sigma \to -\sigma, \tag{16.26}$$

(where $\gamma^5 = \gamma^0\gamma^1$) besides the usual U(1) symmetry.

16.3.2 The Lattice Effective Potential

The electron-lattice interaction in polyacetylene produces profound effects both on the electronic properties and on the lattice itself. Let us first investigate the latter. For this purpose we are going to integrate over the fermion field in order to determine the effective potential of the σ-field (or, equivalently the Δ-field):

$$e^{iS_{eff}[\sigma]} = \exp\left\{i\int d^2x\left[-\frac{1}{2}\sigma^2\right]\right\}$$
$$\times \frac{1}{Z_0}\int D\psi\,D\overline{\psi}\,\exp\left\{i\int d^2x\left[\overline{\psi}_a(i\slashed{\partial} - g\sigma)\psi_a\right]\right\}. \tag{16.27}$$

Evaluating the quadratic fermionic functional integral to leading order in $1/N$, we get the effective potential

$$V_{eff}(\sigma) = \frac{1}{2}\sigma^2 + NTr \ln\left[1 - \frac{g\sigma}{i\partial}\right]$$

$$V_{eff}(\sigma) = \frac{1}{2}\sigma^2 + 2N \int \frac{d^2k}{(2\pi)^2} \ln\left[\frac{k^2 + g^2\sigma^2}{k^2}\right]. \tag{16.28}$$

The k-integral, according to (16.10) sweeps a region delimited by the cutoff, where $|k| \le \Lambda$. The result is

$$V_{eff}(\sigma) = \frac{1}{2}\sigma^2 + \frac{g^2N}{4\pi}\sigma^2\left[\ln\frac{g^2\sigma^2}{\Lambda^2} - 1\right]. \tag{16.29}$$

Polyacetylene is an example where the momentum cutoff Λ is unphysical; therefore, physical quantities should not depend on it. In order to get rid of the cutoff dependence, we introduce the renormalized potential by subtracting from the expression above, the same expression at a certain fixed value σ_0, namely

$$V_{R,eff}(\sigma) = V_{eff}(\sigma) - V_{eff}(\sigma_0). \tag{16.30}$$

The finite part of the renormalized effective potential is fixed by the renormalization condition

$$\left(\frac{\partial^2 V_{R,eff}}{\partial\sigma^2}\right)_{\sigma=\sigma_0} = 1 + \frac{g^2N}{2\pi}. \tag{16.31}$$

This implies

$$\Lambda = g\sigma_0. \tag{16.32}$$

Inserting this result in (16.29), we obtain the renormalized effective potential

$$V_{eff}(\sigma) = \frac{1}{2}\sigma^2 + \frac{g^2N}{4\pi}\sigma^2\left[\ln\frac{\sigma^2}{\sigma_0^2} - 1\right]. \tag{16.33}$$

Observe that this is invariant under the operation $\sigma \to -\sigma$.

16.3.3 Renormalization Group Analysis

The finite effective potential above was obtained by a subtraction made at the finite point σ_0, which thereby fixes an energy-momentum renormalization scale. Then the renormalization group equations tell us $V_{eff}(\sigma)$ does not depend on the σ_0 scale, provided we modify the coupling constant and the field itself by the corresponding quantities g_R and σ_R as we change this scale.

The effective potential, indeed, will satisfy the renormalization group equation

$$\left(\sigma_0\frac{\partial}{\partial\sigma_0} + \beta(g_R)\frac{\partial}{\partial g_R} - \gamma(g_R)\sigma_R\frac{\partial}{\partial\sigma_R}\right) V_{eff}(\sigma_R, g_R, \sigma_0) = 0, \tag{16.34}$$

where the $\beta(g_R)$ function is defined by (6.75) and the so-called anomalous dimension γ, by (6.76). In order to convince ourselves that this is the proper renormalization group equation for the effective potential, observe that, by inserting (6.47) in (16.34), we do obtain, according to (6.74), the correct renormalization group equation for the proper vertices $\Gamma^{(n)}$, which appear in the effective potential expansion (6.47).

We conclude that a change in the renormalization point σ_0 would be promptly compensated by corresponding finite renormalizations in the coupling parameter and in the field in such a way that the physical effective potential remains unchanged. Inserting (16.33) in (16.34), we obtain the explicit form of the β and γ functions, namely,

$$\beta(g) = -\frac{g^3 N}{2\pi + g^2 N}$$

$$\gamma(g) = \frac{\beta(g)}{g}. \tag{16.35}$$

A negative β-function implies, according to (6.75), that as we increase the energy scale, the effective coupling parameter of the theory becomes ever smaller, in such a way that asymptotically it tends to zero and, consequently, the theory becomes free in the regime of high momentum-energy scales. This phenomenon, known as "asymptotic freedom," also occurs in non-abelian gauge theories such as QCD. It is remarkable we identify it in a field theory model for polyacetylene. This, however, is not the only similarity between polyacetylene and the Standard Model.

16.4 The Peierls–Yukawa Mechanism

Having established the physical effective potential for the lattice field σ that corresponds to Δ in polyacetylene, namely (16.33), let us determine what are the minima of such potential. We have

$$V'_{eff}(\sigma) = \sigma \left[1 + \frac{g^2 N}{2\pi} \ln \frac{\sigma^2}{\sigma_0^2} \right]. \tag{16.36}$$

Imposing $V' = 0$ yields the solutions

$$\overline{\sigma} = \pm \sigma_M = \pm \sigma_0 \exp\left[-\frac{\pi}{g^2 N} \right]. \tag{16.37}$$

The second derivative calculated at σ_M is

$$V''(\sigma_M) = \frac{g^2 N}{\pi} > 0, \tag{16.38}$$

thus implying $\sigma = \pm \sigma_M$ are the minima of the effective potential (16.33).

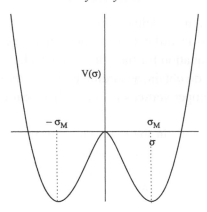

Figure 16.4 The effective potential for the lattice field in polyacetylene

The existence of nontrivial minima $\pm\sigma_M$ characterizes the occurrence of spontaneous dynamical breakdown of the symmetry $\sigma \to -\sigma$ presented by the original system.

In the original TLM model for polyacetylene, $\sigma(x)$ corresponds to $\Delta(x)$. According to (16.19), a constant $\Delta = \pm\Delta_M$, where $\Delta_M = \sigma_M$, given by (16.37) implies a constant y_n, namely, a constant dimerization parameter with two possible, degenerate values, $\bar{y}_n = \pm y_0$, for which the CH radicals will be at the positions $u_n = \pm(-1)^n y_0$.

For polyacetylene, the lattice parameter is $a \simeq 0.122\ nm$ and the dimerization parameter at the ground state is $y_0 \simeq 0.004\ nm$. Each of the minima of the effective potential $\pm\sigma_M$ corresponds to a pattern of the polyacetylene chain, where double $\sigma\pi$ bonds alternate with single σ bonds, either at the left or at the right of each Bravais lattice site.

For stability reasons, we must shift the Δ-field in (16.22) around the vacuum value $|\sigma_M|$. This will generate an extra term, such that the quadratic part of the electronic Hamiltonian becomes

$$\mathcal{H} = -iv_F\psi_\sigma^\dagger\partial_x\sigma^z\psi_\sigma + M\psi_\sigma^\dagger\sigma^x\psi_\sigma. \tag{16.39}$$

Diagonalizing this Hamiltonian, we find the energy eigenvalues

$$\epsilon(k) = \pm\sqrt{v_F^2 k^2 + M^2} \quad ; \quad M = \frac{\alpha}{\sqrt{4Ka}}|\sigma_M|, \tag{16.40}$$

where σ_M is given by (16.37). We see that the energy spectrum opens gap $2M$ precisely at the Fermi points. For polyacetylene, the gap is $2M \simeq 1.5\ eV$. This mechanism of generating a gap for the electrons by coupling them to the lattice through a $\bar{\psi}\psi\Delta$ coupling is known as the Peierls mechanism. The lattice distortion produced by the dimerization is directly responsible for opening a gap in the electronic spectrum. The system, consequently, is actually not a metal, but an insulator. This kind of a coupling was first introduced by Yukawa in order to describe the interaction of protons and neutrons through the exchange of π-mesons [107] by the hypothetic new massive particles proposed by him to be the mediators of the strong interaction among nucleons. The π-mesons were soon observed experimentally by Lattes, Occhialini and Powell [108], and later on it became clear they were but a quark-antiquark bound state.

It is remarkable that the masses of quarks and leptons in the Standard Model (SM) of the fundamental interactions are generated by introducing precisely the same coupling, involving fermions and a scalar field, exactly as in (16.22) and (16.23) for polyacetylene. In the SM case, however, the scalar field is neither the dimerization field Δ nor the pion field, but rather the Higgs field. The mechanism works as the scalar field develops a nonzero vacuum expectation value, which becomes proportional to the mass or, equivalently, to the electronic gap. In the case of the Higgs field, the classical potential is already chosen to possess such nontrivial minima. In the case of the lattice field, Δ, conversely, it acquires the nonzero vacuum expectation value dynamically through quantum corrections generated by the electron-lattice interaction. It is remarkable that essentially the same mechanism works in systems being that different and with energy scales so far apart.

16.5 Solitons in Polyacetylene

The dimerization of the CH chain and the corresponding opening of a gap in the electronic spectrum are not the only interesting effects produced by the electron-lattice interaction in polyacetylene. Let us examine here some further, quite appealing consequences of such interplay. Let us consider the Hamiltonian eigenstates. From (16.22) and (16.25), we may extract the one-particle Hamiltonian

$$h = \begin{pmatrix} -iv_F\partial_x & g\Delta(x) \\ g\Delta(x) & iv_F\partial_x \end{pmatrix}. \tag{16.41}$$

This has the following eigenvalues equation:

$$\begin{pmatrix} -iv_F\partial_x & g\Delta(x) \\ g\Delta(x) & iv_F\partial_x \end{pmatrix} \begin{pmatrix} u_n(x) \\ v_n(x) \end{pmatrix} = E_n \begin{pmatrix} u_n(x) \\ v_n(x) \end{pmatrix}, \tag{16.42}$$

which corresponds to the Bogoliubov–de Gennes equation [109].

This admits the zero-energy ($E_n = 0$) solution [104, 106]

$$\Delta_S(x) = \Delta_M \tanh\left(\frac{x}{\xi}\right)$$

$$\begin{pmatrix} u_0(x) \\ v_0(x) \end{pmatrix} = N_0 \text{sech}\left(\frac{x}{\xi}\right) \begin{pmatrix} 1 \\ -i \end{pmatrix}, \tag{16.43}$$

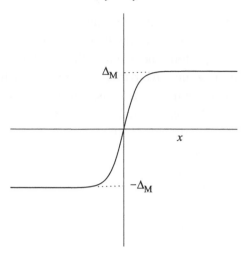

Figure 16.5 The soliton lattice defect in polyacetylene

and the negative energy ($E_n = -\epsilon(k) < 0$) solutions [106, 104],

$$
\begin{pmatrix} u_n(x) \\ v_n(x) \end{pmatrix} = N_- e^{ikx} \begin{pmatrix} \tanh\left(\frac{x}{\xi}\right) + i \left(\frac{\epsilon(k)-v_F k}{\Delta_M}\right) \\ -i \tanh\left(\frac{x}{\xi}\right) - i \left(\frac{\epsilon(k)+v_F k}{\Delta_M}\right) \end{pmatrix},
\tag{16.44}
$$

where

$$
\xi = \frac{v_F}{g\Delta_M} \quad ; \quad N_0 = \sqrt{\frac{\xi}{4}}
\tag{16.45}
$$

and

$$
\epsilon(k) = \sqrt{v_F^2 k^2 + M^2} \quad ; \quad N_- = \frac{\Delta_M}{2\epsilon(k)\sqrt{2\pi}}.
\tag{16.46}
$$

The physical value for polyacetylene is $\xi \simeq 7a$.

We identify the solution $\Delta_S(x)$ with the topological soliton configuration (8.26), corresponding to the nontrivial mapping $\Pi_0([0, \pm 1]) = Z(2)$ and possessing a unit of the topological charge (8.5), which classify such mappings. The structure of the lattice effective potential minima in polyacetylene provides the same kind of mapping between the asymptotic spatial behavior of the classical solutions and the vacuum manifold as in the spontaneously broken φ^4-theory in $d = 1$, hence the similarity of the topological field configurations. We will see that the topological nature of the soliton solution will have a profound influence on the electronic properties of polyacetylene.

Observe that the soliton solution $\Delta_S(x)$ is a defect separating two portions of the dimerized polyacetylene chain corresponding to different minima of the lattice effective potential. The parameter ξ measures the size of the distortion connecting

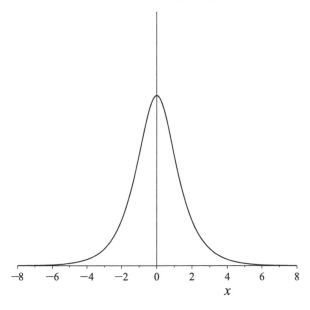

Figure 16.6 The wave-function for the soliton midgap state in polyacetylene

the two opposite dimerizations. We see that it extends for about seven lattice parameters. Let us determine now the soliton energy in polyacetylene. For this, notice that equation (16.42) must be supplemented by

$$g\Delta(x) = \overline{\psi}\psi = \sum_n \left[u_n^*(x)v_n(x) + v_n^*(x)u_n(x) \right],\qquad(16.47)$$

which is obtained from (16.22) and (16.25) by functional derivation with respect to Δ. The sum above is supposed to sweep all the occupied states up to the Fermi level, which is assumed to be $E_F = 0$.

The soliton energy, then, is given by the difference between the total energy of the occupied states in the presence of a soliton configuration and the same total energy in the presence of dimerized ground state, namely

$$E_S = E[\Delta_S] - E[\Delta_M],\qquad(16.48)$$

where, for an arbitrary lattice configuration Δ, we have

$$E[\Delta] = \sum_{E_n<0} E_n(\Delta) + \frac{1}{2}\int dx\,\Delta^2,\qquad(16.49)$$

where $E_n(\Delta)$ are the eigenvalues of (16.42).

Inserting (16.43) and (16.44), as well as the corresponding expressions for the constantly dimerized ground state [106], in (16.48), we get the soliton energy [103, 104, 106]

$$E_S = \frac{2}{\pi} \Delta_M. \tag{16.50}$$

We see that it costs less energy to create a soliton state than to excite a negative electron from the valence to the conduction band, which would require an energy of $2\Delta_M$. For topological reasons, however, we cannot create a single soliton, since this would imply a change in the boundary conditions. Soliton-antisoliton pairs, nevertheless, can be created.

16.6 Polarons in Polyacetylene

We now consider another kind of combined electron-lattice solution for the energy eigenvalue equation (16.42) and consistency gap equation (16.47) derived from the quantum field theory for polyacetylene. These are the polaron solutions [106], which consist in a non-topological, local deformation of the basic, uniformly dimerized ground state. The energy eigenvalues are symmetrically located in the midgap and each of the corresponding localized eigenstates can accommodate up to two electrons with opposite spins. These are, therefore, yoked to this lattice deformation, hence characterizing a polaron. The analytic form of the polaron solution is surprisingly simple [106], namely

$$\Delta_P(x) = \Delta_M \left\{ 1 - \frac{1}{\sqrt{2}} \left\{ \tanh \left[\frac{(x + x_0)}{\sqrt{2}\xi} \right] - \tanh \left[\frac{(x - x_0)}{\sqrt{2}\xi} \right] \right\} \right\}, \tag{16.51}$$

where the relation

$$\tanh \left[\frac{\sqrt{2}x_0}{\xi} \right] = \frac{1}{\sqrt{2}}$$

fixes x_0.

The corresponding energy cigenvalues and eigenfunctions are $\epsilon_+ = \frac{\Delta_M}{\sqrt{2}}$, for

$$u_+(x) = v_+(x) = N_+ \text{sech} \left(\frac{x - x_0}{\sqrt{2}\xi} \right) \tag{16.52}$$

and $\epsilon_- = -\frac{\Delta_M}{\sqrt{2}}$, for

$$u_-(x) = v_-(x) = N_- \text{sech} \left(\frac{x + x_0}{\sqrt{2}\xi} \right), \tag{16.53}$$

where $N_+ = N_- = \sqrt{\frac{\xi}{4\sqrt{2}}}$.

The total polaron energy can be determined in a similar way as for the soliton, namely,

$$E_P = E[\Delta_P] - E[\Delta_M], \tag{16.54}$$

where $E[\Delta]$ is given by (16.49).

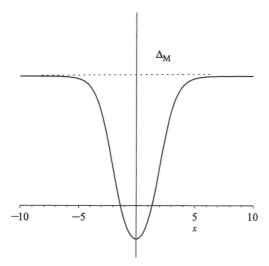

Figure 16.7 The polaron lattice defect in polyacetylene

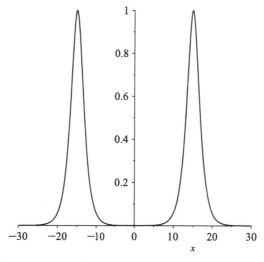

Figure 16.8 The wave-functions for the two-polaron midgap state in polyacetylene

The total polaron solution, then, can be found by considering, apart from the midgap solutions (16.51)–(16.53), also the negative solutions forming the whole valence band [106]. By inserting all these in (16.54), we get

$$E_P = \frac{2\sqrt{2}}{\pi} \Delta_M. \tag{16.55}$$

Notice that the polaron solution (16.51) is essentially a soliton-antisoliton bound state separated a distance $2x_0$ apart. As in the case of solitons, it is energetically more favorable to create an electron in a polaron state than to promote it from the valence to the conduction band. In the case of polarons, we do not have the topological restrictions that applied to the creation of solitons. These observations have a deep impact in the process of doping polyacetylene and the associated consequences in the conductivity of this polymer.

16.7 The Charge and Spin of Solitons

Charged particles and quasi-particles such as electrons and holes usually present a charge-spin relation in which a charge e corresponds to a spin $s = 1/2$ and vice versa. We will see here that this relation gets modified in polyacetylene, in the presence of soliton excitations, a phenomenon that was first predicted in [113, 114, 115]. This is another remarkably interesting consequence of the electron-lattice interaction in this material. We start by observing that polyacetylene is invariant under charge conjugation symmetry. By this we mean invariance under the discrete operation that exchanges particles by antiparticles or, equivalently, the fields that create the former by those that create the latter. Both the quantum field theory model for polyacetylene, given by (16.22) and (16.25), and the one-particle Hamiltonian (16.42), derived from these, present charge conjugation symmetry. For a system with the Fermi level chosen to be at $E_F = 0$, this implies the spectrum of states is symmetric by reflecting the energy about $E = 0$.

Consider now a system defined on a lattice containing N sites. The corresponding valence and the conduction bands will contain together, counting the two spin orientations, twice as many states as the number of sites, namely, $2N$. In charge conjugation symmetric systems, the number of states in the valence and conduction bands must be the same. Each band, therefore, would contain N states in this case.

In polyacetylene, each site has a radical CH, which, when depleted from the carbon p-electron, becomes positively charged. There is, consequently, a background of N positive charges, the same number of available electronic states. In the absence of a soliton, therefore, a full valence band would contain as many electrons as positive charges in the background, hence the whole system is neutral. Moreover, in a completely filled band the total spin is also equal to zero.

Let us examine what happens in the presence of a soliton. For general reasons, in a lattice with N sites, even when there exists a soliton, we must have the same total number of $2N$ states. Hence, the sum of the states in the two bands plus the two states ($\sigma = \uparrow, \downarrow$) in the midgap must be equal to $2N$. The total number of states in the two bands, thus, amounts to $2N - 2$. Charge conjugation symmetry implies, therefore, that valence and conduction bands will contain $N - 1$ states each. The

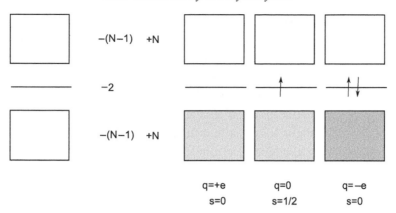

Figure 16.9 The unusual spin-charge relation for soliton excitations in polyacetylene. Each band provides $N - 1$ electronic states, while the ion cores have a total
$+Ne$ charge.

soliton supplies the midgap $E = 0$ two states deficit, thus making the total number
of states equal to $2N$.

Let us examine now the different ways the available states can be occupied in a
polyacetylene chain, in the presence of a soliton. Consider first a state where the
valence band is completely filled and the two midgap states are empty. We have just
seen that in the presence of a soliton, the valence band can take $N - 1$ electrons,
whereas the background charge is $+Ne$. It follows that a completely filled valence
band will have a charge $q = +e$. Since the total spin of a fully occupied band is
zero, this state has $(q, s) = (+e, 0)$.

Now consider a state with a fully occupied valence band and one of the midgap
states occupied. Now there is a total of N electrons, hence the total charge is
zero. The midgap spin, however, is unpaired, hence $(q, s) = (0, \pm 1/2)$ in this
case. Finally, consider a state with a fully occupied valence band and both midgap
states occupied. Now there is a total of $N + 1$ electrons, hence the total charge
is $q = -e$. The midgap spin now is paired, hence $(q, s) = (-e, 0)$ in this case.
Solitons with this unusual charge-spin relation have been observed experimentally
in polyacetylene. Optical absorption experiments, in particular, reveal the midgap
states associated with solitons and polarons [110, 111, 112].

For polarons, the charge-spin relation is the conventional one.

16.8 Conductivity in Polyacetylene

We have seen that, by virtue of the Peierls–Yukawa mechanism, pure polyacetylene
is an insulator, with a gap of approximately 1.5 eV. This quality, however, may be
changed by doping polyacetylene, say with halogen, receptor atoms such as F or

with alkali, donor atoms such as Na. It can be shown that at a doping on the order of 1 percent, the system suffers an insulator-metal transition. Actually, the electrical conductivity in polyacetylene exhibits a dramatic behavior as we dope it. Indeed, up to a doping percentage of approximately 10 percent, the conductivity can be increased by eleven orders of magnitude [116].

The study of soliton and polaron excitations in polyacetylene presented in the previous sections strongly suggests that the doped electrons in this material will form complex combined lattice-electronic excitations, either solitons or polarons, depending on the boundary conditions and on the number of such excitations. Polarons and solitons, consequently, must play a fundamental role in the transport properties of doped polyacetylene. In that respect, we may identify three different regimes where such excitations would contribute for the conductivity in this system.

We have seen that solitons or polarons invariably imply the existence of midgap electronic states. At a low-doping regime, below approximately 1 percent, the solitons or polarons are fixed. Electrons in the midgap states then can contribute to conductivity by thermal activation, very much like a semiconductor. In an intermediate regime of doping in between 1 and 10 percent, the solitons or polarons would become dynamical, themselves moving and being the carriers of electric charge. For dopings above 10 percent, we would have a metallic regime, with the formation of a conduction band with the corresponding Fermi surface.

The first regime can be understood within the theory of semiconductors with a classical description of solitons or polarons. The third regime, on the other hand, can also be understood with the theory of metals. The intermediate regime, however, would require a full quantum theory of solitons or polarons in order to account for the quantum properties of these dynamical excitations.

An evaluation of the contribution of fully quantized solitons to the conductivity of polyacetylene, in the intermediate doping regime, was performed in [17]. This study reveals that the contribution of quantum solitons to the conductivity in polyacetylene vanishes. Quantum polarons are most probably the relevant carriers of charge in the intermediate doping regime.

16.9 Index Theorem and Fermion Fractionalization

The existence of zero-energy eigenstates of the Dirac Hamiltonian in the presence of a background soliton configuration Δ_S is a particular case of the famous Atiyah–Singer theorem [117], which identifies the number of zero-energy eigenstates of an elliptic differential operator with the topological class of the background manifold where it is defined. The theorem is remarkable for connecting two different areas of mathematics, namely Analysis and Topology.

In the case of polyacetylene, it essentially states that the number of zero-energy eigenstates of the Dirac Hamiltonian operator

$$h = -i\hbar v_F \sigma_z \partial_x + g\sigma_x \Delta(x) \tag{16.56}$$

is equal to the topological charge of the background Δ-field configuration. The midgap zero-energy state associated to the soliton, therefore, is a reflection of the fact that this has topological charge equal to one.

Interestingly, in the case of polarons, which have zero topological charge, the Dirac operator has no zero-energy eigenstates, in agreement with the theorem.

17

The Kondo Effect

There are physical systems that lack a natural intrinsic energy scale to which we could refer in order to determine whether the system is in a high-energy or in a low-energy regime. For those systems, the physical properties would have the same character at any energy scale. Interestingly, however, there are systems which albeit not having *a priori* such a characteristic energy scale, yet are able to spontaneously generate it. This, consequently, usually provides a separation between strong and weak interaction regimes. A well-known example is QCD, which dynamically generates an energy scale $\Lambda \simeq 160~MeV$, such that for energies well above this the theory is essentially free, whereas for energies way below it the theory is strongly interacting and exhibiting a confining behavior.

The Kondo system is an archetypal condensed matter system presenting a similar situation. It consists of a good, non-magnetic metal in which some atoms with a permanent magnetic dipole moment are diluted at a very low concentration. The initial description of the system does not contain any energy scale, which could allow one to distinguish among different coupling regimes; nevertheless, an energy scale $T_0 \simeq 0.7~meV$ is dynamically self-generated, clearly separating a high-energy regime where the system behaves as a good metal and the electrons are essentially free, from a low-energy regime where the electrons are strongly coupled to the diluted magnetic moment. As the temperature is varied across the generated scale, a series of anomalies, which are referred to as the Kondo effect, are observed noticeably in the metal resistivity and in the impurity's effective magnetization and magnetic susceptibility.

The Kondo system can be mapped into a quantum field theory closely related to the chiral Gross–Neveu model, and thereafter its Hamiltonian can be exactly diagonalized by using the technique known as the Bethe Ansatz. This solution of the Kondo problem is a beautiful milestone in the field of applications of QFT in condensed matter.

17.1 The Kondo Model

The Kondo model [142] describes a magnetic impurity, highly diluted, in a non-magnetic metal, such as Mn in Au. Due to the low concentration, we may consider a single impurity at the origin, which thereby produces an effective spherical symmetry. The electrons of the host metal interact with the introduced impurity but are otherwise free. Expanding the electron momentum around the Fermi surface of a parabolic dispersion relation,

$$E = \frac{(\mathbf{k}_F + \mathbf{k})^2}{2m} \simeq E_F + \frac{1}{m}\mathbf{k}_F \cdot \mathbf{k}. \tag{17.1}$$

Since the Fermi momentum is radial, we have, up to a constant, $E \simeq v_F k$, where $k = \pm|\mathbf{k}|$. The linearization is valid within a region delimited by a cutoff Λ, such that $|\mathbf{k}| \ll \Lambda$.

The electron creation operator is $c_{k\sigma}^\dagger$, where $-\Lambda \le k \le \Lambda$ is a one-dimensional variable. This is possible due to the spherical symmetry, which decouples all non-radial electronic degrees of freedom. $\sigma = \uparrow, \downarrow$ are the two possible orientations of the electron spin.

The Kondo Hamiltonian [142, 150] contains a kinetic part and two interaction terms, one magnetic and one non-magnetic, between the electrons and the localized impurity:

$$H_K = \sum_k k c_{k\sigma}^\dagger c_{k\sigma} + J \sum_{k,k'} \mathbf{S}_I \cdot \left[c_{k\alpha}^\dagger \vec{\sigma}_{\alpha\beta} c_{k'\beta} \right] + J' \sum_{k,k'} \left[c_{k\alpha}^\dagger c_{k'\alpha} \right], \tag{17.2}$$

where $\mathbf{S}_I$ is the localized impurity spin operator. Summation over all spin orientations is understood. Notice that the second term involves a change in both the electron spin and momentum, whereas the third one, just in momentum.

Now, introducing the Schrödinger picture electron field

$$\psi_\sigma(x) = \sum_k e^{ikx} c_{k\sigma}, \tag{17.3}$$

we can re-write the Kondo Hamiltonian as the 1D field Hamiltonian

$$H_K = \int dx \left\{ i\psi_\sigma^\dagger(x)\partial_x \psi_\sigma + J\delta(x)\mathbf{S}_I \cdot \left[\psi_\alpha^\dagger \vec{\sigma}_{\alpha\beta} \psi_\beta \right] + J'\delta(x)\psi_\alpha^\dagger \psi_\alpha \right\}, \tag{17.4}$$

where the Dirac delta represents the localized impurity density.

Now, assuming the impurity has spin 1/2, we introduce the impurity fermion field [150] $\chi_\sigma(x)$ in such a way that the total number of impurities is equal to one, namely,

$$\delta(x) \longrightarrow \chi_\sigma^\dagger(x)\chi_\sigma(x)$$
$$\delta(x)\mathbf{S}_I \longrightarrow \chi_\alpha^\dagger(x)\vec{\sigma}_{\alpha\beta}\chi_\beta(x). \tag{17.5}$$

The Kondo Hamiltonian, then, becomes

$$H_K = \int dx \left\{ i\psi_\sigma^\dagger(x)\partial_x\psi_\sigma + J\left[\chi_\alpha^\dagger\vec{\sigma}_{\alpha\beta}\chi_\beta\right]\cdot\left[\psi_\alpha^\dagger\vec{\sigma}_{\alpha\beta}\psi_\beta\right] + J'\chi_\sigma^\dagger\chi_\sigma\psi_\alpha^\dagger\psi_\alpha\right\}. \quad (17.6)$$

Notice that the impurity field χ_σ does not have a kinetic term.

It is convenient to express both the electron and impurity fields as the two components of a fermion field, namely

$$\Psi_{a,\sigma}(x) \Leftrightarrow \begin{cases} \Psi_{1,\sigma}(x) = \psi_\sigma(x) \\ \Psi_{0,\sigma}(x) = \chi_\sigma(x). \end{cases} \quad (17.7)$$

In terms of this field, we may express the Kondo Hamiltonian density as

$$\mathcal{H}_K = \sum_{a=0,1} i\Psi_{a,\alpha}^\dagger a \partial_x \Psi_{a,\alpha} + J\left[\Psi_{0,\alpha}^\dagger\vec{\sigma}_{\alpha\beta}\Psi_{0,\beta}\right]\cdot\left[\Psi_{1,\alpha}^\dagger\vec{\sigma}_{\alpha\beta}\Psi_{1,\beta}\right]$$
$$+ J'\Psi_{0,\sigma}^\dagger\Psi_{0,\sigma}\Psi_{1,\alpha}^\dagger\Psi_{1,\alpha}. \quad (17.8)$$

Now, using (3.41), we can write, for $J = J'$,

$$\mathcal{H}_K = \sum_{a=0,1} i\Psi_{a,\alpha}^\dagger a \partial_x \Psi_{a,\alpha}$$
$$+ 2J\Psi_{0,\alpha}^\dagger\Psi_{0,\beta}\left[\delta_{\alpha\beta'}\delta_{\alpha'\beta}\right]\Psi_{1,\alpha'}^\dagger\Psi_{1,\beta'} \quad (17.9)$$

or, equivalently,

$$\mathcal{H}_K = \sum_{a=0,1} i\Psi_{a,\alpha}^\dagger a \partial_x \Psi_{a,\alpha} + 2J\left[\Psi_{0,\alpha}^\dagger\Psi_{1,\alpha}\right]\left[\Psi_{1,\beta}^\dagger\Psi_{0,\beta}\right]. \quad (17.10)$$

We will see that, in the same way the physics of polyacetylene was closely related to the Gross–Neveu model, accordingly, the Kondo system physics will be described by the chiral Gross–Neveu model. In the next section, we introduce this QFT, extract its main physical properties and relate them to the Kondo problem.

17.2 The Chiral Gross–Neveu Model

The chiral Gross–Neveu (CGN) model is defined by the following Lagrangean density in two-dimensional spacetime [152]:

$$\mathcal{L}_{CGN} = i\overline{\psi}_\alpha\partial\!\!\!/\psi_\alpha - \frac{g^2}{2}\left[\left(\overline{\psi}_\alpha\psi_\alpha\right)^2 - \left(\overline{\psi}_\alpha\gamma^5\psi_\alpha\right)^2\right], \quad (17.11)$$

where ψ is a Dirac field,

$$\psi_\sigma = \begin{pmatrix} \psi_{1,\sigma} \\ \psi_{2,\sigma} \end{pmatrix}, \quad (17.12)$$

where $\alpha = 1, \ldots, N$ is an arbitrary flavor index and the convention for the Dirac matrices is the same we used in (16.23).

Interestingly, this model is closely related to the Kondo model. In order to see this, we insert (17.12) into (17.11), whereby

$$
\mathcal{L}_{CGN} = i \left[\psi_{1,\alpha}^{\dagger} \left(\partial_0 + \partial_1 \right) \psi_{1,\alpha} + \psi_{2,\alpha}^{\dagger} \left(\partial_0 - \partial_1 \right) \psi_{2,\alpha} \right]
$$
$$
- \frac{g^2}{2} \left[\left(\psi_{1,\alpha}^{\dagger} \psi_{2,\alpha} + \psi_{2,\alpha}^{\dagger} \psi_{1,\alpha} \right)^2 - \left(\psi_{1,\alpha}^{\dagger} \psi_{2,\alpha} - \psi_{2,\alpha}^{\dagger} \psi_{1,\alpha} \right)^2 \right], \quad (17.13)
$$

which corresponds to the Hamiltonian density

$$
\mathcal{H}_{CGN} = i \left[\psi_{1,\alpha}^{\dagger} \partial_1 \psi_{1,\alpha} - \psi_{2,\alpha}^{\dagger} \partial_1 \psi_{2,\alpha} \right] + 2g^2 \left[\psi_{1,\alpha}^{\dagger} \psi_{2,\alpha} \right] \left[\psi_{2,\beta}^{\dagger} \psi_{1,\beta} \right]. \quad (17.14)
$$

The above Hamiltonian is invariant both under the U(1) and chiral U(1) continuous global symmetry operations

$$
\psi_{\alpha} \to e^{i\theta} \psi_{\alpha} \; ; \quad \psi_{\alpha} \to e^{i\theta \gamma^5} \psi_{\alpha}, \quad (17.15)
$$

where the latter can be written as

$$
\psi_{1\alpha} \to e^{i\theta} \psi_{1\alpha} \; ; \quad \psi_{2\alpha} \to e^{-i\theta} \psi_{2\alpha}
$$
$$
\psi_{2,\beta}^{\dagger} \psi_{1,\beta} \to e^{i2\theta} \psi_{2,\beta}^{\dagger} \psi_{1,\beta}
$$
$$
\psi_{1,\beta}^{\dagger} \psi_{2,\beta} \to e^{-i2\theta} \psi_{1,\beta}^{\dagger} \psi_{2,\beta}. \quad (17.16)
$$

We see that, for a number of flavors $N = 2$, this is identical to the Kondo Hamiltonian (17.10), except for the fact that in the chiral Gross–Neveu model we have right-movers and left-movers as the two components of the Dirac field whereas in the Kondo model we have only right-movers, the component associated to the impurity being static. Observe also that, $g^2 = J$ being dimensionless, neither the CGN nor the Kondo Hamiltonians have any intrinsic energy scale, as announced.

We can closely follow now the steps taken in Section 16.3 in order to extract the properties of the CGN model. These will then apply to the Kondo system [150].

We firstly transform the quartic interactions into trilinear ones by means of two Hubbard–Stratonovitch fields: σ and π. The first one is introduced by (16.24), whereas the second, by

$$
\exp \left\{ -i \int d^2 x (-) \frac{g^2}{2} \left(\overline{\psi}_a \gamma^5 \psi_a \right)^2 \right\} = \int D\pi \exp \left\{ i \int d^2 x \left[\frac{\pi^2}{2} - ig\pi \overline{\psi}_a \gamma^5 \psi_a \right] \right\}.
$$
$$
(17.17)
$$

The resulting Lagrangean is

$$
\mathcal{L}_{CGN} = i \overline{\psi}_{\alpha} \partial\!\!\!/ \psi_{\alpha} - \frac{1}{2} \left(\sigma^2 + \pi^2 \right) - g \left[\sigma \overline{\psi}_{\alpha} \psi_{\alpha} + i\pi \overline{\psi}_{\alpha} \gamma^5 \psi_{\alpha} \right]. \quad (17.18)
$$

Integrating on the fermion fields, we obtain the effective potential, which generalizes (16.33)

$$V_{eff}(\sigma, \pi) = \frac{1}{2}\left(\sigma^2 + \pi^2\right) + \frac{g^2 N}{4\pi}\left(\sigma^2 + \pi^2\right)\left[\ln\frac{\left(\sigma^2 + \pi^2\right)}{\sigma_0^2} - 1\right], \quad (17.19)$$

where σ_0 is a finite renormalization scale. The chiral U(1) symmetry manifests at
this level as an O(2) rotation symmetry in the (σ, π) space.

The minima of $V_{eff}(\sigma, \pi)$ occur at

$$\sqrt{\sigma_M^2 + \pi_M^2} = \sigma_0 \exp\left[-\frac{\pi}{g^2 N}\right]. \quad (17.20)$$

Choosing

$$\sigma_M = \sigma_0 \exp\left[-\frac{\pi}{g^2 N}\right] \quad ; \quad \pi_M = 0 \quad (17.21)$$

dynamically generates a gap for the fermions and an energy scale for the system.
We see that the scalar field π has zero mass, thus constituting the expected Goldstone boson. It has been shown, however [153], that π completely decouples from
the system in such a way that the physical fermions are not affected by the chiral
rotation. Thereby we can reconcile the dynamical mass generation with the preser-
vation of the chiral symmetry, which being continuous cannot be spontaneously
broken in d = 1, according to Coleman's theorem [171]. The renormalized effective
potential satisfies the renormalization group equation

$$\left(\sigma_0\frac{\partial}{\partial\sigma_0} + \beta(g_R)\frac{\partial}{\partial g_R}\right.$$

$$\left. - \gamma_\sigma(g_R)\sigma_R\frac{\partial}{\partial\sigma_R} - \gamma_\pi(g_R)\pi_R\frac{\partial}{\partial\pi_R}\right) V_{eff}(\sigma_R, \pi_R, g_R, \sigma_0) = 0, \quad (17.22)$$

where the $\beta(g_R)$ and $\gamma(g_R)$ functions are defined by (6.75) and (6.76).

Inserting (17.19) in (17.22), we obtain the explicit form of the β and γ functions,
namely,

$$\beta(g) = -\frac{g^3 N}{2\pi + g^2 N}$$

$$\gamma_\sigma(g) = \gamma_\pi(g) = \frac{\beta(g)}{g}. \quad (17.23)$$

A negative β function implies that at high energy scales the effective coupling
will vanish, thus rendering the system effectively free. At low energies, conversely,
it will be strongly interacting. The energy scale separating the two regimes is
dynamically generated by the system itself and is given by σ_M. The spontaneous

generation of such a characteristic energy scale is the essence of the Kondo effect. Its existence has been revealed in an early treatment of the problem [151].

17.3 Exact Solution and Phenomenology

The Kondo and the chiral Gross–Neveu Hamiltonians share the remarkable property of being exactly diagonalizable by means of the technique known as Bethe Ansatz [143]. The first was solved by Adrei and Wiegmann [144, 145], whereas the latter, by Andrei and Lowenstein [146, 147]. Once the exact energy eigenvalues are determined [148], one can derive a set of coupled integral equations for the exact free energy [149, 150]. From these, several properties of the system can be inferred in the asymptotically free, scaling and crossover regimes.

Analogous to the chiral Gross–Neveu model, which dynamically generates an energy scale given by (17.21), the Kondo system also spontaneously generates a scale T_0, given by [150]

$$T_0 = \Lambda \exp\left[-\frac{\pi}{2J}\right], \tag{17.24}$$

where $\Lambda \gg T_0$ is the energy/momentum cutoff. From this, one universally obtains temperature and magnetic crossover scales, T_K and T_H, given, respectively, by [150]

$$\frac{T_K}{T_0} = 4\pi \times 0.102676 \quad ; \quad \frac{T_H}{T_0} = \sqrt{\frac{\pi}{e}}. \tag{17.25}$$

The results for the impurity magnetization at $T = 0$ and magnetic susceptibility are [150], respectively,

$$\mathcal{M}_i = \begin{cases} \mu\left[1 - \frac{1}{2\ln\frac{\mu H}{T_H}}\right] & H \gg T_H \\ \frac{\mu}{\pi T_0} H & H \ll T_H \end{cases} \tag{17.26}$$

and

$$\chi = \begin{cases} \frac{\mu^2}{T}\left[1 - \frac{1}{\ln\frac{T}{T_K}}\right] & T \gg T_K \\ \frac{\mu^2}{\pi T_0} & T \ll T_K \end{cases}. \tag{17.27}$$

In the above expressions, the upper result is valid in the asymptotically free high-energy regime, whereas the lower, in the strongly coupled low-energy regime.

The above expressions should be compared with the free ones

$$\mathcal{M}_i^{(0)} = \mu$$

$$\chi^{(0)} = \frac{\mu^2}{T}. \tag{17.28}$$

Observe that both the magnetization and the magnetic susceptibility tend to the free expressions in the high-temperature regime. Physically, this corresponds to a limit where the electrons' interaction with the magnetic impurity asymptotically vanishes, thereby making the electrons essentially free. The opposite effect at low temperatures is produced by the strong antiferromagnetic interaction between the electrons and the impurity spin, which has the effect of completely screening its magnetic moment.

18

Quantum Magnets in 1D: Fermionization, Bosonization, Coulomb Gases and "All That"

Few systems in physics present such a rich variety of interconnections and unexpected equivalences as the quantum one-dimensional magnetic systems. These turn out to be closely related to strongly interacting one-dimensional electronic systems as well as to classical two-dimensional gases of Coulomb-interacting charged particles. Surprising mappings bridge these different systems, whose equivalence would otherwise be very difficult to anticipate. Interesting and important issues such as the Berezinskii–Kosterlitz–Thouless (BKT) transition, bosonization, supersymmetry and the thermodynamics of interacting gases are brought together in this fascinating subject.

18.1 From Spins to Fermions

18.1.1 The XYZ-Model

Consider a general quantum spin system on a one-dimensional lattice, with Hamiltonian given by

$$H_{XYZ} = \sum_n \left[J_x S_n^x S_{n+1}^x + J_y S_n^y S_{n+1}^y + J_z S_n^z S_{n+1}^z \right], \tag{18.1}$$

the so-called XYZ-Model. Important particular cases are the Heisenberg model ($J_x = J_y = J_z$), the XXZ-Model ($J_x = J_y$; $J_z \neq 0$) and the XY-Model ($J_x = J_y$; $J_z = 0$).

Here S_n^i, $i = x, y, z$ are spin operators acting on a Hilbert space associated to the site n and satisfying the angular momentum algebra

$$[S_n^i, S_m^j] = i \delta_{nm} \epsilon^{ijk} S_n^k. \tag{18.2}$$

We consider that the spin operators at each site are associated to a spin $s = 1/2$ physical system (atom, radical, etc.), thus a spin $1/2$ representation in terms of Pauli matrices is implicitly assumed for each site:

$$S_n^i = \frac{1}{2}\sigma^i. \tag{18.3}$$

We introduce now the operators

$$S_n^\pm = S_n^x \pm i S_n^y, \tag{18.4}$$

such that

$$S_n^+ S_n^- = \frac{1}{2} + S_n^z$$

$$S_n^- S_n^+ = \frac{1}{2} - S_n^z, \tag{18.5}$$

and

$$S_n^x S_{n+1}^x = \frac{1}{4}\left[S_n^+ S_{n+1}^- + S_n^- S_{n+1}^+ + S_n^+ S_{n+1}^+ + S_n^- S_{n+1}^-\right]$$

$$S_n^y S_{n+1}^y = \frac{1}{4}\left[S_n^+ S_{n+1}^- + S_n^- S_{n+1}^+ - S_n^+ S_{n+1}^+ - S_n^- S_{n+1}^-\right]. \tag{18.6}$$

In terms of these, we may rewrite the Hamiltonian as

$$H_{XYZ} = \sum_n \left[\left(\frac{J_x + J_y}{2}\right)\left[S_n^+ S_{n+1}^- + S_n^- S_{n+1}^+\right]\right.$$

$$\left. + \left(\frac{J_x - J_y}{2}\right)\left[S_n^+ S_{n+1}^+ + S_n^- S_{n+1}^-\right] + J_z S_n^z S_{n+1}^z\right]. \tag{18.7}$$

18.1.2 The Jordan–Wigner Transformation

We now derive a transformation that will map the spin operators into fermion operators. This is the Jordan–Wigner transformation [118], obtained already in the early days of quantum mechanics. As we will see, this transformation contains some basic elements later used in the process of bosonization and can be considered the precursor of the order-disorder methods introduced by Kadanoff and Ceva [30].

We start by defining the operator

$$K(n) = \exp\left\{i\pi \sum_{m=-\infty}^{n-1} S_m^+ S_m^-\right\}$$

$$K(n) = \exp\left\{i\pi \sum_{m=-\infty}^{n-1} \left(S_m^z + \frac{1}{2}\right)\right\}. \tag{18.8}$$

In terms of this, we introduce the new operators

$$c_n = K(n)S_n^- = \exp\left\{i\pi \sum_{m=-\infty}^{n-1} S_m^+ S_m^-\right\} S_n^-$$

$$c_n^\dagger = S_n^+ K^\dagger(n) = S_n^+ \exp\left\{-i\pi \sum_{m=-\infty}^{n-1} S_m^+ S_m^-\right\}. \tag{18.9}$$

It follows that these operators are genuine fermion operators, as one can infer from the anticommutation relations that follow directly from the above definition:

$$\{c_n, c_m\} = \{c_n^\dagger, c_m^\dagger\} = 0$$
$$\{c_n, c_m^\dagger\} = \delta_{nm}. \tag{18.10}$$

Moreover, we have, from (18.9),

$$c_n^\dagger c_n = S_n^+ S_n^- = S_n^z + \frac{1}{2}$$

$$S_n^z = c_n^\dagger c_n - \frac{1}{2}. \tag{18.11}$$

Inserting this result in (18.9) and inverting those expressions, we are able to obtain a mapping of pure spin operators into pure fermion operators, namely, the Jordan–Wigner transformation,

$$S_n^+ = c_n^\dagger \exp\left\{i\pi \sum_{m=-\infty}^{n-1} c_m^\dagger c_m\right\}$$

$$S_n^- = \exp\left\{-i\pi \sum_{m=-\infty}^{n-1} c_m^\dagger c_m\right\} c_n$$

$$S_n^z = c_n^\dagger c_n - \frac{1}{2}. \tag{18.12}$$

From this, we obtain

$$S_n^+ S_{n+1}^- = c_n^\dagger c_{n+1}$$
$$S_n^- S_{n+1}^+ = -c_n c_{n+1}^\dagger$$
$$S_n^+ S_{n+1}^+ = c_n^\dagger c_{n+1}^\dagger$$
$$S_n^- S_{n+1}^- = -c_n c_{n+1}. \tag{18.13}$$

Using these relations, we may express the XYZ Hamiltonian entirely in terms of fermions,

$$H_{XYZ} = \sum_n \left[\left(\frac{J_x + J_y}{2}\right)\left[c_n^\dagger c_{n+1} + c_{n+1}^\dagger c_n\right] + \left(\frac{J_x - J_y}{2}\right)\left[c_n^\dagger c_{n+1}^\dagger + c_{n+1} c_n\right]\right.$$

$$\left. + J_z \left(c_n^\dagger c_n - \frac{1}{2}\right)\left(c_{n+1}^\dagger c_{n+1} - \frac{1}{2}\right)\right], \tag{18.14}$$

We recognize in the first term above the familiar tight-binding Hamiltonian. The first surprising connection thus follows: the XY-Model quantum spin system in one dimension ($J_x = J_y = J$; $J_z = 0$) is equivalent to a free fermion on a Bravais lattice. The energy spectrum for this would be $\epsilon(k) = -2t \cos ka$, with hopping parameter $-J$. An occupation of one fermion per site would imply, according to (18.12), a $+1/2$ eigenvalue for the S_z-component of the associated spin system.

Before we consider the general case, let us perform a canonical transformation, introducing the new fermion operator d_n through $c_n \equiv i^n d_n$. Inserting in (18.14), we get

$$H_{XYZ} = \sum_n \left[i \left(\frac{J_x + J_y}{2} \right) \left[d_n^\dagger d_{n+1} - d_{n+1}^\dagger d_n \right] \right.$$
$$\left. + i(-1)^n \left(\frac{J_x - J_y}{2} \right) \left[d_{n+1}^\dagger d_n^\dagger + d_{n+1} d_n \right] + J_z d_n^\dagger d_n d_{n+1}^\dagger d_{n+1} \right], \quad (18.15)$$

where we neglected constant terms.

The Heisenberg equation of motion for d_n will be

$$i\dot{d}_n = [d_n, H_{XYZ}] = i \left(\frac{J_x + J_y}{2} \right) \left[d_{n+1} - d_{n-1} \right]$$
$$- i(-1)^n \left(\frac{J_x - J_y}{2} \right) \left[d_{n+1}^\dagger + d_{n-1}^\dagger \right]$$
$$+ J_z d_n \left[d_{n+1}^\dagger d_{n+1} + d_{n-1}^\dagger d_{n-1} \right]. \quad (18.16)$$

Introducing

$$\psi_n^o = d_n \; ; \quad n = \text{odd}$$
$$\psi_n^e = d_n \; ; \quad n = \text{even} \quad (18.17)$$

and taking the continuum limit

$$\psi_n^{o,e} \longrightarrow \psi_{o,e}(x)$$
$$\frac{\psi_{n+1}^{o,e} - \psi_{n-1}^{o,e}}{2a} \longrightarrow \partial_x \psi_{o,e}(x), \quad (18.18)$$

we obtain

$$i\dot{\psi}_e = i v_0 \partial_x \psi_o - M \psi_o^\dagger + 2g \psi_e \left(\psi_o^\dagger \psi_o \right)$$
$$i\dot{\psi}_o = i v_0 \partial_x \psi_e + M \psi_e^\dagger + 2g \psi_o \left(\psi_e^\dagger \psi_e \right), \quad (18.19)$$

where

$$v_0 = \frac{J_x + J_y}{2} a \quad M = J_x - J_y \quad g = J_z a. \quad (18.20)$$

It will be convenient to express the equation of motion in terms of left- and right-moving degrees of freedom. For this purpose, consider the expansion

$$c_n = i^n d_n = \sum_q c(q) e^{iqna}$$

$$c(q) = \sum_n \left[i^{2n} d_{2n} e^{-iq2na} + i^{2n+1} d_{2n+1} e^{-iq(2n+1)a} \right], \tag{18.21}$$

where, in the last expression, we separated the odd and even components. Assuming we are near the Fermi points $k_F = \pm\pi/2a$ and writing $q = k_F + k$, we have

$$d_R(k) \equiv \frac{1}{\sqrt{2}} c\left(+\frac{\pi}{2a} + k \right) \quad ; \quad d_L(k) \equiv \frac{1}{\sqrt{2}} c\left(-\frac{\pi}{2a} + k \right) \tag{18.22}$$

for the right and left movers. Replacing q for $\pm\pi/2a + k$ in (18.21) and noting that $(i)^{4n} = 1$, $(i)^{4n+2} = -1$, we obtain

$$d_R(k) = \frac{1}{\sqrt{2}} \sum_n \left[d_{2n} e^{-ik2na} + d_{2n+1} e^{-ik(2n+1)a} \right]$$

$$d_L(k) = \frac{1}{\sqrt{2}} \sum_n \left[d_{2n} e^{-ik2na} - d_{2n+1} e^{-ik(2n+1)a} \right]. \tag{18.23}$$

Fourier transforming, we have

$$d_{2n} = \frac{1}{\sqrt{2}} [d_R(n) + d_L(n)] \quad ; \quad d_{2n+1} = \frac{1}{\sqrt{2}} [d_R(n) - d_L(n)]$$

$$d_R(n) = \frac{1}{\sqrt{2}} [d_{2n} + d_{2n+1}] \quad ; \quad d_L(n) = \frac{1}{\sqrt{2}} [d_{2n} - d_{2n+1}] \tag{18.24}$$

or, from the first relation,

$$d_n = \frac{1}{\sqrt{2}} \left[d_R(n) + (-1)^n d_L(n) \right]. \tag{18.25}$$

We now take the right- and left-moving fields $\psi_R(x)$ and $\psi_R(x)$ as the continuum limit of $d_R(n)$ and $d_L(n)$. Using (18.23) and (18.19), we get

$$i\dot\psi_R = iv_0 \partial_x \psi_R + M\psi_L^\dagger + 2g\psi_R \left(\psi_L^\dagger \psi_L \right)$$

$$i\dot\psi_L = -iv_0 \partial_x \psi_L - M\psi_R^\dagger + 2g\psi_L \left(\psi_R^\dagger \psi_R \right). \tag{18.26}$$

This equation of motion corresponds to the Hamiltonian density

$$\mathcal{H} = iv_0 \left[\psi_R^\dagger \partial_x \psi_R - \psi_L^\dagger \partial_x \psi_L \right] + M \left[\psi_R^\dagger \psi_L^\dagger + \psi_L \psi_R \right]$$

$$+ 2g \left(\psi_R^\dagger \psi_R \right) \left(\psi_L^\dagger \psi_L \right). \tag{18.27}$$

Notice that, since the fermion current components are given by $j^0 = \rho_R + \rho_L$ and $j^1 = \rho_R - \rho_L$, then

$$j^\mu j_\mu = (j^0)^2 - (j^1)^2 = 4\left(\psi_R^\dagger \psi_R\right)\left(\psi_L^\dagger \psi_L\right); \qquad (18.28)$$

hence we may write the interaction part of the Hamiltonian as a Thirring interaction:

$$\frac{g}{2}j^\mu j_\mu. \qquad (18.29)$$

Observe that we are assuming a half-filled band, with Fermi points at $k_F = \pm\pi/2a$, which means an average occupation number of $\langle c_n^\dagger c_n \rangle = 1/2$. According to (18.12), this implies an average value $\langle S_z \rangle = 0$ for the z-component of the associated XYZ-spin system. The relations (18.22) therefore imply the operators $d_{R,L}(k)$ already act, creating or annihilating fermions and holes with respect to the Fermi sea, in such a way that $\langle d_n^\dagger d_n \rangle = 0$. Hence, in terms of these operators, we may express the z-component of the spin operator as

$$S_n^z = d_n^\dagger d_n. \qquad (18.30)$$

We can identify here another unexpected connection: the one existing between the XYZ-Model, describing localized quantum spins, and a fermion system with quartic self-interaction.

18.2 From Fermions to Bosons

18.2.1 Bosonization of Localized Quantum Spin Systems

In the previous section, we mapped the chain of localized quantum spins corresponding to the XYZ-Model into a one-dimensional continuum fermion field with Thirring interaction. Now let us proceed by mapping the fermion field itself into a bosonic field.

XXZ Model

Consider firstly the case $M = 0$ ($J_x = J_y \equiv J$). We have seen in Chapter 10 that, in this case, the bosonized Hamiltonian is given by (10.90); therefore, taking into account the velocity $v_0 = Ja$, we have for the XXZ-Model

$$\mathcal{H} = \frac{v}{2}\left(\frac{\Pi^2}{K} + K(\partial_x \phi)^2\right) = \frac{v}{2}\left(\tilde{\Pi}^2 + (\partial_x \tilde{\phi})^2\right). \qquad (18.31)$$

We see that, in this case, the interaction introduces the dimensionless K factors. It can be shown that the exact expression for these is [119]

$$K = \sqrt{\frac{2}{\pi}}\left[1 - \frac{\delta}{\pi}\right]^{1/2} \quad ; \quad \cos\delta = \frac{J_z}{J}, \qquad (18.32)$$

whereas for the velocity,

$$v = \frac{\sin \delta}{\delta} \left(\frac{\pi}{2} \right) Ja. \tag{18.33}$$

The bosonized current, accordingly, is given by

$$j^\mu = \frac{1}{\pi K} \epsilon^{\mu\nu} \partial_\nu \phi. \tag{18.34}$$

This expression is valid for $-1 \leq \frac{J_z}{J} \leq 1$ and corresponds to a ground state with $\langle S^z \rangle = 0$. For $|J_z| > J$, the system suffers a quantum phase transition to a state where $\langle S^z \rangle \neq 0$, being in the same universality class as the Ising model.

From (10.80), we see that the bosonization formula for the fermion fields involves both the bosonic field ϕ and its dual, namely

$$\theta(\mathbf{x}, t) = \frac{1}{v} \int_{-\infty}^{\mathbf{x}} d\xi \, \Pi(\xi, t)$$

$$\partial_x \theta(\mathbf{x}, t) = \frac{1}{v} \dot{\phi}(\mathbf{x}, t) \quad ; \quad \dot{\theta} = \frac{1}{v} \int_{-\infty}^{\mathbf{x}} d\xi \, \partial_t^2 \phi(\xi, t) = v \partial_x \phi. \tag{18.35}$$

In terms of the dual bosonic field, we may express the XXZ-Model Hamiltonian as

$$\mathcal{H} = \frac{v}{2} \left(K \Pi_\theta^2 + \frac{(\partial_x \theta)^2}{K} \right). \tag{18.36}$$

XYZ Model

In this case, we have $J_x \neq J_y$, $M \neq 0$, and there is a mass term in (18.27). According to (10.80), this can be conveniently bosonized in terms of the θ-field. We therefore obtain the following bosonized Hamiltonian for the XYZ-Model:

$$\mathcal{H} = \frac{v}{2} \left(K \Pi_\theta^2 + \frac{(\partial_x \theta)^2}{K} \right) + \alpha \cos 2\theta, \tag{18.37}$$

where $\alpha = M/2\pi$.

The Bosonized Spin Operators

Once we bosonize the fermions emerging from the application of the Jordan–Wigner transformation to the XYZ quantum spin system, it is quite useful to express those spin operators in terms of the bosonic field, thereby obtaining a bosonized form of the spin operators themselves. Using (18.30) and (18.25), we get

$$S_n^z = \left[\psi_R^\dagger \psi_R + \psi_L^\dagger \psi_L + (-1)^n \left(\psi_R^\dagger \psi_L + \psi_L^\dagger \psi_R \right) \right]. \tag{18.38}$$

Now, observing that the first two terms form the zeroth component of the fermion current and using (10.80), we may express the above spin operator in terms of the bosonic field appearing in (18.36) as

$$S_n^z = \frac{\beta}{2\pi} \partial_x \phi + \frac{(-1)^n}{\pi} \cos(\beta\phi), \tag{18.39}$$

where

$$\beta = \frac{2}{K} = \sqrt{\frac{2\pi}{1 - \frac{\delta}{\pi}}}. \tag{18.40}$$

Analogously, we get

$$S_n^+ = e^{-i\frac{2\pi}{\beta}\theta} \left[\cos(\beta\phi) + (-1)^n \right]. \tag{18.41}$$

The previous expressions are very useful, for instance, for the purpose of determining the magnetic properties of systems governed by the XXZ and XYZ models. The uniform, static, magnetic susceptibility, for example, according to (4.19), is given by

$$\chi(0,0) = \lim_{q,\omega \to 0} \int d\tau e^{-i\omega\tau} \sum_n e^{iqna} \langle S_n^z(\tau) S_0^z(0) \rangle$$

$$= \lim_{\omega \to 0} \frac{1}{N} \int d\tau e^{-i\omega\tau} \sum_{n,m} \langle S_n^z(\tau) S_m^z(0) \rangle$$

$$= \lim_{q,\omega \to 0} \frac{1}{N} \int d\tau e^{-i\omega\tau} \sum_{n,m} e^{iq(n-m)a} e^{iqma} \langle S_n^z(\tau) S_m^z(0) \rangle \longrightarrow$$

$$\lim_{\omega \to 0} \frac{1}{N} \int d\tau e^{-i\omega\tau} \sum_{n,m} \langle S_n^z(\tau) S_m^z(0) \rangle$$

$$\chi(0,0) = \int d\tau \sum_n \langle S_n^z(\tau) S_0^z(0) \rangle, \tag{18.42}$$

where N is the total number of sites. Now, taking the continuum limit and considering that the last term in (18.39) cancels out when summed over n, we get

$$\chi(0,0) = \lim_{q,\omega \to 0} \frac{\beta^2}{4\pi^2} \int d\tau dx e^{iqx} e^{-i\omega\tau} (-\partial_x^2) \langle \phi(x,\tau)\phi(0,0) \rangle. \tag{18.43}$$

The ϕ-field two-point correlation function corresponds to the theory given by (18.31) in the case of the XXZ-Model and to the one given by (18.37), in the case of the XYZ-Model. In the first case, the exact result in Euclidean space is

$$G^{(2)}(\omega, q) = \frac{1}{v(\omega^2 + q^2)}. \tag{18.44}$$

In the second case we have just an approximate result, which for small momentum behaves as

$$G^{(2)}(\omega, q) \approx \frac{1}{v(\omega^2 + q^2 + M^2)}. \tag{18.45}$$

We can write the static magnetic susceptibility as

$$\chi(0,0) = \lim_{q,\omega \to 0} \frac{\beta^2}{4\pi^2 v} \frac{q^2}{\omega^2 + q^2 + M^2}. \tag{18.46}$$

In the presence of a gap $(M \neq 0)$, we can see from the above expression that $\chi(0,0) \to 0$. This is a reasonable result from the physical point of view, since, in the presence of a finite gap, the application of a weak external magnetic field is not enough to excite a magnetic mode.

In the absence of a gap$(M = 0)$, however, as is the case for the XXZ-Model, using (18.32) we obtain the following general expression for the static, uniform susceptibility at $T = 0$

$$\chi_{XXZ}(0,0) = \frac{\beta^2}{4\pi^2 v} = \frac{\delta}{\pi J a (\pi - \delta) \sin \delta}. \tag{18.47}$$

There are two important special cases: the XY-Model, where $J_z = 0$; $\delta = \pi/2$, $\beta = 2\sqrt{\pi}$, $v = Ja$ and

$$\chi_{XY}(0,0) = \frac{1}{\pi J a}, \tag{18.48}$$

and the Heisenberg Model, where $J_z = J$; $\delta = 0$, $\beta = \sqrt{2\pi}$, $v = \frac{\pi}{2} Ja$ and

$$\chi_H(0,0) = \frac{1}{\pi^2 J a}. \tag{18.49}$$

18.2.2 Bosonization of Itinerant Quantum Spins (Interacting Fermions)

In the previous subsection, we considered localized quantum spin systems in one spatial dimension and studied the bosonization of the associated fermion systems obtained by the application of the Jordan–Wigner transformation to those systems. Now we consider one-dimensional interacting fermionic systems with itinerant quantum spins and once again apply the bosonization method in order to map these into familiar bosonic systems.

Basic Bosonization Formulas

Let us consider spin $s = 1/2$ fermion excitations associated to a field $\psi_{\alpha,\sigma}(x)$, where $\alpha = R, L$ denotes, respectively, right-movers and left-movers and $\sigma = \uparrow, \downarrow$, the two possible values of the z-component of their spin. We will focus on quartic interactions, in which, typically, an initial pair of fermions with given (α, σ) shall scatter into a final pair with (α', σ').

We bosonize each $\sigma = \uparrow, \downarrow$ component of the fermion field $\psi_{\alpha,\sigma}(x)$ through the bosonization formulas (10.80) with the bosonic field $\phi_\sigma(x)$, corresponding to each spin component. According to that, we have

$$j_\sigma^0 = \rho_{R,\sigma} + \rho_{L,\sigma} = \frac{1}{\sqrt{\pi}} \partial^1 \phi_\sigma$$

$$j_\sigma^1 = \rho_{R,\sigma} - \rho_{L,\sigma} = -\frac{1}{\sqrt{\pi}} \partial^0 \phi_\sigma$$

$$\rho_{R,\sigma} = -\frac{1}{2\sqrt{\pi}} \left(\partial^0 - \partial^1 \right) \phi_\sigma$$

$$\rho_{L,\sigma} = \frac{1}{2\sqrt{\pi}} \left(\partial^0 + \partial^1 \right) \phi_\sigma. \tag{18.50}$$

Analogously,

$$\psi_{R,\sigma}^\dagger \psi_{L,\sigma} = \frac{1}{4\pi} e^{-2ib\phi_\sigma} \quad ; \quad \psi_{L,\sigma}^\dagger \psi_{R,\sigma} = \frac{1}{4\pi} e^{2ib\phi_\sigma}, \tag{18.51}$$

and for the charge density

$$\rho = \psi_\sigma^\dagger \psi_\sigma = \psi_\uparrow^\dagger \psi_\uparrow + \psi_\downarrow^\dagger \psi_\downarrow = \frac{1}{\sqrt{2\pi}} \partial^1 \phi_c \tag{18.52}$$

and spin operators

$$S^z = \psi_\sigma^\dagger \sigma_{\sigma\lambda}^z \psi_\lambda = \psi_\uparrow^\dagger \psi_\uparrow - \psi_\downarrow^\dagger \psi_\downarrow = \frac{1}{\sqrt{2\pi}} \partial^1 \phi_s, \tag{18.53}$$

where the charge and spin bosonic fields are

$$\phi_c = \frac{1}{\sqrt{2}} \left(\phi_\uparrow + \phi_\downarrow \right)$$

$$\phi_s = \frac{1}{\sqrt{2}} \left(\phi_\uparrow - \phi_\downarrow \right). \tag{18.54}$$

Response Functions

Bosonization is an extremely powerful method for determining basic response functions of the system. The magnetic susceptibility, for instance, can be derived from (4.19), (18.42) and (18.53), namely

$$\chi(\omega, q) = \frac{1}{2\pi T} \int dx e^{iqx} \int d\tau e^{-i\omega\tau} \partial_x \partial_y \langle \phi_s(x, \tau) \phi_s(y, 0) \rangle. \tag{18.55}$$

The electric conductivity, according to the Kubo formula, is given by

$$\sigma(\omega, q) = \frac{i}{\omega} \int dx e^{iqx} \int d\tau e^{-i\omega\tau} \langle j(x, \tau) j(0, \tau') \rangle$$

$$= \frac{i}{\omega} \int dx e^{iqx} \int d\tau e^{-i\omega\tau} \partial_\tau \partial_\tau' \langle \phi_c(x, \tau) \phi_c(0, \tau') \rangle. \tag{18.56}$$

Finally, we consider the compressibility, which is defined as the ratio between the average density variations under a certain applied pressure. This is given by the correlator

$$\kappa(\omega, q) = \int dx e^{iqx} \int d\tau e^{-i\omega\tau} \langle n(x,\tau)n(y,0)\rangle$$

$$n(x,\tau) = \sum_{\sigma=\uparrow,\downarrow} \left[\rho_{R,\sigma} + \rho_{L,\sigma}\right] = \sqrt{\frac{2}{\pi}}\partial_x\phi_c. \qquad (18.57)$$

Free Hamiltonian

We start with the free Hamiltonian

$$\mathcal{H}_0 = \sum_\sigma i v_0 \overline{\psi}_\sigma \slashed{\partial} \psi_\sigma. \qquad (18.58)$$

Using the bosonization formula (10.80), we find

$$\mathcal{H}_0 = \sum_\sigma \frac{1}{2}\partial_\mu\phi_\sigma\partial^\mu\phi_\sigma, \qquad (18.59)$$

which can be cast in the form

$$\mathcal{H}_0 = \frac{1}{2}\partial_\mu\phi_c\partial^\mu\phi_c + \frac{1}{2}\partial_\mu\phi_s\partial^\mu\phi_s, \qquad (18.60)$$

or

$$\mathcal{H} = \frac{v_0}{2}\left(\Pi_c^2 + (\partial_x\phi_c)^2\right) + \frac{v_0}{2}\left(\Pi_s^2 + (\partial_x\phi_s)^2\right) \qquad (18.61)$$

where

$$\Pi_{c,s} = \frac{\dot{\phi}_{c,s}}{v_0}. \qquad (18.62)$$

Notice that the free Hamiltonian separates in charge and spin degrees of freedom; nevertheless, both have the same velocity v_0, hence they are not truly separated in the free theory.

Small Momentum Interactions

Let us consider first interactions involving a momentum exchange, which is small in comparison to the Fermi momentum $|k_F| = \pi/2a$. These are interactions where an R-mover is scattered into an R-mover and an L-mover into an L-mover. The corresponding Hamiltonians are subdivided in two groups, namely

$$\mathcal{H}_{1c} = g_{1c}\left[\left(\rho_{R,\uparrow} + \rho_{R,\downarrow}\right)^2 + \left(\rho_{L,\uparrow} + \rho_{L,\downarrow}\right)^2\right]$$
$$\mathcal{H}_{1s} = g_{1s}\left[\left(\rho_{R,\uparrow} - \rho_{R,\downarrow}\right)^2 + \left(\rho_{L,\uparrow} - \rho_{L,\downarrow}\right)^2\right] \qquad (18.63)$$

and

$$\mathcal{H}_{2c} = g_{2c}\left[\rho_{R,\uparrow} + \rho_{R,\downarrow}\right]\left[\rho_{L,\uparrow} + \rho_{L,\downarrow}\right]$$
$$\mathcal{H}_{2s} = g_{2s}\left[\rho_{R,\uparrow} - \rho_{R,\downarrow}\right]\left[\rho_{L,\uparrow} - \rho_{L,\downarrow}\right]. \qquad (18.64)$$

Using the bosonization formulas (18.50), we get

$$\mathcal{H}_{1c} = \frac{g_{1c}}{2\pi} \left[\Pi_c^2 + (\partial^1 \phi_c)^2 \right]$$

$$\mathcal{H}_{1s} = \frac{g_{1c}}{2\pi} \left[\Pi_s^2 + (\partial^1 \phi_s)^2 \right] \tag{18.65}$$

and

$$\mathcal{H}_{2c} = -\frac{g_{2c}}{2\pi} \partial_\mu \phi_c \partial^\mu \phi_c = \frac{g_{2c}}{2\pi} \left[-\Pi_c^2 + (\partial^1 \phi_c)^2 \right]$$

$$\mathcal{H}_{2s} = -\frac{g_{2s}}{2\pi} \partial_\mu \phi_s \partial^\mu \phi_s = \frac{g_{2c}}{2\pi} \left[-\Pi_s^2 + (\partial^1 \phi_s)^2 \right]. \tag{18.66}$$

Large Momentum Interactions

Now, we will consider interactions where the momentum exchanged is comparable to the Fermi momentum $|k_F| = \pi/2a$. These are interactions where an R-mover is scattered into an L-mover and an L-mover into an R-mover. The first interaction of such a kind is the "Backscattering," where an initial pair of R and L movers with opposite spins is scattered into another pair of R and L movers in such a way that the initial L-mover becomes a final R-mover, without changing its spin, and vice versa. The underlying interaction is

$$\mathcal{H}_3 = g_3 \sum_\sigma \psi_{R,\sigma}^\dagger \psi_{L,\sigma} \psi_{L,-\sigma}^\dagger \psi_{R,-\sigma}. \tag{18.67}$$

Using (18.51), it is not difficult to infer that the bosonized form of the Backscattering interaction Hamiltonian is

$$\mathcal{H}_3 = \frac{g_3}{2\pi^2} \cos\left(2\sqrt{2}b\phi_s \right). \tag{18.68}$$

The second type of large momentum interactions is the Umklapp process. In this, a pair of R-movers (or L-movers) with opposite spins scatter into a pair of L-movers (or R-movers), without changing their spin. This is a process that occurs at the first Brillouin zone boundaries, hence its name. The Hamiltonian corresponding to this interaction is

$$\mathcal{H}_4 = \frac{g_4}{2} \sum_\sigma \left[\psi_{R,\sigma}^\dagger \psi_{R,-\sigma}^\dagger \psi_{L,\sigma} \psi_{L,-\sigma} + \psi_{L,-\sigma}^\dagger \psi_{L,\sigma}^\dagger \psi_{R,-\sigma} \psi_{R,\sigma} \right]. \tag{18.69}$$

Again, using (18.51), we find the corresponding bosonized version of the above interaction to be

$$\mathcal{H}_4 = \frac{g_4}{2\pi^2} \cos\left(2\sqrt{2}b\phi_c \right). \tag{18.70}$$

18.3 From Bosons to Gases

We have seen that two classes of bosonic Hamiltonians emerge in the process of bosonization, both of localized spin systems and itinerant, strongly interacting fermions. The first category involves essentially free gapless systems that can be solved exactly for the spectrum, correlation functions and response functions. The second one comprises gapped systems associated to a sine-Gordon interaction, for which a full exact solution is, so far, not available. There exists, nevertheless, an extremely useful and instructive mapping of such systems into a two-dimensional classical gas of charged point particles, interacting through the Coulomb potential, in the grand-canonical ensemble [121, 122, 123, 124, 131].

18.3.1 The Sine-Gordon Theory: 2D Coulomb Gas

The Grand-Partition Function

Let us start by considering the sine-Gordon vacuum functional corresponding to (10.100) in Euclidean space, namely,

$$Z_{SG} = Z_0^{-1} \int D\phi \exp \left\{ -\int d^2z \left[\frac{1}{2} \partial_\mu \phi \partial_\mu \phi + \alpha \cos \beta \phi \right] \right\}. \qquad (18.71)$$

Then, expanding the cosine term,

$$
\begin{aligned}
Z_{SG} &= \sum_{m=0}^{\infty} \frac{\alpha^m}{m!} \int d^2z_1 \ldots d^2z_m \\
&\quad \times Z_0^{-1} \int D\phi \exp \left\{ -\int d^2z \left[\frac{1}{2} \partial_\mu \phi \partial_\mu \phi \right] \right\} \cos \beta \phi(z_1) \ldots \cos \beta \phi(z_m) \\
&= Z_0^{-1} \sum_{m=0}^{\infty} \frac{\left(\frac{\alpha}{2} \right)^m}{m!} \int d^2z_1 \ldots d^2z_m \\
&\quad \times \sum_{\{\lambda\}, \lambda = \pm 1} \int D\phi \exp \left\{ -\int d^2z \left[\frac{1}{2} \phi(-\nabla^2)\phi + \rho(z; z_1 \ldots z_m)\phi(z) \right] \right\},
\end{aligned}
$$
$$(18.72)$$

where $\lambda_i = \pm 1$,

$$\rho(z; z_1 \ldots z_m) = i\beta \sum_{i=1}^{m} \lambda_i \delta(z - z_i), \qquad (18.73)$$

and the sum over $\{\lambda\}$ sweeps all possible configurations of λ_i values in the set $\{\lambda_1 = \pm 1, \ldots, \lambda_m = \pm 1\}$.

Notice that the functional integral in Z_{SG} now becomes quadratic and can be performed with the help of (5.49):

$$I(z_1 \ldots z_m) = Z_0^{-1} \int D\phi \exp \left\{ -\int d^2z \left[\frac{1}{2}\phi(-\nabla^2)\phi + \rho(z; z_1 \ldots z_m)\phi(z) \right] \right\}$$

$$= \exp \left\{ \frac{1}{2} \int d^2z d^2z' \rho(z; z_1 \ldots z_m) D(z - z') \rho(z'; z_1 \ldots z_m) \right\},$$
$$(18.74)$$

where $D(z - z')$ is the Euclidean Green function of the $-\nabla^2$ operator, which, by definition, is the Coulomb potential created by a point charge in any dimension. According to (10.72), it is given by

$$D(z - z') = -\frac{1}{4\pi} \ln \mu^2 \left[|z - z'|^2 + |\epsilon|^2 \right]|_{\mu,\epsilon \to 0}, \qquad (18.75)$$

where μ and ϵ are, respectively, infrared and ultraviolet regulators, which must be removed at the end.

Inserting (18.73) and (18.75) in (18.74), we get

$$I(z_1 \ldots z_m) = \exp \left\{ \frac{\beta^2}{8\pi} \sum_{i,j=1}^{m} \lambda_i \lambda_j \ln \mu^2 \left[|z_i - z_j|^2 + |\epsilon|^2 \right] \right\}. \qquad (18.76)$$

Now, the IR regulator μ-dependent term, which is additive in the exponent, yields an overall multiplicative factor:

$$[\mu]^{\frac{\beta^2}{4\pi} \left[\sum_{i=1}^{m} \lambda_i \right]^2}. \qquad (18.77)$$

Conversely, the $i = j$ terms of the μ-independent terms of the sum above yield another overall multiplicative factor,

$$|\epsilon|^{m \frac{\beta^2}{4\pi}}, \qquad (18.78)$$

depending on the UV regulator.

The μ-dependent factor in (18.77) is equal to one whenever $\sum_{i=1}^{m} \lambda_i = 0$, and vanishes, otherwise, when we remove the IR cutoff, by taking the limit $\mu \to 0$. We conclude, therefore, that the only nonzero contribution to Z_{SG} occurs when $\sum_{i=1}^{m} \lambda_i = 0$. This last condition implies there are n $\lambda_i = +1$ and n $\lambda_i = -1$ in such a way that $m = 2n$.

The ϵ-dependent factor in (18.78), conversely, can be absorbed in a renormalization of α in (18.72):

$$\alpha_R = \alpha |\epsilon|^{\frac{\beta^2}{4\pi}}. \qquad (18.79)$$

This is Coleman's renormalization [121].

Inserting (18.78) in (18.72) and taking into account the fact that there are n positive and n negative λ_i, values, it follows that the sum over the possible λ_i configurations yields

$$\sum_{\{\lambda\}} = \frac{m!}{n!n!} \quad ; \quad m = 2n.$$

We can therefore write

$$Z_{SG} = \sum_{n=0}^{\infty} \frac{\zeta^{2n}}{n!n!} Z(n; n), \tag{18.80}$$

where

$$Z(n; n) = \int d^2 z_1 \ldots d^2 z_{2n} \exp \left\{ \frac{\beta^2}{8\pi} \sum_{i,j=1, i\neq j}^{2n} \lambda_i \lambda_j \ln \left[|z_i - z_j|^2 + |\epsilon|^2 \right] \right\}$$

$$\tag{18.81}$$

is the partition function of a two-dimensional neutral, classical gas of $2n$ point charges, n of which having positive and the remaining n having negative signs and interacting through the Coulomb potential. Z_{SG}, conversely, is identified as the grand-partition function of this neutral Coulomb gas, with the fugacity given by

$$\zeta = \frac{\alpha_R}{2}.$$

Coleman's renormalization (18.79), in this language, corresponds to the removal of the unphysical self-energies of the point charges.

The result above is just one more remarkable, unexpected connection exhibited by the systems studied in the present chapter.

Ultraviolet Divergences: Non-Perturbative Renormalization

The self-energies of the $2n$ point charges are the only divergences we find in the Coulomb gas for $0 \leq \beta^2 < 4\pi$. These are eliminated by a fugacity renormalization, as we have seen. For $\beta^2 \geq 4\pi$, however, a new class of short-distance divergences starts to show up [125, 126, 127, 128, 129, 130], requiring further consideration. In the Coulomb gas language, these new divergences correspond to the coalescence of neutral multipoles (p positive and p negative charges) in the integration region over the charges' positions in (18.81). Dipoles ($p = 1$) start to diverge at $\beta^2 = 4\pi$, quadrupoles ($p = 2$) at $\beta^2 = 6\pi$ and ∞-poles, ultimately, start to diverge at $\beta^2 = 8\pi$. In general, p-poles diverge for [130]

$$\beta^2 \geq \left(\frac{2p-1}{2p} \right) 8\pi \quad ; \quad p = 1, 2, 3, \ldots \tag{18.82}$$

A detailed analysis of the short-distance divergences occurring in the Coulomb gas for $4\pi \leq \beta^2 < 8\pi$ has been presented in [130]. By subdividing the total integration region of the integral in (18.81) both in subregions corresponding to

divergent configurations of coalescing p-charges as well as subregions corresponding to finite contributions to the partition function, one is capable of factoring out
all the divergent pieces in an overall multiplicative term. This is achieved by means
of a resummation of the series appearing in (18.80) [130] that leads to

$$Z_{SG} = e^{V\Lambda(\epsilon)} \sum_{n=0}^{\infty} \frac{\zeta^{2n}}{n!n!} \tilde{Z}(n; n), \tag{18.83}$$

where $\tilde{Z}(n; n)$ is finite and $\Lambda(\epsilon) \to \infty$, for $\epsilon \to 0$.

All the UV divergences, therefore, may be eliminated by a renormalization

$$Z_{SG}^R = Z_{SG} e^{-V\Lambda(\epsilon)}, \tag{18.84}$$

which, according to (18.71) amounts to the subtraction

$$\mathcal{L}_{SG}^R = \mathcal{L}_{SG} - \Lambda(\epsilon). \tag{18.85}$$

Exact Equation-of-State

We have just demonstrated the equivalence between the sine-Gordon theory and
a two-dimensional neutral Coulomb gas in the grand-canonical ensemble. Let us
derive now the exact equation of state of such a gas. For this purpose, let us recall
that the grand-canonical potential $\Omega(T, V, \zeta)$, where ζ is the fugacity and V is the
volume, is defined as

$$\Omega(T, V, \zeta) = -k_B T \ln Z_{SG}. \tag{18.86}$$

The pressure, then, is expressed as

$$p = -\left(\frac{\partial \Omega(T, V, \zeta)}{\partial V}\right)_{T,\zeta} = k_B T \frac{1}{Z_{SG}} \frac{\partial Z_{SG}(T, V, \zeta)}{\partial V}. \tag{18.87}$$

In order to extract the volume dependence of (18.80), let us rescale the z_i
variables and the UV regulator in (18.81) as

$$z_i = V^{1/2}\hat{z}_i \quad ; \quad \epsilon = V^{1/2}\hat{\epsilon}. \tag{18.88}$$

We can, then, write (18.81) as

$$Z(n; n) = V^{2n(1-\beta^2/8\pi)} \int d^2\hat{z}_1 \dots d^2\hat{z}_{2n}$$

$$\times \exp\left\{\frac{\beta^2}{8\pi} \sum_{i,j=1,i\neq j}^{2n} \lambda_i\lambda_j \ln\left[|\hat{z}_i - \hat{z}_j|^2 + |\hat{\epsilon}|^2\right]\right\}, \tag{18.89}$$

where we used the fact that, in view of the neutrality of the gas,

$$\left(\sum_{i=1}^{2n}\lambda_i\right)^2 = \sum_{i,j=1}^{2n}\lambda_i\lambda_j = \sum_{i,j=1,i\neq j}^{2n}\lambda_i\lambda_j + 2n = 0. \tag{18.90}$$

It follows from (18.87) that

$$p = \left(1 - \frac{\beta^2}{8\pi}\right)\frac{k_B T}{V}\left[\frac{1}{Z_{SG}}\sum_{n=0}^{\infty} 2n\frac{\zeta^{2n}}{n!n!}Z(n;n)\right]$$

$$pV = \left(1 - \frac{\beta^2}{8\pi}\right)k_B T \,\langle 2n\rangle$$

$$pV = \left(1 - \frac{\beta^2}{8\pi}\right)Nk_B T, \tag{18.91}$$

where $N \equiv \langle 2n\rangle$ is the average number of particles in the grand-canonical ensemble. This is the exact equation of state of the Coulomb gas associated to the sine-Gordon equation.

Observe that the pressure vanishes at $\beta^2 = 8\pi$ and would become negative for larger values of β. This indicates that the system becomes unstable and undergoes a phase transition at this value of the coupling. Interestingly, precisely at this coupling, further UV divergences set in, namely, the ones associated to the coalescence of non-neutral groups of charges in the integration region in (18.81) [130]

Notice, moreover, that for $\beta^2 = 4\pi$, according to the energy equipartition theorem, the above equation of state becomes that of an ideal gas. This is related to the fact that, at this point, the sine-Gordon equation describes a free massive fermion field.

18.3.2 The Supersymmetric 2D Sine–Gordon/Coulomb Gas

It is quite interesting and instructive to examine the supersymmetric extension of the usual sine-Gordon theory. As it turns out, it is equivalent to the supersymmetric extension of the usual Coulomb gas in D = 2: the supersymmetric Coulomb gas, a Coulomb gas in the superspace. We start by introducing some basic concepts and notation.

Notation and Basic Concepts

The 2D space with coordinates $x^\mu = (x^0, x^1)$ is Euclidean. The γ-matrices are hermitian: $\gamma^0 = \sigma^x$, $\gamma^1 = \sigma^y$, $\gamma^5 = \sigma^z$. $\overline{\Psi} \equiv \Psi^\dagger\gamma^5$. A point in superspace is associated to a pair (x, Θ), where

$$\Theta = \begin{pmatrix} \theta \\ \theta* \end{pmatrix} \tag{18.92}$$

and θ is a Grassmann variable. The volume element in Θ-space is $d\theta^* d\theta$ and the delta function: $\delta^2(\theta) = \theta\theta^*$.

The displacement vector corresponding to two points i and j in superspace is then expressed as

$$R_\mu^{ij} = x_\mu^i - x_\mu^j + \overline{\Theta}_i \gamma^\mu \Theta_j. \qquad (18.93)$$

The square of the distance between such points, $|R^{ij}|^2 = R_\mu^{ij} R_\mu^{ij}$, is invariant under the supersymmetry (Susy) transformation

$$x_\mu^i \rightarrow x_\mu^i + \overline{\epsilon}\gamma^\mu \Theta_i \quad ; \quad \Theta_i \rightarrow \Theta_i + \epsilon, \qquad (18.94)$$

where ϵ, the transformation parameter, is a two-component Grassmann variable spinor.

The real scalar superfield is defined as

$$\Phi = \phi + \overline{\Theta}\Psi + \frac{1}{2}\overline{\Theta}\Theta F, \qquad (18.95)$$

where ϕ and F are real scalar fields and Ψ is a Majorana fermion field (defined as $\Psi = \gamma^0 \Psi^*$). Any function of the superfield Φ, upon expansion in the Grassmann variables, may be expressed in terms of the component fields as

$$V(\Phi) = V(\phi) + V'(\phi)\overline{\Theta}\Psi + \frac{1}{2}\overline{\Theta}\Theta \left[V'(\phi)F - V''(\phi)\overline{\Psi}\Psi \right]. \qquad (18.96)$$

The covariant derivative and its dual are defined as

$$D = \partial_{\theta^*} - \theta^* \partial_- \quad ; \quad \overline{D} = \partial_\theta + \theta \partial_+, \qquad (18.97)$$

where $\partial_\pm = \partial_0 \pm \partial_1$.

Defining the Green function of the $\overline{D}D$ operator as

$$\overline{D}D\Delta(x_1, \theta_1; x_2, \theta_2) = \delta^2(x_1 - x_2)\delta^2(\theta_1 - \theta_2), \qquad (18.98)$$

it is not difficult to show that

$$\Delta(x_1, \theta_1; x_2, \theta_2) = -\frac{1}{4\pi} \ln \mu^2 \left[|R^{12}|^2 + |\epsilon|^2 \right]|_{\mu,\epsilon \to 0}, \qquad (18.99)$$

where as in the non-Susy case, μ and ϵ are, respectively, IR and UV regulators.

The Model Lagrangean

The Susy sine-Gordon theory [132] is defined by the following action, written in superspace, in terms of the scalar superfield Φ:

$$S_{SG} = \int d^2x d^2\theta \left[\frac{1}{2}D\Phi D\Phi - \alpha \cos \beta\Phi \right]. \qquad (18.100)$$

Fugacity Expansion

We can now go through the same steps that led to the Coulomb gas description in the usual sine-Gordon theory. As before, we start with the expansion

$$
Z_{Susy} = Z_0^{-1} \sum_{m=0}^{\infty} \frac{\left(\frac{\alpha}{2}\right)^m}{m!} \int \Pi_{i=1}^m d^2 z_i d^2 \theta_i \sum_{\{\lambda\}, \lambda = \pm 1} \int D\Phi
$$
$$
\times \exp\left\{ -\int d^2 z d^2 \theta \left[\frac{1}{2} \Phi(\overline{D}D)\Phi + \rho(z, \theta; z_1\theta_1, \ldots, z_m\theta_m)\Phi(z, \theta) \right] \right\}
$$

$$(18.101)$$

where $\lambda_i = \pm 1$,

$$
\rho(z, \theta; z_1\theta_1, \ldots, z_m\theta_m) = i\beta \sum_{i=1}^m \lambda_i \delta^2(z - z_i)\delta^2(\theta - \theta_i) \tag{18.102}
$$

and the sum over $\{\lambda\}$ sweeps all possible λ_i configurations in the set $\{\lambda_1 = \pm 1, \ldots, \lambda_m = \pm 1\}$.

Performing the quadratic functional integral over the scalar superfield, with the help of the Green function (18.99), we arrive at the Susy Coulomb gas grand-partition function, which is given by (18.80), with $Z(n; n)$ replaced by

$$
Z(n; n)_{Susy} = \int \Pi_{i=1}^{2n} d^2 z_i d^2 \theta_i \exp\left\{ \frac{\beta^2}{8\pi} \sum_{i,j=1, i \neq j}^{2n} \lambda_i \lambda_j \ln\left[|R^{ij}|^2 + |\epsilon|^2 \right] \right\},
$$

$$(18.103)$$

where $|R_{ij}|$ is the distance in superspace, given in terms of (18.93). The fugacity is defined in terms of α_R, precisely as in (18.79). As in the non-Susy Coulomb gas, the IR regulator imposes the neutrality, so we have n positive and n negative point charges in the Susy Coulomb gas.

Divergence Structure and Renormalization

The divergence structure is very similar to the one found in the usual sine-Gordon/Coulomb gas system [133, 130]. For $0 \leq \beta^2 < 2\pi$ the system is completely finite, except for the fugacity renormalization that eliminates the charges' self-energies. For $2\pi \leq \beta^2 < 4\pi$, additional divergences appear, associated to the coalescence of neutral agglomerates, as before. Again, after a resummation made in the fugacity expansion of the grand-partition function [133], we are able to factorize all the divergences as in (18.83) and eliminate them by a subtraction analogous to (18.84) and (18.85). This subtraction, however, has profound consequences in a supersymmetric theory, as we show below.

Dynamical Supersymmetry Breaking

A supersymmetric system is invariant under transformations in which both the parameter and the generator are fermionic. This can be seen, for instance, in (18.94). Denoting by Q the Susy generator, assuming there is only one, a supersymmetric transformation can be expressed as

$$U(\epsilon) = e^{i[\epsilon^\dagger Q + Q^\dagger \epsilon]}. \tag{18.104}$$

Then, a fundamental property of any supersymmetric theory follows. This is that the Hamiltonian can be written as

$$H = Q^\dagger Q. \tag{18.105}$$

For the vacuum to be invariant under the operation in (18.104), a necessary and sufficient condition for Susy not to be spontaneously broken, we must have $Q|0\rangle = 0$. The immediate conclusion is that a necessary and sufficient condition for preventing Susy being spontaneously broken is that the vacuum state is an energy eigenstate with eigenvalue precisely equal to zero: $H|0\rangle = 0$.

This observation has strong consequences when we, in order to eliminate divergences in a Susy theory, are forced to make an overall subtraction in the Hamiltonian, as we did in (18.85). As a matter of fact, we can no longer guarantee, after subtracting a constant from the system Hamiltonian, that the condition $H|0\rangle = 0$ still holds; hence, supersymmetry is spontaneously broken for $\beta^2 \geq 2\pi$ where this subtraction is unavoidable.

A general theorem states, however, that spontaneous supersymmetry breakdown cannot possibly occur at any finite order in perturbation theory [303]. The spontaneous Susy breaking we find for this value of the coupling complies with this theorem. This becomes clear when we notice that the overall subtraction made in the Hamiltonian was only possible after a full resummation of the fugacity series. This, however, was only feasible by taking into account all terms of the series, in agreement with the mentioned theorem.

18.4 Applications: Magnetic Systems

18.4.1 Strontium Cuprate: Sr_2CuO_3

Strontium cuprate is, for different reasons, a remarkable compound. It displays linear arrays of Cu^{++} ions in a $3d^9$ electronic configuration, which has spin $s = 1/2$, surrounded by four O^{--} ions, which have a noble gas configuration, $2p^6$. As we can see in Fig. 18.1, the orbitals arrange in such a way that the oxygen p-orbitals overlap the copper d-orbitals, thereby creating an antiferromagnetic

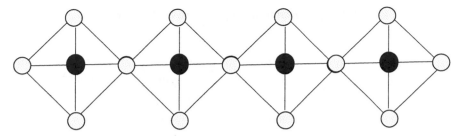

Figure 18.1 Strontium cuprate forms linear chains of Cu^{++} ions (black dots). These have a localized magnetic dipole moment, presenting an antiferromagnetic exchange coupling with its nearest neighbors, produced by a superexchange mechanism, which is mediated by O^{--} ions (white dots).

coupling between the spins of neighbor copper ions through the superexchange mechanism.

Interestingly, the same structure extended to CuO_2 planes on a square lattice forms the base of the high-Tc superconducting cuprates such as La_2CuO_4, as we will see in Chapter 22.

The coupling between neighboring chains is negligible ($J' \simeq 10^{-5}J$), hence the system is a natural realization (actually one of the best) of the isotropic, one-dimensional Heisenberg Hamiltonian given by (18.7), with $J_x = J_y = J_z = J > 0$. This is bosonized as in (18.31) or (18.36), for $\delta = 0$.

The magnetic susceptibility of Sr_2CuO_3 at $T = 0$ is given by (18.49). The correction to this result for $T \neq 0$ has been evaluated in [135] using conformal perturbation and renormalization group techniques. It reads

$$\chi_H(0, 0; T) = \frac{1}{\pi^2 Ja} \left[1 - \frac{1}{2 \ln \frac{T}{T_0}} \right], \tag{18.106}$$

where the reference temperature T_0 is chosen as $T_0 \simeq 7.7J$.

This result is in excellent agreement with the experiment, provided we choose the superexchange coupling as $J = 2200K$ [136].

Another interesting result, obtained by similar methods, is the $T = 0$ magnetic susceptibility in the presence of a uniform magnetic field H. This is given, in the weak field regime, by [137]

$$\chi_H(0, 0; H) = \frac{1}{\pi^2 Ja} \left[1 - \frac{1}{2 \ln \frac{\mu_B g H}{J}} \right], \tag{18.107}$$

where g is the gyromagnetic factor.

18.4.2 Copper Benzoate: $Cu(C_6H_5COO)_2 \bullet 3H_2O$

Hamiltonian

Copper benzoate is another extremely interesting material. In the same way as Sr_2CuO_3, it presents linear chains of magnetic, spin $s = 1/2$, Cu^{++} ions. Each copper ion now is in the center of an oxygen octahedron. The peculiar feature, however, is that the octahedron main axis alternates its direction at every next site along the chain. This produces a site-dependent gyromagnetic tensor of the copper ions, possessing a uniform and a staggered component (this is caused in part by the so-called Dzyaloshinskii–Morya interaction [137, 138]). The magnetic interaction along the chain is governed by an isotropic Heisenberg interaction such that, in the presence of an applied magnetic field $\mathbf{H}$, the copper benzoate Hamiltonian is

$$\mathcal{H}_{CB} = J \sum_n \mathbf{S}_n \cdot \mathbf{S}_{n+1} + \mu_B \sum_n \mathbf{H}^a g_{ab} S_n^b, \tag{18.108}$$

where the gyromagnetic tensor g_{ab} possesses both a uniform and a staggered component, namely

$$g_{ab} = g_{ab}^u + (-1)^n g_{ab}^s. \tag{18.109}$$

The two components of the gyromagnetic tensor produce, respectively, a uniform field $\mathbf{H}^u$ and a staggered field $\mathbf{H}^s = (-1)^n \mathbf{h}$. As it turns out [137], $\mathbf{H}^u$ and $\mathbf{H}^s$ are almost perpendicular and $h \ll H^u$. Consequently, we can write the copper benzoate effective Hamiltonian as [137]

$$\mathcal{H}_{CB} = J \sum_n \mathbf{S}_n \cdot \mathbf{S}_{n+1} + \mu_B \sum_n H^u S_n^z + \mu_B \sum_n (-1)^n h S_n^x. \tag{18.110}$$

Bosonization

The first term above can be written in bosonic form as (18.31) or (18.36). The second term in the expression above can be bosonized as

$$H \frac{\beta}{2\pi} \partial_x \phi. \tag{18.111}$$

This can be absorbed in a redefinition of the bosonic field, namely,

$$\phi \to \phi - H \frac{\beta}{2\pi} x. \tag{18.112}$$

Using the bosonization formula (18.41) for the last term above, as well as the necessary rescalings, we get

$$\mathcal{H}_{CB}(h) = \frac{v}{2} \left(\Pi_\theta^2 + (\partial_x \theta)^2 \right) + M(h) \cos \tilde{\beta} \theta, \tag{18.113}$$

where $\tilde{\beta} = \frac{2\pi}{\beta}$ and we have neglected a rapidly oscillating term, which vanishes upon integration. This is the usual sine-Gordon theory, with $M(h) = Ch$.

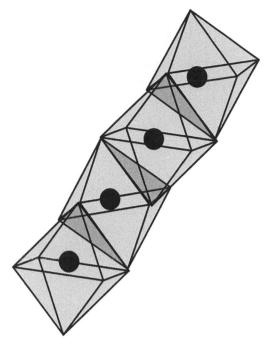

Figure 18.2 The crystal structure of copper benzoate showing Cu^{++} ions at the center of oxygen octahedra with alternating axes' orientation

The Gap, Soliton Mass, Breather Mass

An important issue is the determination of the sine-Gordon gap, which is twice the soliton mass (denoted by Δ) as a function of the staggered field h. Dimensional analysis [137, 139] imposes the following dependence on h:

$$\frac{\Delta(h)}{J} = \mathcal{F}\left(\frac{\mu_B g H}{J}\right)\left(\frac{\mu_B h}{J}\right)^{\frac{\tilde{\beta}^2}{2\tilde{\beta}^2 - \pi}}, \tag{18.114}$$

where logarithmic corrections have been neglected. Then, renormalization group techniques associated with Bethe Ansatz results allow for the explicit determination of the pre-factor function of the uniform field, yielding [137]

$$\frac{\Delta(h)}{J} = 1.50416\left(\ln\frac{\mu_B g H}{J}\right)^{1/6}\left(\frac{\mu_B g h}{J}\right)^{\frac{\tilde{\beta}^2}{2\tilde{\beta}^2 - \pi}}. \tag{18.115}$$

A simplified approach that works quite well consists in bosonizing the sine-Gordon system at $H = 0$ or, equivalently, using the value $\tilde{\beta}^2 = \beta^2 = 2\pi$, but otherwise including the effect of H. The result is [137]

$$\frac{\Delta(h)}{J} = 1.50416 \left(\ln \frac{\mu_B g H}{J} \right)^{1/6} \left(\frac{\mu_B g h}{J} \right)^{2/3}. \tag{18.116}$$

Now, specific heat measurements in copper benzoate for different values of the applied field ranging from $0T$ up to $7T$ reveal the onset of a gapped regime with a gap that scales with the applied field as [140]

$$\frac{\Delta(H)}{J} = \left(\frac{\mu_B g H}{J} \right)^{0.65 \pm 0.03}. \tag{18.117}$$

Considering that $h \propto H$ and that $2/3 \simeq 0.67$, we conclude that the gap predicted by the sine-Gordon description of copper benzoate under the action of an external field is in excellent agreement with the experiments.

The sine-Gordon theory predicts, in the range of couplings to which $\tilde{\beta}$ belongs the occurrence of soliton bound-states, known as "breathers," with masses given by (10.104). The mass of the first breather, then, is given by

$$M_1 = 2\Delta \sin \left(\xi \frac{\pi}{2} \right) \quad ; \quad \xi = \frac{\tilde{\beta}^2}{8\pi - \tilde{\beta}^2}. \tag{18.118}$$

Now, for determining $\tilde{\beta}$ as a function of the magnetic field, we use the exact Bethe Ansatz result, [137, 139]. This gives, for a field $H = 7T$:

$$\frac{\tilde{\beta}^2}{2\pi} (H = 7T) = 0.82. \tag{18.119}$$

This results in a value of $\xi = 0.258$, which implies a breather/soliton mass ratio of $(M_1/\Delta)_{th} \simeq 0.788$.

Neutron scattering experiments performed in copper benzoate for an applied field of $7T$ reveal the presence of sharp peaks [140] corresponding to the soliton at $\Delta = 0.22 meV$ and to a first breather at $M_1 = 0.17 meV$. The ratio $(M_1/\Delta)_{exp} \simeq 0.773$, hence, is in good agreement with the theoretical prediction of the sine-Gordon theory.

Specific Heat

The use of thermal Bethe Ansatz in the sine-Gordon theory has allowed the obtainment of an explicit expression for the copper benzoate free energy, out of which the specific heat can be obtained [139]. At small T, the free energy reads

$$F(T) = -\frac{2T\Delta}{\pi} K_1 \left(\frac{\Delta}{T} \right) - \frac{2T M_1}{\pi} K_1 \left(\frac{M_1}{T} \right), \tag{18.120}$$

where K_1 is a modified Bessel function. The specific heat is, then, given by

$$c = T \frac{\partial^2 F(T)}{\partial T^2}. \tag{18.121}$$

In [139], the theoretical expression derived from (18.120) is compared with the experimental data. The agreement is excellent.

We conclude that sine-Gordon theory provides an accurate description of copper benzoate physical properties, constituting a beautiful example of an application of QFT to a condensed matter system.

18.4.3 The 2D XY-Model and the BKT Transition

The XY-Model

We now consider a third application of bosonization and, in particular, of the Coulomb gas expansion, this time to a classical two-dimensional magnetic system. This is the XY-Model on a square lattice [155, 156, 157]. In this extremely beautiful example of deep interconnection between classical statistical mechanics and quantum field theory of topological excitations, the thermal correlation functions of the 2D XY-spins is exactly mapped into the quantum soliton correlators of the sine-Gordon theory. The Coulomb gas expansion thereof provides the derivation of the exact critical behavior of such correlators, thus allowing for the exact obtainment [141] of the famous $\eta = 1/4$ critical exponent of the Berezinskii–Kosterlitz–Thouless (BKT) phase transition.

The ferromagnetic XY Model is defined by the Hamiltonian (Heisenberg)

$$H_{XY} = -J \sum_{\langle ij \rangle} \mathbf{n}_i \cdot \mathbf{n}_j, \tag{18.122}$$

where the sum runs over nearest neighbors of a square lattice. The 2D classical spins are conveniently parametrized as

$$\mathbf{n}_i = \left(\cos \tilde{\theta}_i, \sin \tilde{\theta}_i \right) \tag{18.123}$$

in such a way that

$$H_{XY} = -J \sum_{\langle ij \rangle} \cos \left(\tilde{\theta}_i - \tilde{\theta}_j \right). \tag{18.124}$$

Assuming a small variation of the spins direction between neighboring sites, we may expand the cosine, obtaining up to a constant

$$H_{XY} = \frac{J}{2} \sum_{\langle ij \rangle} \left(\tilde{\theta}_i - \tilde{\theta}_j \right)^2. \tag{18.125}$$

Taking the continuum limit and considering that each site has two neighbors in each direction, we get

$$H_{XY} = J \int d^2x \nabla \tilde{\theta} \cdot \nabla \tilde{\theta}. \tag{18.126}$$

Considering that the temperature enters the partition function as $H/k_B T$, we choose

$$\tilde{\theta}_i \equiv \sqrt{\frac{T}{2J}} \theta_i$$

$$\mathbf{n}(\mathbf{r}) = \left(\cos \sqrt{\frac{T}{2J}} \theta(\mathbf{r}), \sin \sqrt{\frac{T}{2J}} \theta(\mathbf{r}) \right), \qquad (18.127)$$

such that

$$\frac{H_{XY}}{T} = \frac{1}{2} \int d^2 r \nabla \theta \cdot \nabla \theta. \qquad (18.128)$$

The Vortices and the Sine-Gordon/Coulomb Gas System

The XY-Model in two dimensions possesses topological excitations that carry a nonzero topological charge Q, which is given by

$$Q = \frac{1}{2\pi} \oint d\mathbf{l} \cdot \mathbf{n}. \qquad (18.129)$$

An example of such topological excitations is the vortex configuration

$$\mathbf{n}_v(\mathbf{r} - \mathbf{r}_i) = \lambda_i \nabla \arg(\mathbf{r} - \mathbf{r}_i), \quad ; \quad \lambda_i = \pm 1 \qquad (18.130)$$

centered at $\mathbf{r}_i$, for which $Q = \lambda_i$.

Let us evaluate now the energy associated to a configuration containing n vortices. Inserting (18.130) into (18.122), for each of the vortices we get, after taking again the continuum limit,

$$\frac{H_V}{T} = -\frac{J}{2T} \sum_{i,j}^{n} \lambda_i \lambda_j \int d^2 r \nabla \arg(\mathbf{r} - \mathbf{r}_i) \cdot \nabla \arg(\mathbf{r} - \mathbf{r}_j), \qquad (18.131)$$

where $\lambda_i = \pm 1$ is the topological charge of each (anti)vortex. Now, since $\arg(\mathbf{r})$ is the imaginary part of the analytical function $\ln z$, $z = |\mathbf{r}| e^{i \arg(\mathbf{r})}$, it follows that it must satisfy the Cauchy–Riemann equation with the real part thereof, namely

$$\nabla^i \arg(\mathbf{r} - \mathbf{r}_j) = \epsilon^{ik} \nabla^k \ln \left| \frac{\mathbf{r} - \mathbf{r}_j}{R} \right|, \qquad (18.132)$$

where R is the spatial radius of the system.

The above equation implies we may write (18.131) as

$$\frac{H_V}{T} = -\frac{J}{T} \sum_{i,j}^{n} \lambda_i \lambda_j \int d^2 r \nabla \ln \left| \frac{\mathbf{r} - \mathbf{r}_i}{R} \right| \cdot \nabla \ln \left| \frac{\mathbf{r} - \mathbf{r}_j}{R} \right|. \qquad (18.133)$$

Integrating by parts and considering that

$$- \nabla^2 \ln \left| \frac{\mathbf{r} - \mathbf{r}_i}{R} \right| = 2\pi \delta(\mathbf{r} - \mathbf{r}_i), \tag{18.134}$$

we obtain, after integration in d^2r,

$$\frac{H_V}{T} = -\frac{2\pi J}{T} \sum_{i \neq j}^{n} \lambda_i \lambda_j \ln |\mathbf{r}_i - \mathbf{r}_j|$$

$$- n \frac{2\pi J}{T} \ln |\epsilon| + \frac{2\pi J}{T} \left(\sum_{i=1}^{n} \lambda_i \right)^2 \ln R. \tag{18.135}$$

The partition function being $Z(n) = \mathrm{Tr} e^{-\frac{H}{T}}$, we can infer that in the infinite volume limit, $R \to \infty$ the only vortex configurations that contribute to it are the neutral ones, for which $\sum_{i=1}^{n} \lambda_i = 0$. Hence, since $n = 2m$, we have

$$Z(m; m) = \int d^2r_1 \ldots d^2r_{2m} \exp \left\{ \frac{J\pi}{T} \sum_{i \neq j}^{2m} \lambda_i \lambda_j \ln |\mathbf{r}_i - \mathbf{r}_j|^2 \right\}, \tag{18.136}$$

where m of the λ_i values are $+1$, and the remaining m, -1.

Comparing with (18.81) we see that the vortex partition function coincides with that of the two-dimensional Coulomb gas, provided we make the identifications

$$\frac{J\pi}{T} = \frac{\beta^2}{8\pi} \quad ; \quad \beta = 2\pi \sqrt{\frac{2J}{T}} \quad ; \quad \frac{2\pi}{\beta} = \sqrt{\frac{T}{2J}}. \tag{18.137}$$

One immediately concludes, using (18.80) and (18.128), that the XY-Model partition function in the presence of a neutral configuration with an arbitrary number of vortices is nothing but the vacuum functional of the sine-Gordon theory of the field ϕ, dual to θ, defined so as to satisfy

$$\nabla^i \theta = \epsilon^{ij} \nabla^j \phi,$$

namely,

$$Z_{XY} = \sum_{m=0}^{\infty} \frac{\zeta^{2m}}{m! m!} Z(m; m)$$

$$= Z_{SG} = Z_0^{-1} \int D\phi \exp \{-S_{SG}[\phi]\}, \tag{18.138}$$

where Z_{SG} is given by (18.71), with $\zeta = \frac{\alpha_R}{2}$ being the fugacity, which is renormalized with the second term in the exponent in (18.135).

The sine-Gordon soliton creation operator, $\mu(x, t)$, which carries one unit of the topological charge

$$Q = \frac{\beta}{2\pi} \int_{-\infty}^{\infty} dx \partial_x \phi(x, y), \tag{18.139}$$

then, can be expressed in terms of both ϕ and θ as

$$\mu(\mathbf{r}) = \exp\left\{i\frac{2\pi}{\beta} \int_{-\infty}^{x} d\xi \dot{\phi}(\xi, y)\right\} = \exp\left\{i\frac{2\pi}{\beta}\theta(\mathbf{r})\right\}, \tag{18.140}$$

where $\mathbf{r} = (x, y)$. The SG quantum soliton correlation function, therefore, is given by

$$\langle\mu(\mathbf{r})\mu^\dagger(0)\rangle_{SG} = \left\langle\exp\left\{i\frac{2\pi}{\beta}\theta(\mathbf{r})\right\}\exp\left\{-i\frac{2\pi}{\beta}\theta(0)\right\}\right\rangle_{SG}$$

$$= \left\langle\exp\left\{i\frac{2\pi}{\beta}[\theta(\mathbf{r}) - \theta(0)]\right\}\right\rangle_{SG} = \left\langle\cos\left[\frac{2\pi}{\beta}(\theta(\mathbf{r}) - \theta(0))\right]\right\rangle_{SG}, \tag{18.141}$$

where the last part follows from the invariance of the SG action under $\theta \leftrightarrow -\theta$.

Now, let us consider the thermal correlation function of the XY-Model. According to (18.137) and (18.138) this is given by

$$\langle\mathbf{n}(\mathbf{r}) \cdot \mathbf{n}(0)\rangle_{XY} = \left\langle\cos\left[\frac{2\pi}{\beta}(\theta(\mathbf{r}) - \theta(0))\right]\right\rangle_{SG}. \tag{18.142}$$

We see that this coincides with the quantum correlator of a sine-Gordon operator, namely

$$\langle\mathbf{n}(\mathbf{r}) \cdot \mathbf{n}(0)\rangle_{XY} = \left\langle\exp\left\{i\frac{2\pi}{\beta}\theta(\mathbf{r})\right\}\exp\left\{-i\frac{2\pi}{\beta}\theta(0)\right\}\right\rangle_{SG} = \langle\mu(\mathbf{r})\mu^\dagger(0)\rangle_{SG}. \tag{18.143}$$

The BKT Transition

This is another remarkable and unexpected connection: the thermal correlation functions of the XY-Model spins coincide with the Euclidean quantum correlators of the sine-Gordon soliton operators. In what follows we are going to explore the Coulomb gas description of the latter in order to obtain the exact critical behavior of such functions, in particular at the BKT transition point, at $\beta^2 = 8\pi$ or $T_{BKT} = \pi J$. Indeed, at this temperature the XY-Model undergoes a continuous phase transition, which has already manifested in the Coulomb gas exact equation of state (18.91).

We saw that the XY-vortex corresponds to the sine-Gordon soliton and this, by its turn, corresponds through bosonization to the massive Thirring fermion. Then, from (10.103), it follows that for $\beta^2 < 4\pi$ the soliton-antisoliton interaction is repulsive (soliton-soliton attractive), whereas for $\beta^2 > 4\pi$ the soliton-antisoliton interaction is attractive (soliton-soliton repulsive). Precisely the same properties apply to XY-vortices at the corresponding temperatures related to β through (18.137). As we approach the temperature $T_{BKT} = \pi J$ from above, which means approaching $\beta^2 = 8\pi$ from below, we are in an attractive regime for vortices-antivortices (solitons-antisolitons), where, according to (18.91), the pressure is positive. As we go through the critical point, the pressure becomes negative. This indicates that the gas collapses through the formation of vortex-antivortex (soliton-antisoliton) bound states. This phase transition is the celebrated Berezinskii–Kosterlitz–Thouless (BKT) [156, 157] transition.

One of the benchmarks of the BKT transition is the fact that the XY-spin correlator scales with a critical exponent $\eta = 1/4$ at the transition. This result has been obtained by approximate scaling arguments in the low-temperature (large β) phase [157, 158, 159, 160] and by heuristic arguments in the high-temperature (small β) phase [154].

We now evaluate the XY spin thermal correlator by means of the Coulomb gas expansion of the quantum soliton correlation function, namely [159]

$$
\left\langle \exp\left\{ i\frac{2\pi}{\beta}\theta(\mathbf{x}) \right\} \exp\left\{ -i\frac{2\pi}{\beta}\theta(\mathbf{y}) \right\} \right\rangle_{SG} = |\mathbf{x} - \mathbf{y}|^{-\frac{2\pi}{\beta^2}} Z_{SG}^{-1} \sum_{n=0}^{\infty} \frac{\alpha^{2n}}{n!n!} \int d^2 r_1 \ldots d^2 r_{2n}
$$

$$
\times \exp\left\{ \frac{\beta^2}{8\pi} \sum_{i\neq j=1}^{2n} \lambda_i\lambda_j \ln|\mathbf{r}_i - \mathbf{r}_j|^2 + i\sum_{i=1}^{2n} \lambda_i \left[\arg(\mathbf{r}_i - \mathbf{y}) - \arg(\mathbf{r}_i - \mathbf{x}) \right] \right\}
$$

$$
\equiv \frac{F(\mathbf{x}, \mathbf{y})}{|\mathbf{x} - \mathbf{y}|^{\frac{2\pi}{\beta^2}}}. \tag{18.144}
$$

We are going to demonstrate now that, at the critical point $\beta^2 \to 8\pi$, the function $F(\mathbf{x}, \mathbf{y}) \to 1$, hence the famous scaling with the exponent $\eta = 1/4$ follows as an exact result [141].

In order to derive this result, we shall make use of the so-called bipolar coordinate system, the main properties of which we describe below.

Bipolar Coordinates

Given the position vector on the plane, $\mathbf{r}$ and two "poles," at the positions $\mathbf{x}$ and $\mathbf{y}$, the bipolar coordinate system (ξ, η) is defined as [161]

$$
\xi = \arg(\mathbf{r} - \mathbf{y}) - \arg(\mathbf{r} - \mathbf{x}) \quad ; \quad 0 \leq \xi \leq 2\pi
$$

$$
\eta = \ln|\mathbf{r} - \mathbf{y}| - \ln|\mathbf{r} - \mathbf{x}| \quad ; \quad -\infty \leq \eta \leq \infty. \tag{18.145}
$$

We may express the position vector as

$$\mathbf{r} = |\mathbf{x} - \mathbf{y}| \frac{(\sinh \eta, \sin \xi)}{2[\cosh \eta - \cos \xi]} \equiv |\mathbf{x} - \mathbf{y}| \hat{\mathbf{r}}(\eta, \xi) \tag{18.146}$$

and the integration volume element as

$$d^2 r = \frac{|\mathbf{x} - \mathbf{y}|^2}{4[\cosh \eta - \cos \xi]^2} d\eta d\xi. \tag{18.147}$$

Using this coordinate system, we can rewrite (18.144) as

$$\left\langle \exp\left\{ i \frac{2\pi}{\beta} \theta(\mathbf{x}) \right\} \exp\left\{ -i \frac{2\pi}{\beta} \theta(\mathbf{y}) \right\} \right\rangle_{SG} = |\mathbf{x} - \mathbf{y}|^{-\frac{2\pi}{\beta^2}} Z_{SG}^{-1} \sum_{n=0}^{\infty} \frac{\alpha^{2n}}{n!n!}$$

$$\times \int \Pi_{i=1}^{2n} d\xi_i d\eta_i \frac{|\mathbf{x} - \mathbf{y}|^{4n}}{4[\cosh \eta_i - \cos \xi_i]^2}$$

$$\times \exp\left\{ \frac{\beta^2}{8\pi} \sum_{i \neq j = 1}^{2n} \lambda_i \lambda_j \ln\left[|\mathbf{x} - \mathbf{y}|^2 |\hat{\mathbf{r}}_i - \hat{\mathbf{r}}_j|^2 \right] + i \sum_{i=1}^{2n} \lambda_i \xi_i \right\}. \tag{18.148}$$

We see that the use of bipolar coordinates allows a remarkable factorization of the $|\mathbf{x} - \mathbf{y}|$ factor. Considering the neutrality of the gas and (18.90), we obtain

$$\left\langle \exp\left\{ i \frac{2\pi}{\beta} \theta(\mathbf{x}) \right\} \exp\left\{ -i \frac{2\pi}{\beta} \theta(\mathbf{y}) \right\} \right\rangle_{SG} = \frac{|\mathbf{x} - \mathbf{y}|^{-\frac{2\pi}{\beta^2}}}{Z_{SG}} \sum_{n=0}^{\infty} C_n |\mathbf{x} - \mathbf{y}|^{2n\left(2 - \frac{\beta^2}{4\pi}\right)},$$

$$\tag{18.149}$$

where C_n are constants.

It is clear that at the BKT-point, $\beta^2 = 8\pi$, we have

$$\langle \mathbf{n}(\mathbf{r}) \cdot \mathbf{n}(0) \rangle_{XY} = \left\langle \exp\left\{ i \frac{2\pi}{\beta} \theta(\mathbf{r}) \right\} \exp\left\{ -i \frac{2\pi}{\beta} \theta(0) \right\} \right\rangle_{SG} = \frac{K}{|\mathbf{r}|^{1/4}}, \tag{18.150}$$

which is the well-known BKT critical exponent. Notice that all powers in the series (18.149) become negative for $\beta^2 > 8\pi$, indicating that the large-distance behavior of the spin/soliton correlator is completely determined by the free term, just confirming that the cosine term in the sine-Gordon theory becomes irrelevant in this region.

One can use the Coulomb gas description in order to obtain additional results in the XY-Model. The XY-vortex correlators, for instance, are mapped into the sine-Gordon correlation functions of the $\sigma(x) = \exp\{i \frac{\beta}{2} \phi(x)\}$ operators, which are dual to the soliton operators $\mu(x)$. Following the same steps, then, it is not difficult to show that [141]

$$\langle \mathbf{n}_v(\mathbf{r}) \cdot \mathbf{n}_v(0) \rangle_{XY} = \left\langle \exp\left\{ i \frac{\beta}{2} \phi(\mathbf{r}) \right\} \exp\left\{ -i \frac{\beta}{2} \phi(0) \right\} \right\rangle_{SG} = \frac{K}{|\mathbf{r}|}. \quad (18.151)$$

From the large-distance behavior of the correlation functions (18.149) and (18.151), we infer that under the BKT transition the expectation value of the complex exponential of the phase field changes as

$$\left\langle \exp\left\{ i \frac{\beta}{2} \phi(\mathbf{r}) \right\} \right\rangle_{SG} = \begin{cases} 0 & \beta^2 \geq 8\pi \\ \neq 0 & \beta^2 < 8\pi. \end{cases} \quad (18.152)$$

Consequently, the BKT transition can be seen as one where phase coherence of a complex order parameter sets in, rather than in a Landau–Ginzburg type transition, where the modulus of such parameter becomes different from zero.

The present example shows how powerful is the use of the Coulomb gas picture in order to extract useful information as well as to establish unexpected illuminating interconnections among systems in statistical mechanics, quantum field theory and condensed matter physics that otherwise would seem to be completely unrelated. The inter-relation of the 2D Coulomb gas with bosonization and magnetic systems is a rich and beautiful chapter of theoretical physics. Surprisingly, there are also interesting consequences of the use of the logarithmic gas in 3D [162].

18.5 Applications: Strongly Correlated Systems

18.5.1 The Tomonaga–Luttinger Model

Bosonization

One of the most used models for describing one-dimensional strongly correlated electronic systems is the Tomonaga–Luttinger Model. This is a combination of the low-momentum transfer interactions introduced in subsection 18.2.2, namely

$$\mathcal{H}_{TL} = \mathcal{H}_0 + \mathcal{H}_{1c} + \mathcal{H}_{1s} + \mathcal{H}_{2c} + \mathcal{H}_{2s}. \quad (18.153)$$

Now, using (18.61), (18.65) and (18.66), we can see that, upon bosonization, the Hamiltonian separates into charge and spin bosonic degrees of freedom,

$$\mathcal{H}_{TL} = \mathcal{H}_{TL}^c + \mathcal{H}_{TL}^s$$

$$\mathcal{H}_{TL}^c = \frac{v_c}{2} \left[K_c \Pi_c^{\,2} + \frac{1}{K_c} (\partial_x \phi_c)^2 \right]$$

$$\mathcal{H}_{TL}^s = \frac{v_s}{2} \left[K_s \Pi_s^{\,2} + \frac{1}{K_s} (\partial_x \phi_s)^2 \right], \quad (18.154)$$

where

$$
v_c = \sqrt{\left(v_0 + \frac{g_1^c}{\pi} \right)^2 - \left(\frac{g_2^c}{\pi} \right)^2}
$$

$$
v_s = \sqrt{\left(v_0 + \frac{g_1^s}{\pi} \right)^2 - \left(\frac{g_2^s}{\pi} \right)^2}
\tag{18.155}
$$

and

$$
K_c = \sqrt{\frac{\pi v_0 + g_1^c - g_2^c}{\pi v_0 + g_1^c + g_2^c}}
$$

$$
K_s = \sqrt{\frac{\pi v_0 + g_1^s - g_2^s}{\pi v_0 + g_1^s + g_2^s}}.
\tag{18.156}
$$

We may absorb the constants K_c and K_s through a canonical transformation

$$
\phi_c \to \frac{\phi_c}{\sqrt{K_c}} \quad ; \quad \Pi_c \to \sqrt{K_c} \Pi_c
$$

$$
\phi_s \to \frac{\phi_s}{\sqrt{K_s}} \quad ; \quad \Pi_s \to \sqrt{K_s} \Pi_c.
\tag{18.157}
$$

The exact dispersion relation of the charge and spin modes is

$$
\omega_c = v_c |k| \quad ; \quad \omega_s = v_s |k|,
\tag{18.158}
$$

where the velocities v_c, v_s, are given by (18.155).

We can see that in general there will be a separation of the spin and charge degrees of freedom, as a consequence of the difference between the two associated velocities.

Correlation Function

Let us consider now the one-particle correlation function $\langle \psi_{R,\uparrow}(x) \psi_{R,\uparrow}^\dagger(0) \rangle$. Using the fact that $\psi_{R,\uparrow} = \psi_c \psi_s$, where

$$
\psi_\alpha(x) = \sqrt{\frac{-i}{4\pi}} \exp \left\{ -i \sqrt{\frac{\pi}{2K_\alpha}} \phi_\alpha(x, t) + i \sqrt{\frac{\pi K_\alpha}{2}} \int_{-\infty}^{x} d\xi \, \Pi_\alpha(\xi, t) \right\}
\tag{18.159}
$$

and $\alpha = c, s$ refers to the charge and spin degrees of freedom. Taking (18.154) into account, one readily obtains

$$
\langle \psi_{R,\uparrow}(x) \psi_{R,\uparrow}^\dagger(0) \rangle = \langle \psi_c(x_c) \psi_c^\dagger(0) \rangle \langle \psi_s(x_s) \psi_s^\dagger(0) \rangle,
\tag{18.160}
$$

where $x_\alpha = x - v_\alpha t$, for $\alpha = c, s$.

Then we can use (10.170) in order to determine an exact Euclidean expression for the above function, namely,

$$\langle \psi_{R,\uparrow}(x,t)\psi_{R,\uparrow}^{\dagger}(0)\rangle = \frac{1}{2\pi}\frac{1}{(x-iv_ct)^{1/2}}\frac{1}{2\pi}\frac{1}{(x-iv_st)^{1/2}}$$
$$\times \frac{1}{(x^2+v_c^2t^2)^{\theta_c/2}}\frac{1}{(x^2+v_s^2t^2)^{\theta_s/2}}, \qquad (18.161)$$

where

$$\theta_c = \frac{1}{4}\left[K_c + \frac{1}{K_c} - 2\right] \; ; \quad \theta_s = \frac{1}{4}\left[K_s + \frac{1}{K_s} - 2\right].$$

Notice that the effects of the interaction manifest in the θ_α terms. $K_\alpha = 1$ represents the free case.

The spectral density $A(\omega,k)$ can be obtained, according to (13.53), from the Fourier transform of the above correlator, both in space and time. It has been shown that [120] it has the following properties (for $K_s = 1$):

$$A(\omega,k) = 0 \; ; \quad -v_ck < \omega < v_sk$$
$$A(\omega \simeq v_sk, k) \sim \theta(\omega - v_sk)(\omega - v_sk)^{\theta_c-1/2}$$
$$A(\omega \simeq -v_ck, k) \sim \theta(-\omega + v_ck)(\omega + v_ck)^{\theta_c}$$
$$A(\omega \simeq v_ck, k) \sim \theta(\omega - v_ck)(\omega - v_ck)^{\frac{\theta_c-1}{2}}. \qquad (18.162)$$

The spectral density shows some remarkable features. For instance, it does not contain any delta peak that would indicate the existence of one-particle fermion excitations. Rather, it reveals that the only excitations present in the spectrum are the collective bosonic excitations associated to the ϕ_c and ϕ_s fields, related, respectively, with charge and spin degrees of freedom. The characteristic velocities of these collective excitations, however, become different as a consequence of the interaction. We observe, therefore, the occurrence of the phenomenon of spin-charge separation as a dramatic consequence of the strong interactions.

Response Functions

Using the bosonization formulas, we can straightforwardly determine basic response functions such as the magnetic susceptibility, the electric conductivity and the compressibility. These are, respectively, given by

$$\chi(\omega,q) = \frac{K_s}{2\pi}\frac{q^2}{\omega^2 + v_s^2q^2}$$
$$\sigma(\omega,q) = \frac{K_c}{2\pi}\frac{i\omega}{\omega^2 + v_c^2q^2}$$
$$\kappa(\omega,q) = \frac{K_c}{2\pi}\frac{q^2}{\omega^2 + v_c^2q^2} \qquad (18.163)$$

according to (18.55), (18.56) and (18.57)

18.5.2 The Hubbard Model

As a second application of bosonization for describing strongly correlated fermion systems in one spatial dimension, let us consider the Hubbard model, given by (3.34), on a linear chain, namely

$$H_H = -t \sum_{n,\sigma} \left[c^\dagger_{n,\sigma} c_{n+1,\sigma} + c^\dagger_{n+1,\sigma} c_{n,\sigma} \right] + U \sum_n n_{n,\uparrow} n_{n,\downarrow}, \tag{18.164}$$

where $n_{n,\sigma} = c^\dagger_{n,\sigma} c_{n,\sigma}$. The first term is the familiar tight-binding Hamiltonian, and in the continuum limit is bosonized as in (18.60).

Considering that, in the continuum limit

$$n_{n,\sigma} \to j^0_\sigma = \rho_{R,\sigma}(x) + \rho_{L,\sigma}(x) \quad ; \quad \sigma = \uparrow, \downarrow \tag{18.165}$$

and using (18.50), we find the Hubbard interaction term is bosonized as

$$U \sum_n n_{n,\uparrow} n_{n,\downarrow} \to \frac{U}{\pi} \partial_x \phi_\uparrow \partial_x \phi_\downarrow. \tag{18.166}$$

Combining this expression with the bosonized form of the tight-binding term, we get the following bosonized Hamiltonian density for the Hubbard model:

$$\mathcal{H}_H = \frac{v_F}{2} \left[\Pi_c^2 + \left(1 + \frac{U}{v_F \pi}\right)(\partial_x \phi_c)^2 \right] + \frac{v_F}{2} \left[\Pi_s^2 + \left(1 - \frac{U}{v_F \pi}\right)(\partial_x \phi_s)^2 \right], \tag{18.167}$$

where $v_F = 2ta$ and we are assuming half-filling occupation.

The above Hamiltonian can be expressed, equivalently, as

$$\mathcal{H}_H = \frac{v_c}{2} \left[K_c \Pi_c^2 + \frac{1}{K_c}(\partial_x \phi_c)^2 \right] + \frac{v_c}{2} \left[K_s \Pi_s^2 + \frac{1}{K_s}(\partial_x \phi_s)^2 \right], \tag{18.168}$$

where

$$v_c = v_F \sqrt{1 + \frac{U}{v_F \pi}} \quad ; \quad v_s = v_F \sqrt{1 - \frac{U}{v_F \pi}}$$

$$K_c = \left[1 + \frac{U}{v_F \pi} \right]^{-1/2} \quad ; \quad K_s = \left[1 - \frac{U}{v_F \pi} \right]^{-1/2}. \tag{18.169}$$

We see that also in the d = 1 Hubbard model, the only physical excitations are the collective bosonic excitations ϕ_c and ϕ_s, respectively associated to charge and spin. From the expression above we can infer that the velocities of charge and spin excitations are split by the interaction and the system exhibits, similarly to the Tomonaga–Luttinger model, the remarkable phenomenon of spin-charge separation. Correlation functions and response functions can be obtained in the same way as we did in that model.

19

Quantum Magnets in 2D: Nonlinear Sigma Model, CP^1 and "All That"

We saw in the last chapter how convenient it was to map one-dimensional quantum magnetic systems into bosonic quantum field theories. This procedure has frequently led to unsuspected connections among systems that otherwise seemed to be completely unrelated. The sequence of steps leading from quantum spins to fermions, from fermions to bosons and, finally, from bosons to classical gases, has proved to be extremely useful, revealing a rich variety of interesting physical results. Now we turn to quantum magnetic systems in two spatial dimensions. In this case we will see that, once again, the use of ingenuous methods will enable one to map these systems into bosonic quantum field theories, such as the nonlinear sigma model (NLSM) or its equivalent CP^1 model (CP1M). Exploring the physical properties of such field theory models will allow the obtainment of nontrivial results about such quantum magnetic systems. Then, in the next two chapters, we extend the range of application of the same methodology in two different ways. Firstly we apply it to systems enlarged by the presence of electrons embedded in that magnetic background, the so-called spin-fermion model, and secondly we introduce magnetic disorder and study the onset of the phase known as spin-glass.

Later on, we will see that two-dimensional quantum magnetic systems play a central role in the physics of high-Tc superconductors, both the cuprates and iron-based pnictides. The mappings of such magnetic systems into the NLSM and CP1M consequently will prove to be quite useful in the description of those materials.

19.1 From Heisenberg Model to Nonlinear Sigma Model

Our starting point is the Heisenberg Hamiltonian, describing a magnetic system with nearest-neighbor interactions of quantum spins on a square lattice,

$$H = J \sum_{\langle ij \rangle} \mathbf{S}_i \cdot \mathbf{S}_j, \tag{19.1}$$

where $J > 0$ corresponds to the antiferromagnetic (AF) case and $J < 0$, to the ferromagnetic one. The above Hamiltonian corresponds to the homogeneous case, where the exchange coupling J is uniform. This can be generalized to a non-uniform exchange, where the coupling changes from link to link: $J \to J_{ij}$.

The main purpose is to evaluate the partition function

$$Z = \mathrm{Tr} e^{-\beta H} = \sum_n \langle n | e^{-\beta H} | n \rangle \tag{19.2}$$

for some complete set of states $|n\rangle$. The present derivation is close to the ones found in [163, 164].

19.1.1 Coherent Spin States and Continuum Limit

A very convenient set of states is the so-called coherent spin states [166]. These are quantum states $|\mathbf{N}\rangle$, labeled by the classical unit vector $\mathbf{N}$, having the property

$$\langle \mathbf{N} | \mathbf{S} | \mathbf{N} \rangle = s\mathbf{N}, \tag{19.3}$$

where s is the spin quantum number, labeling the angular momentum eigenstates $|sm\rangle$, $m = -s, -s+1, \ldots, s-1, s$.

One can easily verify that

$$|\mathbf{N}_0\rangle = |s, m = s\rangle, \quad ; \quad \mathbf{N}_0 = \hat{z}. \tag{19.4}$$

Also, it is not difficult to show that an arbitrary state $|\mathbf{N}\rangle$ is obtained out of $|\mathbf{N}_0\rangle$ by applying the rotation operator corresponding to a rotation of an angle θ about the unit vector $\hat{t}(\varphi) = (-\sin\varphi, \cos\varphi)$, tangent to a circle in the xy-plane, namely,

$$|\mathbf{N}\rangle = U(\theta, \varphi) |\mathbf{N}_0\rangle, \tag{19.5}$$

where

$$U(\theta, \varphi) = \exp\left\{-i\theta \hat{t}(\varphi) \cdot \mathbf{S}\right\}. \tag{19.6}$$

The U-operator satisfies

$$U(\theta, \varphi) \sigma_z U^\dagger(\theta, \varphi) = Q = \mathbf{N} \cdot \sigma. \tag{19.7}$$

The states $|\mathbf{N}\rangle$ have the following property,

$$\int \frac{d\mathbf{N}}{2\pi} |\mathbf{N}\rangle \langle \mathbf{N}| = 1, \tag{19.8}$$

and we will use them for evaluating the trace in (19.2).

For this purpose, let us write

$$e^{-\beta H[\mathbf{S}]} = \underbrace{e^{-\epsilon H[\mathbf{S}]} \ldots e^{-\epsilon H[\mathbf{S}]}}_{M} \tag{19.9}$$

with $\beta = M\epsilon$.

Then, using (19.2) repeatedly, we get

$$
Z = \int \frac{d\mathbf{N}}{2\pi} \langle \mathbf{N} | e^{-\beta H[\mathbf{S}]} | \mathbf{N} \rangle
$$

$$
= \lim_{M \to \infty; \epsilon = \frac{\beta}{M}} \int \frac{d\mathbf{N}}{2\pi} \int \prod_{i=1}^{M-1} \frac{d\mathbf{N}_i}{2\pi} \underbrace{\langle \mathbf{N} | e^{-\epsilon H[\mathbf{S}]} | \mathbf{N}_1 \rangle \dots \langle \mathbf{N}_{M-1} | e^{-\epsilon H[\mathbf{S}]} | \mathbf{N} \rangle}_{M} .
$$

$$(19.10)$$

Now, considering the interval $[0, \beta]$ of the real variable τ and introducing the continuous function $\mathbf{N}(\tau)$ in such a way that, making a partition of this interval in M pieces of uniform size $\epsilon = \frac{\beta}{M} = \tau_{i+1} - \tau_i$, $i = 0, \dots, M - 1$, we have $\mathbf{N}(\tau_i) = \mathbf{N}_i$. Consequently, we may write each of the M factors in (19.10) as

$$
\langle \mathbf{N}(\tau_i) | e^{-\epsilon H[\mathbf{S}]} | \mathbf{N}(\tau_{i+1}) \rangle \simeq \langle \mathbf{N}(\tau_i) | [1 - \epsilon H[\mathbf{S}]] | \mathbf{N}(\tau_{i+1}) \rangle. \tag{19.11}
$$

Now, making a Taylor expansion

$$
|\mathbf{N}(\tau_{i+1})\rangle = |\mathbf{N}(\tau_i)\rangle + \epsilon \frac{d}{d\tau} |\mathbf{N}(\tau_i)\rangle + O(\epsilon^2), \tag{19.12}
$$

we have, up to $O(\epsilon)$,

$$
\langle \mathbf{N}(\tau_i) | e^{-\epsilon H[\mathbf{S}]} | \mathbf{N}(\tau_{i+1}) \rangle \simeq 1 + \epsilon \langle \mathbf{N}(\tau_i) | \frac{d}{d\tau} |\mathbf{N}(\tau_i)\rangle - \epsilon \langle \mathbf{N}(\tau_i) | H[\mathbf{S}] | \mathbf{N}(\tau_i) \rangle
$$

$$
\simeq \exp \left\{ \epsilon \left[\langle \mathbf{N}(\tau_i) | \frac{d}{d\tau} | \mathbf{N}(\tau_i) \rangle - H[s\mathbf{N}] \right] \right\}, \tag{19.13}
$$

where we used the fact that

$$
\langle \mathbf{N}(\tau_i) | H[\mathbf{S}] | \mathbf{N}(\tau_i) \rangle = H[s\mathbf{N}]. \tag{19.14}
$$

Now, inserting (19.13) in (19.10) and taking the continuum limit $\epsilon \to 0$; $M \to \infty$, we finally obtain

$$
Z = \int D\mathbf{N} \exp \left\{ \int_0^\beta d\tau \left[\langle \mathbf{N}(\tau) | \frac{d}{d\tau} | \mathbf{N}(\tau) \rangle - H[s\mathbf{N}] \right] \right\}. \tag{19.15}
$$

We see that the use of the coherent spin states allowed us to express the partition function as a functional integral over the unit classical vector field $\mathbf{N}(\tau)$.

In the next two subsections, we consider in detail the two terms in the exponent above, both for $J > 0$ and $J < 0$.

19.1.2 The Antiferromagnetic Case

The Hamiltonian

We first consider the second term in (19.15). Let us start with the AF case, $J > 0$. We have

$$H[s\mathbf{N}] = Js^2 \sum_{\langle ij \rangle} \mathbf{N}_i \cdot \mathbf{N}_j = \frac{Js^2}{2} \sum_{\langle ij \rangle} \left[\mathbf{N}_i + \mathbf{N}_j \right]^2, \tag{19.16}$$

where we used the fact that $|\mathbf{N}|^2 = 1$ and neglected a constant term.

We introduce now a continuum field $\mathbf{n}(\mathbf{r}, \tau)$ such that $|\mathbf{n}(\mathbf{r}, \tau)|^2 = 1$ and a transverse field $\mathbf{L}(\mathbf{r}, \tau)$ such that $\mathbf{L} \cdot \mathbf{n} = 0$. These represent, respectively, the antiferromagnetic and ferromagnetic components of the field $\mathbf{N}(\mathbf{r}, \tau)$. We then write, in terms of the lattice parameter a of a square lattice,

$$\mathbf{N}(\mathbf{r}_i, \tau) = (-1)^i \mathbf{n}(\mathbf{r}_i, \tau)\sqrt{1 - a^4 \mathbf{L}^2(\mathbf{r}_i, \tau)} + a^2 \mathbf{L}(\mathbf{r}_i, \tau)$$
$$= (-1)^i \mathbf{n}(\mathbf{r}_i, \tau) + a^2 \mathbf{L}(\mathbf{r}_i, \tau) + O(a^4) \tag{19.17}$$

in such a way that $|\mathbf{N}|^2 = |\mathbf{n}|^2 = 1$. Then, making the Taylor expansion,

$$\mathbf{N}(\mathbf{r}_j, \tau) = \mathbf{n}(\mathbf{r}_i, \tau) + a\nabla_j \mathbf{n}(\mathbf{r}_i, \tau) + a^2 \mathbf{L}(\mathbf{r}_i, \tau) + O(a^4). \tag{19.18}$$

Using (19.17), we get, for i, j nearest neighbors,

$$\left[\mathbf{N}_i + \mathbf{N}_j \right]^2 = a^2 \nabla_j \mathbf{n}(\mathbf{r}_i, \tau) \cdot \nabla_j \mathbf{n}(\mathbf{r}_i, \tau)$$
$$+ 4a^4 |\mathbf{L}(\mathbf{r}_i, \tau)|^2 + 4a^3 \nabla_j \mathbf{n}(\mathbf{r}_i, \tau) \cdot \mathbf{L}(\mathbf{r}_i, \tau). \tag{19.19}$$

Inserting this in (19.16) and considering that, for each site i of a square lattice, the two opposite nearest neighbors along each direction will produce contributions of opposite sign for the last term, we conclude that only the first two terms above will contribute to the sum. Hence, taking the continuum limit, we obtain

$$H[s\mathbf{N}] = \frac{1}{2} \int d^2r \left[Js^2 \nabla_i \mathbf{n} \cdot \nabla_i \mathbf{n} + 4Js^2 a^2 |\mathbf{L}|^2 \right]. \tag{19.20}$$

The Berry Phase

Let us consider now the first term in (19.15), namely

$$\zeta = \int_0^\beta d\tau \langle \mathbf{N}(\tau)| \frac{d}{d\tau} |\mathbf{N}(\tau)\rangle. \tag{19.21}$$

This is a pure imaginary number because the fact that

$$\frac{d}{d\tau} \langle \mathbf{N}(\tau)|\mathbf{N}(\tau)\rangle = 0$$

implies

$$\langle \mathbf{N}(\tau)|\frac{d}{d\tau}|\mathbf{N}(\tau)\rangle = -\frac{d}{d\tau}\left[\langle \mathbf{N}(\tau)|\right]|\mathbf{N}(\tau)\rangle,$$

hence $\zeta^* = -\zeta$ and, consequently, it appears in (19.15) as a phase.

This is an example of the famous Berry phase, which manifests in many areas of physics when the Hamiltonian of a given system depends on some parameters that are varied adiabatically [167]. We study the Berry phase from a general point of view in subsection 26.2.3 and in connection to the quantum Hall effect in subsection 26.2.4.

Observe that $\tau \in [0, \beta]$ parametrizes a closed curve $S(C)$ on the spherical surface $|\mathbf{N}|^2 = 1$, as a consequence of the periodic boundary conditions in β.

We can write ζ as a line integral on this curve, namely

$$\zeta = \int_0^\beta d\tau \frac{d\mathbf{N}}{d\tau} \cdot \langle \mathbf{N}(\tau)|\nabla_\mathbf{N}|\mathbf{N}(\tau)\rangle$$

$$\zeta = \oint_{S(C)} d\mathbf{N} \cdot \langle \mathbf{N}(\tau)|\nabla_\mathbf{N}|\mathbf{N}(\tau)\rangle. \tag{19.22}$$

Calling

$$\vec{\mathcal{A}} = i\langle \mathbf{N}(\tau)|\nabla_\mathbf{N}|\mathbf{N}(\tau)\rangle, \tag{19.23}$$

we may use Stokes' theorem in order to express ζ as a surface integral. It happens, however, that there are two surfaces $S_1(C)$ and $S_2(C)$ bound by C, such that their union is the sphere. Then, for consistency, these two integrals must be equal. We can, therefore, express ζ as an integral over the whole sphere,

$$\zeta = \frac{i}{2}\int_{S_2} d^2\mathbf{N} \cdot \vec{\mathcal{B}}, \tag{19.24}$$

where $\vec{\mathcal{B}} = \nabla_\mathbf{N} \times \vec{\mathcal{A}}$. This, however, is the "magnetic flux" of $\vec{\mathcal{B}}$ across the whole sphere, which must be quantized as $2\pi n$ for $n \in \mathbb{Z}$ (see subsection 26.2.3). We see that the Berry phase ζ is a topological invariant because it does not change by local transformations.

For the homotopy class with $n = 1$, we can write

$$\zeta = i2\pi \int_{S_2} d^2\mathbf{N} \cdot \mathbf{N}. \tag{19.25}$$

Now, parametrizing the spherical surface by $\tau \in [0, \beta]$ and $x \in [0, 1]$, where the two limits in the x-interval correspond to the north and south poles, and changing the integration variables from $d^2\mathbf{N}$ to $d\tau\, dx$, we obtain

$$\zeta = i\int_0^1 dx \int_0^\beta d\tau \left[\mathbf{N} \cdot \left(\frac{\partial \mathbf{N}}{\partial x} \times \frac{\partial \mathbf{N}}{\partial \tau}\right)\right]. \tag{19.26}$$

In the above expression $J^a dx d\tau = d^2 N^a$, where J^a is the Jacobian for the change of variables $dx d\tau \leftrightarrow d^2 \mathbf{N}^a$, given by

$$J^a = \frac{1}{2} \left[\frac{\partial \mathbf{N}}{\partial x} \times \frac{\partial \mathbf{N}}{\partial \tau} \right]^a. \tag{19.27}$$

We have, therefore,

$$H_B = s\zeta. \tag{19.28}$$

This phase expresses the solid angle described by the vector $\mathbf{N}$ on the unit radius sphere $|\mathbf{N}| = 1$, as the parameters (x, τ) vary in the intervals $x \in [0, 1]$, $\tau \in [0, \beta]$. It is an example of the well-known Berry phase [167].

We now want to express the Berry phase in terms of the fields $\mathbf{n}$ and $\mathbf{L}$. For this, we use (19.17) and obtain, up to the order a^2 in the lattice parameter,

$$H_B = is \sum_{i \in \mathbb{Z}} \int_0^1 dx \int_0^\beta d\tau \left[(-1)^i \mathbf{n} \cdot \left(\frac{\partial \mathbf{n}}{\partial x} \times \frac{\partial \mathbf{n}}{\partial \tau} \right) \right.$$
$$\left. + a^2 \mathbf{n} \cdot \left(\frac{\partial \mathbf{n}}{\partial x} \times \frac{\partial \mathbf{L}}{\partial \tau} \right) + a^2 \mathbf{n} \cdot \left(\frac{\partial \mathbf{L}}{\partial x} \times \frac{\partial \mathbf{n}}{\partial \tau} \right) \right]. \tag{19.29}$$

The first term above vanishes for smooth configurations of the sublattice magnetization field $\mathbf{n}$. The remaining terms can be written as

$$H_B = is \int d^2 r \int_0^1 dx \int_0^\beta d\tau \left\{ \frac{\partial}{\partial \tau} \left[\mathbf{n} \cdot \left(\frac{\partial \mathbf{n}}{\partial x} \times \mathbf{L} \right) \right] - \frac{\partial}{\partial x} \left[\mathbf{n} \cdot \left(\frac{\partial \mathbf{n}}{\partial \tau} \times \mathbf{L} \right) \right] \right\}, \tag{19.30}$$

where we used the fact that $\mathbf{L}$, $\frac{\partial \mathbf{n}}{\partial x}$, $\frac{\partial \mathbf{n}}{\partial \tau}$ are coplanar vectors. Now, considering that we have periodic boundary conditions in the τ integral and that $\frac{\partial \mathbf{n}}{\partial \tau}|_{x=0} = 0$, we get

$$H_B = -is \int d^2 r \int_0^\beta d\tau \mathbf{n} \cdot \left[\frac{\partial \mathbf{n}}{\partial \tau} \times \mathbf{L} \right]$$
$$H_B = -is \int d^2 r \int_0^\beta d\tau \mathbf{L} \cdot \left[\mathbf{n} \times \frac{\partial \mathbf{n}}{\partial \tau} \right]. \tag{19.31}$$

Combining with (19.20), we can express the partition function (19.2) as

$$Z = \int D\mathbf{n} D\mathbf{L} \delta(|\mathbf{n}|^2 - 1)$$
$$\times \exp \left\{ \frac{1}{2} \int d^2 r \int_0^\beta d\tau \left[J s^2 \nabla_i \mathbf{n} \cdot \nabla_i \mathbf{n} + 4 J s^2 a^2 |\mathbf{L}|^2 \right] - is \mathbf{L} \cdot \left[\mathbf{n} \times \frac{\partial \mathbf{n}}{\partial \tau} \right] \right\}. \tag{19.32}$$

Performing the integration over the ferromagnetic fluctuations, **L**, we obtain

$$Z = \int D\mathbf{n}\delta(|\mathbf{n}|^2 - 1)$$

$$\times \exp\left\{\frac{1}{2}\int d^2r \int_0^\beta d\tau \left[Js^2\nabla_i\mathbf{n}\cdot\nabla_i\mathbf{n} + \frac{1}{4Ja^2}\left[\mathbf{n}\times\frac{\partial\mathbf{n}}{\partial\tau}\right]\cdot\left[\mathbf{n}\times\frac{\partial\mathbf{n}}{\partial\tau}\right]\right]\right\}.$$

$$(19.33)$$

Finally, using the fact that, for $|\mathbf{n}|^2 = 1$,

$$\left[\mathbf{n}\times\frac{\partial\mathbf{n}}{\partial\tau}\right]\cdot\left[\mathbf{n}\times\frac{\partial\mathbf{n}}{\partial\tau}\right] = \frac{\partial\mathbf{n}}{\partial\tau}\cdot\frac{\partial\mathbf{n}}{\partial\tau}. \tag{19.34}$$

This ultimately leads to the relativistic $O(3)$ nonlinear sigma model, namely,

$$Z = \int D\mathbf{n}\delta(|\mathbf{n}|^2 - 1)$$

$$\times \exp\left\{\frac{\rho_s}{2}\int d^2r \int_0^\beta d\tau \left[\nabla_i\mathbf{n}\cdot\nabla_i\mathbf{n} + \frac{1}{c^2}\partial_\tau\mathbf{n}\cdot\partial_\tau\mathbf{n}\right]\right\}, \tag{19.35}$$

where $\rho_s = Js^2$ is the spin stiffness and $c = 2Jsa$ is the spin-waves velocity.

From the above expression, we can immediately infer that the AF spin-wave will have a dispersion relation given by $\omega = c|\mathbf{k}|$.

19.1.3 The Ferromagnetic Case

The Hamiltonian

Now, let us consider the second term in (19.15), for $J < 0$. It is clear that we can in this case write, up to a constant,

$$H[s\mathbf{N}] = -|J|s^2\sum_{\langle ij\rangle}\mathbf{N}_i\cdot\mathbf{N}_j = \frac{|J|s^2}{2}\sum_{\langle ij\rangle}[\mathbf{N}_i - \mathbf{N}_j]^2. \tag{19.36}$$

Now, to use the same notation, we identify, in the continuum limit,

$$\mathbf{N}(\mathbf{r}_i, \tau) = \mathbf{n}(\mathbf{r}_i, \tau). \tag{19.37}$$

Now, for nearest neighbors ij,

$$\mathbf{N}(\mathbf{r}_j, \tau) = \mathbf{n}(\mathbf{r}_i, \tau) + a\nabla_j\mathbf{n}(\mathbf{r}_i, \tau) + O(a^4)$$
$$\mathbf{N}(\mathbf{r}_j, \tau) - \mathbf{N}(\mathbf{r}_i, \tau) = a\nabla_j\mathbf{n}(\mathbf{r}_i, \tau)$$
$$[\mathbf{N}(\mathbf{r}_j, \tau) - \mathbf{N}(\mathbf{r}_i, \tau)]^2 = a^2\nabla_j\mathbf{n}\cdot\nabla_j\mathbf{n}. \tag{19.38}$$

Now, inserting the above result in (19.16) and taking the continuum limit, we get, for $J < 0$,

$$H[s\mathbf{N}] = \frac{1}{2} \int d^2r |J| s^2 \nabla_i \mathbf{n} \cdot \nabla_i \mathbf{n}. \tag{19.39}$$

The Berry Phase

Let us turn now to the Berry phase in the ferromagnetic case. From (19.26) and (19.37), we conclude that the Berry phase of the magnetization field $\mathbf{n}$ no longer cancels.

The partition function in this case is given by

$$Z = \int D\mathbf{n} D\lambda \, \exp \left\{ \int d^2r \int_0^\beta d\tau \left[\frac{\rho_s}{2} \nabla_i \mathbf{n} \cdot \nabla_i \mathbf{n} + i\lambda(|\mathbf{n}|^2 - 1) \right. \right.$$
$$\left. \left. + i\frac{s}{a^2} \int_0^1 dx \mathbf{n} \cdot \left(\frac{\partial \mathbf{n}}{\partial x} \times \frac{\partial \mathbf{n}}{\partial \tau} \right) \right] \right\}, \tag{19.40}$$

where we introduced the functional integral representation of the functional delta and $\rho_s = |J| s^2$.

The field equation satisfied by $\mathbf{n}$ in real time is

$$\rho_s \nabla^2 \mathbf{n} = \lambda \mathbf{n} + \frac{s}{a^2} \mathbf{n} \times \partial_t \mathbf{n}, \tag{19.41}$$

Taking the rotational of the above equation, we get the relation

$$\partial_t \mathbf{n} = |J| a^2 s \mathbf{n} \times \nabla^2 \mathbf{n}, \tag{19.42}$$

which is known as the Landau–Lifshitz equation [168] and implies a quadratic dispersion relation for the ferromagnetic spin-waves, as it should.

19.2 From the $NL\sigma$ Model to the CP^1 Formulation

In this section, we are going to describe the so-called CP^1 formulation of the non-linear sigma model. Consider two complex scalar fields z_1 and z_2, in terms of which we express the NLSM field as

$$\mathbf{n} = z_i^* \sigma_{ij} z_j. \tag{19.43}$$

The use of (3.41) reveals the fact that $|\mathbf{n}|^2 = 1$ implies $z_i^* z_i = 1$. Then, using the expression above, one finds

$$\partial_\mu \mathbf{n} \cdot \partial_\mu \mathbf{n} = 4 \left[\partial_\mu z_i^* \partial_\mu z_i + (z_i^* \partial_\mu z_i)(z_i^* \partial_\mu z_i) \right]. \tag{19.44}$$

19.2.1 The Antiferromagnetic Case

Starting from the NLSM action for the AF case, namely,

$$S_{NLSM} = \frac{\rho_s}{2} \int d^3x \, \partial_\mu \mathbf{n} \cdot \partial_\mu \mathbf{n}, \tag{19.45}$$

and using (19.44), one ends up with the CP^1 action,

$$S_{CP^1} = 2\rho_s \int d^3x \left[\partial_\mu z_i^* \partial_\mu z_i + (z_i^* \partial_\mu z_i)(z_i^* \partial_\mu z_i) \right]. \tag{19.46}$$

Now, using the auxiliary Hubbard–Stratonovitch field A_μ, we can express the above action as

$$e^{-S_{CP^1}} = \int DA_\mu \exp\left\{ -2\rho_s \int d^3x \left[\partial_\mu z_i^* \partial_\mu z_i - 2i(z_i^* \partial_\mu z_i) A_\mu + A_\mu A^\mu \right] \right\}. \tag{19.47}$$

Then, using the fact that $z_i^* z_i = 1$, we may write the partition function of the AF Heisenberg model on a square lattice as

$$Z = \int Dz_i \, Dz_i^* \, DA_\mu \, D\lambda \exp\left\{ -2\rho_s \int d^3x (D_\mu z_i)^*(D^\mu z_i) - i\lambda[z_i^* z_i - 1] \right\}, \tag{19.48}$$

where $D_\mu = \partial_\mu + iA_\mu$ is a covariant derivative and we introduced the functional delta representation. The field equation for A_μ yields

$$A_\mu = i z_i^* \partial_\mu z_i. \tag{19.49}$$

19.2.2 The Ferromagnetic Case

The relation (19.44) is valid component-wise, hence we may express the Hamiltonian part of the NLSM action, namely,

$$S_{NLSM} = \frac{\rho_s}{2} \int d^3x \nabla_i \mathbf{n} \cdot \nabla_i \mathbf{n} \tag{19.50}$$

as

$$S_{H,CP^1} = 2\rho_s \int d^3x \left[\partial_j z_i^* \partial_j z_i + (z_i^* \partial_j z_i)(z_i^* \partial_j z_i) \right]. \tag{19.51}$$

As in the AF case, we can use the auxiliary vector field A_i in the action

$$S_{H,CP^1}[A_i] = 2\rho_s \int d^3x \left[D_j z_i^* D_j z_i \right], \tag{19.52}$$

where, upon integration on A_i, we recover (19.51).

In the ferromagnetic case, the Berry phase term can be expressed, in CP^1 language, as [169]

$$S_{B,CP^1} = i2s \int d^3x z_i^* \partial_0 z_i. \tag{19.53}$$

Consequently, combining (19.52) and (19.53), we can write the full partition of the Heisenberg model on a square lattice, in the ferromagnetic case, as

$$Z = \int Dz_i Dz_i^* DA_i \exp\left\{-\int d^3x \left[i2sz_i^*\partial_0 z_i + 2\rho_s(D_j z_i)^*(D_j z_i)\right]\right\} \delta[z_i^* z_i - 1]. \tag{19.54}$$

19.3 Quantum and Thermal Fluctuations: the Antiferromagnetic Case

19.3.1 Quantum Fluctuations

Let us consider now the effects of quantum fluctuations in the determination of the possible different phases of the AF magnetic system associated with the partition function given by (19.35). Calling the different components of the NLSM field as $\mathbf{n} = (\pi, \sigma)$ and rescaling the fields as

$$\sigma \longrightarrow \sqrt{\rho_s}\sigma \quad ; \quad \pi \longrightarrow \sqrt{\rho_s}\pi \quad ; \quad \lambda \longrightarrow \frac{\lambda}{\rho_s}, \tag{19.55}$$

we may write the partition function, at $T = 0$, as

$$Z = \int D\pi \, D\sigma \, D\lambda$$
$$\times \exp\left\{\int d^3r \left[\frac{1}{2}\partial_\mu\pi \cdot \partial_\mu\pi + \frac{1}{2}\partial_\mu\sigma\partial_\mu\sigma + i\lambda\left[\sigma^2 + |\pi|^2 - \rho_s\right]\right]\right\}. \tag{19.56}$$

Now, performing the quadratic functional integral on π, namely, on the transverse components of the sublattice magnetization $\mathbf{n}$, we obtain

$$Z = \int D\sigma \, D\lambda$$
$$\times \exp\left\{\int d^3r \left[\frac{1}{2}\partial_\mu\sigma\partial_\mu\sigma + i\lambda\left[\sigma^2 - \rho_s\right]\right] + \text{Tr}\ln\left[1 + 2i\frac{\lambda}{-\Box}\right]\right\}. \tag{19.57}$$

We are going to look for constant minima of the above effective action corresponding to $\langle\sigma\rangle$ and $\langle\lambda\rangle$. Functional differentiating with respect to σ and λ, respectively, we get

$$\langle \sigma \rangle m^2 = 0$$

$$\langle \sigma \rangle^2 = \rho_s - \int \frac{d^3 k}{(2\pi)^3} \frac{1}{k^2 + m^2}. \tag{19.58}$$

19.3.2 The Ordered Phase

We look for solutions with $\langle \sigma \rangle \neq 0$, hence we must have necessarily $m = 0$. Evaluating the integral in (19.58), for $m = 0$, with the help of a momentum cutoff $\Lambda = \frac{2\pi}{a}$, where a is the lattice parameter, we get a nonzero sublattice magnetization

$$M \equiv \langle \sigma \rangle = \sqrt{\rho_s - \rho_c} \qquad \rho_c = \frac{\Lambda}{2\pi^2} = \frac{1}{\pi a}, \tag{19.59}$$

provided $\rho_s > \rho_c$. Then, $m^2 = 0$.

Defining the renormalized spin-stiffness as

$$\rho_R = \sqrt{\rho_s - \rho_c} \qquad \rho_R = Z^{-1}\rho_s \; ; \quad Z = 1 + \frac{\Lambda}{2\pi^2 \rho_R}. \tag{19.60}$$

Introducing the renormalized fields

$$\mathbf{n}_R = Z^{1/2}\mathbf{n} \; ; \quad \lambda_R = Z^{-1}\lambda$$

$$\rho_R = \sqrt{\rho_s - \rho_c} \qquad \rho_R = Z^{-1}\rho_s \; ; \quad Z = 1 + \frac{\Lambda}{2\pi^2 \rho_R} \tag{19.61}$$

and action

$$S_R = S + i(Z - 1) \int d^3 x \lambda_R, \tag{19.62}$$

we obtain for the renormalized action

$$S_R = \int d^3 r \left[\frac{\rho_R}{2} |\partial_\mu \mathbf{n}_R|^2 + i\lambda_R \left[|\mathbf{n}_R|^2 - 1 \right] \right]. \tag{19.63}$$

Inserting the saddle-point solution $\langle \lambda_R \rangle = 0$ and shifting the σ_R field around the vacuum value, namely, $\eta = \sigma_R - \sqrt{\rho_R}$, we find the physical excitations about the ground state

$$S_R = \frac{1}{2} \int d^3 r \, \partial_\mu \eta \partial_\mu \eta. \tag{19.64}$$

The Green function

$$G_\eta(\omega, \mathbf{k}) = \frac{1}{\omega^2 - c^2 |\mathbf{k}|^2} \tag{19.65}$$

reveals that the basic excitations, AF magnons, possess a dispersion relation $\omega = c|\mathbf{k}|$. In coordinate space, it is given by

$$G_\eta(t, \mathbf{x}) = \frac{1}{4\pi \left[c^2 t^2 - |\mathbf{x}|^2 \right]^{1/2}}, \tag{19.66}$$

which corresponds to (4.27) for $m \propto 1/\xi = 0$ in the ordered phase.

The existence of gapless excitations in the ordered phase of the NLSM, as a matter of fact, is imposed by the Goldstone theorem. Indeed, the original SO(3) symmetry of the system is spontaneously broken in the ordered phase, hence, the occurrence of gapless magnons is a consequence of that theorem.

19.3.3 The Disordered Phase

Now $m \neq 0$, implying the sublattice magnetization vanishes: $M \equiv \langle \sigma \rangle = 0$. Evaluating the integral in (19.58), now with nonzero m, we get, for a large cutoff,

$$M^2 \equiv \langle \sigma \rangle^2 = 0 = \rho_s - \rho_c + \frac{m}{4\pi}$$

$$\frac{m}{4\pi} = \rho_c - \rho_s, \tag{19.67}$$

for $\rho_s < \rho_c$.

It is customary to introduce the coupling $g = \frac{1}{\rho_s}$. Then, for $g < g_c$ we have the ordered, Néel phase, whereas for $g > g_c$, we have the quantum disordered phase. The phase transition at $g_c = \pi a$ is driven exclusively by quantum fluctuations, which thereby destroy the Néel state when the spin stiffness is less than a critical value.

In the disordered phase $\langle \lambda \rangle \propto m^2 \neq 0$, and therefore the basic excitations will be governed by the Euclidean action,

$$S = \frac{1}{2} \int d^3 r \left[\partial_\mu \eta \partial_\mu \eta + m^2 \eta^2 \right]. \tag{19.68}$$

The Euclidean Green function now is given by

$$G_\eta(\omega, \mathbf{k}) = \frac{1}{\omega^2 + c^2 |\mathbf{k}|^2 + m^2}. \tag{19.69}$$

The basic excitations, now, possess a dispersion relation $\omega = \sqrt{c^2 |\mathbf{k}|^2 + m^2}$.

In coordinate space, the Green function reads

$$G_\eta(t, \mathbf{x}) = \frac{e^{-m[c^2 t^2 - |\mathbf{x}|^2]^{1/2}}}{4\pi \left[c^2 t^2 - |\mathbf{x}|^2 \right]^{1/2}}, \tag{19.70}$$

which is the result announced in (4.27).

Notice that a gapped spectrum implies the symmetries can be implemented by unitary operators, hence the vacuum must be invariant and the symmetry, preserved [14].

19.3.4 Temperature Effects

We now take into account the effects of temperature fluctuations on the phase diagram. The saddle-point equations now read

$$\langle\sigma\rangle m^2 = 0$$

$$\langle\sigma\rangle^2 = \rho_s - T \sum_{n=-\infty}^{\infty} \int \frac{d^2k}{(2\pi)^2} \frac{1}{\omega_n^2 + k^2 + m^2}, \tag{19.71}$$

where ω_n are the bosonic Matsubara frequencies. Performing the Matsubara sum and integrating over momentum, with the help of a cutoff Λ, we get

$$\langle\sigma\rangle^2 = \rho_s - \frac{\Lambda}{4\pi} + \frac{T}{2\pi} \ln\left[2\sinh\left(\frac{m}{2T}\right)\right]. \tag{19.72}$$

Observe that for finite temperature, no solution with a nonzero sublattice magnetization is allowed, since this would require $m = 0$. In this case, however, the last term of the equation above tends to $-\infty$, thereby precluding a $\langle\sigma\rangle^2 \neq 0$ solution. Assuming $\langle\sigma\rangle = 0$, we have two phases.

Considering firstly the case $\rho_s > \rho_c$, $\rho_c = \frac{\Lambda}{4\pi}$, we have, in this case

$$2\sinh\left(\frac{m}{2T}\right) = e^{-\frac{2\pi\rho_R}{T}}, \tag{19.73}$$

where $\rho_R = \rho_s - \rho_c$. In the limit $T \ll \rho_R$, the rhs must be small, and so must be the lhs, hence we have

$$m(T) = T e^{-\frac{2\pi\rho_R}{T}}. \tag{19.74}$$

We see that even for $\rho_s > \rho_c$ for any $T \neq 0$, we would have $m(T) \neq 0$, which implies $\langle\sigma\rangle = 0$. This result is a manifestation of the Hohenberg–Mermin–Wagner theorem, which prohibits the occurrence of spontaneous breakdown of a continuous symmetry in two-dimensional space [170]. A related theorem is Coleman's theorem, which precludes the spontaneous breaking of a continuous symmetry even at $T = 0$ in one spatial dimension [171]. A summary of the two theorems is: only discrete symmetries can be spontaneously broken in $d = 1$, whereas in $d = 2$ continuous symmetries can be only possibly spontaneously broken at $T = 0$.

Now, consider the case $\rho_s < \rho_c$, $\rho_c = \frac{\Lambda}{4\pi}$. In this case, (19.72) implies,

$$m(T) = \Delta + 2T e^{-\frac{\Delta}{T}}, \tag{19.75}$$

where $\Delta = 4\pi(\rho_c - \rho_s)$ and $T \ll \Delta$.

19.4 Quantum and Thermal Fluctuations: the Ferromagnetic Case

In the ferromagnetic case it is more convenient to use the CP^1 formulation in order to extract the effect of the quantum and thermal fluctuations and thereby establish the phase diagram. For this purpose, we write the partition function (19.55) as

$$Z = \int Dz_i Dz_i^* DA_i D\lambda$$

$$\times \exp\left\{-\int d^3x \left[i2sz_i^*\partial_0 z_i + 2\rho_s(D_j z_i)^*(D_j z_i) + i\lambda[z_i^* z_i - \rho_s]\right]\right\} \quad (19.76)$$

after rescaling the fields as

$$z_i \longrightarrow \frac{z_i}{\sqrt{\rho_s}} \quad ; \quad \lambda \longrightarrow \lambda\rho_s. \quad (19.77)$$

The saddle-point equations will now read

$$m^2\langle z_i\rangle = 0$$

$$\langle z_i z_i^*\rangle = \rho_s. \quad (19.78)$$

The second equation above can be written as

$$|\langle z_i\rangle|^2 = \rho_s - T\sum_{n=-\infty}^{\infty}\int \frac{d^2k}{(2\pi)^2}\frac{1}{i\omega_n + k^2 + m^2}, \quad (19.79)$$

where $m^2 \propto \langle\lambda\rangle$.

Evaluating the Matsubara sum and integrating on the momentum, we obtain

$$|\langle z_i\rangle|^2 = \rho_s - \rho_c + \frac{m}{4\pi} + \frac{T}{4\pi}\ln\left[1 - e^{-\frac{m}{T}}\right]. \quad (19.80)$$

Again, we find two regimes, according to whether $\rho_s > \rho_c$ or $\rho_s < \rho_c$. In the first case the equation above admits a solution with $m = 0$ for $T = 0$ and nonzero magnetization $M = |\langle z_i\rangle|^2 = \rho_R = \rho_s - \rho_c$. For finite temperature, $T \neq 0$, a solution with $m = 0$ clearly does not exist. In the regime where $m \ll T$, we have

$$m(T) = Te^{-\frac{4\pi\rho_R}{T}} \quad ; \quad m(0) = 0 \quad (19.81)$$

with magnetization $M = 0$.

For $\rho_s < \rho_c$, conversely, we have

$$m(T) = \Delta + Te^{-\frac{\Delta}{T}} \quad ; \quad m(0) = \Delta \quad (19.82)$$

where $\Delta = 4\pi(\rho_s - \rho_c)$.

We see that for $T = 0$, we have $M \neq 0$; $m = 0$ for $\rho_s > \rho_c$; and $M = 0$; $m = \Delta \neq 0$ for $\rho_s < \rho_c$, hence the ferromagnetic Heisenberg model, similarly to its antiferromagnetic counterpart, undergoes a quantum phase transition at $\rho_s = \rho_c$. At finite temperature, both phases have zero magnetization.

Let us turn now to the spin fluctuation correlation functions. For this, assuming we are in a phase with magnetization $M \neq 0$ along the z-direction, let us decompose the **n** field as

$$\mathbf{n} = M\hat{z} + \sigma \quad ; \quad \sigma \in \mathbb{R}^2_{xy}. \quad (19.83)$$

The excitations around a uniformly magnetized state along the z-direction are contained in the σ-field, which belongs to the xy-plane. They are known as spin waves, while their associated quanta are the magnons. The field equations governing such excitations are given by (19.41) and (19.42). Inserting (19.83) in that equation, it follows that

$$\partial_t \sigma^{\pm} = \pm \left[|J| a^2 s \nabla^2 + m \right] \sigma^{\pm}, \tag{19.84}$$

where $\sigma^{\pm} = \sigma_x \pm i\sigma_y$.

From the above expression we may infer the form of the magnon quantum correlation function. This is so because such correlator is the inverse of the operator appearing in the quadratic part of the action describing these excitations, whereas the field equation is precisely this operator acting on the field. It follows, consequently, that the relevant Green function in momentum-frequency space is

$$G_\sigma(\omega, \mathbf{k}) = \frac{1}{\omega - ic^2 |\mathbf{k}|^2 - im}. \tag{19.85}$$

In this expression, $m = 1/\xi$ is the inverse correlation length, ξ, which diverges in the ordered phase. We conclude that in the ferromagnetic (ordered) phase the magnons dispersion relation is $\omega = c^2 |\mathbf{k}|^2$.

19.5 The Topological Charge and the Hopf Term

We have seen in Chapter 8 that the nonlinear sigma field is classified according to two topological invariants, the topological charge Q and the Hopf invariant, given respectively by (8.18) and (8.20), in terms of the topological current (8.19). The topological current (8.19) can be expressed in terms of the CP^1 vector field as [172, 173]

$$J^\mu = \frac{1}{2\pi} \epsilon^{\mu\alpha\beta} \partial_\alpha A_\beta \quad ; \quad A_\beta = iz_i^* \partial_\beta z_i. \tag{19.86}$$

The topological charge (8.18), then, becomes

$$Q = \frac{1}{2\pi} \epsilon^{ij} \partial_i A_j, \tag{19.87}$$

which is the magnetic flux associated to the field $\mathbf{A}$ along the xy-plane.

The Hopf term, conversely, can also be expressed in a simple form in terms of the CP^1 field A_μ. Indeed, inserting (19.86) in (8.20) and using the identity

$$\frac{x^\mu}{|x|^3} = 4\pi \partial^\mu \left(\frac{1}{-\Box} \right), \tag{19.88}$$

we obtain

$$H = \frac{1}{8\pi^2} \int d^3x \epsilon^{\mu\alpha\beta} A_\mu \partial_\alpha A_\beta. \tag{19.89}$$

We see that the Hopf term becomes a Chern–Simons term in the CP^1 formulation.

19.6 Classic and Quantum Skyrmions

Classic Skyrmions

In Section 8.5, we showed that the relativistic O(3) nonlinear sigma model possesses excitations carrying a nonzero topological charge, $Q = 1$. These are known as skyrmions and the corresponding classical static solutions are given by (8.48). In the associated AF quantum magnetic system, the skyrmion appears as a defect in each of the sublattices of the perfectly ordered Néel ground state. The defect consists of one reversed (-1) spin at the origin and, as we recede from it, the spins belonging to circles centered at the origin progressively bend in such a way that at $r = \lambda$ the z-component vanishes and at $r \to \infty$ it returns to the ground state value $+1$.

It is instructive to look at the skyrmion excitations from the perspective of the CP^1 formulation of the NLSM. The skyrmion solution (8.48), when expressed in terms of CP^1 fields [37], becomes

$$z_1 = \cos \frac{f(r)}{2} e^{-\frac{i}{2} \arg(\mathbf{r})}$$

$$z_2 = \sin \frac{f(r)}{2} e^{\frac{i}{2} \arg(\mathbf{r})}$$

$$A_i = \cos f(r) \partial_i \arg(\mathbf{r}) \quad ; \quad A_0 = 0,$$

$$f(r) = 2 \arctan \left(\frac{\lambda}{r} \right), \tag{19.90}$$

where λ is a parameter determining the skyrmion size. It is clear that the topological charge (in this language, magnetic flux) of the skyrmion solution is $Q = 1$. The classical skyrmion energy is given by (8.49).

In this language, the quantum AF magnetic system resembles a superconductor described by the Landau–Ginzburg theory, such that the ordered Néel phase corresponds to the SC phase, with a nonvanishing order parameter. The skyrmion topological excitations correspond to magnetic vortices with quantized flux.

Quantum Skyrmions

We have just seen that in the CP^1 language, skyrmions become vortices. Therefore, the most convenient method for describing quantum skyrmion excitations in two-dimensional AF quantum magnetic systems is to use the CP^1 formulation and apply to it the vortex quantization method developed in Chapter 9

Calling $\mu(x)$ the quantum skyrmion creation operator, we have

$$Q|\mu(x)\rangle = |\mu(x)\rangle. \tag{19.91}$$

In order to obtain the skyrmion correlation function, we start from (19.48) and directly obtain the CP^1 model expression that is related to (9.79) and (9.92):

$$
\begin{aligned}
\langle \mu(x)\mu^\dagger(y)\rangle_{CP^1} &= Z_0^{-1} \int DA_\mu \exp\left\{-\int d^3z\right. \\
&\quad \left.\times \frac{1}{4}\left[W^{\mu\nu} + \tilde{B}^{\mu\nu}\right]\left[\frac{4\rho_s^2}{-\Box}\right]\left[W_{\mu\nu} + \tilde{B}_{\mu\nu}\right]\right\}, \\
&= \exp\left\{\Lambda(x, y; L) - \int d^3z \frac{1}{4}\tilde{B}^{\mu\nu}\left[1 + \frac{4\rho_s^2}{-\Box}\right]\tilde{B}_{\mu\nu}\right\},
\end{aligned}
\tag{19.92}
$$

where $\tilde{B}_{\mu\nu}$ is given by (9.65). This corresponds to (9.63), less the Maxwell term, which is absent in the CP^1 model. Notice also the $4\rho_s^2$ factor that appears here. Going through the same steps that led us from (9.92) to (9.103), we arrive at the vortex correlation function. Nevertheless, we must remove the second term in (9.103), which was produced by the Maxwell term.

The quantum skyrmion correlation function, in the ordered phase of the AF two-dimensional quantum Heisenberg magnetic system, which is described by the CP^1-NLSM, therefore is given by

$$\langle \mu(x)\mu^\dagger(y)\rangle_{CP^1} = \exp\left\{-2\pi\rho_s^2|x - y|\right\}. \tag{19.93}$$

From this we infer the quantum skyrmion mass (energy) is given by $\mathcal{M} = 2\pi\rho_s^2$, which is a half of the classical skyrmion energy, given by (8.49) [213].

19.6.1 Duality between Magnons and Skyrmions

Notice that there exists a duality relation between spin waves (magnons) and skyrmions. Indeed, in the ordered phase, the magnon correlation function is given by (19.69), which implies $\langle \eta \rangle = 0$ or $\langle \sigma \rangle \neq 0$. The skyrmion correlator, conversely, is expressed by (19.93). This implies, $\langle \mu \rangle = 0$.

In the disordered phase, on the other hand, we have $\langle \sigma \rangle = 0$ and $\langle \mu \rangle \neq 0$. The behavior of the magnon and skyrmion excitations in each of the different phases of the NLSM reveals the duality relation existing between such excitations. The magnons are Hamiltonian excitations, namely degrees of freedom appearing explicitly in the Hamiltonian, while skyrmions are topological excitations, which therefore require treatment as such.

20

The Spin-Fermion System: a Quantum Field Theory Approach

We have seen in the previous chapter how the dynamics of a two-dimensional quantum magnetic system on a square lattice can be described in the framework of the CP^1/nonlinear sigma model. It is quite appealing, not only from the standpoint of basic principles but also from the point of view of modelling real materials, to investigate the behavior of electrons in the presence of this magnetic background. From this perspective, it would be interesting to study, among other issues: what would be the effective electronic interactions generated by the magnetic background, how these would depend on the phase transitions undergone by the underlying magnetic system, according to the values of different control parameters; what would be the role of skyrmion topological defects on the physical properties of the associated electrons; and how would electron or hole doping affect this interplay.

On the other hand, from the experimental point of view, several advanced materials have been obtained recently, the phase diagram of which present a very rich set of phases displaying different types of order. These are typically superconducting, magnetic or charge orders. Among these materials, we find heavy fermions such as $CeCoIn_5$, high-Tc cuprates such as $La_{2-x}Sr_xCuO_4$ and iron pnictides, such as $Sr_{1-x}K_xFe_2As_2$. The richness of phases observed in such materials suggests there could be an underlying interaction responsible for the observed output, depending on the values of internal as well as external control parameters such as coupling constants and temperature, respectively. The spin-fermion model [174] describes this kind of system. Here we develop a quantum field theory approach the spin-fermion system and investigate the possible effective interactions that are induced among the electrons by the (AF) magnetically ordered substrate.

20.1 Itinerant Electrons and Ordered Localized Spins

The Hamiltonian

We envisage a system containing both localized and itinerant electrons, the former belonging to atomic orbitals fixed to the sites of a square lattice. These generate localized magnetic dipole moments, which interact with nearest neighbors according to the SO(3), AF Heisenberg model. The itinerant electrons, conversely, have their kinematics determined by a tight-binding Hamiltonian, containing a hopping between nearest neighbors. The itinerant electrons typically could visit the same orbitals that contain the localized ones or, alternatively, could circulate along some extra orbitals that could be available at the square lattice links. It does not look like there will be a significant difference between both in a continuum field theoretical description. The picture is completed by introducing a magnetic interaction between the itinerant and localized spins.

The Hamiltonian is given by

$$H_{SF} = J \sum_{\langle ij \rangle} \mathbf{S}_i \cdot \mathbf{S}_j - t \sum_{\langle ij \rangle} \left(c_{i\alpha}^\dagger c_{j\alpha} + c_{j\alpha}^\dagger c_{i\alpha} \right) + J_K \sum_i \mathbf{S}_i \cdot \left(c_{i\alpha}^\dagger \boldsymbol{\sigma}_{\alpha\beta} c_{i\beta} \right), \quad (20.1)$$

where $\mathbf{S}_i$ is the spin operator of an electron localized at the site i of a square lattice and $c_{i\alpha}^\dagger$ is the creation operator of an itinerant electron of spin $\alpha = \uparrow, \downarrow$ at the site i. The spin operator of the latter is given by

$$\mathbf{s}_i = c_{i\alpha}^\dagger \boldsymbol{\sigma}_{\alpha\beta} c_{i\beta}. \quad (20.2)$$

The Continuum QFT

In order to obtain the continuum limit of the Hamiltonian above, we go through the same steps leading to (19.32), but now including the fermion field $\psi_\alpha(\mathbf{x})$, which is the continuum limit of $c_{i\alpha}$.

After employing the spin coherent states $|\mathbf{N}\rangle$, where $\mathbf{N}$ is given by (19.17), we can write the Kondo coupling between the itinerant and localized spins as

$$J_K \mathbf{S}_i \cdot \left(c_{i\alpha}^\dagger \boldsymbol{\sigma}_{\alpha\beta} c_{i\beta} \right) \to J_K S \left(a^2 \mathbf{L} + (-1)^{|\mathbf{x}|} \mathbf{n} \right) \cdot \mathbf{s}, \quad (20.3)$$

where S is the spin quantum number of the localized spin operators.

Assuming the electrons have a dispersion relation $\omega_{\alpha\beta}(\mathbf{k})$, we therefore get

$$Z = \int D\mathbf{n} D\psi_\alpha D\psi_\alpha^\dagger D\mathbf{L} \delta(|\mathbf{n}|^2 - 1)$$

$$\times \exp \left\{ -\int d^2r \int_0^\beta d\tau \left[\psi_\alpha^\dagger \partial_\tau \psi_\beta \psi_\alpha^\dagger \omega_{\alpha\beta}(-i\nabla)\psi_\beta + \frac{JS^2}{2}|\nabla \mathbf{n}|^2 + \frac{4JS^2a^2}{2}|\mathbf{L}|^2 \right] \right.$$

$$\left. - \mathbf{L} \cdot \left[is\mathbf{n} \times \frac{\partial \mathbf{n}}{\partial \tau} + J_K Sa^2 \mathbf{s} \right] + \frac{J_K S}{a^2}(-1)^{|\mathbf{x}|}\mathbf{n} \cdot \mathbf{s} \right\}. \quad (20.4)$$

We now integrate over the ferromagnetic fluctuations, $\mathbf{L}$, obtaining

$$Z = \int Dn D\psi_\alpha D\psi_\alpha^\dagger \delta(|\mathbf{n}|^2 - 1)$$

$$\times \exp\left\{-\int d^2r \int_0^\beta d\tau \left[\psi_\alpha^\dagger \partial_\tau \psi_\beta - \psi_\alpha^\dagger \omega_{\alpha\beta}(-i\nabla)\psi_\beta + \frac{JS^2}{2}\left(|\nabla\mathbf{n}|^2 + \frac{1}{c^2}|\partial_\tau\mathbf{n}|^2\right)\right.\right.$$

$$\left.\left. - a^2 J_K S\left[\frac{1}{4Ja^2}\mathbf{n} \times \frac{\partial\mathbf{n}}{\partial\tau} + a^{-4}(-1)^{|\mathbf{x}|}\mathbf{n}\right]\cdot\mathbf{s} + \frac{a^2 J_K^2}{8J}\mathbf{s}\cdot\mathbf{s}\right]\right\}, \qquad (20.5)$$

where $\rho_s = JS^2$ is the spin stiffness and $c = 2JSa$ is the spin-wave velocity.

At this point, we shall assume the electrons, by virtue of the Fermi surface shape, which is determined by several different factors, have a Dirac dispersion relation

$$\omega_{\alpha\beta}(-i\nabla) = \left[-i\gamma^0\gamma^i\nabla_i\right]_{\alpha\beta}, \qquad (20.6)$$

where $\gamma^0\gamma^i = \sigma^i$ (Pauli matrix) and $(\gamma^0)^2 = \mathbb{I}$.

Also, from now on, we switch to the CP^1 formulation, which will be more convenient for our purposes. In this language, the Lagrangean density corresponding to (20.5) in real-time Minkowski space is

$$\mathcal{L} = \overline{\psi}\gamma^\mu\partial_\mu\psi + 2\rho_s|D_\mu z_i|^2 + \frac{J_K}{4J}\mathbf{s}\cdot\mathbf{n}\times\frac{\partial\mathbf{n}}{\partial\tau} + J_K S(-1)^{|\mathbf{x}|}\mathbf{n}\cdot\mathbf{s} + \frac{a^2 J_K^2}{8J}\mathbf{s}\cdot\mathbf{s}, \qquad (20.7)$$

where

$$\mathbf{s}(x) = \psi_\alpha^\dagger \vec{\sigma}_{\alpha\beta}\psi_\beta \quad ; \quad \mathbf{n}(x) = z_i^*\vec{\sigma}_{ij}z_j \qquad (20.8)$$

with $\sum_i |z_i|^2 = 1$ and $D_\mu = \partial_\mu + i\Lambda_\mu$.

Let us first concentrate on the third term of the Lagrangean above, namely

$$\psi_\alpha^\dagger\vec{\sigma}_{\alpha\beta}\psi_\beta \cdot \mathbf{n} \times \frac{\partial\mathbf{n}}{\partial\tau} = \psi_\alpha^\dagger\psi_\beta z_\mu^* z_\nu \partial_\tau(z_\lambda^* z_\rho)\epsilon^{ijk}\sigma^i\sigma^j\sigma^k. \qquad (20.9)$$

Using the identity

$$\sigma_{\alpha\beta}^i\sigma_{\mu\nu}^j = \frac{\delta^{ij}}{3}\left[2\delta_{\alpha\nu}\delta_{\beta\mu} - \delta_{\alpha\beta}\delta_{\mu\nu}\right] + i\epsilon^{ijk}\left[\delta_{\beta\mu}\sigma_{\alpha\nu}^k - \delta_{\alpha\mu}\sigma_{\beta\nu}^k\right] \qquad (20.10)$$

and the polar representation of the complex fields z_i,

$$z_i = \frac{1}{\sqrt{2}}\rho_i e^{i\theta_i} \quad ; \quad i = 1, 2 \quad ; \quad \frac{1}{2}\left(\rho_1^2 + \rho_2^2\right) = 1, \qquad (20.11)$$

we may write (20.9) as

$$\psi_\alpha^\dagger\vec{\sigma}_{\alpha\beta}\psi_\beta \cdot \mathbf{n} \times \frac{\partial\mathbf{n}}{\partial\tau} = F(\rho_i) + G(\rho_i, e^{i\theta_j}) = \psi_\alpha^\dagger\psi_\alpha z_\mu^* z_\mu \partial_\tau(z_\lambda^* z_\lambda) + G(\rho_i, e^{i\theta_j}). \qquad (20.12)$$

The last term is proportional to $e^{i\theta_i}$, therefore rapidly oscillating, and consequently does not contribute to the functional integral when we integrate over the phase field θ_i. The first, phase-independent term, by its turn, vanishes because $z_\lambda^* z_\lambda = 1$, hence the derivative is zero.

The QFT associated with the spin-fermion system is, therefore,

$$\mathcal{L} = \overline{\psi}\gamma^\mu \partial_\mu \psi + 2\rho_s |D_\mu z_i|^2 + J_K S(-1)^{|\mathbf{x}|}\mathbf{n}\cdot\mathbf{s} + \frac{a^2 J_K^2}{8J}\mathbf{s}\cdot\mathbf{s}. \tag{20.13}$$

We are just left the two last terms of (20.8). One of them corresponds to the spin interaction of the itinerant electrons with the localized spins and the other, the spin-spin interaction among the itinerant electrons themselves. In the next section, we will show that the former can be replaced by a gauge coupling with the CP^1 field A_μ.

20.2 The Gauge Coupling Replaces Magnetic Coupling

In CP^1 formulation, the Heisenberg magnetic coupling among the localized spins manifests as the minimal gauge coupling of the complex scalar fields, in terms of which the localized spins are expressed, with the CP^1 gauge field A_μ. We will now show that the magnetic coupling between the itinerant electrons with the localized spins, analogously, can be expressed as a minimal gauge coupling of the electrons with the same field.

For this purpose, let us perform a canonical transformation on the electron fields, given by

$$\psi_\alpha \to U_{\alpha\beta}\psi_\beta, \quad ; \quad U = \begin{pmatrix} z_1 & -z_2^* \\ z_2 & z_1^* \end{pmatrix}, \tag{20.14}$$

where α, β are spin indices of the Dirac field.

The matrix U has the property

$$U^\dagger \vec{\sigma}\cdot\mathbf{n}U = \sigma^z. \tag{20.15}$$

Using (20.10), one can verify that the last term in (20.13) is invariant under (20.14). The third term, however, is transformed into

$$J_K S(-1)^{|\mathbf{x}|}\left[\psi_\uparrow^\dagger \psi_\uparrow - \psi_\downarrow^\dagger \psi_\downarrow\right]. \tag{20.16}$$

Assuming a uniform electron density, this term will be washed out by the rapid oscillations when spatially integrated.

Since the canonical transformation (20.14) is local, it follows that, under it, the first term in (20.13) will produce an additional term given by

$$i\overline{\psi}\gamma^\mu U^\dagger \partial_\mu U\psi. \tag{20.17}$$

Now, from (20.14), we have

$$U^\dagger \partial_\mu U = i\sigma^z A_\mu + \begin{pmatrix} 0 & z_2^* \partial_\mu z_1^* - z_1^* \partial_\mu z_2^* \\ z_1 \partial_\mu z_2 - z_2 \partial_\mu z_1 & 0 \end{pmatrix}, \qquad (20.18)$$

where $A_\mu = -iz^* \partial_\mu z$.

The off-diagonal terms above contain the phases θ_i of the complex fields z_i. Consequently, in the same way as the last term in (20.12), they will not contribute to the partition functional when functionally integrated.

After the canonical transformation (20.14), therefore, we may cast the field theory Lagrangean associated to the spin-fermion system in the form

$$\mathcal{L} = \overline{\psi}\gamma^\mu \partial_\mu \psi + 2\rho_s |D_\mu z_i|^2 + \overline{\psi}_\alpha \gamma^\mu \sigma_{\alpha\beta}^z \psi_\beta A_\mu + \frac{a^2 J_K^2}{8J} \mathbf{s} \cdot \mathbf{s}, \qquad (20.19)$$

which is invariant under the gauge transformation

$$\psi \to e^{i\Lambda}\psi$$
$$z_i \to e^{i\Lambda} z_i \;\; ; \;\; \theta_i \to \theta_i + \Lambda$$
$$A_\mu \to A_\mu - \partial_\mu \Lambda. \qquad (20.20)$$

We see that the magnetic interaction between the itinerant electrons and the localized spins is expressed by a gauge coupling of the electrons with the CP^1 gauge field.

20.3 Competing Electronic Interactions

We shall now integrate over all the CP^1 fields in order to obtain an effective electronic interaction. For this, we start by introducing gauge invariant phase fields

$$\chi_i = \theta_i + \frac{\partial_\mu A^\mu}{\Box} \;\; ; \;\; i = 1, 2, \qquad (20.21)$$

which decouple from the vector gauge field A_μ. Indeed, in the constant ρ_i approximation, and noting that as usual $F_{\mu\nu} = \partial_\mu A_\nu - \partial_\nu A_\mu$, we get

$$\mathcal{L} = \overline{\psi}\gamma^\mu \partial_\mu \psi + \frac{1}{4} F_{\mu\nu} \left[\frac{2\rho_s}{-\Box} \right] F_{\mu\nu} + \overline{\psi}_\alpha \gamma^\mu \sigma_{\alpha\beta}^z \psi_\beta A_\mu$$
$$+ \frac{1}{2} \sum_{i=1,2} \rho_i^2 \partial_\mu \chi_i \partial_\mu \chi_i + \frac{a^2 J_K^2}{8J} \mathbf{s} \cdot \mathbf{s}, \qquad (20.22)$$

which is explicitly gauge invariant.

Integration over χ_i and ρ_i just produces a trivial constant multiplicative factor. The nontrivial contribution for the electronic effective interaction comes, in fact, from integration over the gauge field A_μ. This yields [175]

$$\mathcal{L}_{eff} = \overline{\psi}\gamma^\mu \partial_\mu \psi + \frac{1}{2\rho_s} \left(\overline{\psi}_\alpha \gamma^\mu \sigma_{\alpha\beta}^z \psi_\beta \right) \left(\overline{\psi}_\alpha \gamma^\mu \sigma_{\alpha\beta}^z \psi_\beta \right) + \frac{a^2 J_K^2}{8J} \mathbf{s} \cdot \mathbf{s}, \quad (20.23)$$

Writing explicitly the Dirac field components $\psi_{i,\sigma}$, with $i = 1, 2;, \sigma =\uparrow, \downarrow$, one obtains, after some algebra [175], the Lagrangean describing the effective electronic interaction

$$\mathcal{L}_{eff} = \overline{\psi}\gamma^{\mu}\partial_{\mu}\psi + \frac{1}{4\rho_s}\left(\psi^{\dagger}_{1\uparrow}\psi^{\dagger}_{2\downarrow} + \psi^{\dagger}_{2\uparrow}\psi^{\dagger}_{1\downarrow}\right)\left(\psi_{2\downarrow}\psi_{1\uparrow} + \psi_{1\downarrow}\psi_{2\uparrow}\right)$$

$$+ \frac{1}{8\rho_s}\left[\left(\overline{\psi}_{\sigma}\psi_{\sigma}\right)^2 - \left(\overline{\psi}_{\sigma}\gamma^0\psi_{\sigma}\right)^2\right] + \frac{a^2 J^2_K}{8J}\mathbf{s}\cdot\mathbf{s} - \frac{a^2 J}{2}s^2_z. \quad (20.24)$$

The second term in the effective Lagrangean above is a BCS-type interaction that tends to produce a superconducting ground state, whereas the next term is a Nambu–Jona–Lasinio type interaction [176], which would rather produce a charge-gapped insulating ground state. The last term, finally, is a spin interaction that combines with the spin-spin magnetic interaction term of the itinerant electrons in the form

$$\mathcal{L}_{eff,mag} = \frac{a^2}{8J}\left[\left(J^2_K - 4J^2\right)s^2_z + J^2_K\mathbf{s}_{\perp}\cdot\mathbf{s}_{\perp}\right]. \quad (20.25)$$

20.4 Phases

In order to determine the phase diagram of the system, we introduce Hubbard–Stratonovich auxiliary fields, with the help of which we may transform the quadratic effective electronic interactions into trilinear ones. These are

$$\Delta = \psi_{2\downarrow}\psi_{1\uparrow} + \psi_{1\downarrow}\psi_{2\uparrow}$$
$$M = \overline{\psi}_{\sigma}\psi_{\sigma}$$
$$\sigma = \psi_{\alpha}\sigma^z_{\alpha\beta}\psi_{\beta} \quad (20.26)$$

and their vacuum expectation values constitute the relevant order parameters for each phase. Their nonzero vacuum expectation values indicate the onset of each of such different phases.

In order to determine the phase diagram, we must obtain the free energy (effective potential) as a function of the order parameters and see what are the nonzero minima as a function of temperature, chemical potential and the coupling parameters $\lambda_{SC} = \frac{1}{4\rho_s}$, $\lambda_{EXC} = \frac{1}{8\rho_s}$ and $\lambda_{MAG} = \frac{a^2}{8J}\left(J^2_K - 4J^2\right)$.

In conclusion, we saw that different effective electronic interactions can be generated out of the original magnetic interactions existing among the localized spins as well as from the ones occurring between itinerant electrons and localized spins. The phase diagram of the system will be very rich, on account of the different effective interactions. Nevertheless the source of all those distinct interactions can be traced back to the original magnetic interaction involving localized and itinerant spins.

21

The Spin Glass

Real systems are often discrepant from ideal situations. A real crystal lattice, for instance, is never perfect and differs from a Bravais lattice. The exchange couplings of magnetic systems, conversely, are never identical for all pairs of nearest lattice sites, a random distribution thereof being a closer picture of reality. In the latter case, situations arise where the exchange magnetic couplings are not only random but a subset of them possess an opposite sign. Dramatic effects can, then, occur. A competition between two opposite tendencies sets in, each one trying to push the system, respectively, into an ordered state, either of ferromagnetic or antiferromagnetic (Néel) nature. Since the two tendencies cannot be fulfilled simultaneously, a situation which is called frustration describes the ground-state features. Under these circumstances a new phase, known as spin glass, frequently occurs. This phase presents features, some of which are common to the paramagnetic phase, while others are shared with the ordered Néel or ferromagnetic phase. The absence of spatial long-range order is an example of the former, whereas the breakdown of ergodicity is one of the latter.

A common feature of spin glasses is the fact that the characteristic time scale of the disordered background is much larger than the corresponding time scale of the dynamical degrees of freedom. This fact leads us to use the quenched approach for determining the thermodynamic properties of a spin glass. In this, the free energy is evaluated at a fixed disordered configuration of exchange couplings, and subsequently it is averaged over these random configurations with a certain probability distribution. Pioneering exploration of spin glasses was conducted by Edwards and Anderson, who proposed a model and an order parameter for characterizing a spin glass [94]. This measures the presence of infinite time correlations in the system, which would be also present in an ordered state. This, however, would also exhibit infinite spatial correlations, which would be absent in a spin glass.

21.1 The Quantum SO(3) Spin Glass

The system we are going to investigate is characterized by the Hamiltonian [177]

$$H = \sum_{\langle ij \rangle} J_{ij} \mathbf{S}_i \cdot \mathbf{S}_j, \tag{21.1}$$

where the $\mathbf{S}_i$ operators represent localized spins occupying the sites of a square lattice and interacting with nearest neighbors with exchange couplings J_{ij}. These are supposed to be random and characterized by a Gaussian probability distribution of variance ΔJ and centered at an antiferromagnetic coupling $J_0 > 0$, namely

$$P[J_{ij}] = \frac{1}{\sqrt{2\pi(\Delta J)^2}} \exp\left[\frac{[J_{ij} - J_0]^2}{2(\Delta J)^2}\right]. \tag{21.2}$$

We assume, for convenience that $\Delta J \ll J_0$.

The present model is similar to the Edwards–Anderson (EA) model [94], however, a crucial difference is the fact that the distribution of exchange couplings is centered at a nonzero coupling J_0, whereas in that case the Gaussian is centered at zero. This feature will allow us to treat the spin-glass as a perturbation of the non-random system, something that would not be possible in the EA model. In addition, the fact that $J_0 > 0$ is an AF coupling will guarantee the cancellation of the topological term originated from the Berry phase, when summed on the whole square lattice.

In the quenched situation, we evaluate the free energy as a functional of the background configuration J_{ij}

$$F[J_{ij}] = -k_B T \ln Z[J_{ij}] \tag{21.3}$$

and subsequently average over it, namely

$$\overline{F} = \int \prod_{\langle ij \rangle} P[J_{ij}] F[J_{ij}] . d J_{ij}. \tag{21.4}$$

In order to facilitate the averaging process, we introduce the replica method by making use of the identity

$$\ln Z[J_{ij}] = \lim_{n \to 0} \frac{Z^n[J_{ij}] - 1}{n}. \tag{21.5}$$

It follows that

$$\overline{F} = -\frac{1}{\beta} \lim_{n \to 0} \frac{\overline{Z^n} - 1}{n}. \tag{21.6}$$

The replicated partition function is given by

$$
Z^n[J_{ij}] = \text{Tr} \exp\left[-\beta \sum_{\alpha=1}^{n} \sum_{\langle ij \rangle} J_{ij} \mathbf{S}_i^\alpha \cdot \mathbf{S}_j^\alpha\right]. \tag{21.7}
$$

In order to evaluate the above trace, it will be convenient to use the spin coherent states we used in Chapter 17, properly adapted to the present system:

$$
|\mathbf{N}_i^\alpha\rangle \quad ; \quad \langle \mathbf{N}_i^\alpha | \mathbf{S}_i^\alpha | \mathbf{N}_i^\alpha \rangle = S \mathbf{N}_i^\alpha, \tag{21.8}
$$

where $|\mathbf{N}_i^\alpha| = 1$.

As we did in Chapter 17, we can express the replicated partition function (21.7) as a functional integral over the $\mathbf{N}_i^\alpha$ field,

$$
Z^n[J_{ij}] = \int D\mathbf{N}_i^\alpha \exp\left\{-\sum_{\alpha=1}^{n} \int_0^\beta d\tau \left[L_B^\alpha + \sum_{\langle ij \rangle} J_{ij} \mathbf{N}_i^\alpha \cdot \mathbf{N}_j^\alpha\right]\right\}, \tag{21.9}
$$

where

$$
L_B^\alpha = \sum_i \langle \mathbf{N}_i^\alpha(\tau) | \frac{d}{d\tau} | \mathbf{N}_i^\alpha(\tau)\rangle \tag{21.10}
$$

is the Berry phase.

We can now average the replicated partition function over the background configurations. Inserting (21.9) and (21.1) into (21.4), we get a Gaussian integral on the variable J_{ij} at each link. Performing this integration, we obtain a functional integral representation for the averaged replicated partition function, which is expressed as

$$
\overline{Z^n}[J_0, \Delta] = \int D\mathbf{N}_i^\alpha \exp\left\{-S[\mathbf{N}_i^\alpha; J_0, \Delta J]\right\}, \tag{21.11}
$$

where the effective action in the previous expression is given by

$$
S[\mathbf{N}_i^\alpha; J_0, \Delta J] = \int_0^\beta d\tau \sum_{\alpha=1}^{n} \left[L_B^\alpha - J_0 S^2 \sum_{\langle ij \rangle} \mathbf{N}_i^\alpha \cdot \mathbf{N}_j^\alpha\right]
$$
$$
+ \frac{S^4(\Delta J)^2}{2} \int_0^\beta d\tau \int_0^\beta d\tau' \sum_{\alpha,\beta=1}^{n} \sum_{\langle ij \rangle} N_{ia}^\alpha(\tau) N_{ib}^\beta(\tau') N_{ja}^\alpha(\tau) N_{jb}^\beta(\tau'). \tag{21.12}
$$

In the above expression, $a, b = 1, 2, 3$ are SO(3) internal indices and the sum over them is implicitly assumed. We can write the sum over nearest neighbors as a sum sweeping the whole lattice, by using the connectivity matrix K_{ij}. This is defined by

$$
K_{ij} = \begin{cases} 1 & \text{nearest neighbors} \\ 0 & \text{otherwise.} \end{cases} \tag{21.13}
$$

Then, we can introduce a Hubbard–Stratonovitch transformation with the auxiliary tensor variable $Q_{ab}^{\alpha\beta}(i, \tau, \tau')$ in such a way that the effective action becomes

$$
S[\mathbf{N}_i^\alpha; J_0, \Delta J] = \int_0^\beta d\tau \sum_{\alpha=1}^n \left[L_B^\alpha - J_0 S^2 \sum_{\langle ij \rangle} \mathbf{N}_i^\alpha \cdot \mathbf{N}_j^\alpha \right]
$$

$$
+ S^4 (\Delta J)^2 \int_0^\beta d\tau \int_0^\beta d\tau' \sum_{\alpha,\beta=1}^n \sum_{\langle ij \rangle} \left[\frac{1}{2} Q_{ab}^{\alpha\beta}(i, \tau, \tau') Q_{ab}^{\alpha\beta}(j, \tau, \tau') \right.
$$

$$
\left. - N_{ia}^\alpha(\tau) Q_{ab}^{\alpha\beta}(j, \tau, \tau') N_{ib}^\beta(\tau') \right]. \tag{21.14}
$$

Observe that this is no longer a random system. The parameters characterizing the probability distribution, namely the variance ΔJ and the average coupling J_0, become coupling constants of the effective, non-randomic theory. In the limit of zero variance, the last two terms above vanish and we retrieve the usual AF system with an exchange coupling J_0. The effect of disorder, reflected through a nonzero variance, manifests as the two last interaction terms above. Notice that, as announced, we can do perturbation around the non-random AF Heisenberg model by expanding in the variance ΔJ.

21.2 Quantum Field Theory Approach to the SO(3) Spin Glass

21.2.1 Nonlinear Sigma Formulation

Before taking the continuum limit, we now introduce the decomposition (19.17) into (21.14). Only the antiferromagnetic fluctuation field contributes to the last term above in the small a limit. The first term, conversely, may be treated precisely as we did in Chapter 17 for the pure Heisenberg/NLSM system. The topological phases in particular cancel out when summed all over the square lattice. After taking the continuum limit and integrating over the ferromagnetic fluctuation field, we get the following effective Lagrangean density:

$$
\mathcal{L} = \frac{1}{2} |\nabla \mathbf{n}^\alpha|^2 + \frac{1}{2c^2} |\partial_\tau \mathbf{n}^\alpha|^2 + i\lambda_\alpha \left(|\mathbf{n}^\alpha|^2 - \rho_s \right)
$$

$$
+ D \int_0^\beta d\tau' \left[\frac{1}{2} Q_{ab}^{\alpha\beta}(\mathbf{r}, \tau, \tau') Q_{ab}^{\alpha\beta}(\mathbf{r}, \tau, \tau') - \frac{1}{\rho_s} n_a^\alpha(\mathbf{r}, \tau) Q_{ab}^{\alpha\beta}(\mathbf{r}, \tau, \tau') n_b^\beta(\mathbf{r}, \tau') \right],
$$

$$
\tag{21.15}
$$

where $D = \frac{S^4 (\Delta J)^2}{a^2}$, $\rho_s = S^2 J_0$ and a, the lattice parameter.

The Hubbard–Stratonovitch field, according to (21.8) and (21.12), corresponds to

$$
Q_{ab}^{\alpha\beta}(\mathbf{r}, \tau, \tau') = \langle S_a^\alpha(\mathbf{r}, \tau) S_b^\beta(\mathbf{r}, \tau') \rangle. \tag{21.16}
$$

Our next step is to decompose such a field in replica diagonal and off-diagonal components:

$$Q_{ab}^{\alpha\beta}(\mathbf{r}, \tau, \tau') = \left[\delta^{\alpha\beta}\chi(\mathbf{r}, \tau, \tau') + q^{\alpha\beta}(\mathbf{r}, \tau, \tau')\right]\delta_{ab}, \qquad (21.17)$$

where $q^{\alpha\beta} = 0$, for $\alpha = \beta$ and we assume the isotropy of the system, which implies a delta-dependence on the SO(3) indices. The resulting effective Lagrangean density becomes

$$\begin{aligned}
\mathcal{L} = &\frac{1}{2}|\nabla \mathbf{n}^\alpha|^2 + \frac{1}{2c^2}|\partial_\tau \mathbf{n}^\alpha|^2 + i\lambda_\alpha\left(|\mathbf{n}^\alpha|^2 - \rho_s\right) \\
&+ \frac{3D}{2}\int_0^\beta d\tau'\left[n\chi^2(\tau, \tau') - q^{\alpha\beta}(\tau, \tau')q^{\alpha\beta}(\tau, \tau') - \frac{D}{\rho_s}\mathbf{n}^\alpha(\tau)\chi(\tau, \tau')\mathbf{n}^\alpha(\tau')\right. \\
&\left.- \frac{D}{\rho_s}\mathbf{n}^\alpha(\tau)q^{\alpha\beta}(\tau, \tau')\mathbf{n}^\beta(\tau')\right].
\end{aligned} \qquad (21.18)$$

21.2.2 CP^1 Formulation

We want to fully include quantum-mechanical effects in our description of the SO(3) AF spin glass system. These effects are more transparently seen under the CP^1 formulation. Thus, introducing the CP^1 fields through

$$\mathbf{n}^\alpha(\tau) = \frac{1}{\sqrt{\rho_s}}\left[z_i^{*\alpha}(\tau)\vec{\sigma}_{ij}z_j^\alpha(\tau)\right], \qquad (21.19)$$

with

$$|z_1^\alpha|^2 + |z_2^\alpha|^2 = \rho_s, \qquad (21.20)$$

we obtain

$$\frac{1}{2}|\nabla \mathbf{n}^\alpha|^2 + \frac{1}{2c^2}|\partial_\tau \mathbf{n}^\alpha|^2 \leftrightarrow 2\sum_{i=1,2}|D_\mu z_i^\alpha|^2. \qquad (21.21)$$

In this language, the effective Lagrangean, in terms of which the averaged replicated partition function is expressed, becomes

$$\begin{aligned}
\mathcal{L} = &2\sum_{i=1,2}|D_\mu z_i^\alpha|^2 + i\lambda_\alpha\left(\sum_{i=1,2}|z_i^\alpha|^2 - \rho_s\right) \\
&+ \frac{3D}{2}\int_0^\beta d\tau'\left[n\chi^2(\tau, \tau') - q^{\alpha\beta}(\tau, \tau')q^{\alpha\beta}(\tau, \tau') - \frac{2}{3\rho_s^2}\chi(\tau, \tau')\right. \\
&\left.- \frac{2}{3\rho_s^2}\left[z_i^{*\alpha}z_j^\alpha\right](\tau)\left[\delta^{\alpha\beta}\chi(\tau, \tau') - q^{\alpha\beta}(\tau, \tau')\right]\left[z_i^{*\beta}z_j^\beta\right](\tau')\right]. \qquad (21.22)
\end{aligned}$$

The averaged replicated partition function is given by the following functional integral

$$\overline{Z^n} = \int Dz_i^\alpha Dz_i^{*\alpha} D\lambda_\alpha D\chi Dq^{\alpha\beta} DA_\mu e^{-S[z_i^\alpha, z_i^{*\alpha}, \lambda_\alpha, \chi, q^{\alpha\beta}, A_\mu]}, \tag{21.23}$$

where S is the action corresponding to the effective Lagrangean (21.22).

21.3 The Quenched Free Energy

In order to determine the thermodynamic phase diagram, we need the quenched free energy, which is given by (21.6) in terms of $\overline{Z^n}$. For obtaining the latter, we evaluate the previous functional integral by expanding around the stationary point and integrating the quadratic fluctuations of the z_i fields. The result is [178]

$$\overline{Z^n} = \exp\left\{ -S_0\left[z_{i,s}^\alpha, z_{i,s}^{*\alpha}, \lambda_s^\alpha, \chi_s, q_s^{\alpha\beta}, A_{\mu,s} \right] - \ln \mathrm{Det}\mathbb{M} \right\}, \tag{21.24}$$

where the subscript s indicates the quantity is evaluated at the stationary point and the matrix $\mathbb{M}$ is the coefficient of the quadratic form of the fluctuations in such an expansion, also evaluated at the same point.

The classic fields evaluated at the stationary points coincide with the vacuum expectation value of the corresponding quantum operators, namely

$$\chi_s(\tau - \tau') = \langle \chi(\tau, \tau') \rangle$$
$$q_s^{\alpha\beta}(\tau - \tau') = \langle q^{\alpha\beta}(\tau, \tau') \rangle$$
$$\lambda_s^\alpha = \langle \lambda^\alpha(\tau) \rangle \quad ; \quad m^2 = 2i\lambda_s^\alpha, \quad \forall \alpha$$
$$A_{\mu,s} = \langle A_\mu(\tau) \rangle = 0$$
$$z_{i,s}^\alpha = \langle z_i^\alpha(\tau) \rangle \quad ; \quad |z_{1,s}^\alpha|^2 + |z_{2,s}^\alpha|^2 \equiv \sigma_\alpha^2$$
$$\sigma^2 \equiv \frac{1}{n} \sum_{\alpha=1}^n \sigma_\alpha^2. \tag{21.25}$$

It will be convenient, for later use, to introduce the "replica average" quantity

$$\tilde{q} = \frac{1}{n(n-1)} \sum_{\alpha\beta} q^{\alpha\beta}. \tag{21.26}$$

The (Euclidean) time-dependent quantities are conveniently expanded as Matsubara sums, namely

$$\chi_s(\tau - \tau') = \frac{1}{\beta} \sum_{\omega_n} \chi(\omega_n) e^{-i\omega_n(\tau-\tau')}$$
$$q_s^{\alpha\beta}(\tau - \tau') = \frac{1}{\beta} \sum_{\omega_n} q^{\alpha\beta}(\omega_n) e^{-i\omega_n(\tau-\tau')}$$
$$\tilde{q}(\tau - \tau') = \frac{1}{\beta} \sum_{\omega_n} \tilde{q}(\omega_n) e^{-i\omega_n(\tau-\tau')}. \tag{21.27}$$

Notice that, according to (21.16) and (21.17), $\chi_0 \equiv \chi(\omega_n = 0)$ is the static magnetic susceptibility.

We also introduce now, in the present framework, the Edwards–Anderson parameter, which detects the presence of a spin glass phase. This is basically given by the large-time limit of the spin-spin correlation function on the same site, which tell us essentially how much a spin on a certain site is correlated to itself at a large later time. Using the Riemann–Lebesgue lemma we have

$$q_s^{\alpha\beta}(\tau - \tau') \xrightarrow{\tau - \tau' \to \infty} T q^{\alpha\beta}(\omega_n = 0)$$

$$\tilde{q}(\tau - \tau') \xrightarrow{\tau - \tau' \to \infty} T \tilde{q}(\omega_n = 0)$$

$$q_{EA} \equiv T \tilde{q}(\omega_n = 0). \tag{21.28}$$

The last line defines the Edwards–Anderson parameter.

The stationary point quantities, in particular, are spatially uniform. In terms of these, the effective action at the stationary points reads

$$S_0 \left[z_{i,s}^\alpha, z_{i,s}^{*\alpha}, \lambda_s^\alpha, \chi_s, q_s^{\alpha\beta}, A_{\mu,s} \right] = \frac{nV\beta}{2} \left\{ m^2 \left[\sigma^2 - \rho_s \right] \right\}$$

$$+ nV \int_0^\beta d\tau \int_0^\beta d\tau' \left\{ \frac{3D}{2} \left[\chi^2(\tau, \tau') - \frac{1}{n} q^{\alpha\beta}(\tau, \tau') q^{\alpha\beta}(\tau, \tau') \right] \right.$$

$$\left. - \frac{D}{n\rho_s} \left[\delta^{\alpha\beta} \chi(\tau, \tau') q^{\alpha\beta}(\tau, \tau') \right] \sigma^\alpha \sigma^\beta \right\}. \tag{21.29}$$

The quantum-mechanical contribution to the averaged replicated partition function is given by the logarithm of the determinant in (21.24). The determinant involves three parts, related respectively to the CP^1 indices (ij), the replica indices $(\alpha\beta)$ and the frequency-momentum ω_n, $\mathbf{k}$. The first two can be made exactly, in the limit $n \to 0$, provided we replace $q^{\alpha\beta}$ with its replica average $\tilde{q}$. Taking the trace over frequency-momentum in the logarithm of $\mathbb{M}$, we obtain ultimately [178]

$$\ln \mathrm{Det}\mathbb{M} = nV \sum_{\omega_n} \int \frac{d^2k}{(2\pi)^2} \left[\ln\left(k^2 + M_n\right) - \frac{A\tilde{q}(\omega_n)}{k^2 + M_n} \right], \tag{21.30}$$

where $A = \frac{2D}{\rho_s}$ and

$$M_n = m^2 + \omega_n^2 + A \left[\chi(\omega_n) - \tilde{q}(\omega_n) \right]. \tag{21.31}$$

Fourier transforming (21.29) to frequency space, in which (21.30) already is, and inserting both in (21.24), we get the averaged replicated partition function. Then, using (21.6), we obtain the quenched free-energy density, namely

$$\bar{f}\left[\sigma^\alpha, m^2, \chi(\omega_n), q^{\alpha\beta}(\omega_n)\right] = \frac{1}{2}m^2[\sigma^2 - \rho_s]$$

$$+ \frac{D}{n\rho_s}\left[\chi(\omega_n = 0)\delta^{\alpha\beta} + q^{\alpha\beta}(\omega_n = 0)\right]\sigma^\alpha\sigma^\beta$$

$$+ 3DT\sum_{\omega_n}\left[\chi(\omega_n)\chi(-\omega_n) - \frac{1}{n}q^{\alpha\beta}(-\omega_n)q^{\alpha\beta}(-\omega_n)\right]$$

$$+ \sum_{\omega_n}\int\frac{d^2k}{(2\pi)^2}\left[\ln\left(k^2 + M_n\right) - \frac{A\tilde{q}(\omega_n)}{k^2 + M_n}\right],$$

$$(21.32)$$

where σ^2 was defined in (21.25).

21.4 The Phase Diagram

The Stationary Point Equations

Let us determine now the phase diagram of the system. For this, we consider the equations obtained by imposing the first derivatives of the free-energy density vanish. The stability of the phases will be guaranteed by verifying that the eigenvalues of the Hessian matrix of the free energy has only positive eigenvalues. We have

1) $\dfrac{1}{n}\left[[m^2 - A\chi_0]\delta^{\alpha\beta} - Aq_0^{\alpha\beta}\right]\sigma^\beta = 0$

2) $\sigma^2 = \rho_s - \dfrac{T}{2\pi}\sum_{\omega_r}\ln\left(1 + \dfrac{\Lambda^2}{M_r}\right) + 2A\sum_{\omega_r}\tilde{q}(\omega_r)G_r$

3) $3DT\chi(-\omega_n) = \dfrac{TA}{4\pi}\ln\left(1 + \dfrac{\Lambda^2}{M_n}\right) + A^2\tilde{q}(\omega_n)G_n - A\sigma^2\delta_{\omega_n 0}$

4) $3DTq^{\alpha\beta}(-\omega_n) = A^2\tilde{q}(\omega_n)G_n + \dfrac{A}{2n}\sigma^\alpha\sigma^\beta\delta_{\omega_n 0},$ (21.33)

where $\chi_0 \equiv \chi(\omega_n = 0)$, $q_0^{\alpha\beta} \equiv q^{\alpha\beta}(\omega_n = 0)$ and $\tilde{q}_0 \equiv \tilde{q}(\omega_n = 0)$, Λ is a high-momentum cutoff and

$$G_n = \frac{T}{4\pi}\left[\frac{1}{M_n} - \frac{1}{\Lambda^2 + M_n}\right].$$ (21.34)

Notice that in the absence of disorder ($\Delta J = D = A = 0$), the above equations reduce to (19.58), which determine the phase structure of the pure AF magnetic system described by the NLSM.

Preliminary Results

Let us consider firstly the phases for which $\sigma^\alpha = 0$. These could be either paramagnetic (PM) or spin glass (SG) phases. Then, summing both sides of Eq. 4) in α, β, we get

$$\tilde{q}(\omega_n)G_n = \tilde{q}(-\omega_n)\Gamma, \tag{21.35}$$

where

$$\Gamma = \frac{T}{4\pi}\frac{\gamma}{\Lambda^2}$$

$$\gamma = 3\pi\left(\frac{J_0}{\Delta J}\right)^2. \tag{21.36}$$

Let us take (21.35) for $\omega_n = 0$ and insert it in Eq. 3). It yields

$$M_0 = \frac{\Lambda^2}{e^{6\pi\rho_s(\chi_0-\tilde{q}_0)} - 1}, \tag{21.37}$$

which can also be expressed as

$$F(\chi_0 - \tilde{q}_0) = m^2$$

$$F(x) = \frac{\Lambda^2}{e^{6\pi\rho_s x} - 1} + Ax. \tag{21.38}$$

The function $F(x)$ has an absolute minimum $F(x_0) = m_0^2$ such that x_0 satisfies

$$\gamma = [2\sinh(3\pi\rho_s x_0)]^2$$

$$m_0^2 = \frac{\Lambda^2}{\gamma}\ln\gamma. \tag{21.39}$$

It follows that

$$G_0(M_0(x_0)) = \Gamma. \tag{21.40}$$

Now, since $G_0(M_0)$ and $M_0(x)$ are both monotonically decreasing functions and the physical solutions of (21.38) (left branch in Fig. 21.1) occur for $x < x_0$, we have $G_0 \le \Gamma$, the equality holding only at $x = x_0$ or $m^2 = m_0^2$. In this case, (21.35) implies $\tilde{q}_0 = 0$. We will only have $\tilde{q}_0 \ne 0$ for $G_0 = \Gamma$.

On the other hand, it has also been shown [178] that $\tilde{q}(\omega_n \ne 0) = 0$. This follows from the fact that $|G_n| < \Gamma$, for $\omega_n \ne 0$ and that $\tilde{q}(-\omega_n \ne 0) = \tilde{q}^*(\omega_n \ne 0)$, because $\tilde{q}(\tau - \tau') \in \mathbb{R}$.

The Paramagnetic Phase

The paramagnetic phase is characterized by: $\sigma = 0$, $\tilde{q} = 0$, $m^2 > m_0^2$, $G_0 < \Gamma$.

Inserting Eq. 3) in Eq. 2), we readily find the integrated magnetic susceptibility is given by

$$\chi_I \equiv \sum_{\omega_n} \chi(\omega_n) = \frac{1}{3T} \tag{21.41}$$

and, hence, satisfies the Curie Law. The static magnetic susceptibility is given by

$$\chi_0 \equiv \chi_I - \sum_{\omega_n \neq 0} \chi(\omega_n) = \chi_I - \Upsilon(J_0, \Delta J, T), \tag{21.42}$$

where the last sum, evaluated in [178], yields a function $\Upsilon(J_0, \Delta J, T)$, such that,

$$\Upsilon(J_0, \Delta J, T) \overset{T \gg \Lambda}{\to} 0, \tag{21.43}$$

hence χ_0 also satisfies the Curie Law.

The Spin-Glass Phase

An SG phase is characterized by $\tilde{q}_0 \neq 0$. It occurs for $m^2 < m_0^2$, where a PM solution with $\tilde{q}_0 = 0$ does not exist. From (21.35), a nonzero $\tilde{q}_0$ implies $G_0 = \Gamma$, for which $M_0(\chi_0 - \tilde{q}_0) = M_0(x_0)$. It follows that

$$m^2 - A(\chi_0 - \tilde{q}_0) = m_0^2 - A x_0$$

$$\tilde{q}_0 = \chi_0 - x_0 + \frac{1}{A}(m_0^2 - m^2). \tag{21.44}$$

From Eq. 2), conversely, we get

$$\tilde{q}_0 = \frac{1}{3T} - \Upsilon(J_0, \Delta J, T) - x_0. \tag{21.45}$$

From the two previous equations, we get, in the SG phase, the static susceptibility, namely

$$\chi_0 = \frac{1}{3T} - \Upsilon(J_0, \Delta J, T) - \frac{1}{A}(m_0^2 - m^2), \tag{21.46}$$

and the integrated susceptibility,

$$\chi_I = \frac{1}{3T} - \frac{1}{A}(m_0^2 - m^2). \tag{21.47}$$

From the above equation and the general relation valid in an SG phase [180]

$$\chi_I = \frac{1}{3T} - \frac{1}{3} \sum_{\omega_n} \tilde{q}(\omega_n)$$

$$\chi_I = \frac{1}{3T} - \frac{1}{3}\tilde{q}_0, \tag{21.48}$$

where we used the fact that $\tilde{q}(\omega_n \neq 0) = 0$, we infer that

$$\tilde{q}_0 = \frac{3}{A}(m_0^2 - m^2) \tag{21.49}$$

in the SG phase.

Critical Behavior

The transition between the SG and PM phases occurs for $m^2 = m_0^2$; $\tilde{q}_0 = 0$. Using this in (21.45), we obtain the critical condition that will determine the curve separating these phases:

$$\frac{1}{3T_c} - \Upsilon(J_0, \Delta J, T_c) = x_0, \qquad (21.50)$$

where T_c is the critical temperature.

Also, for $T \simeq T_c$

$$\frac{1}{3T} - \Upsilon(J_0, \Delta J, T) \sim \frac{T}{T_c} x_0$$

$$m^2 - m_0^2 \sim 4\pi \Lambda \left[\frac{T - T_c}{T_c} \right] [\rho_{s,0} - \rho_s]. \qquad (21.51)$$

Now, how do we determine T_c? For $T_c \ll \Lambda$, we get, from (21.50)

$$T_c \left[\ln \left(\frac{\Lambda}{T_c} \right) - \frac{1}{2} \ln (1 + \gamma) \right] = \pi [\rho_{s,0} - \rho_s], \qquad (21.52)$$

where

$$\rho_{s,0} = \frac{\Lambda}{2\pi} \left[1 + \frac{1}{\gamma} \left[1 + \frac{1}{2} \ln [1 + \gamma] \right] \right] \qquad (21.53)$$

is the critical spin stiffness. Notice that the T_c curve touches the horizontal axis ($T_c = 0$) at $\rho_s = \rho_{s,0}$, and only for $\rho_s < \rho_{s,0}$ we find a finite critical temperature $T_c(\rho_s)$. For $\rho_s > \rho_{s,0}$, there is no SG phase at any temperature.

From (21.46), (21.47) and (21.51), we get the critical behavior ($T \simeq T_c$) of the static and integrated susceptibilities, namely

$$\chi_0 \sim \frac{T}{T_c} x_0 - \frac{4\pi}{A} \Lambda \left[\frac{T_c - T}{T_c} \right] [\rho_{s,0} - \rho_s] \qquad (21.54)$$

and the integrated susceptibility

$$\chi_I \sim \frac{1}{3T} - \frac{4\pi}{A} \Lambda \left[\frac{T_c - T}{T_c} \right] [\rho_{s,0} - \rho_s]. \qquad (21.55)$$

This is depicted in Fig. 21.1 and exhibit the characteristic cusp at the transition.

The Néel Phase

Let us consider now the ordered AF phase, for which $\sigma \neq 0$. In this case, the quantity between brackets in Eq. 1) vanishes. Then, summing in α and β, we get $M_0 = 0$, which implies

$$\chi_0 - \tilde{q}_0 = \frac{m^2}{A}, \qquad (21.56)$$

where m^2 is the spin-gap.

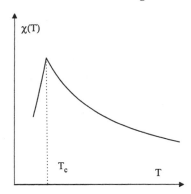

Figure 21.1 The integrated magnetic susceptibility exhibiting the characteristic sharp cusp at the transition temperature T_c

Now, Eq. 2) implies that a phase with $M_0 = 0$ can only occur at $T = 0$, otherwise the sublattice magnetization σ would have an unphysical infinite imaginary value. The fact that an ordered AF phase can only occur at zero temperature is in agreement with the Mcrmin–Wagner–Hohenberg theorem [170], which precludes the spontaneous breakdown of a continuous symmetry at a finite temperature in two spatial dimensions.

In the Néel phase, Eq. 2) and Eq. 3) imply

$$\chi_I = \frac{1}{3T} \left(1 - \frac{2\sigma^2}{\rho_s} \right). \tag{21.57}$$

From this, using (21.48), we infer

$$\tilde{q}_0 = \frac{2\sigma^2}{T\rho_s}. \tag{21.58}$$

Also, from (21.42) and

$$\Upsilon(J_0, \Delta J, T) \stackrel{T \to 0}{\to} \frac{\rho_{s,0}}{3T\rho_s}, \tag{21.59}$$

we conclude that, for $T \to 0$

$$\chi_0 = \frac{1}{3T} \left[\left(\frac{\rho_s - \rho_{s,0}}{\rho_s} \right) - \frac{2\sigma^2}{\rho_s} \right]. \tag{21.60}$$

From (21.56), (21.58) and (21.60), we can solve for χ_0, $\tilde{q}_0$ and σ, obtaining

$$\chi_0 = \tilde{q}_0 = \frac{1}{4T\rho_s} (\rho_s - \rho_{s,0})$$

$$\sigma^2 = \frac{1}{8} (\rho_s - \rho_{s,0}), \tag{21.61}$$

which imply the spin-gap vanishes: $m = 0$, in compliance with the Goldstone theorem. Notice that the static susceptibility χ_0 diverges in the Néel phase, as it should, whereas the Edwards–Anderson parameter, $q_{EA} = T\tilde{q}_0$, remains finite but nonzero. The AF ordered phase sets in for $\rho_s > \rho_{s,0}$ at $T = 0$.

Summary

We have found three phases in the disordered SO(3) quantum AF Heisenberg system with nearest neighbor interactions: a Néel phase ($\sigma \neq 0, \tilde{q}_0 \neq 0$) at $T = 0$, $\rho_s > \rho_{s,0}$, $m^2 = 0$, with $\rho_{s,0}$ given by (21.53); an SG phase ($\sigma = 0, \tilde{q}_0 \neq 0$), for $T < T_c$, $\rho_s < \rho_{s,0}$, $0 < m^2 < m_0^2$, with T_c given by (21.52) and m_0, by (21.39); and a PM phase ($\sigma = 0, \tilde{q}_0 = 0$), for $T > T_c$, $\rho_s < \rho_{s,0}$, $m^2 > m_0^2$ and for $T > 0$ for $\rho_s > \rho_{s,0}$.

Notice that a smaller average coupling will favor the onset of a spin-glass phase, whereas a large one will eliminate it. This is in agreement with the fact that the frustration region of the distribution function, namely, the tail on the negative coupling side, will disappear as we increase the average coupling (see Fig. 21.4).

Observe that, as the disorder is removed, $\Delta J \to 0$, the critical spin stiffness, $\rho_{s,0}$, reduces to the quantum-critical coupling $\rho_c = \Lambda/2\pi$, which we found for the pure AF Heisenberg model in Section 19.3, separating the paramagnetic phase from the Néel phase at $T = 0$. The spin-glass phase is removed accordingly because the Edwards–Anderson parameter, needed for the onset of a spin-glass phase, vanishes for $\Delta J \to 0$, as we can infer from (21.49). This observation is corroborated by

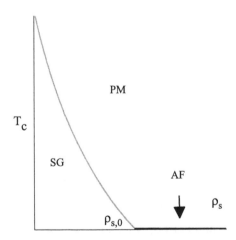

Figure 21.2 The phase diagram of the quantum SO(3) disordered system defined by (21.1) and (21.2), exhibiting an SG phase for $T < T_c$, $\rho_s < \rho_{s,0}$, an AF phase for $T = 0$, $\rho_s > \rho_{s,0}$ and a PM phase for $T > T_c$, $\rho_s < \rho_{s,0}$ and for $T > 0$ for $\rho_s > \rho_{s,0}$

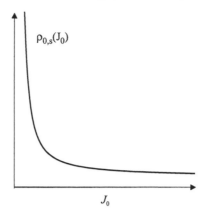

Figure 21.3 The critical spin stiffness, $\rho_{s,0}$, as a function of the average coupling J_0, taken from (21.53)

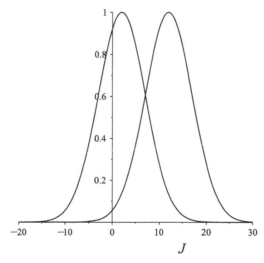

Figure 21.4 The coupling distribution function (21.2) for different average couplings (J_0). Frustration is produced by the tail on the negative coupling side.

the fact that the frustration area of the distribution function, namely, the tail on the negative coupling side, will disappear as we make $\Delta J \to 0$ (see Fig. 21.6).

21.5 Thermodynamic Stability

The thermodynamic stability of a spin glass is an issue of crucial importance. Historically, after the problem of quenched disordered magnets was addressed by Edwards and Anderson [94], a spin glass model was proposed by Sherrington and Kirkpatrick (SK) [181], who presented a mean-field solution exhibiting a

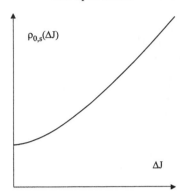

Figure 21.5 The critical spin stiffness, $\rho_{s,0}$, as a function of the amount of disorder, ΔJ, taken from (21.53)

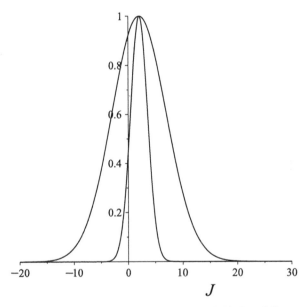

Figure 21.6 The coupling distribution function (21.2) for different amounts of disorder (ΔJ). Frustration is produced by the tail on the negative coupling side.

spin-glass phase, apart from the usual phases. The application of the mean-field method was significantly facilitated by the fact that the SK Hamiltonian contained an infinite range interaction, where each spin would interact with every other spin on the lattice. The SK solution was soon proved to be unstable by Almeida and Thouless [182] and the instability was ascribed to the replica symmetry of the SK solution. A stable replica symmetry breaking solution was subsequently found by Parisi [183].

We have studied a disordered quantum magnet model with SO(3) symmetry, introduced in [177], which contains only nearest-neighbor interactions, and obtained the quenched free energy

$$\overline{f}\left[\sigma^{\alpha}, q^{\alpha\beta}(\omega_n), m^2, \chi(\omega_n)\right] \equiv \overline{f}\left[\varphi_i\right] \tag{21.62}$$

by taking into account quadratic quantum fluctuations. In the above expression, φ_i is a generic symbol for the fields appearing as arguments of $\overline{f}$. We then extracted the phase diagram of the system by finding which solutions minimize the above free energy for different ranges of the parameters and temperature. A crucial condition for a given solution of the extremant condition

$$\frac{\partial \overline{f}\left[\varphi_i\right]}{\partial \varphi_i} = 0 \tag{21.63}$$

to be a minimum is that all eigenvalues of the Hessian matrix of the free energy, namely

$$\mathbb{H}_{ij} = \frac{\partial^2 \overline{f}\left[\varphi_i\right]}{\partial \varphi_i \partial \varphi_j}\Big|_{\varphi_{i,0}}, \tag{21.64}$$

are strictly positive.

A theorem from linear algebra, then, states that a sufficient condition for all the eigenvalues of a given matrix to be positive is that all principal minor determinants of such a matrix be positive. A very careful evaluation of all principal minors of the quenched free energy Hessian was presented in [178]. As it turns out, in the limit $n \to 0$, all the principal minors except the one corresponding to the $\sigma^{\alpha\beta}$ sector, which is equal to one, can be written in the form

$$\xi\left[(\Gamma - G_0) + H_0\right] + \zeta\left[(\Gamma - G_0]^2\right., \tag{21.65}$$

where

$$H_0 = \tilde{q}_0 A \frac{T}{4\pi}\left[\frac{1}{M_0^2} - \frac{1}{(M_0 + \Lambda^2)^2}\right], \tag{21.66}$$

where ξ and ζ are real factors, such that $\xi > 0$ and $\zeta \geq 0$.

In a paramagnetic (PM) phase, $H_0 = 0$ and $G_0 < \Gamma$, whereas in a spin-glass (SG) phase, $H_0 > 0$, with $G_0 = \Gamma$. We conclude, therefore, that both in the PM and SG phases, all the principal minors of the free energy Hessian are always positive, thereby demonstrating unequivocally the stability of such phases. Furthermore, these principal minors vanish right at the phase transition connecting such phases, so we can clearly see the phase transition occurring directly in the Hessian. The stability of the Néel phase can be demonstrated accordingly.

21.6 Duality and the Nature of the Spin-Glass Phase

The Skyrmion-Spin Duality

It is very instructive to analyze the different phases we found in the disordered quantum SO(3) Heisenberg system in the light of the order-disorder duality. For this purpose the quantum skyrmion correlation function was evaluated, applying the methods of quantization of topological excitations exposed in Section 17.6 in each of the phases encountered in the system [179]. The results, obtained in the Néel, spin-glass and paramagnetic phases are, respectively,

$$\langle \mu(\mathbf{x}, 0)\mu^\dagger(\mathbf{y}, 0)\rangle_N \overset{|\mathbf{x}-\mathbf{y}|\to\infty}{\longrightarrow} \exp\left\{-2\pi\sigma^2|\mathbf{x} - \mathbf{y}|\right\} \tag{21.67}$$

where σ^2 is given by (21.61);

$$\langle \mu(\mathbf{x}, 0)\mu^\dagger(\mathbf{y}, 0)\rangle_{SG} \overset{|\mathbf{x}-\mathbf{y}|\to\infty}{\longrightarrow} \frac{1}{|\mathbf{x} - \mathbf{y}|^{\nu\tilde{q}_0}}, \tag{21.68}$$

where $\tilde{q}_0$ is given by (21.49) and ν is real and positive; and

$$\langle \mu(\mathbf{x}, 0)\mu^\dagger(\mathbf{y}, 0)\rangle_{PM} \overset{|\mathbf{x}-\mathbf{y}|\to\infty}{\longrightarrow} \exp\left\{\frac{\kappa}{|\mathbf{x} - \mathbf{y}|}\right\}, \tag{21.69}$$

where κ is real and positive. The above expressions imply $\langle\mu\rangle_N = 0$, $\langle\mu\rangle_{SG} = 0$ and $\langle\mu\rangle_{PM} = 1$.

Notice that the power-law decay of the skyrmion correlators in the SG phase implies these excitations are massless, while $\langle 0|\mu|0\rangle_N = 0$ implies they are genuine nontrivial excitations, since the skyrmion state is orthogonal to the ground-state. In the PM phase, conversely, $\langle 0|\mu|0\rangle_{PM} = 1$ means skyrmions are not genuine excitations. Acting on a disordered ground state with the skyrmion operator produces essentially the same ground state, hence $\langle\mu\rangle_{PM} = 1$. In an SG phase, conversely, the ground state is also disordered but frozen; hence, acting on this with the skyrmion operator creates another frozen disordered state which is orthogonal to the former, despite having the same energy – in other words, a gapless excitation. We have, therefore, a power-law behavior of the skyrmion quantum correlator.

Stability of the SG Phase and the BKT Mechanism

In the CP^1 language, skyrmions are vortices. The fact that they have a long-range power-law behavior characterizes the SG phase as the low-temperature phase of a two-dimensional BKT system [155, 156, 157]. This explains the stability we found for the SG phase at a finite temperature in a quantum system with short-range interaction and a continuous symmetry.

The massless quantum spin-waves states of the AF phase, conversely, not possessing a topological protection, are washed-out to $T = 0$ in the Néel phase by virtue of the Hohenberg–Mermin–Wagner mechanism. This explains the asymmetry existing between the dual phases: AF and SG. Both contain gapless excitations, respectively spin-waves and skyrmions, the former occurring only at $T = 0$, whereas the latter, by means of the BKT mechanism, surviving below a finite temperature T_c.

A New Characterization of the SG Phase

A new way of characterizing an SG phase emerges from the previous analysis. The system realizes the three phases allowed by the dual algebra existing between the spin-wave and skyrmion operators, namely: ($\langle \sigma \rangle \neq 0$; $\langle \mu \rangle = 0$; Ordered AF phase); ($\langle \sigma \rangle = 0$; $\langle \mu \rangle = 0$; SG phase) and ($\langle \sigma \rangle = 0$; $\langle \mu \rangle = 1$; PM phase).

In this framework, taking advantage of order-disorder duality, an SG phase would be characterized in general as one for which both $\langle \sigma \rangle = 0$ and $\langle \mu \rangle = 0$.

22

Quantum Field Theory Approach to Superfluidity

Superfluidity and superconductivity are two extremely interesting twin phenomena, the discovery of which was made possible by the development of the helium liquefaction technique by Kamerlingh Onnes in 1908. Superconductivity was discovered by Onnes himself in 1911, whereas superfluidity was discovered by Kapytza, Allen and Misener in 1938. While superconductivity entails the flow of charge carriers without any resistance, superfluidity involves the frictionless flow of a fluid, without any viscosity. Superconductivity, as we have seen, was observed when mercury was cooled down to approximately $4\ K$, in contact with a liquid helium bath. Superfluidity, conversely, was firstly observed in liquid ^4He itself, when it was cooled down below $2.17\ K$. Later on, it was also observed in liquid ^3He below $2.7\ mK$. In the first case, the phenomenon was related to the Bose–Einstein condensation of the bosonic ^4He atoms, whereas in the second case, it involves the formation of bound states of the fermionic ^3He atoms, very much like Cooper pairs, before the superfluid phase can set in.

In this chapter, after a brief introduction of superfluidity, we describe the use of a field theory approach for explaining its most important features. The quantum field theory approach to superconductivity is described in the next chapter.

22.1 Basic Features of Superfluidity

Superfluidity is most conveniently described in the framework of a Landau–Ginzburg approach, similar to the one employed in the case of superconductivity in (4.68). An important point to be mentioned is that superfluidity concerns the transport of neutral matter, whereas superconductivity, that of charged carriers. It follows that, for superfluids, the free energy would be given by (4.68), but without the electromagnetic gauge field. The Landau–Ginzburg theory in this case would be replaced by the corresponding theory with a global continuous U(1) symmetry, namely, the Gross–Pitaevskii free energy [187, 188],

$$F[\Psi] = F_0 + \int d^3r \left\{ \frac{\hbar^2}{2M} |\nabla\Psi|^2 + a(T)|\Psi|^2 + \frac{1}{2}b|\Psi|^4 \right\}, \tag{22.1}$$

where $b > 0$ and $a(T) = a_0(T - T_c)$, with T_c being the critical temperature for the system to undergo the superfluid transition and $a_0 > 0$.

Writing the order parameter wave-function as $\Psi_{SF} = \sqrt{n_{SF}}e^{i\theta}$ where n_{SF} is the density of superfluid matter, which is assumed to be constant, we have

$$n_{SF} = -\frac{a(T)}{b} \quad ; \quad T < T_c. \tag{22.2}$$

Then we can infer that the velocity eigenvalue of the operator $\frac{\mathbf{P}}{M}$ is

$$\mathbf{v} = \frac{\hbar}{M}\nabla\theta, \tag{22.3}$$

where M is the mass of the atoms forming the superfluid matter. Also,

$$\frac{\mathbf{P}^2}{2M}\Psi_{SF} = E_0\Psi_{SF}$$

$$E_0 = \frac{\hbar^2}{2M}\nabla\theta \cdot \nabla\theta. \tag{22.4}$$

It follows that the superfluid current is given by

$$\mathbf{j}_{SF} = n_{SF}\mathbf{v}$$

$$\mathbf{j}_{SF} = n_{SF}\frac{\hbar}{M}\nabla\theta, \tag{22.5}$$

which satisfies the continuity equation

$$\nabla \cdot \mathbf{j}_{SF} = -\frac{\partial n_{SF}}{\partial t} = n_{SF}\frac{\hbar}{M}\nabla^2\theta = 0. \tag{22.6}$$

Similarly to the case of superconductors, we see that the superfluid velocity at different points is coherently locked by the fact that it is proportional to the gradient of the order parameter phase. This coherence rules out the individual scattering phenomena responsible for friction and viscosity, and therefore explains superfluidity. Bose–Einstein condensation is responsible for producing the condensate, which is described by the wave-function Ψ_{SF} [186].

22.2 Classical Vortices

From (22.3), we find

$$\nabla \times \mathbf{v} = \frac{\hbar}{M}\nabla \times \nabla\theta$$

$$\epsilon^{ijk}\partial_j v_k = \frac{\hbar}{M}\epsilon^{ijk}\partial_j\partial_k\theta. \tag{22.7}$$

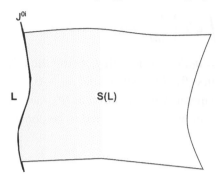

Figure 22.1 The vortex along the line L describes the universe-surface S(L) as it evolves in time.

The right-hand side is equal to zero everywhere except at the vortex cores, implying the superfluid flow is irrotational except when vortices are present [189]. The vortex, being an extended object that exists along a line L inside the superfluid, has a tensor density

$$J^{0i} = \frac{\hbar}{M}\epsilon^{ijk}\partial_j\partial_k\theta = \int_{S(L)} d^2\xi^{i0}\delta^4(x-\xi), \tag{22.8}$$

also described in (10.22). Here, $S(L)$ is the universe-surface of the vortex, namely, the surface it describes as it evolves in time.

Vortices are, in fact, infinite string excitations that, having a cylindrical symmetry, produce a mapping $\Pi_1(U(1))$, which is topologically nontrivial. The topological charge is the velocity circulation, or vorticity, along a closed loop pierced by the vortex line:

$$\Gamma = \oint_C d\mathbf{l}\cdot\mathbf{v} = 2\pi\frac{\hbar}{M}n \quad ; \quad n = 0, \pm1, \pm2, \ldots \tag{22.9}$$

We see that the topological charge is an integer multiple of the vorticity quantum $\phi_0 = h/M$.

We can also express Γ as the flux of the vortex current density across the surface $R(C)$ bounded by C,

$$\Gamma = \int_{R(C)} d^2S^i\, J^{0i} = \frac{\hbar}{M}\int_{R(C)} d\mathbf{S}\cdot(\nabla\times\nabla\theta)$$

$$= \frac{\hbar}{M}\oint_C d\mathbf{l}\cdot\nabla\theta. \tag{22.10}$$

It is a well-known property of superfluids that, when contained in a rotating vessel, the fluid remains at rest in the laboratory frame when the vessel's angular velocity is less than a certain threshold. Above the threshold, the fluid bulk remains steady, however vortices are created in such a way that there is a nonzero number of circulation quanta.

Notice, at this point, the strong similarity with the physics of superconductors. Irrotational flow corresponds to the Meissner effect, namely, the absence of a magnetic field in the superconducting bulk; the rotating vessel corresponds to the external magnetic field; the threshold angular velocity corresponds to the lower critical magnetic field in a type-II superconductor; and the vortex lines correspond to the magnetic flux lines that pierce through in such superconducting materials.

There is, consequently, an important physical interest in the study of both classical and quantum properties of vortices in superfluids. For this purpose, we will see that a field theory approach shall be quite convenient.

22.3 The Goldstone Mode

Consider a complex field Ψ with Lagrangean

$$\mathcal{L} = \partial_\mu \Psi^* \partial_\mu \Psi + a(T)|\Psi|^2 - \frac{1}{2}b|\Psi|^4, \tag{22.11}$$

which describes a field theory related to the Gross–Pitaevskii free energy (22.1). Using the polar representation $\Psi = \frac{\rho}{\sqrt{2}}e^{i\varphi}$, we can write this as

$$\mathcal{L} = \frac{\rho^2}{2}\partial_\mu \varphi \partial_\mu \varphi + \frac{1}{2}\partial_\mu \rho \partial_\mu \rho + a(T)\rho^2 - \frac{1}{2}b\rho^4. \tag{22.12}$$

This Lagrangean possesses a conserved current $j^\mu = \rho^2 \partial^\mu \varphi$ as a consequence of the global U(1) symmetry it has.

In the constant density regime, which applies to a superfluid, $\rho = \rho_0 = \sqrt{n_{SF}}$ is a constant and we have the superfluid density given by

$$n_{SF} = \frac{j^0}{\partial_t \varphi}$$
$$\mathbf{j} = n_{SF}\nabla\varphi. \tag{22.13}$$

In this regime the only dynamical degree of freedom is the Goldstone mode φ, corresponding to

$$\mathcal{L} = \rho_0^2 \partial_\mu \varphi \partial_\mu \varphi, \tag{22.14}$$

and we see that

$$\nabla\varphi = \frac{\hbar}{M}\nabla\theta = \frac{\phi_0}{2\pi}\nabla\theta. \tag{22.15}$$

We will see in the next section that it is possible to reformulate the theory in terms of a rank-two antisymmetric tensor gauge field, the Kalb–Ramond field. This formulation of the theory will be particularly convenient for the full description of quantum vortices.

22.4 The Kalb–Ramond Field

In order to derive the Kalb–Ramond formulation of superfluidity [190], consider the following functional integral representation of the vacuum functional Z, corresponding to (22.14) [192]:

$$Z = \int D\varphi \exp\left\{i \int d^4x \frac{\rho_0^2}{2} \partial_\mu \varphi \partial_\mu \varphi\right\}.$$

This can be written, up to a multiplicative constant, as

$$Z = \int D\varphi DH_\mu \exp\left\{i \int d^4x \left[\frac{1}{2}H_\mu H^\mu + \rho_0 H^\mu \partial_\mu \varphi\right]\right\}, \tag{22.16}$$

where H_μ is a real vector field.

Now let us decompose the scalar field φ as

$$\varphi = \varphi_r + \varphi_{mv}$$
$$D\varphi = D\varphi_r, \tag{22.17}$$

where φ_r is regular and φ_{mv}, multivalued. Inserting in (22.16) and functional integrating on φ_r, we get, once again up to a multiplicative constant,

$$Z = \int DH_\mu \delta\left[\partial_\mu H^\mu\right] \exp\left\{i \int d^4x \left[\frac{1}{2}H_\mu H^\mu + \rho_0 H^\mu \partial_\mu \varphi_{mv}\right]\right\}. \tag{22.18}$$

We solve the identity $\partial_\mu H^\mu \equiv 0$, which is imposed by the functional delta functional, by expressing H_μ as

$$H_\mu = \frac{1}{2}\epsilon^{\mu\nu\alpha\beta} \partial_\nu B_{\alpha\beta}$$
$$\frac{1}{2}H_\mu H^\mu = \frac{1}{12}H^{\mu\alpha\beta} H_{\mu\alpha\beta}, \tag{22.19}$$

in terms of the Kalb–Ramond field $B_{\alpha\beta}$, [191], which has the field intensity tensor given by $H_{\mu\alpha\beta} = \partial_\mu B_{\alpha\beta} + \partial_\alpha B_{\beta\mu} + \partial_\beta B_{\mu\alpha}$.

Inserting (22.19) in (22.18) and integrating by parts the last term, we get

$$\frac{1}{2}\epsilon^{\mu\nu\alpha\beta} \partial_\alpha \partial_\beta \varphi_{mv} = J^{\mu\nu} = \int_{S(L)} d^2\xi^{\mu\nu} \delta^4(x - \xi), \tag{22.20}$$

where $S(L)$ is the universe-surface of a vortex string along the line L, according to (22.8) and (10.22).

We therefore can write (22.18) as

$$Z = \int DB_{\mu\nu} \exp\left\{i \int d^4x \left[\frac{1}{12} H^{\mu\alpha\beta} H_{\mu\alpha\beta} + \rho_0 J^{\mu\nu} B_{\mu\nu}\right]\right\}. \tag{22.21}$$

We see that the vortex string is the source of the Kalb–Ramond field, namely

$$\partial_\alpha H^{\mu\nu\alpha} = \rho_0 J^{\mu\nu} \tag{22.22}$$

with the generalized Gauss' law being expressed as

$$\partial_j \Pi^{ij} = \rho_0 J^{0i}$$
$$\Pi^{ij} = H^{0ij}, \tag{22.23}$$

where Π^{ij} is the momentum canonically conjugate to B_{ij}.

In Kalb–Ramond language, the topological charge or vorticity piercing a surface $R(C)$ can be written as

$$\Gamma_R = \frac{1}{\rho_0} \int_{R(C)} d^2 S^i \partial_j \Pi^{ij}, \tag{22.24}$$

in agreement with (10.25) and (10.26).

As we remarked in the first section, the physical properties of vortices become a central issue in the physics of superfluids. In the next section we approach these basic excitations from the quantum-mechanical point of view.

22.5 Quantum Vortices

Superfluids are essentially quantum fluids; therefore, a full quantum-mechanical treatment of superfluid vortices is unavoidably required. Once we identify the vorticity with the vector charge of a Kalb–Ramond field, namely (10.26), we are enabled to use the operator (10.24),

$$\sigma(S(C), t) = \exp\left\{-i\phi_0 \int_{S(C)} d^2\xi^{ij} B_{ij}(\xi, t)\right\}, \tag{22.25}$$

as the quantum vortex creation operator [192, 42]. Indeed, according to (10.28) and (10.30), it follows that the operator above, when acting on the vacuum, creates quantum states that are eigenvectors of the quantum vorticity operator, the eigenvalue of which is one unit of vorticity, ϕ_0, along the curve C, namely,

$$\Gamma_R|\sigma(S(C), t)\rangle = \phi_0|\sigma(S(C), t)\rangle, \tag{22.26}$$

provided the curve C pierces the surface R once.

The physical properties of quantum vortices are encoded in the correlation functions of the above operator. The energy required to create a quantum vortex line, for instance, is an important quantity, which may be extracted from the large-distance behavior of the vortex, two-point correlation function. This has been obtained after a detailed calculation presented in [192, 42, 193].

For a long straight vortex line of length L along the z-direction, located at the position $(\mathbf{r}, 0)$ in the xy-plane, this is given by

$$\langle \sigma_L(\mathbf{x}, t)\sigma_L^\dagger(\mathbf{y}, t)\rangle \overset{|\mathbf{y}-\mathbf{y}| \to \infty}{\sim} \exp\left\{-L\frac{\phi_0^2 \rho_0^2}{8\pi}|\mathbf{y} - \mathbf{y}|\right\}. \qquad (22.27)$$

From this we may infer the vortex energy per unit length,

$$\epsilon(L) = \frac{E(L)}{L} = \frac{\phi_0^2 \rho_0^2}{8\pi}. \qquad (22.28)$$

This would be, accordingly, the energy cost for creating a quantum vortex excitation in a superfluid.

The quantum vortex operator and the respective correlation functions are the basic required tools for describing any physical process involving vortices in a superfluid. In conclusion, we can only emphasize how useful it was to use quantum field theory language and methods in order to provide a full quantum-mechanical description of vortices in a superfluid.

In the next chapter, we present a full account of superconductivity, both of regular and Dirac electrons. The same approach used here to describe quantum-mechanical vortices could be applied there; however, in the case of a (type II) superconductor, the magnetic vortices correspond to relatively strong external magnetic fields, which are most naturally described within a classical approach.

Quantum Field Theory Approach to Superconductivity

The basic features of superconductivity were introduced in Chapter 4. There we saw that the key condition for the occurrence of superconductivity is the onset of a phase containing an incompressible fluid of the charge carriers, such that their velocities are all coherently locked by the gradient of the complex order parameter phase. The Landau–Ginzburg theory provided a phenomenological framework where this situation would occur below a certain critical temperature. Here, we employ quantum field theory methods in order to demonstrate that, under appropriate conditions, this follows from the electronic interactions. We derive in particular an effective potential, generated by the microscopic dynamics of the system, which will replace the quartic potential postulated in the Landau–Ginzburg phenomenological approach as being responsible for generating such incompressible fluid.

We shall first consider the case of electrons with a non-relativistic dispersion relation, which is the situation usually found in regular metals. Then, we will move to the case where the electrons in a crystal, in spite of having a speed much less than the speed of light, exhibit a dispersion relation that would befit a relativistic particle. This is the situation found in some advanced materials.

23.1 Superconductivity of Regular Electrons

23.1.1 The BCS Quantum Field Theory

As we have seen, the dynamics behind the phenomenon of superconductivity is synthesized by the quartic Hamiltonian (3.62), where the interaction potential may be chosen as in (4.82). Then, using (3.12) and (3.19), we can express the total Hamiltonian corresponding to the interaction (3.62) in terms of the electron field $\psi_\sigma(\mathbf{r})$:

$$H = \int d^3r \left[\psi_\sigma^\dagger(\mathbf{r}) \left(-\frac{\hbar^2}{2m}\nabla^2 \right) \psi_\sigma(\mathbf{r}) - \frac{\lambda}{2} \psi_{\sigma'}^\dagger(\mathbf{r})\psi_\sigma^\dagger(\mathbf{r})\psi_\sigma(\mathbf{r})\psi_{\sigma'}(\mathbf{r}) \right], \quad (23.1)$$

where summation over repeated indices is understood. The corresponding Lagrangean density is

$$\mathcal{L}_{BCS} = \psi_\sigma^\dagger(\mathbf{r}) \left(i\hbar \frac{\partial}{\partial t} - \frac{\hbar^2}{2m} \nabla^2 \right) \psi_\sigma(\mathbf{r}) + \lambda \, \psi_\downarrow^\dagger \psi_\uparrow^\dagger \psi_\uparrow \psi_\downarrow, \qquad (23.2)$$

where we took into account the fact that terms containing products of fermionic fields with the same spin vanish. We call the theory described by this Lagrangean the BCS field theory.

Both in (23.1) and (23.2), the key condition, expressed by (4.82) and necessary for the occurrence of an attractive interaction leading to a superconducting phase, is implicitly understood. Such a condition is that the electrons described by the Lagrangean and Hamiltonian above belong to a shell of width of the order $\hbar\omega_D$ around the Fermi surface, where ω_D is the Debye frequency. We are going to explicitly impose such condition below.

The corresponding vacuum-functional is given by

$$\mathcal{Z} = \frac{1}{\mathcal{Z}_{0,\psi}} \int \mathcal{D}\psi_\sigma^\dagger \mathcal{D}\psi_\sigma \exp \left\{ i \int d^4x \mathcal{L}_{BCS}[\psi_\sigma] \right\}. \qquad (23.3)$$

It is convenient to transform the quartic interaction into a trilinear one in the functional integral above. This is achieved by means of the Hubbard–Stratonovitch transformation. For this purpose, consider a complex field η, having a quadratic action, and multiply and divide Eq. (23.3) by the η field vacuum-functional as follows:

$$\mathcal{Z} = \frac{1}{\mathcal{Z}_\eta \mathcal{Z}_{0,\psi}} \int \mathcal{D}\psi_\sigma^\dagger \mathcal{D}\psi_\sigma \mathcal{D}\eta^* \mathcal{D}\eta \, \exp \left\{ i \int d^4x \, \mathcal{L}_0[\psi_\sigma, \eta] \right\}, \qquad (23.4)$$

where

$$\mathcal{L}_0[\psi_\sigma, \eta] = \mathcal{L}_{BCS}[\psi_\sigma] - \frac{1}{\lambda} \, \eta^* \eta. \qquad (23.5)$$

The elimination of the quartic term then follows upon shifting the η functional integration variable in (23.4) as

$$\begin{aligned} \eta &\to \eta - \lambda \psi_\uparrow \psi_\downarrow \\ \eta^* &\to \eta^* - \lambda \psi_\downarrow^\dagger \psi_\uparrow^\dagger, \end{aligned} \qquad (23.6)$$

an operation that leaves the functional integration measure invariant. The final Lagrangean appearing in the exponent of the integrand in (23.4) thus becomes

$$\mathcal{L}[\psi_\sigma, \eta] = \psi_\sigma^\dagger(\mathbf{r}) \left(i\hbar \frac{\partial}{\partial t} - \frac{\hbar^2}{2m} \nabla^2 \right) \psi_\sigma(\mathbf{r}) - \frac{1}{\lambda} \, \eta^* \eta + \eta^* \psi_\uparrow \psi_\downarrow + \eta \psi_\downarrow^\dagger \psi_\uparrow^\dagger. \quad (23.7)$$

From this Lagrangean, varying with respect to the auxiliary field and its complex conjugate, we obtain the field equations for such field, namely

$$\eta = \lambda \, \psi_\uparrow \psi_\downarrow$$
$$\eta^* = \lambda \, \psi_\downarrow^\dagger \psi_\uparrow^\dagger. \tag{23.8}$$

From this we may infer that η is the Cooper pair field. Consequently, its vacuum expectation value expresses the Cooper pair density in the ground state and therefore serves as an order parameter for superconductivity.

We now take advantage of the fact that the Lagrangean became quadratic in the fermion fields and integrate over these, thereby obtaining an effective action in terms of the Cooper Pair field η. For this purpose, we introduce the Nambu fermion field $\Phi^\dagger = (\psi_\downarrow^\dagger \; \psi_\uparrow)$. In terms of this we can rewrite (23.7) as

$$\mathcal{L}[\Psi, \eta] = -\frac{1}{\lambda} \, \eta^* \eta + \Phi^\dagger \mathcal{A} \Phi, \tag{23.9}$$

where the matrix $\mathcal{A}$ is given, in momentum space, by

$$\mathcal{A} = \begin{pmatrix} \xi(\mathbf{k}) + \hbar\omega & \eta \\ \eta^* & \xi(\mathbf{k}) - \hbar\omega \end{pmatrix} \tag{23.10}$$

with $\xi(\mathbf{k}) = \frac{\hbar^2 \mathbf{k}^2}{2m} - \mu$. Notice that we have subtracted the chemical potential μ in order to comply with the fact that the kinetic energy of the electrons playing an active role in the mechanism of superconductivity is expressed with respect to the Fermi level.

Upon integration over the fermion fields, we obtain the Cooper pair field vacuum functional, namely,

$$\mathcal{Z} = \frac{1}{\mathcal{Z}_{0,\eta}} \int \mathcal{D}\eta^* \mathcal{D}\eta \, e^{iS_{eff}[\eta]}, \tag{23.11}$$

where

$$S_{eff}[\eta] = \int d^4x \left(-\frac{1}{\lambda} |\eta|^2 \right) - i \ln \mathrm{Det} \left[\frac{\mathcal{A}[\eta]}{\mathcal{A}[0]} \right]. \tag{23.12}$$

The determinant of the matrix $\mathcal{A}$ is

$$\det \mathcal{A}[\eta] = \xi^2 - (\hbar\omega)^2 - |\eta|^2, \tag{23.13}$$

hence the above expression becomes

$$S_{eff}[\eta] = \int d^4x \left(-\frac{1}{\lambda} |\eta|^2 \right) - i \mathrm{Tr} \ln \left[1 + \frac{|\eta|^2}{\partial_t^2 + \left(-\frac{\hbar^2 \nabla^2}{2m} - \mu \right)^2} \right]. \tag{23.14}$$

23.2 A Dynamically Generated Effective Potential

From (23.14), we may extract the effective potential. This is a function of the vacuum expectation value of the Cooper pair field, namely $\Delta = \langle 0|\eta|0\rangle$, which is the order parameter for the superconducting phase. We get

$$V_{\text{eff}}\left(|\Delta|, T\right) = \frac{|\Delta|^2}{\lambda} - T \int \frac{d^3k}{(2\pi)^3} \sum_{n=-\infty}^{\infty} \left\{ \ln \left[1 + \frac{|\Delta|^2}{\omega_n^2 + \xi(\mathbf{k})^2} \right] \right\}, \quad (23.15)$$

where the sum runs over fermionic Matsubara frequencies, corresponding to the fermionic functional integration. This effective potential derives from the electronic interaction generated by the electron-phonon coupling and replaces the phenomenological quartic potential contained in the Landau–Ginzburg theory.

We now impose the condition that the only electron states participating of the above summation and integral are the ones located within a distance of the order $\hbar\omega_D$ from the Fermi surface. Calling $N(\xi)$ the density of states at ξ, it follows that integration on the momenta around the Fermi surface is written as

$$\int \frac{d^3k}{(2\pi)^3} = \int_{-\hbar\omega_D}^{\hbar\omega_D} d\xi\, N(\xi) \simeq N(E_F) \int_{-\hbar\omega_D}^{\hbar\omega_D} d\xi, \quad (23.16)$$

where $N(E_F)$ is the density of states at the Fermi surface and the last step follows from the fact that $\hbar\omega_D \ll E_F$. The way of evaluating the momentum integrals in (23.15), outlined above, guarantees that the condition for having a phonon-mediated attractive interaction, effectively described by (4.82), has been enforced.

The different phases of our system correspond to stable equilibrium points of the effective potential. Hence, the phase diagram is determined by

$$\frac{\partial}{\partial|\Delta|} V_{\text{eff}}\left(|\Delta|, T\right) = 0 \qquad \frac{\partial^2}{\partial|\Delta|^2} V_{\text{eff}}\left(|\Delta|, T\right) > 0. \quad (23.17)$$

The first condition implies (from now on, we take $\hbar = 1$)

$$0 = 2|\Delta| \left\{ \frac{1}{\lambda} - T N(E_F) \int_{-\omega_D}^{\omega_D} d\xi \sum_{n=-\infty}^{\infty} \left[\frac{1}{\omega_n^2 + \xi^2 + |\Delta|^2} \right] \right\}. \quad (23.18)$$

Carrying on the Matsubara sum, we get

$$0 = 2|\Delta| \left\{ \frac{1}{\lambda} - N(E_F) \int_0^{\omega_D} d\xi \frac{\tanh\left(\frac{\sqrt{\xi^2 + |\Delta|^2}}{2T}\right)}{\sqrt{\xi^2 + |\Delta|^2}} \right\}. \quad (23.19)$$

Let us investigate the possibility for the onset of a superconducting phase, namely, a phase with $|\Delta| \neq 0$. In this case the expression between brackets in (23.19) must vanish, thereby leading to the gap equation:

$$1 = \lambda N(E_F) \int_0^{\omega_D} d\xi \, \frac{\tanh\left(\frac{\sqrt{\xi^2 + |\Delta|^2}}{2T}\right)}{\sqrt{\xi^2 + |\Delta|^2}}. \tag{23.20}$$

Now, the second derivative (23.17) evaluated at the $|\Delta| \neq 0$ solution is

$$\frac{\partial^2}{\partial |\Delta|^2} V_{\text{eff}}(|\Delta|, T) = |\Delta| N(E_F) I(|\Delta|) > 0$$

$$I(|\Delta|) = \int_0^{\omega_D} \frac{d\xi}{f(\xi)} \left[\sinh\left(\frac{\sqrt{\xi^2 + |\Delta|^2}}{2T}\right) - \frac{\sqrt{\xi^2 + |\Delta|^2}}{2T} \right] > 0, \tag{23.21}$$

where

$$f(\xi) = \cosh\left(\frac{\sqrt{\xi^2 + |\Delta|^2}}{2T}\right) \left(\frac{\sqrt{\xi^2 + |\Delta|^2}}{2T}\right)^{3/2} > 0. \tag{23.22}$$

Notice that the second derivative in (23.21) is positive whenever $|\Delta| \neq 0$, because the integrand in the expression of $I(|\Delta|)$ is always positive. The superconducting phase therefore will occur below a critical temperature T_c above which the gap Δ vanishes. T_c, therefore, can be determined from (23.20), by imposing $\Delta = 0$. Then, performing the change of variable $y = \frac{\xi}{2T}$ and integrating by parts, one readily gets (re-instating $\hbar$)

$$\frac{1}{\lambda N_{E_F}} = \ln\left(\frac{2\gamma \hbar \omega_D}{\pi T_c}\right), \tag{23.23}$$

or

$$T_c = \frac{2\gamma}{\pi} \hbar \omega_D e^{-\frac{1}{\lambda N_{E_F}}}, \tag{23.24}$$

where $\gamma = e^C$ is the exponential of the Euler constant, $C \simeq 0.577$.

An interesting feature of BCS superconductivity is that any system describable by the BCS Hamiltonian will be in a superconducting phase for $T < T_c$, irrespective of the value of the coupling parameter λ. As we will see, this is no longer valid for Dirac-BCS systems. The gap at zero temperature, namely $|\Delta_0|$, in particular will be always non-vanishing for conventional BCS systems. It can be easily determined from the gap equation by making $T = 0$ in (23.20). The result is

$$|\Delta_0| = 2\hbar \omega_D e^{-\frac{1}{\lambda N_{E_F}}}, \tag{23.25}$$

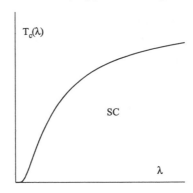

Figure 23.1 The BCS critical temperature as a function of the coupling strength. For any $\lambda > 0$, there will be a finite T_c, below which the system is in an SC phase.

in such a way that the famous ratio

$$\frac{T_c}{|\Delta_0|} = \frac{\gamma}{\pi} \tag{23.26}$$

is a universal number for conventional BCS superconductors.

The exponential factor in the expression for T_c is responsible for the fact that conventional BCS superconductivity occurs at considerably low temperatures. The same factor tends to preclude the onset of this form of superconductivity at high temperatures. Expression (23.24) for the critical temperature also explains the so-called isotopic effect, namely, the observed modification of T_c upon substitution of the material constituents by isotopes of different mass, according to

$$\frac{T_c}{T_c'} = \left(\frac{M'}{M}\right)^{1/2}. \tag{23.27}$$

The above relation follows directly from (23.24), if we realize that the Debye frequency is inversely proportional to the square-root of the ion mass, namely $\omega_D \propto M^{-1/2}$.

23.3 Superconductivity of Dirac Electrons

Let us consider in this section the interesting situation of electrons possessing kinematical properties described by the massless Dirac equation, and yet undergoing an effective interaction described by the BCS Lagrangean. The former property derives from the peculiar form of the crystal structure on which the electrons move and is known to occur in many advanced materials such as graphene, iron pnictides and transition metal dichalcogenides, among others. The BCS interaction, by its turn, can be mediated by the exchange of phonons, as we saw before; however, we do not exclude the possible existence of a different mechanism producing such effective interaction. We shall actually see that many features of this form of superconductivity strongly suggest that its underlying mechanism does not involve phonons.

23.3.1 The BCS-Dirac Lagrangean

The system we have in mind contains electrons with kinematics described by the Dirac massless equation in two-dimensional space and that are therefore associated to a Dirac field, which we consider to have two components:

$$\psi_{\sigma,a} = \begin{pmatrix} \psi_{1,\sigma,a} \\ \psi_{2,\sigma,a,} \end{pmatrix}, \tag{23.28}$$

where $\sigma = \uparrow, \downarrow$ are the two spin components and $a = 1, \ldots, N$ is a "flavor" index specifying other electronic attributes, such as the layer, band or valley to which the electron belongs.

The so-called relativistic BCS Lagrangean density combining relativistic kinematics with the BCS interaction, analogous to (23.2), is given by [15]

$$\mathcal{L}_{RBCS} = i\overline{\psi}_{\sigma a} \, \not{\partial} \, \psi_{\sigma a} + \frac{\lambda}{N} \left(\psi_{1\uparrow a}^\dagger \psi_{2\downarrow a}^\dagger + \psi_{2\uparrow a}^\dagger \psi_{1\downarrow a}^\dagger \right) \left(\psi_{2\downarrow b} \psi_{1\uparrow b} + \psi_{1\downarrow b} \psi_{2\uparrow b} \right). \tag{23.29}$$

In this expression, $\not{\partial} = \gamma^\mu \partial_\mu$ and $\overline{\psi}_{\sigma a} = \psi_{\sigma a}^\dagger \gamma^0$, with the two-dimensional Dirac matrices given by

$$\gamma^0 = \sigma^z, \quad \gamma^0 \gamma^1 = \sigma^x, \quad \gamma^0 \gamma^2 = \sigma^y, \tag{23.30}$$

where the σs are Pauli matrices. We envisage a system in two spatial dimensions, and for that reason, consider only three Dirac gamma-matrices.

As in the non-relativistic case, we introduce an auxiliary complex scalar field η and carry out a Hubbard–Stratonovitch transform leading to the Lagrangean

$$\mathcal{L}\left[\psi_{\sigma,a}, \eta\right] = \overline{\psi}_{\sigma a} \, \not{\partial} \, \psi_{\sigma a} - \frac{1}{\lambda} \eta^* \eta + \eta^* \left(\psi_{2\downarrow b} \psi_{1\uparrow b} + \psi_{1\downarrow b} \psi_{2\uparrow b} \right)$$
$$+ \eta \left(\psi_{1\uparrow a}^\dagger \psi_{2\downarrow a}^\dagger + \psi_{2\uparrow a}^\dagger \psi_{1\downarrow a}^\dagger \right). \tag{23.31}$$

This implies the following field equation for the auxiliary field η

$$\eta = \lambda \left(\psi_{2\downarrow a} \psi_{1\uparrow a} + \psi_{1\downarrow a} \psi_{2\uparrow a} \right)$$
$$\eta^* = \lambda \left(\psi_{1\uparrow a}^\dagger \psi_{2\downarrow a}^\dagger + \psi_{2\uparrow a}^\dagger \psi_{1\downarrow a}^\dagger \right). \tag{23.32}$$

As we did in the non-relativistic case, we now perform the quadratic functional integration over the fermionic fields, thereby obtaining an effective action for the Cooper pair field η, which includes the effects of the interaction, as well as those produced by quantum fluctuations. We, therefore arrive at a vacuum partition functional similar to (23.11), but with the effective action given by

$$S_{eff}[\eta] = \int d^3x \left(-\frac{N}{\lambda}|\eta|^2 \right) - i2N \mathrm{Tr} \ln \left[1 + \frac{|\eta|^2}{\partial_t^2 - v_F^2 \nabla^2} \right]. \qquad (23.33)$$

Notice that we are working with zero chemical potential, $\mu = 0$, in compliance to the fact that the Fermi level is located at the vertex of the Dirac cones at $E = 0$. Also, observe that we are working in two spatial dimensions, for the purpose of modelling most of materials exhibiting Dirac electrons, which are essentially two-dimensional. We also assume the system has a natural energy cutoff Λ, which is provided by the lattice itself. Indeed, we have $\Lambda \simeq \hbar v_F / a$, where a is the lattice parameter. Observe that in conventional BCS theory a natural cutoff also exists, which is the Debye energy, which, by the way, is of the same order of Λ.

23.4 The Effective Potential at $T \neq 0$: the Phase Diagram

The effective potential corresponding to (23.33) is given by [15]

$$V_{\mathrm{eff}}(|\Delta|, T) = \frac{|\Delta|^2}{\lambda} - 2T \int \frac{d^2k}{(2\pi)^2} \sum_{n=-\infty}^{\infty} \left\{ \ln \left[1 + \frac{|\Delta|^2}{\omega_n^2 + v_F^2 \mathbf{k}^2} \right] \right\}, \qquad (23.34)$$

where, again, the sum runs over fermionic Matsubara frequencies.

Again the stable phases will be determined by Eqs. (23.17), hence, after performing the Matsubara sum, we conclude that the superconducting gap Δ must satisfy

$$\frac{\partial V_{\mathrm{eff}}}{\partial |\Delta|} = 2|\Delta| \left\{ \frac{1}{\lambda} - \frac{1}{\alpha} \int_{|\Delta|}^{\sqrt{|\Delta|^2 + \Lambda^2}} dy \tanh \left(\frac{y}{2T} \right) \right\} = 0, \qquad (23.35)$$

where $\alpha = 2\pi v_F^2$ and we introduced the physical cutoff Λ.

For a stable state, characterizing a phase, the second derivative at the nonzero solution of the equation above must be positive, hence

$$\frac{\partial^2 V_{\mathrm{eff}}}{\partial |\Delta|^2} = \frac{2|\Delta|^2}{\alpha} \left\{ \frac{\tanh \left(\frac{|\Delta|}{2T} \right)}{|\Delta|} - \frac{\tanh \left(\frac{\sqrt{|\Delta|^2 + \Lambda^2}}{2T} \right)}{\sqrt{|\Delta|^2 + \Lambda^2}} \right\} > 0. \qquad (23.36)$$

This is always satisfied, provided $|\Delta| \neq 0$, because the function $\frac{\tanh x}{x}$ is monotonically decreasing.

Solving the gap equation (23.35), we obtain an implicit equation for the gap, namely [15]

$$|\Delta|(T) = 2T \cosh^{-1}\left[e^{-\frac{\alpha}{2T\lambda}} \cosh\left[\frac{\sqrt{\Delta^2(T) + \Lambda^2}}{2T} \right] \right]. \qquad (23.37)$$

From this we can derive an implicit equation determining the critical temperature above which the gap would vanish. Indeed, taking $\Delta \to 0$ in the above equation, we get

$$\cosh\left(\frac{\Lambda}{2T_c} \right) = e^{\frac{\alpha}{2T_c\lambda}}, \qquad (23.38)$$

from which we can extract T_c as a function of the coupling parameter λ.

We can now obtain an expression for the gap as a function of the temperature in the regime where $|\Delta|, T_c \ll \Lambda$. Indeed, by inserting (23.38) in (23.37), one straightforwardly obtains

$$|\Delta|(T) = 2T \cosh^{-1}\left\{ \left[\frac{2}{1 + e^{-\Lambda/T_c}} \right]^{\frac{T_c}{T} - 1} \right\}. \qquad (23.39)$$

The fact that a nonzero Δ exists is the sign that the system is in a superconducting phase, both for Dirac and regular electrons.

23.5 The Onset of Superconductivity at $T = 0$: A Quantum Phase Transition

Let us investigate now the superconducting transition at $T = 0$. Taking (23.37) in the limit $T \to 0$, we obtain the gap $\Delta_0 = \Delta(T = 0)$:

$$|\Delta_0| = \frac{\alpha\lambda}{2}\left[\frac{1}{\lambda_c^2} - \frac{1}{\lambda^2} \right], \qquad (23.40)$$

where $\lambda_c = \frac{\alpha}{\Lambda}$. This should be compared with (23.25). Observe that for a superconducting gap to occur at zero temperature, the coupling parameter λ must be larger than a threshold value λ_c. This characterizes a quantum phase transition, namely a transition that is produced by quantum fluctuations rather than by thermal fluctuations and therefore occurs already at $T = 0$.

We can actually see that, below this threshold, a superconducting phase will not occur at any temperature. Indeed, from (23.38) we obtain

$$T_c = \frac{\Lambda\left(1 - \frac{\lambda_c}{\lambda}\right)}{2\ln\left[\frac{2}{1 + e^{-\Lambda/T_c}} \right]} \xrightarrow{\Lambda \gg T_c} \frac{\Lambda\left(1 - \frac{\lambda_c}{\lambda}\right)}{2\ln 2}, \qquad (23.41)$$

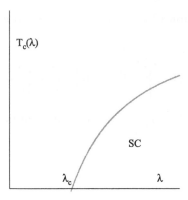

Figure 23.2 The SC transition temperature, T_c for Dirac electrons, showing the existence of a threshold coupling λ_c below which there is no SC phase, contrary to the case of usual electrons. To be compared with Fig. 23.1

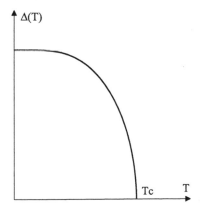

Figure 23.3 The SC gap as a function of temperature, for Dirac electrons, showing the critical temperature T_c, above which it vanishes. Notice that a nonzero T_c requires a coupling $\lambda > \lambda_c$.

which should be compared with (23.24). Since the denominator is always positive, we will only have a positive solution for the critical temperature when $\lambda > \lambda_c$.

Inserting (23.41) in (23.39), we obtain a clean expression for the superconducting gap of Dirac electrons as a function of the cutoff Λ, the critical coupling λ_c and the critical temperature T_c, namely

$$|\Delta|(T) = 2T \cosh^{-1}\left\{ \exp\left[\frac{\Lambda}{2T_c}\left(1 - \frac{\lambda_c}{\lambda}\right)\left(\frac{T_c}{T} - 1\right)\right]\right\}. \tag{23.42}$$

Notice that the gap will vanish at the critical points $\lambda = \lambda_c$ and $T = T_c$. Conversely it will only be different from zero when the argument of the inverse hyperbolic cosine is larger than one, namely at $\lambda > \lambda_c$ and $T < T_c$.

Near the transition ($T < T_c$; $\lambda > \lambda_c$) the gap will behave as

$$|\Delta|(T) \xrightarrow{T \sim T_c} 2\sqrt{\Lambda T_c}\left(1 - \frac{\lambda_c}{\lambda}\right)^{\frac{1}{2}}\left(1 - \frac{T}{T_c}\right)^{\frac{1}{2}}. \qquad (23.43)$$

This implies the superconducting transition is of second order with a critical exponent 1/2. Notice that we obtain, from the microscopic theory, the same critical behavior that was advanced by the phenomenological Landau–Ginzburg approach in Eq. (4.72).

Now, from (23.39) we can obtain a relation between the critical temperature and the gap at $T = 0$. Indeed, taking the zero temperature limit in that equation, we get [15]

$$\frac{T_c}{|\Delta_0|} = \frac{1}{2\ln\left[\frac{2}{1+e^{-\Lambda/T_c}}\right]} \xrightarrow{\Lambda \gg T_c} \frac{1}{2\ln 2}, \qquad (23.44)$$

which should be compared with (23.26). Notice that also for Dirac electrons, we obtain an universal relation between the critical temperature and the gap at $T = 0$, provided, of course this in non-vanishing.

The behavior of Dirac electrons with respect to superconducting properties is radically different from that found in regular metals, where a positive value of T_c would be always found for the whole range of the coupling parameter λ, as we can clearly see in (23.24). Another crucial difference is the fact that in regular BCS superconductors, we have a finite density of states N_{E_F} at the Fermi level. These play an important role in the mechanism of conventional BCS superconductivity, as we can infer from (23.24) and also from the form of the effective electron-phonon interaction responsible for this phenomenon: Eqs. (3.62) and (4.82). This is a manifestation of the Cooper theorem, according to which, for electrons around a Fermi surface there will always exist a superconducting phase below a certain critical temperature T_c, no matter how weak the attractive interaction might be, or, equivalently, no matter how small the coupling parameter λ might be. In the case of Dirac electrons, (23.24) must be replaced by (23.41) and we immediately see that a superconducting phase will only occur when the coupling parameter is larger than a threshold value given by λ_c.

The above study strongly indicates that, even though, for the description of superconductivity of Dirac electrons, we used same effective quartic BCS interaction usually employed for describing this phenomenon in regular metals, the underlying mechanism must be completely different in each case. For conventional electrons forming a Fermi surface, we have seen that the phonon-mediated interaction does indeed lead to the BCS effective interaction. In the case of systems with Dirac electrons, conversely, the absence of a Fermi surface, the existence of a lower coupling threshold, showing that the Cooper theorem does not apply, among other features, unequivocally points to a novel mechanism, not involving phonons, behind the attractive electronic effective interaction.

23.6 The Effect of an External Magnetic Field

We considered in Chapter 4 the consequences of the onset of a superconducting phase in a material under the action of an external magnetic field. It was seen that the magnetic field is deeply affected by the superconducting state of the sample, actually being completely or partially expelled from the region where a persistent current exists. By the same token, one should expect that the presence of a magnetic field will strongly affect the superconducting properties of the material.

For this purpose, let us investigate in this subsection the effects an applied external magnetic field will produce in the superconducting state itself [16]. We shall pursue this study in the framework of Dirac electrons constrained to move on a plane, hence we take as our starting point the BCS-Dirac Lagrangean, represented by (23.29).

We shall consider here that system in the presence of a constant and uniform external magnetic field $\mathbf{B} = \nabla \times \mathbf{A}$, applied perpendicularly to the plane where the electrons move, which corresponds to

$$\mathcal{L}_{RBCS}[\mathbf{B}] = i\,\overline{\psi}_{\sigma a}\left[\gamma^0 \partial_0 + v_F \gamma^i \left(\partial_i + i\frac{e}{c}A_i\right)\right]\psi_{\sigma a} - \mu_B \psi_{\sigma a}^\dagger\,(\mathbf{B}\cdot\vec{\sigma})\,\psi_{\sigma a}$$
$$+ \frac{\lambda}{N}\left(\psi_{1\uparrow a}^\dagger\,\psi_{2\downarrow a}^\dagger + \psi_{2\uparrow a}^\dagger\,\psi_{1\downarrow a}^\dagger\right)\left(\psi_{2\downarrow b}\,\psi_{1\uparrow b} + \psi_{1\downarrow b}\,\psi_{2\uparrow b}\right),\ (23.45)$$

where μ_B is the magneton-Bohr.

As we did before, it is convenient to perform a Hubbard–Stratonovitch transformation in order to transform the quartic interaction into a trilinear one, involving two fermion fields and the Cooper pair complex scalar field η. Thereafter we may integrate over the fermion fields, thereby obtaining an effective action for the η-field. From this, we may derive the expression of the effective potential, which is a function of the vacuum expectation value $\Delta = \langle 0|\eta|0\rangle$ and the magnetic field $\mathbf{B}$. We choose the vector potential corresponding to this magnetic field as $\mathbf{A} = B(0, x)$.

Proceeding as we did in Subsection 19.2.1, we obtain the effective action [16]

$$S_{eff}[\eta, B] = -\frac{N}{\lambda}\int d^3x |\eta|^2$$
$$- i2N\,\text{Tr}\ln\left[1 + \frac{|\eta|^2}{(\partial_t + \mu_B B)^2 - v_F^2(\nabla + i\frac{e}{c}\mathbf{A})^2}\right],\qquad (23.46)$$

and out of this we derive the effective potential

$$V_{\text{eff}}(|\Delta|, B) = \frac{|\Delta|^2}{\lambda} - 2T \int \frac{d^2k}{(2\pi)^2} \sum_{n=-\infty}^{\infty} \left\{ \ln\left[(\omega_n + \mu_B B)^2 \right. \right.$$

$$\left. \left. + v_F^2 \left[k_x^2 + \left(k_y + \frac{e}{c} B\langle x \rangle \right)^2 \right] + |\Delta|^2 + gB \right] - \ln\left[\omega_n^2 + v_F^2 |k|^2 \right] \right\},$$

$$(23.47)$$

where $g = v_F^2(e/c)$ and the sum runs over fermionic Matsubara frequencies.

In the above expression, we have replaced the y-component of the applied vector potential, $A_y = Bx$, by its average value, $\langle A_y \rangle = B\langle x \rangle$. The details of how this average is computed actually are not important since, being a constant, it can be shifted away through a change of variable in the k_y integral.

In order to determine the phase diagram in the presence of the magnetic field, we must look for the minima of the effective potential. Taking the $|\Delta|$-derivative and performing the Matsubara sum, we get

$$\frac{\partial V_{\text{eff}}}{\partial |\Delta|} = 2|\Delta| \left\{ \frac{1}{\lambda} - \int \frac{d^2k}{(2\pi)^2} \frac{1}{E} \left[\frac{\sinh(\beta E)}{\cosh(\beta E) + \cosh(\beta \mu_B B)} \right] \right\} = 0, \quad (23.48)$$

where $E = \sqrt{v_F^2 k^2 + |\Delta|^2 + gB}$ and $\beta = 1/T$

From (23.48) we derive the gap equation [16]

$$1 = \frac{\lambda}{\alpha} \int_{\sqrt{|\Delta|^2 + gB}}^{\Lambda} dy \, \frac{\sinh(\beta y)}{\cosh(\beta y) + \cosh(\beta \mu_B B)}, \quad (23.49)$$

where Λ/v_F is the high-momentum cutoff, assumed to be much larger than the lower integration limit of the above integral.

The condition for a stable minimum is given by

$$\frac{\partial^2 V_{\text{eff}}}{\partial |\Delta|^2} = \frac{2|\Delta|^2}{\alpha\beta\sqrt{|\Delta|^2 + gB}} \left[\frac{\sinh\beta\sqrt{|\Delta|^2 + gB}}{\cosh\beta\sqrt{|\Delta|^2 + gB} + \cosh\beta\mu_B B} \right] > 0, \quad (23.50)$$

which is always satisfied for a non-vanishing solution $|\Delta|$.

Performing the y-integral in (23.49), we get an equation for the gap, namely [16]

$$\Delta^2(T, B) = T^2 \left\{ \cosh^{-1} \left\{ e^{-\frac{\alpha}{T\lambda}} \left[\cosh\left(\frac{\Lambda}{T} \right) + \cosh\left(\frac{\mu_B B}{T} \right) \right] \right. \right.$$

$$\left. \left. - \cosh\left(\frac{\mu_B B}{T} \right) \right\} \right\}^2 - gB. \quad (23.51)$$

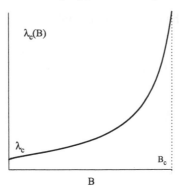

Figure 23.4 The critical coupling increases with a magnetic field and eventually diverges at a critical upper field B_c, above which, consequently, the SC phase is destroyed.

From this equation we can obtain the gap at zero temperature. Indeed, making $T = 0$ above, we obtain

$$|\Delta_0| = \sqrt{\alpha^2 \left(\frac{1}{\lambda_c} - \frac{1}{\lambda}\right)^2 - gB}. \qquad (23.52)$$

This will vanish for $\lambda < \lambda_c(B)$, where the new, field-dependent critical coupling is given by

$$\lambda_c(B) = \frac{\lambda_c}{1 - \lambda_c \frac{\sqrt{gB}}{\alpha}}. \qquad (23.53)$$

Notice that the applied magnetic field increases the critical coupling or, in other words, in the presence of an external magnetic field, a stronger interaction is required for producing a superconducting phase.

By making $|\Delta| = 0$ in (23.51), we can determine the critical temperature [16]

$$T_c(B) = \frac{\sqrt{|\Delta_0|^2 + gB}}{\ln\left\{2\left[\cosh\left(\frac{\sqrt{gB}}{T_c}\right) + \cosh\left(\frac{\mu_B B}{T_c}\right)\right]\right\}}. \qquad (23.54)$$

Notice that for $B = 0$, we have $\Delta_0/T_c = 2\ln 2$, in agreement with the result obtained in the previous subsection. Observe that $T_c(B) < T_c(0)$, and we conclude that the magnetic field reduces the critical temperature for the onset of a superconducting phase. In the above expression, we may clearly distinguish the contributions provenient of the coupling of the magnetic field to orbital and spin degrees of freedom.

Expressing the gap in terms of T_c, we have

$$\Delta^2(T, B) = \left\{ T \cosh^{-1} \left\{ 2^{(\frac{T_c}{T}-1)} \left[\cosh \left(\frac{\sqrt{gB}}{T_c} \right) + \cosh \left(\frac{\mu_B B}{T_c} \right) \right]^{T_c/T} \right. \right.$$
$$\left. \left. - \cosh \left(\frac{\mu_B B}{T} \right) \right\} \right\}^2 - gB. \qquad (23.55)$$

Observe that the right-hand side of the above equation vanishes for $T \to T_c$, implying that the superconducting gap will be zero above this temperature, as it should.

We can now determine the critical magnetic field, namely the field threshold above which the gap would vanish. For temperatures near T_c, where the critical field is expected to be small, a B-expansion in (23.55) leads to [16]

$$B_c(T) \sim 8 \ln 2 \, T_c(0)^2 \left(1 - \frac{T}{T_c(0)} \right), \qquad (23.56)$$

where $T_c(0)$ is given by (23.41). The critical field shows a linear dependence on the temperature near the transition point. This is in clear contrast to conventional BCS theory, where the critical field decays quadratically with the temperature.

23.7 Overview

We have demonstrated in this chapter, using the efficient methodology of quantum field theory, the occurrence of a nonvanishing quantum average of Cooper pair density, in different systems, under different regimes of temperature, interaction strength, coupling parameter and magnetic field. This charge carrier density, which was the central target of the calculations pursued here, is represented mathematically by the gap parameter Δ.

Notice that, whenever $|\Delta|$ is non-vanishing, in the many examples seen above, it is always a constant. This implies the charge carriers form the incompressible fluid required for the existence of a persistent current, or equivalently, for a superconducting phase.

We have examined in detail the specific conditions required for a phase displaying a nonzero Δ to occur, both in conventional and unconventional Dirac-BCS superconductors, determining in particular the critical expressions of temperature, coupling strength and magnetic field leading to such a phase.

Several results, such as the critical temperature dependence on the coupling parameter, the existence of a lower coupling threshold, the linear dependence of the critical magnetic field on the temperature, the occurrence of a superconducting phase even in the absence of a Fermi surface, among others, strongly indicate that for Dirac electrons a new mechanism not involving phonons must be responsible for superconductivity.

Furthermore, the linear dependence of the critical temperature on the inverse coupling parameter, which we found in Dirac systems, rather than the exponentially decaying dependence that was found in the conventional BCS case, points toward the concrete possibility of attaining ever higher critical temperatures in these novel superconductors.

23.8 The Anderson–Higgs–Meissner Mechanism

We cannot close a chapter on quantum field theory methods in superconductivity without describing the close relationship existing between this phenomenon, more specifically the Meissner effect, and the Anderson–Higgs mechanism, which plays a fundamental role in the Standard Model of the fundamental interactions.

The Landau–Ginzburg Theory Revisited

We have derived in this chapter, starting from a basic microscopic theory, results which were also described by the phenomenological Landau–Ginzburg theory introduced in Chapter 4. These included the existence of a constant gap parameter $|\Delta|$ corresponding to the Landau–Ginzburg order parameter $|\Psi|^2$, namely, the density of charge carriers.

The accurate description of the phenomenon of superconductivity, described here to a large extent relying on the properties of $|\Delta|$, is a consequence of the dynamics underlying the electronic interaction. The dynamics, by itself, leads to the resulting superconducting phase, provided the temperature and perhaps coupling are in a certain range.

The same features, however, can emerge from the phenomenological Landau–Ginzburg theory as a consequence of the judicious choice of the parameters found in the quartic potential appearing in the free-energy Eq. (4.68). Yet, the microscopic theory, of course, offers a deeper and further-reaching comprehension of the system, which will certainly allow a higher degree of control.

A side effect of the existence of a nonzero average of charge carriers, which by the way is responsible for the occurrence of persistent currents, as we have seen, is that the gauge field that is coupled to these charge carriers becomes exponentially damped, with the damping parameter being precisely proportional to the charge carrier average density. The exponential damping of the gauge field, by its turn, was the explanation for the Meissner effect, namely, the expulsion of a magnetic field from a superconducting region.

Anderson–Higgs–Meissner Effect

The Anderson–Higgs mechanism is a procedure by which a gauge field coupled to a scalar field becomes exponentially damped, provided this scalar field possesses a nonzero vacuum expectation value. The damping parameter, then, is proportional to this vacuum expectation value. From a quantum-mechanical point of view, the particles consisting in the quantized version of this gauge field acquire a mass, which is also proportional to that vacuum expectation value.

This mechanism was required in the Standard Model of the fundamental interactions to account for the extremely short range of the weak interactions and for the massive nature of the intermediate bosons $W^\pm$ and Z.

We immediately see that the Anderson–Higgs mechanism is precisely the Meissner effect, except perhaps for the fact that the gauge group is $SU(2) \times U(1)$ in the former and $U(1)$ in the latter.

Let us look a bit closer to the mechanism. Assuming a non-abelian gauge field A_μ^a, $a = 1, \ldots, N$ corresponding to a symmetry group with generators T_{ij}^a, $a = 1, \ldots, N$, $i, j = 1, \ldots, M$ and a scalar field Φ_i, $i = 1, \ldots, M$, both transforming in the fundamental representation of this group. A "mass" term is given by

$$\frac{1}{2} M_{ab}^2 A_\mu^a A_\mu^b. \tag{23.57}$$

Such a term, at a classical level, would produce an exponential damping of the field, and at a quantum-mechanical level would produce particles with a mass given by the eigenvalues of the mass matrix M_{ab}.

The problem with this term is that it breaks gauge invariance and, without gauge invariance, we lose unitarity. Hence, we cannot explicitly introduce a mass term for the gauge field. Can we generate such a mass term without breaking gauge invariance? The Anderson–Higgs mechanism is the positive answer.

Consider a gauge invariant Lagrangean of the form

$$\mathcal{L} = (D_\mu \Phi)_i (D_\mu \Phi)_i - V(\Phi), \tag{23.58}$$

where the covariant derivative is given by

$$(D_\mu)_{ij} = \delta_{ij} \partial_\mu - ig A_\mu^a T_{ij}^a. \tag{23.59}$$

This Lagrangean will clearly contain a term

$$\frac{1}{2} \left[2 T_{ik}^a (\Phi)_k T_{il}^b (\Phi)_l \right] A_\mu^a A_\mu^b, = \frac{1}{2} \left[2 T_{ik}^a T_{il}^b \right] \Phi_k \Phi_l A_\mu^a A_\mu^b, \tag{23.60}$$

which in principle describes a quartic interaction involving the scalar and gauge fields. Now, suppose we choose the potential in (23.58) in such a way that the minimum occurs at a nonzero constant value $\langle \Phi_i \rangle = \varphi_i^0$. For the sake of stability, then, we must shift the scalar field around this constant value,

$$\Phi_i \to \Phi_i - \langle \Phi_i \rangle, \tag{23.61}$$

thereby generating a mass matrix

$$M_{ab}^2 = 2T_{ik}^a(\varphi^0)_k T_{il}^b(\varphi^0)_l. \tag{23.62}$$

In the U(1) case, Φ is the scalar field Ψ, the modulus of which represents the density of charge carriers. The mass matrix reduces to a scalar: the exponential damping factor, which is precisely the inverse penetration length responsible for the Meissner effect! We may therefore safely assert that the Meissner effect is the Anderson–Higgs mechanism for a U(1) symmetric field theory, which is the Landau–Ginzburg theory.

Let us now look into the past and inquire about the future. All results obtained originally from the Landau–Ginzburg theory were derived later on from a more complete microscopic theory, as we have shown in detail in this chapter. The Landau–Ginzburg or Cooper pair field, in particular, was shown to be composite of two electron fields. Should the same happen with the Anderson–Higgs boson?

It seems appealing, from the esthetic point of view, and natural from the historic perspective that the Higgs sector of the Standard Model constitutes a kind of phenomenological Landau–Ginzburg theory, which should be replaced by a more fundamental microscopic theory, where the mass generation mechanism would be produced by the dynamics, as it happen in the BCS theory.

24

The Cuprate High-Temperature Superconductors

Until 1986, the highest temperature at which superconductivity had been observed was $23.2K$ in Nb_3Ge. There was an issue as to whether BCS theory would impose an upper limit on the critical temperature for the onset of superconductivity. Actually, as remarked before, the exponential factor in (23.24) may be viewed as responsible for the relatively low temperatures of the BCS superconductors. That expression for T_c has in fact a mathematical upper limit, of the order of the Debye temperature, which is typically about a few hundred Kelvin. This, however, would correspond to an infinite value of the product $\lambda N(E_F)$ of coupling parameter and Fermi level density-of-states. The mathematical upper bound, consequently, is way above the temperatures corresponding to realistic physical values of parameters. In 1986, $La_{2-x}Ba_xCuO_4$ and $La_{2-x}Sr_xCuO_4$ (LSCO) where shown to present a superconducting phase at temperatures up to about $40K$. Soon after, $YBa_2Cu_3O_{6+x}$ (YBCO) was shown to have a superconducting phase up to $92K$. These were the first members of the cuprate family, which contains materials such as $HgBa_2Ca_2Cu_3O_{8+x}$, which exhibits a superconducting phase up to a temperature of the order of $130K$ without pressure and of about $160K$ under high pressure. Many indications suggest that the mechanism leading to superconductivity in cuprates is not the phonon-mediated BCS mechanism. There is general consensus on this point. Nevertheless, there is so far no agreement about the specific mechanism leading to the formation of Cooper pairs in cuprates, despite the many proposals that have been made. In this chapter, we keep mostly in the region where consensus has been reached, except for Section 24.4, which contains our own contribution to the subject.

24.1 Crystal Structure, CuO_2 Planes and Phase Diagram

A common feature of all materials belonging to the cuprates family is the presence of CuO_2 planes, consisting in a square lattice, which, before doping, contains

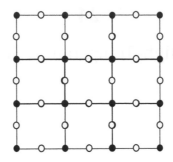

Figure 24.1 The CuO_2 planes, a common feature of all the high-Tc cuprate superconductors. Black dots and white dots are, respectively, Cu^{++} and O^{--} ions.

Cu^{++} ions on the sites and O^{--} ions on the links. The latter has a $2p^6$ electronic configuration, whereas the former is in a $3d^9$ configuration. The oxygen ions are in a noble gas configuration, while the copper ions have an unpaired electron, which creates a localized magnetic moment at each site. This crystal structure produces a super-exchange interaction that leads to an antiferromagnetic coupling among the unpaired electrons of the copper ions on the sites of the square lattice, which is mediated by the electrons in the oxygen ions.

The parent, undoped cuprates are therefore accurately described by a two-dimensional AF Heisenberg model on a square lattice. The unpaired electrons of the copper ion are the main actors at this stage. The oxygen ions only role, at this level, is to provide the super-exchange interaction between the neighboring copper electrons. The localized magnetic moments settle in the ordered Néel ground state, below a certain critical temperature $T_N(0)$. There is a clear dependence of T_c on the number of adjacent CuO_2 planes that are found in the cuprates unit cell. In the Hg sub-family, for instance, $HgBa_2CuO_{4+x}$, $HgBa_2CaCu_2O_{6+x}$ and $HgBa_2Ca_2Cu_3O_{8+x}$, which possess, respectively, one, two and three planes in the unit cell, the highest transition temperatures are, respectively, $94K$, $128K$ and $134K$. The same happens in the Bi and Tl sub-families.

Hole Doping

In between the CuO_2 planes there are atoms, such as La in LSCO, which in the undoped compounds have just a structural function. Most of the interesting physics of the cuprates, however, actually occurs when we dope the system by replacing a certain fraction of the atoms in between the planes with others having a different stoichiometry. In LSCO, for instance, we replace a fraction of the La atoms with Sr, which is an electron receptor. Charge balance is, then, achieved by the Sr atoms absorbing one electron from the CuO_2 planes. Experimental evidence [194] indicates that this electron comes from the O^{--} ions of the CuO_2 planes. By the

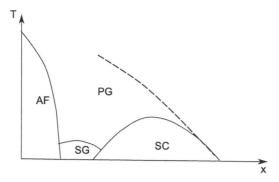

Figure 24.2 The schematic phase diagram of the hole-doped high-Tc cuprate superconductors, exhibiting the SC and AF domes, along with the spin-glass phase (SG) and the pseudogap line

doping process, therefore, we create holes in such planes, or more precisely, in the O^{--} ions located in the links of the square lattice. The presence of these quickly destroys the Néel order, as we can see from the doping-dependent Néel temperature $T_N(x)$ in the phase diagram.

By further doping, the cuprates eventually undergo a superconducting phase transition with a critical temperature $T_c(x)$, which starts at zero for some critical doping x_c, increases up to an optimal doping and eventually decays to zero, thereby forming a superconducting dome in the $T(x)$ phase diagram. This and the AF dome are two of the most important features of the phase diagram of the cuprates. There is clear experimental evidence that superconductivity in the cuprates occurs by the formation of Cooper pairs out of the holes created by the doping process in the O^{--} ions. The supercurrent, therefore, flows along the CuO_2 planes. The crucial issue is, what interaction produces the formation of Cooper pairs out of the holes?

The so-called pseudogap (PG) region of the phase diagram is characterized by a depletion of the electronic density of states and has been interpreted as a precursor to the SC phase. Below a temperature $T^*(x)$, the bound-state Cooper pairs would form in such a way that the complex SC order parameter would have the form

$$\langle \psi^\dagger \psi^\dagger \rangle = \Delta e^{i\theta}$$

with $\Delta \neq 0$. The phase θ, then, decouples and, for temperatures below T_{KT}, according to (18.152), the system undergoes a Berezinskii–Kosterlitz–Thouless transition to a regime where phase coherence sets in and the following thermal averages become different from zero:

$$\langle \cos \theta \rangle = \langle \sin \theta \rangle = 0 \Rightarrow \langle \cos \theta \rangle \neq 0 \; ; \; \langle \sin \theta \rangle \neq 0.$$

Effectively, therefore, the SC transition occurs at $T_c = T_{KT}$, below which

$$\Delta \langle e^{i\theta} \rangle \neq 0.$$

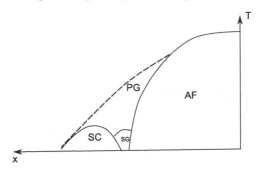

Figure 24.3 The schematic phase diagram of the electron-doped high-Tc cuprate superconductors, exhibiting a smaller SC dome and a larger AF one. Despite the existence of a clear asymmetry between hole- and electron-doped cuprates, yet the qualitative features of the phase diagram look the same for both.

The first term sets in below $T^*(x)$, while the second does it below $T_{KT} < T^*$, which is therefore the actual $T_c(x)$.

The shortage of available electrons, due to the pair formation, accounts for the depletion of the density of states.

Electron Doping

For some materials in the cuprates family, doping introduces electron donors in between the CuO_2 planes. Such is the case of Nd_2CuO_4, doped with Ce: $Nd_{2-x}Ce_xCuO_4$, which has a transition temperature $T_c = 24K$ at $x = 0.15$. Further examples are $Pr_{2-x}Ce_xCuO_4$ and $Sm_{2-x}Ce_xCuO_4$. In the case of electron doping, an extra electron is pumped into the CuO_2 planes, occupying the vacancy available in the Cu^{++} ions. These extra electrons pair with the copper ions' unpaired electron, thus destroying the local magnetic moment and, eventually, the Néel state. Upon further doping, these extra electrons form Cooper pairs and a superconducting phase sets in. Despite the qualitative symmetry existing between hole- and electron-doped cuprates, clearly the latter seem to exhibit lower transition temperatures and a smaller SC dome.

In the next section we analyze phenomenological features that characterize the cuprates.

24.2 Phenomenology

24.2.1 Superconducting Phase

Quite a few features of the SC phase of cuprates differ from conventional BCS superconductors. Among these, we have the isotopic effect, the coherence length and the symmetry of the order parameter.

First of all, there is no isotopic effect in the cuprates. Then, the coherence length, which roughly measures the Cooper pair size, ranges from $\xi = 1nm$ to $\xi = 4nm$ in cuprates, whereas in conventional superconductors it ranges from $\xi = 50nm$ to $\xi = 1000nm$. As a consequence, the ratio between penetration length and coherence length, λ/ξ, is considerably higher in cuprates than in conventional superconductors, which makes of all of them Type-II superconductors. Another distinguished feature of the cuprate superconductors is the order parameter's symmetry. The analysis of ARPES photoemission experiments indicates a momentum dependence of the superconducting order parameter, contrary to the case of conventional BCS superconductors, where the SC gap is uniform. Fig. 24.4 shows the results for $Bi_2Sr_2CaCu_2O_{8+x}$ [196]. The experiment is performed by sweeping different paths along the first Brillouin zone, both above and below T_c. Path B clearly shows different results for the two cases, whereas Path A shows no difference at all. This reveals the anisotropy of the SC order parameter $\Delta(\mathbf{k}) = \langle c^\dagger(-\mathbf{k})c^\dagger(\mathbf{k})\rangle$, which is compatible with a d-wave symmetry, namely

$$\Delta(\mathbf{k}) = \langle c^\dagger(-\mathbf{k})c^\dagger(\mathbf{k})\rangle = \Delta_0\left[\cos k_x a - \cos k_y a\right], \qquad (24.1)$$

where $a = 0.38nm$ is the lattice parameter of the CuO_2 square lattice.

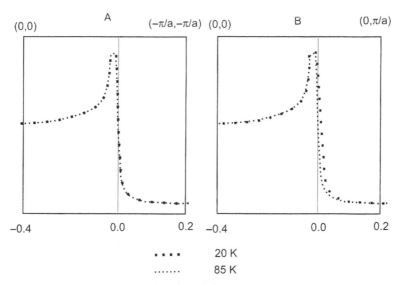

Figure 24.4 Schematic result of ARPES experiment showing the photoemission intensity versus the energy relative to the Fermi level in eV for a sample of $T_c = 78\ K$ along different paths in the Brillouin zone. Path A shows no temperature dependence, whereas path B shows clearly different results above and below T_c. Different temperature-sensible results for paths A and B show evidence for the gap anisotropy [196].

24.2.2 Normal Phase

This is a feature where there exists a strong difference between conventional BCS superconductors and cuprates. In the former, the system, when in the normal phase, is a metal. This is closely related to the fact that the effective electron-electron interaction, responsible for the Cooper pair formation, is attractive whenever the electrons' energy difference is less than the phonon energy, $\hbar\omega$, according to (4.82). This condition is fulfilled when most of the active electrons are precisely on the Fermi surface or within a shell of width $\hbar\omega$ around it. This, however, just happens at low temperatures; otherwise the electrons are far away from the Fermi surface. Hence we can figure out why conventional, phonon-mediated, BCS superconductivity usually occurs at very low temperatures. They build up around a Fermi surface.

Cuprate superconductors are neither regular metals nor Fermi liquids in the normal phase. An evidence for this is a temperature linear dependence of the resistivity, whereas for a metal we would have a T^5 dependence and for a Fermi liquid, T^2.

There is no evidence indicating that Cooper pairs should form around a Fermi surface in the normal phase of cuprates, as is the case in BCS superconductivity. The inexistence of the isotopic effect in the SC phase and the d-wave symmetry of the SC gap are additional features that strongly suggest that the mechanism responsible for superconductivity in cuprates is not mediated by phonons.

An interesting feature of some high-T_c cuprates is the organization of the doped holes in completely anisotropic form in the CuO_2 planes. In lightly doped LSCO, for instance, there is experimental evidence [214] that the totality of doped holes assemble along one of the directions of the O^{--} square lattice, which is rotated by 45° with respect to the Cu^{++} lattice. This creates direction-dependent doping effects, as we shall see below.

24.3 The Undoped System

The pure parent compounds of the cuprates family are accurately described by an antiferromagnetic Heisenberg model on a square lattice [195]. The localized spins thereof correspond to the unpaired electrons of the copper ions located at each site of such lattice. We start, therefore, with the Hamiltonian

$$H = J \sum_{\langle ij \rangle} \mathbf{S}_i \cdot \mathbf{S}_j, \tag{24.2}$$

where $J > 0$ is the super-exchange coupling and $\mathbf{S}_i$ is the spin of the unpaired electron of the $3d^9$ orbital of the copper ion Cu^{++} placed at the site i of a square lattice corresponding to the CuO_2 planes of La_2CuO_4. We choose this compound because it is one of the most studied and consequently one for which there are the most abundant experimental results.

We have seen in Chapter 17 that, within a quantum field theory approach, we describe the sublattice magnetization of the above magnetic system as the O(3) nonlinear sigma field $\mathbf{n}$, governed by the Lagrangean density

$$\mathcal{L} = \frac{\rho_s}{2} \left[\frac{1}{c^2} \partial_t \mathbf{n} \cdot \partial_t \mathbf{n} - \nabla \mathbf{n} \cdot \nabla \mathbf{n} \right], \tag{24.3}$$

supplemented by the constraint $\mathbf{n} \cdot \mathbf{n} = 1$. In the above expression, ρ_s and c are, respectively, the spin stiffness and the spin-wave velocity.

The present description can be reformulated, in the so-called CP^1 language, involving two complex scalar fields z_i, $i = 1, 2$, which are related to the spin sublattice magnetization field, as follows

$$\mathbf{n} = z_i^* \vec{\sigma}_{ij} z_j. \tag{24.4}$$

The CP^1 partition functional is expressed as

$$Z = Z_0^{-1} \int Dz_i \, Dz_i^* \, DA_\mu \exp \left\{ -2\rho_s \int d^3x (D_\mu z_i)^* (D_\mu z_i) \right\} \delta \left[z_i^* z_i - 1 \right], \tag{24.5}$$

where $D_\mu = \partial_\mu + i A_\mu$.

Notice that we did not include a Hopf term, which in CP^1 language has the form of a Chern–Simons Lagrangean. There is a vast amount of evidence indicating that this term is absent in the pure system corresponding to a Heisenberg antiferromagnet on a square lattice [197, 198, 199, 200, 201]. This, however, is no longer true in the presence of doping, as we shall see below.

24.4 A Mechanism of Doping: Skyrmions

24.4.1 A Model for Doping

The above partition functional correctly describes the pure, half-filled, AF system. We use it as the starting point for the description of the doping process. We will see that the final outcome of the doping procedure will be to replace the spin-stiffness ρ_s by a doping-dependent effective spin-stiffness $\rho_s(x)$, which will be reduced until it eventually vanishes at a critical doping x_c. The Néel temperature at zero doping, $T_N(x = 0) = T_N(\rho_s)$, accordingly, will be replaced by $T_N(x) = T_N(\rho_s(x))$. This will also vanish at the critical doping, $T_N(x_c) = 0$, thus creating the AF dome in the $T_N \times x$ phase diagram. We will consider the specific case of LSCO, because for this material, for each doped Sr atom, there corresponds the creation of a hole in the CuO_2, hence there is a one-to-one correspondence between the amount of charge units doped into the planes, δ, and the stoichiometric parameter x characterizing the compound. This is no longer true, for instance, for YBCO, where part of the doped charge goes into $O - Cu - O$ chains outside of the planes and, consequently, $\delta \neq x$. In this case, any doping theory, which are based on δ, when compared to experiments, which are based on x, requires a detailed modelling of the distribution of the doped charges in between the planes and chains.

We describe the doped holes into the CuO_2 planes by means of a fermion field ψ. According to both numerical estimates [174] and experimental results [204, 205], LSCO has a Fermi surface with four branches, which is represented in Fig. 24.5. This can be conveniently modeled by a dispersion relation

$$\epsilon(\mathbf{k}) = \pm\sqrt{\left(k_x \pm \frac{\pi}{a}\right)^2 v_F^2 + \left(k_y \pm \frac{\pi}{a}\right)^2 v_F^2 + m^{*2}v_F^4}. \qquad (24.6)$$

Each branch above corresponds to the dispersion relation of a Dirac field ψ, but shifted about the points $(\pm\frac{\pi}{a}, \pm\frac{\pi}{a})$, namely, centered on the corners of the first Brillouin zone. Expanding $\epsilon(\mathbf{k})$ around the Fermi energy $\epsilon_F > m^*v_F^2$, we get a linear dispersion relation.

The doping constraint on the charge-current density on the CuO_2 planes, then, is introduced by identifying the Dirac field current with dopants' universe-line [202, 203]:

$$\overline{\psi}\gamma^\mu\psi = \Delta^\mu \equiv \sum_{i=1}^4 \delta_i \int_{X;L_i}^\infty d\xi^\mu \delta^3(\xi - x). \qquad (24.7)$$

The above integral is taken along the universe-line, L_i, of the doped charge, located at $X_i = (\mathbf{X}_i, t)$ and the sum accounts for the four components of the Dirac field. The dopant density would be given by the zeroth component of the above expression, namely,

$$\psi^\dagger\psi \equiv \sum_{i=1}^4 \psi_i^\dagger\psi_i = \sum_{i=1}^4 \delta_i\,\delta^2(\mathbf{x} - \mathbf{X}_i(t)). \qquad (24.8)$$

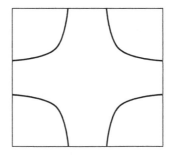

Figure 24.5 The four-branched Fermi surface of LSCO

Notice that there are four branches in the Fermi surface, each one satisfying the relation above, hence the effective doping parameter must be multiplied by a factor four.

The considerations made so far in this subsection enable us to model the kinematics of the holes doped into the CuO_2 planes. The kinematical Lagrangean density would be

$$\mathcal{L}_0 = \overline{\psi}\gamma^\mu \left(i\partial_\mu + \overline{A}_\mu\right)\psi - \frac{m^* v_F}{\hbar}\overline{\psi}\psi, \tag{24.9}$$

where $\overline{A}_\mu$, which in momentum space is given by $\overline{A}_\mu(\mathbf{k}) = (0, \frac{\pi}{a}, \frac{\pi}{a})$, results from shifting the center of the Fermi surface to one of the corners of the first Brillouin zone, according to (24.6), thereby generating one of the four Fermi arcs.

With regard to the interaction of doped holes with the localized spins background, we will assume it is magnetic and directly involving the spin of the holes and the localized spins of the copper ions given by $\mathbf{n} = z_i^\dagger \vec{\sigma}_{ij} z_j$. Inspired by the results of Section 18.3, the interaction of the doped holes with the localized spins background is assumed to be described by the minimal coupling of the Dirac fermion field with the CP^1 vector field A_μ. This also guarantees overall gauge invariance.

We therefore ultimately write the partition functional describing the low doping regime of LSCO as

$$Z = \int Dz_i \, Dz_i^* \, DA_\mu \, D\psi \, D\psi^\dagger \exp\left\{-\int d^3x \left[2\rho_s (D_\mu z_i)^*(D_\mu z_i)\right.\right.$$
$$\left.+ \frac{\theta}{2}\epsilon^{\mu\nu\alpha} A_\mu \partial_\nu A_\alpha + i\overline{\psi}\gamma^\mu\partial_\mu\psi - \frac{m^* v_F}{\hbar}\overline{\psi}\psi - \overline{\psi}\gamma^\mu\psi\left(A_\mu + \overline{A}_\mu\right)\right]\right\}$$
$$\times \delta\left[z_i^* z_i - 1\right]\delta\left[\overline{\psi}\gamma^\mu\psi - \Delta^\mu\right], \tag{24.10}$$

where we made the shift corresponding to (24.6) and (24.9). Notice that we allowed the presence of a residual Hopf term, which should exist in the presence of doping.

We now introduce a functional representation for the last delta above:

$$\delta\left[\overline{\psi}\gamma^\mu\psi - \Delta_\mu\right] = \int D\lambda_\mu \exp\left\{i\int d^3x\lambda_\mu\left[\overline{\psi}\gamma^\mu\psi - \Delta^\mu\right]\right\}. \tag{24.11}$$

We then integrate over the fields $z_i, z_i^*, \psi, \psi^\dagger, \lambda_\mu$, in order to obtain an effective theory for the A_μ field. The z_i's integral can be easily done in the approximation of constant $|z_i|$ and using the functional delta. The fermionic integral is performed in the small fermion mass expansion [271] and the remaining λ_μ-integral is quadratic. The resulting effective Lagrangean for A_μ is [202]

$$\mathcal{L} = \frac{\theta}{2}\epsilon^{\mu\nu\alpha}A_\mu\partial_\nu A_\alpha - \frac{1}{4}F_{\mu\nu}\left[\frac{2\rho_s}{-\Box}\right]F^{\mu\nu} - \Delta^\mu\left(A_\mu + \overline{A}_\mu\right)$$
$$+ \Delta^\mu\left[\Pi_1 P^{\mu\nu} + \Pi_2 C^{\mu\nu}\right]\Delta^\nu. \tag{24.12}$$

In the above expression, $P^{\mu\nu} = -\Box\delta^{\mu\nu} + \partial_\mu\partial_\nu$ and $C^{\mu\nu} = -i\epsilon^{\mu\alpha\nu}\partial_\alpha$ and Π_1 and Π_2 are operators having the Fourier transform

$$\Pi_1(k) = \frac{\alpha}{(k^2)^{3/2}} - \frac{\gamma m^* v_F/\hbar}{k^4}$$
$$\Pi_2(k) = \frac{\beta}{k^2} - \frac{\eta m^* v_F/\hbar}{(k^2)^{3/2}}, \tag{24.13}$$

where

$$\alpha = \frac{8\pi^2}{\pi^2 + 16} \quad ; \quad \gamma = \frac{32\pi(9\pi^2 - 16)}{(\pi^2 + 16)^2}$$
$$\beta = -\frac{32\pi}{\pi^2 + 16} \quad ; \quad \eta = \frac{32\pi^2(24 - \pi^2)}{(\pi^2 + 16)^2}. \tag{24.14}$$

Let us consider now the field equation that results from (24.12) by choosing a transverse gauge. This is given by [202]

$$\theta\epsilon^{\mu\alpha\beta}\partial_\alpha A_\beta = \Delta^\mu + \rho_s A_\mu. \tag{24.15}$$

Taking the zeroth component and applying to the case of the skyrmion excitation, for which

$$\epsilon^{ij}\partial_i A_j^S = \mathcal{B}^S = 2\pi\delta(\mathbf{x} - \mathbf{X}_S) \quad ; \quad A_0^S = 0, \tag{24.16}$$

we find

$$2\pi\theta\delta(\mathbf{x} - \mathbf{X}_i^S) = \delta_i\,\delta(\mathbf{x} - \mathbf{X}_i^h), \tag{24.17}$$

where $\mathbf{X}_i^S$ is the skyrmion position and $\mathbf{X}_i^h$ is the doped hole position. We conclude that a skyrmion topological excitation is created precisely at the same point where the doped hole is introduced in the CuO_2 planes. Summing both sides on i and integrating on $\mathbf{x}$, we infer that

$$\theta(\delta_i) = \frac{1}{2\pi}\sum_{i=1}^{4}\delta_i \tag{24.18}$$

and we see that in the absence of doping, namely, for $\delta_i = 0$, we have $\theta = 0$, as it should.

Our model for doping in cuprates predicts that a skyrmion excitation is created precisely at the hole location, thereby confirming previous proposals of that mechanism [207, 208, 209, 210, 211].

Notice that the A_μ field is screened because of the second term in (24.12). This excludes the presence of long-range in-plane magnetic fields, which have been ruled out by muon relaxation experiments [212].

24.4.2 Skyrmion Correlation Function

In order to describe the phase diagram of the cuprates and, more specifically, that of LSCO, we are now going to apply our method of quantization of topological excitations, described in detail in Chapter 9, to the quantum skyrmions generated through doping in these materials. Specifically, we are going to evaluate the quantum skyrmion two-point correlation function and, out of its large-distance behavior, we shall extract the skyrmion energy as a function of the doping parameter. The method for obtaining the skyrmion quantum field correlator has been developed in [213] and was described in detail in Section 19.6. Applying it to the effective theory of doped LSCO, described in (24.12), we get [202, 203]

$$\langle \mu(\mathbf{x}, t) \mu^\dagger(\mathbf{y}, t) \rangle \stackrel{|\mathbf{x}-\mathbf{y}| \to \infty}{\longrightarrow} \exp \left\{ -\frac{E_S(\delta)}{\hbar c} |\mathbf{x} - \mathbf{y}| \right\}, \qquad (24.19)$$

where the skyrmion energy is

$$E_S(\delta) = 2\pi \rho_s \left(1 - \frac{2\sqrt{2}\hbar c}{a\rho_s}(4\delta) - \frac{\sqrt{2}\gamma \hbar c}{\pi a \rho_s}(4\delta)^2 \right). \qquad (24.20)$$

In the above expression we have assumed all δ_is are identical or, in other words, that the doped holes are uniformly distributed among the four Dirac field components. Furthermore, we have included a factor four corresponding to the four, branches of the Fermi surface.

Observe that the skyrmion energy decreases as we dope the system. This was expected, since the introduction of skyrmion defects spoils the order present in the Néel state.

24.4.3 Sublattice Magnetization × Doping

The skyrmion energy works as a convenient order parameter for the ordered AF state, since the skyrmion energy would vanish in a disordered paramagnetic state.

The first result we may extract from the expression (24.20), derived for the energy cost for creating a skyrmion as a function of the doping percentage is the critical doping for the destruction of the AF order at $T = 0$. This is obtained by imposing $E_S(\delta_c) = 0$ on (24.20). Using the experimental input of $\hbar c = 0.85 \pm 0.03$ eVÅ and $\rho_s = 0.0578$ eV and $a = 3.8$ Å, we obtain $x_c = \delta_c = 0.020 \pm 0.03$, which is in excellent agreement with the experimental value.

An interesting observable quantity we can extract from our model is the doping-dependent sublattice magnetization $M(x)$, which has been measured directly both by muon spin relaxation and nuclear quadrupole resonance techniques [215]. This is given, at zero doping, by $M(0) = \sqrt{\rho_s}$. The dependence on the doping parameter x can be obtained, provided we introduce an effective, doping-dependent spin stiffness $\rho_s(x)$, such that $M(x) = \sqrt{\rho_s(x)}$. This can be obtained naturally from (24.20), however we must keep in mind the presence of stripes in LSCO, which make the doping modify the spin stiffness only along one of the directions on the plane, leaving the other unaffected. We will have, consequently,

$$\rho_{sx}(\delta) = \rho_s \left(1 - \frac{2\sqrt{2}\hbar c}{a\rho_s}(4\delta) - \frac{\sqrt{2}\gamma\hbar c}{\pi a\rho_s}(4\delta)^2 \right)$$

$$\rho_{sy}(\delta) = \rho_s. \tag{24.21}$$

Now, how could one obtain the sublattice magnetization from the above expressions? It was shown in [216] that the presence of stripes creates an anisotropy, such that this is given by the square-root of an effective, doping-dependent spin-stiffness, given by the geometric average of the two direction-dependent spin-stiffnesses in (24.21): $M(\delta) = \sqrt{\bar{\rho}(\delta)}$, where

$$\bar{\rho}_s(\delta) = \sqrt{\rho_{sx}(\delta)\rho_{sy}} = \rho_s \left(1 - \frac{2\sqrt{2}\hbar c}{a\rho_s}(4\delta) - \frac{\sqrt{2}\gamma\hbar c}{\pi a\rho_s}(4\delta)^2 \right)^{1/2}. \tag{24.22}$$

Hence, since $\delta = x$ for LSCO, we finally get

$$\frac{M(x)}{M(0)} = \left(1 - \frac{2\sqrt{2}\hbar c}{a\rho_s}(4x) - \frac{\sqrt{2}\gamma\hbar c}{\pi a\rho_s}(4x)^2 \right)^{1/4}. \tag{24.23}$$

This is plotted in Fig. 24.6, with the input of the above experimental values for $\hbar c$ and ρ_s and a for LSCO. It shows excellent agreement with the experimental data without adjusting any parameter.

From (24.23), we see that, close to the critical doping x_c, the sublattice magnetization behaves as

$$\frac{M(x)}{M(0)} = \left(1 - \frac{x}{x_c} \right)^{0.25}. \tag{24.24}$$

The critical exponent 0.25 is very close to the measured experimental value of 0.236 [215].

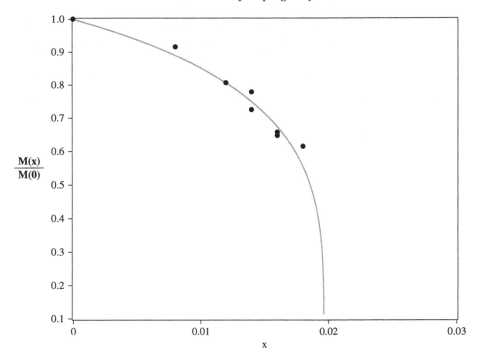

Figure 24.6 Sublattice magnetization as a function of doping for LSCO. Experimental data from μSR and NQR, from [215]. Solid line from (24.23)

24.4.4 The AF Dome: Néel Temperature × Doping

We now turn to the phase diagram $T \times x$; more specifically, to the AF dome in this diagram, which displays the Néel temperature as a function of doping: $T_N(x)$. Now, from the very outset, we must include in our theoretical description the three-dimensional character of the system; otherwise we would have to face the Hohenberg–Mermin–Wagner theorem, which forbids two-dimensional systems from undergoing phase transitions at a finite temperature, $T \neq 0$ [170].

Our strategy will be to start from the expression for the Néel temperature as a function of the spin stiffness ρ_s and then replace this by our doping-dependent effective stiffness $\overline{\rho}_s(\delta)$, given by (24.22). For this purpose, we follow [217], where the interlayer coupling was taken into account, and write the partition functional,

$$Z = \int \prod_i D\mathbf{n}_i \exp\left\{-\frac{1}{\hbar}\int_0^{\hbar\beta} d\tau \int d^2x \sum_i \frac{\rho_s}{2}\left[\frac{1}{c^2}\partial_t\mathbf{n}_i \cdot \partial_t\mathbf{n}_i - \nabla\mathbf{n}_i \cdot \nabla\mathbf{n}_i\right]\right.$$
$$\left.\frac{\alpha}{2}[\mathbf{n}_{i+1} - \mathbf{n}_i]^2 \right\} \delta\left[|\mathbf{n}_i|^2 - 1\right], \tag{24.25}$$

which considers the three-dimensional character of the layered system. The inter-layer coupling, the intensity for which is measured by α, is supposed to be very small.

The authors of Ref. [217] have made a careful determination of the Néel temperature as a function of ρ_s in the model above. The result reads [217]

$$T_N = \frac{4\pi\rho_s}{\ln\left[5.5005\frac{(4\pi\rho_s)^2}{\alpha_R c^2}\right]}, \tag{24.26}$$

where α_R is the renormalized interlayer coupling parameter. We obtain, from this expression, the doping-dependent Néel temperature [203], through replacement of ρ_s by our effective, doping-dependent spin stiffness $\overline{\rho}_s(x)$, given by (24.22), namely,

$$T_N(x) = \frac{4\pi\overline{\rho}_s(x)}{\ln\left[5.5005\frac{(4\pi\overline{\rho}_s(x))^2}{\alpha_R c^2}\right]}. \tag{24.27}$$

This is plotted in Fig. 24.7. The solid line was obtained by adjusting only one parameter, namely, α_R. There is, again, an excellent agreement with the experimental data for LSCO.

The full understanding of the mechanism underlying superconductivity in the cuprates has not yet been achieved. Among other features, this must certainly

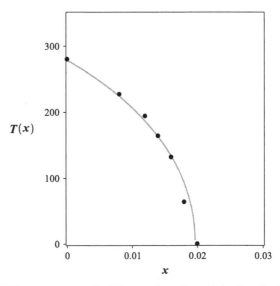

Figure 24.7 Néel temperature (in K) as a function of doping for LSCO. Experimental data of μSR and NQR, from [215]. Solid line from Eqs. (24.27) and (24.22)

include the full description of the $T \times x$ phase diagram and, in particular, the superconducting dome of such diagram. The results obtained for the AF part of the phase diagram suggest that skyrmions, in particular quantum skyrmions, should play an important role in the pairing mechanism.

24.5 Overview

The subject of high-Tc cuprates is one of the beautiful chapters in the history of physics. Nevertheless, it is as yet unfinished and for this reason it is not frequently seen in textbooks. The inclusion of a chapter on the cuprates in this book, therefore, requires a few explanatory words. We tried to provide here a general account of the main features of the subject, which, being one of the most important in contemporary physics, deserves a special place in any book on condensed matter physics. The material included will, hopefully, serve as an introduction to the main features of the subject. In the previous section, however, we took the liberty of including material that is our own contribution to the subject. Consequently, we would like to call the reader's attention to the fact that in doing so, by no means do we try to imply the existence of any kind of recognition of such results by the community. They stand only on their agreement with experimental data.

25

The Pnictides: Iron-Based Superconductors

A new class of superconducting materials with critical temperatures typically ranging from about $40K$ to $60K$ was synthesized in 2008. These materials, known as pnictides, are based in Fe and As and show some similarities with the high-T_c cuprates. Pnictide materials are subdivided in two classes: the so-called 1111 compounds, $RO_xF_{1-x}FeAs$, with $R = La, Ce, Sm, Pr, \ldots$, also known as oxypnictides, and the 122 compounds $M_{1-x}K_xFe_2As_2$, where $M = Ba, Sr, \ldots$ Contrary to the cuprates, which had insulator parent (undoped) compounds, the pnictides' parent compounds are metals. They have, however, analogously to the cuprates, a structure containing layers of square lattices exhibiting nontrivial magnetic properties and interactions. The interlayer magnetic coupling is a bit larger than in the cuprates, namely, $J_\perp/J \sim 10^{-4}$ for 1111 pnictides and $J_\perp/J \sim 10^{-2}$ for 122 pnictides, whereas for cuprates we have $J_\perp/J \sim 10^{-5}$. Despite that fact, a two-dimensional description, both of the parent compounds and of their doped descendants, works quite well. The phase diagram is quite similar to that of the cuprates, exhibiting a SC dome and a magnetically ordered dome as well.

25.1 Crystal Structure, $FeAs$ Planes and Phase Diagram

A common feature among all pnictides is the presence of a layered structure containing square lattices, in which Fe^{++} ions occupy the sites and As atoms are located in the center of each plaquette, in a buckled structure, alternating between the two sides of the lattice.

The Fe^{++} ion is in a $3d^6$ electronic configuration, with two unpaired electrons, respectively occupying orbitals $3d_{xz}$ and $3d_{yz}$, according to the Hund rules. This implies a $S = 1$ spin quantum number for the localized spins. The doped electrons, for energetic reasons, will occupy the same orbitals, exhibiting different hopping parameters, as discussed below. They will undergo a Hund ferromagnetic exchange interaction with the localized electrons.

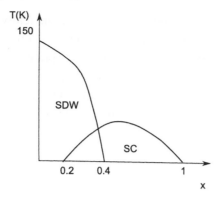

Figure 25.1 The schematic phase diagram of the pnictide superconductor $Ba_{1-x}K_xFe_2As_2$, exhibiting the SC and magnetically ordered domes

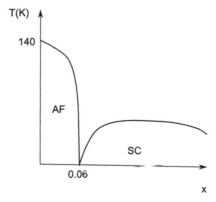

Figure 25.2 The schematic phase diagram of the pnictide superconductor $CeO_{1-x}F_xFeAs$, exhibiting the SC and magnetically ordered (SDW) domes

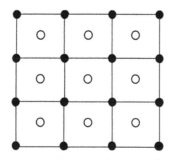

Figure 25.3 The *FeAs* planes, a common feature of all the iron based pnictide superconductors. Black dots and white dots are, respectively, Fe^{++} ions and *As* atoms. The latter alternate between the two sides of the plane in a buckled structure.

The parent compounds invariably present a magnetically ordered ground state, usually a Néel state or a collinear magnetic/spin stripes ordered state. Upon doping, the magnetic order quickly disappears, as in the cuprates, thus forming a magnetic dome. Eventually, as the doping is increased, a superconducting phase sets in, with T_c increasing up to a maximum and subsequently being reduced down to zero, thus forming a SC dome. In some cases, the magnetic and SC domes overlap.

25.2 The $J_1 - J_2$ Localized-Itinerant Model

We will describe the pnictide materials by means of a mixed localized-itinerant model corresponding to the Hamiltonian [218]

$$H = J_1 \sum_{\langle ij \rangle} \mathbf{S}_i \cdot \mathbf{S}_j + J_2 \sum_{\langle\langle ij \rangle\rangle} \mathbf{S}_i \cdot \mathbf{S}_j - K \sum_i (S_i^z)^2$$
$$+ \sum_{ij;\alpha\beta} \left[t_{ij}^{\alpha\beta} (c_{i\sigma}^\alpha)^\dagger c_{j\sigma}^\beta + hc \right] - J_H \sum_{i;\alpha} (c_{i\sigma}^\alpha)^\dagger \vec{\sigma}_{\sigma\sigma'} c_{i\sigma'}^\alpha \cdot \mathbf{S}_i. \quad (25.1)$$

Here $\langle ij \rangle$ runs over nearest neighbors and $\langle\langle ij \rangle\rangle$ over next-nearest neighbors sites of a square lattice and α, β run over the d_{xz}, d_{yz} orbitals of the Fe^{++} ions located at the sites of this lattice. $\mathbf{S}_i$ are the localized spins of the Fe^{++} ions, having eigenvalues with a spin quantum number $S = 1$.

The itinerant doped electrons are described by a two-band minimal model [219], where $(c_{i\sigma}^\alpha)^\dagger$ is the creation operator of the itinerant electron/hole doped into the iron square lattice. The hopping parameters $t_{ij}^{\alpha\beta}$ vary according to the pair of orbitals involved in the hopping.

The constant $J_1 > 0$ is the superexchange coupling between nearest neighbor Fe^{++} ions and $J_2 > 0$, the As-mediated superexchange coupling between next-nearest neighbor Fe^{++} ions in the square lattice formed by these ions. J_H is the

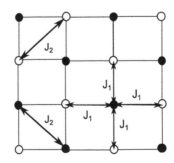

Figure 25.4 The $FeAs$ square lattice with the two sublattices (black and white dots) and the corresponding exchange coupling constants J_1 and J_2. The latter within each sublattice, while the former between the two sublattices. Notice that the lattice in this figure is rotated by 45° with respect to that in Fig. 25.3.

Hund ferromagnetic exchange coupling between the itinerant electron/holes and the localized spins. Finally $K > 0$ is the coupling constant that accounts for the single-ion anisotropy, which occurs for $S > 1/2$, due to spin-orbit and crystal field effects [220].

25.3 The Magnetic Spectrum

Let us start by considering the pure, undoped system, for which only the first three terms of the Hamiltonian (25.1), associated with the localized spins, are relevant. Later on we are going to include the doped itinerant electrons/holes as well.

The experiments suggest that the exchange couplings satisfy $J_1 < 2J_2$, hence we assume this relation between the exchange couplings holds true. In a first approach, also, we will assume $K = 0$. For these couplings the ground state is verified to be such that each sublattice, A and B of the Fe ion square lattice is in an opposite Néel state. This produces an overall ordered state with collinear diagonal spin stripes containing spins that point at the same direction along these stripes.

In order to obtain a functional integral representation for the partition function, we follow the procedure used in Chapter 17 and use the coherent spin states $|\mathbf{N}_\alpha\rangle$, with $\alpha = A, B$, corresponding, respectively, to the two sublattices, such that

$$\langle \mathbf{N}_\alpha | S^\alpha | \mathbf{N}_\alpha \rangle = S \mathbf{N}_\alpha, \quad ; \quad \alpha = A, B. \tag{25.2}$$

We then decompose the field $\mathbf{N}_\alpha$ into ferromagnetic and antiferromagnetic fluctuation fields $\mathbf{n}_\alpha$ and $\mathbf{L}_\alpha$, similarly to (19.17),

$$\mathbf{N}_\alpha(\mathbf{r}_i, \tau) = e^{i\mathbf{Q}\cdot\mathbf{r}_i} \mathbf{n}_\alpha(\mathbf{r}_i, \tau)\sqrt{1 - a^4 \mathbf{L}_\alpha^2(\mathbf{r}_i, \tau)} + a^2 \mathbf{L}_\alpha(\mathbf{r}_i, \tau), \tag{25.3}$$

for $\alpha = A, B$.

Following the same steps as in Chapter 17, we take the continuum limit and integrate over the ferromagnetic components $\mathbf{L}_A$ and $\mathbf{L}_B$, obtaining the resulting action for the antiferromagnetic ones [218]

$$
\begin{aligned}
S = \int_0^{\hbar\beta} d\tau \int d^2x \Bigg\{ & \frac{\rho_s}{2}\left[|\nabla \mathbf{n}_A|^2 + |\nabla \mathbf{n}_B|^2 + \frac{1}{c_0^2}|\partial_\tau \mathbf{n}_A|^2 + \frac{1}{c_0^2}|\partial_\tau \mathbf{n}_B|^2 \right] \\
& + \gamma\left[\mathbf{n}_A \cdot \partial_x \partial_y \mathbf{n}_B + \mathbf{n}_B \cdot \partial_x \partial_y \mathbf{n}_A + \eta \mathbf{n}_A \cdot \mathbf{n}_B \right] \\
& + ib\left[\mathbf{n}_A \cdot \mathbf{n}_B \times \partial_\tau \mathbf{n}_B + \mathbf{n}_B \cdot \mathbf{n}_A \times \partial_\tau \mathbf{n}_A \right] \\
& + \frac{1}{c_1^2}\left[(\partial_\tau \mathbf{n}_A \cdot \mathbf{n}_B)(\mathbf{n}_A \cdot \partial_\tau \mathbf{n}_B) - (\mathbf{n}_A \cdot \mathbf{n}_B)(\partial_\tau \mathbf{n}_A \cdot \partial_\tau \mathbf{n}_B) \right] \Bigg\}.
\end{aligned}
\tag{25.4}
$$

This is supplemented by the constraints $|\mathbf{n}_A|^2 = |\mathbf{n}_B|^2 = 1$.

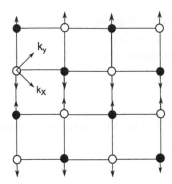

Figure 25.5 The ordered ground state with the collinear stripes: the two sublattices with opposite Néel states. Notice the orientation of the k_x and k_y axes: the lattice in this figure is rotated by 45° with respect to that in Fig. 25.3.

The first line above contains the usual kinematical low-energy fluctuation terms of each of the two sublattices, the second and fourth, their interaction, while the third describes the precession of the spins of one sublattice around the other.

The coupling parameters are given by the following expressions in terms of the original ones:

$$\rho_s = 2J_2^2 s^2 \; ; \; \hbar c_0 = 2\sqrt{2} Sa\sqrt{4J_2^2 - J_1^2} \; ; \; c_1 = c_0 \frac{J_2}{J_1}$$

$$\gamma = \frac{2J_1}{J_2}\left[1 + \frac{J_1^2/4}{4J_2^2 - J_1^2}\right] \; ; \; \eta = \frac{2\hbar}{S^2}\left[\frac{J_1^2}{4J_2^2 - J_1^2}\right] \; ; \; b = \frac{2J_1}{a^2 J_2}\left[\frac{J_1}{4J_2^2 - J_1^2}\right].$$

$$(25.5)$$

We are interested in obtaining the spectrum of magnetic quantum fluctuations around the collinear striped ordered ground state. For this purpose, we write the magnetization field, for each sublattice, as

$$\mathbf{n}_\alpha = \left(\pi_\alpha^x, \pi_\alpha^y, \sigma_\alpha\right) \; ; \; \alpha = A, B. \tag{25.6}$$

In the ordered ground-state $|\sigma_{A,B}| = \sigma$, where $\sigma \in \mathbb{R}$ and $|\pi_{A,B}| = 0$. The magnon quantum excitations are fluctuations in the transverse fields around zero. In order to identify these, we expand the magnetic action (25.4) in powers of the fluctuation fields, namely, π_α^i; $i = x, y$; $\alpha = A, B$ around the ground state,

$$S[\sigma, \pi_\alpha^i] = S_0[\sigma] + \frac{1}{2}\int d^3x d^3y \pi_\alpha^i(x)\left[G_{\alpha\beta}^{ij}\right]^{-1}(x; y)\pi_\beta^j(y) + O(\pi_{i,\alpha}^4). \tag{25.7}$$

In the above expression, $G_{\alpha\beta}^{ij}$ is the Green function of the fluctuating fields π_α^i, which can be derived from (25.4). The dispersion relation of the magnon elementary excitations, finally, may be inferred from the poles of $G_{\alpha\beta}^{ij}$.

We find two magnon modes [218]: one acoustic, with energy

$$\hbar\omega(\mathbf{k}) = \hbar c_- \sqrt{\mathbf{k}^2 + 2\gamma k_x k_y} \tag{25.8}$$

and one optical with energy

$$\hbar\Omega(\mathbf{k}) = \hbar c_+ \sqrt{\mathbf{k}^2 - 2\gamma k_x k_y + \eta}. \tag{25.9}$$

In the above expressions $c+$ and $c-$ are defined by the following relation:

$$\frac{1}{c_\pm^2} = \left(\frac{1}{c_0^2} \pm \frac{\sigma^2}{2c_1^2} \right). \tag{25.10}$$

Notice that the effective spin-wave velocity of the acoustic mode,

$$\mathbf{v}_{SW}(\mathbf{k}) = \nabla_k \omega(\mathbf{k}),$$

is enhanced by a factor $\sqrt{1+\gamma}$ along the direction $k_x = k_y$, when compared to the one along the directions $k_x = 0, k_y \neq 0$ or $k_y = 0, k_x \neq 0$. The enhancement consequently occurs precisely along the spin stripes.

We introduce at this point the effect of the single-ion-anisotropy. Repeating the above derivation for $K \neq 0$, we obtain, in the regime $K \ll J_2$ [218], the following magnon modes:

$$\hbar\omega(\mathbf{k}) = \hbar c_- \sqrt{\mathbf{k}^2 + 2\gamma k_x k_y + \Lambda^2}$$

$$\hbar\Omega(\mathbf{k}) = \hbar c_+ \sqrt{\mathbf{k}^2 - 2\gamma k_x k_y + \Delta^2 + \eta}. \tag{25.11}$$

In the above expression,

$$\Delta^2 = \frac{S - 1/2}{Sa^2} \left(\frac{K}{J_2} \right) \left[1 + \frac{J_1^2/8}{4J_2^2 - J_1^2} \right] \tag{25.12}$$

is the gap introduced by the single-ion-anisotropy.

The remarkable fact here is the emergence of two different gaps in the magnon spectrum. This theoretical result is in agreement with inelastic neutron scattering experiments in $SrFe_2As_2$, which indeed show the existence of two direction-dependent energy thresholds [221].

25.4 The Electronic Spectrum

We now include the effect of doping in our description of pnictides. For this purpose, we take into account the last two terms in (25.1). We want to determine, among other features, what is the effect of the magnetic ground state on the itinerant carriers doped into the system. Here we shall adopt a semiclassical

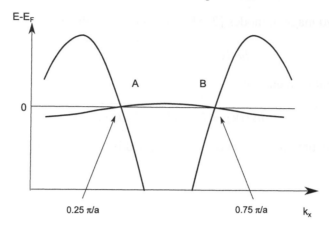

Figure 25.6 Schematic representation of the result obtained in [218] showing that Dirac cones develop in the spectrum of itinerant carriers, in $BaFe_2Sr_2$, in the presence of a striped collinear magnetically ordered spin background that occurs below $T \simeq 160K$

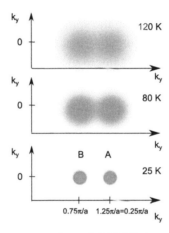

Figure 25.7 Schematic representation of ARPES data [222] showing that Dirac cones develop in $BaFe_2Sr_2$ as the system is cooled down. The result obtained in [218] shows Dirac cones precisely at the same positions.

approach and replace the localized spin background in the Hund interaction term by its magnetically ordered collinear striped ground-state configuration. The part of the Hamiltonian corresponding to the itinerant electrons/holes, then, becomes quadratic and can be easily diagonalized. Our results show that above the magnetic ordering temperature, the Fermi surface is given by the usual tight-binding result. As the temperature lowers and the system orders in the magnetic striped order state, Dirac cones develop along the $\Gamma - M$ line of the first Brillouin zone, symmetrically around the center, at the points $\mathbf{k} = (\pm 0.25\frac{\pi}{a}, 0)$ and $\mathbf{k} = (\pm 0.75\frac{\pi}{a}, 0)$, see Fig. 25.6. This result is confirmed by ARPES experiments [222] in $BaFe_2Sr_2$ [222], see Fig. 25.7.

25.5 Thermodynamic Properties

Let us consider now thermodynamic aspects of the pnictides. We will concentrate more specifically on the specific heat, which is given by

$$C = \frac{1}{V} \frac{\partial U}{\partial T}, \tag{25.13}$$

where U is the internal energy. This contains three contributions, namely, $U = U_L + U_e + U_M$, which correspond, respectively, to the lattice, electronic and magnetic contributions.

25.5.1 Lattice and Electronic Specific Heats

We take the electronic specific heat as given by the Sommerfeld expression

$$C_e = AT. \tag{25.14}$$

The lattice contribution to the specific heat, conversely, can be obtained from the Debye model, and reads [223]

$$C_L = BT^3 \left[1 - \frac{15}{4\pi^4} \left(\frac{\theta_D}{T} \right)^4 \ln Z_D - \frac{15}{4\pi^4} \left(\frac{\theta_D}{T} \right)^3 \left(\frac{T}{\Delta_i} \right) L_1(Z_D) \right.$$
$$\left. - \frac{45}{4\pi^4} \left(\frac{\theta_D}{T} \right)^2 L_2(Z_D) - \frac{90}{4\pi^4} \left(\frac{\theta_D}{T} \right) L_3(Z_D) - \frac{90}{4\pi^4} L_4(Z_D) \right], \tag{25.15}$$

where $L_n(Z)$ are the polylogarithm functions and $Z_D = e^{-\frac{\theta_D}{T}}$, with θ_D the Debye temperature. In the above expressions, A and B are, respectively, constants specified in the models of Sommerfeld and Debye.

25.5.2 Magnetic Specific Heat

We have obtained the acoustic and optical magnon energies starting from a theoretical model for the iron pnictide materials. Let us obtain now their thermodynamic properties, or more specifically, the magnetic contribution to the specific heat. This is obtained from the magnetic internal energy U_M. Assuming that ideal gas conditions apply to the magnons, this is given by

$$U_M = U_{ac} + U_{opt}$$
$$U_{ac} = \mathcal{V} \int \frac{d^2k}{(2\pi)^2} \hbar\omega(\mathbf{k}) n_{BE}(\mathbf{k}, T)$$
$$U_{opt} = \mathcal{V} \int \frac{d^2k}{(2\pi)^2} \hbar\Omega(\mathbf{k}) n_{BE}(\mathbf{k}, T), \tag{25.16}$$

where $\mathcal{V}$ is the unit-cell area, $n_{BE}(\mathbf{k}, T)$ is the Bose–Einstein population of magnons of a given energy at temperature T. $\omega(\mathbf{k})$ and $\Omega(\mathbf{k})$ are, respectively, the acoustic and optical magnon frequencies given by (25.11).

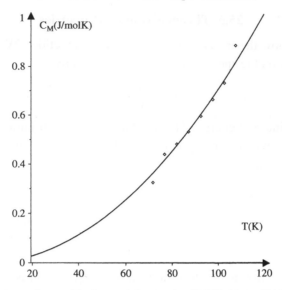

Figure 25.8 Magnetic specific heat of $SrFe_2As_2$ [223]. The solid line is the analytic expression for the magnetic specific heat, Eq. (25.18), with no adjustable parameters. The experimental data from [224, 225] are $C_M = C_T(SrFe_2As_2) - C_T(Sr_{0.6}K_{0.4}Fe_2As_2)$.

From (25.13) and (25.16), we derive the following expressions for the magnetic specific heat,

$$C_M = C_{ac}(\Delta_{ac}, T) + C_{opt}(\Delta_{opt}, T), \tag{25.17}$$

where $\Delta_{ac} = \Delta$ and $\Delta_{opt} = \sqrt{\Delta^2 + \eta}$ are the gaps of the acoustic and optical modes, respectively.

The contribution from each mode is (i = ac, opt) [223]

$$C_i(\Delta_i, T) = \frac{a^2 N_A k_B^3 \Delta^3}{f \pi \hbar^2 c_\pm^2 \sqrt{4 - \gamma^2}} \left(\frac{1}{T}\right) \left[\ln Z_i + 3\left(\frac{T}{\Delta_i}\right) L_1(Z_i) \right.$$
$$\left. + 6\left(\frac{T}{\Delta_i}\right)^2 L_2(Z_i) + 6\left(\frac{T}{\Delta_i}\right)^3 L_3(Z_i)\right]. \tag{25.18}$$

In the above expression, $L_n(Z)$ are the polylogarithm functions and $Z_i = e^{-\frac{\Delta_i}{T}}$. N_A is the Avogadro's number and f is the number of chemical formulas per unit square cell. Comparison of the theoretical expression for the magnetic specific heat, (25.17) and (25.18) is shown in Fig. 25.8. Experimental data are obtained by subtracting the total specific heat of $Sr_{0.6}K_{0.4}Fe_2As_2$ from that of $SrFe_2As_2$. For this level of K doping, the magnetic order of the first compound has already been destroyed, hence its total specific heat is supposed to represent the corresponding sum of lattice and electronic specific heats of $SrFe_2As_2$. Subtraction of the total specific heats of both compounds therefore should express the magnetic-specific heat of $SrFe_2As_2$. The good agreement, without adjusting parameters, attests the accuracy of the model adopted here for describing the magnetic and electronic properties of iron pnictide superconductors.

The total analytic calculation of the specific heat given by from Eqs. (25.14), (25.15) and (25.18) also agrees very well with the experimental data for $SrFe_2As_2$ without any adjustable parameters [223].

25.6 Basic Questions

Despite the mass of information gathered about iron pnictide materials, still, fundamental questions are not yet answered. For instance: what is the mechanism leading to superconductivity in these materials? Is it the same as in the cuprates? If not, why are the phase diagrams so similar? If yes, how can it derive from parent compounds that are metallic in one case and insulating in the other? In due time, the whole picture will be disclosed. Meanwhile, the words found in the last section of the previous chapter also apply here.

26

The Quantum Hall Effect

The Quantum Hall Effect (QHE) is one of the most remarkable, fascinating and, yes, complex phenomena in physics. Its essence, nevertheless, is quite simple: given a steady electric current, whenever we apply a perpendicular uniform magnetic field, a spontaneous electric voltage difference can be measured in the direction perpendicular to the current-magnetic-field plane. Requiring a simple set-up, the classical version of the effect was observed for the first time by Edwin Hall in 1879. The ratio between the transverse voltage and the current yields the "Hall resistance," which increases linearly with the applied magnetic field. On general grounds, the effect is a natural consequence of the Lorentz force acting on moving charges forming the current and it should not come to be a surprise.

One century later, in 1980, von Klitzing [226] repeated the Hall experiment under specific conditions. The electric current was injected in a metal slice $3nm$ wide, squeezed between an insulator and a semiconductor, in a device called MOS-FET, at a temperature of the order of 1 K and under an applied magnetic field of the order of 10 T. The result was stunning. The simple straight line, which represented the magnetic field dependence of the Hall resistance, was replaced by a complex pattern, in which one could observe a sequence of plateaus corresponding to integer multiples of a basic resistance unit.

Two years later, in 1982, Tsui and Störmer [227] repeated the experiment, this time injecting the current in a gas of electrons trapped in the interfaces of a multiple junction alternating $GaAs$ and $GaAs_{1-x}Al_x$, called heterostructure, at a temperature of the order of 0.1 K, under a magnetic field of up to 30 T. The curve representing the magnetic field dependence of the Hall resistance became even more complex, now exhibiting plateaus at rational multiples, mostly with odd denominators, of the same resistance unit.

Both the discoveries described above were laureated with the Nobel Prize and became known, respectively, as the integer and fractional QHE. The theoretical understanding of the two new phenomena required the use of ideas and methods

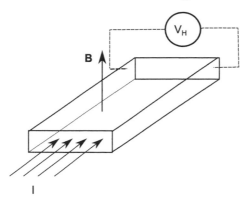

Figure 26.1 Experimental setup for observing the Hall effect

that mobilized the interplay of deep properties of quantum mechanics, the quantum theory of disordered systems, topologically driven physical mechanisms and deep theorems of mathematics. The efforts applied in the attempt to understand the QHE, in addition, have led to the opening of new areas of investigation such as the one of topological insulators and topological quantum computation, for instance.

26.1 The Classical Hall Effect

Let us consider an electron gas confined in a two-dimensional rectangular geometry with an electric current flowing in the x-direction. The conductivity matrix has the form

$$\underline{\sigma} = \begin{pmatrix} \sigma_{xx} & \sigma_{xy} \\ -\sigma_{xy} & \sigma_{xx} \end{pmatrix} = \begin{pmatrix} \sigma_0 & 0 \\ 0 & \sigma_0 \end{pmatrix} \quad ; \quad \sigma_0 = \frac{ne^2\tau}{m} \tag{26.1}$$

where σ_0 is the Drude conductivity, obtained in Section 14.4. Here, n is the electron surface density and τ is the characteristic time interval between the electron scatterings producing the resistance. The resistivity matrix is the inverse of $\underline{\sigma}$.

The above expression can be obtained within the Drude model from the balance between the force exerted by an electric field and the average opposing force due to electron scatterings occurring at an average time interval τ, when the average velocity is steady,

$$\frac{d\langle\mathbf{v}\rangle}{dt} = 0 \Longrightarrow -e\mathbf{E} = \frac{m}{\tau}\langle\mathbf{v}\rangle. \tag{26.2}$$

Then the above expression for the conductivity follows from the fact that the current density is $\mathbf{j} = -ne\langle\mathbf{v}\rangle$ and

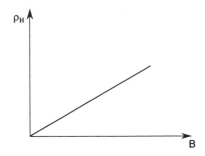

Figure 26.2 The classical Hall effect: transverse (Hall) resistivity as a function of the magnetic field

$$\mathbf{j}_i = \sigma_{ij}\mathbf{E}_j$$
$$\rho_{ij}\mathbf{j}_j = \mathbf{E}_i. \tag{26.3}$$

In the presence of a uniform magnetic field perpendicular to the plane, (26.2) is modified to

$$\frac{d\langle\mathbf{v}\rangle}{dt} = 0 \implies -e\mathbf{E} - \frac{e}{c}\langle\mathbf{v}\rangle \times \mathbf{B} = \frac{m}{\tau}\langle\mathbf{v}\rangle. \tag{26.4}$$

For a magnetic field along the z-direction, the solution for the resistivity matrix becomes

$$\underline{\rho} = \begin{pmatrix} \rho_{xx} & \rho_{xy} \\ -\rho_{xy} & \rho_{xx} \end{pmatrix} = \frac{1}{\sigma_0}\begin{pmatrix} 1 & \omega_C\tau \\ -\omega_C\tau & 1 \end{pmatrix} \quad ; \quad \omega_C = \frac{eB}{mc}. \tag{26.5}$$

The Hall resistivity becomes

$$\rho_H = \rho_{xy} = \frac{\omega_C\tau}{\sigma_0} = \frac{B}{nec}. \tag{26.6}$$

This is the magnetic field dependence observed by Hall in 1879 (see Fig. 26.2). A useful instrument for determining the charge sign and density of the current carriers is the Hall coefficient, defined as $R_H = \frac{\rho_H}{B}$. Notice that it does not depend on the field, but only on intrinsic properties of the material system.

26.2 The Integer Quantum Hall Effect

The Integer QHE manifests through the peculiar magnetic field dependence of the transverse resistivity or conductivity, which completely departs from the linear dependence obtained in the classical effect. The former is schematically represented in Fig. 26.3.

Notice the presence of plateaus in the transverse conductivity such that the corresponding longitudinal conductivity vanishes in the regions where the plateaus form:

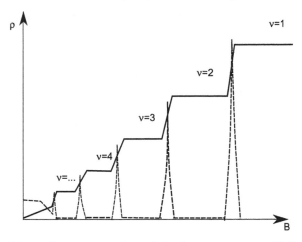

Figure 26.3 Schematic representation of the integer quantum Hall effect [226]. Solid line represents the transverse (Hall) resistivity and dashed line the longitudinal resistivity, both as a function of the magnetic field.

$$\rho = \begin{pmatrix} \rho_{xx} & \rho_{xy} \\ -\rho_{xy} & \rho_{xx} \end{pmatrix} = \underline{\sigma}^{-1} = \frac{1}{\sigma_H} \begin{pmatrix} 0 & 1 \\ -1 & 0 \end{pmatrix}, \tag{26.7}$$

where $\sigma_H = \nu\frac{e^2}{h}$, $\nu = 1, 2, 3, \ldots$, is the transverse part of the conductivity, namely

$$\underline{\sigma} = \begin{pmatrix} \sigma_{xx} & \sigma_H \\ -\sigma_H & \sigma_{xx} \end{pmatrix} = \underline{\rho}^{-1} = \begin{pmatrix} 0 & \sigma_H \\ -\sigma_H & 0 \end{pmatrix}. \tag{26.8}$$

We see that $\sigma_{xx} = \rho_{xx} = 0$ and $\sigma_{xy} = \rho_{xy}^{-1} = \sigma_H = \nu\frac{e^2}{h}$, $\nu = 1, 2, 3, \ldots$

We now describe the several ingredients required for understanding the integer QHE.

26.2.1 Landau Levels

Zero Electric Field

Consider first the Hamiltonian of non-interacting electrons on a plane, in the presence of a perpendicular magnetic field, which, in the Landau gauge, is associated to the vector potential $\mathbf{A} = (0, Bx)$. The Hamiltonian, then, is given by

$$H = \frac{1}{2m}|\mathbf{P} + \frac{e}{c}\mathbf{A}|^2$$

$$H = \frac{\mathbf{P}_x^2}{2m} + \frac{1}{2}m\omega_C^2 (x - x_0)^2$$

$$x_0 = \frac{\mathbf{P}_y c}{eB}. \tag{26.9}$$

We shall neglect the spin degrees of freedom, which at such high magnetic fields at which the effect is observed are frozen in the lowest of the states separated by the Zeeman splitting.

Since $[\mathbf{P}_y, H] = 0$, we may replace $\mathbf{P}_y$ by its eigenvalue $\hbar k_y$ and we see that the Hamiltonian describes a harmonic oscillator shifted by $x_0 = -\hbar k_y c/eB$.

The eigenfunctions with the corresponding eigenvalues

$$\psi_n(x, y) = C_n H_n(x - x_0) e^{-\frac{|x-x_0|^2}{2l_B^2}} e^{ik_y y}$$

$$E_n = \left(n + \frac{1}{2}\right) \hbar \omega_C \tag{26.10}$$

are the famous Landau levels [228]. In the above expression, H_n are Hermite polynomials and $l_B^2 = \hbar c/eB$ is the so-called magnetic length.

These are highly degenerate because the energy does not depend on k_y. For a system with dimensions L_x and L_y, wave functions with the same energy may be centered at a number of different points, given by the ratio between the width of the system and the minimum value of wave function center: $\overline{x}_0$, namely

$$G(n) = \frac{L_x}{\overline{x}_0} = \frac{L_x}{\hbar c \frac{2\pi}{eBL_y}}$$

$$G(n) = \frac{BL_xL_y}{hc/e} = \frac{\Phi}{\phi_0} = N_\Phi. \tag{26.11}$$

We see that the degeneracy of the Landau levels is given by the number of magnetic flux units that pierce through the plane, namely, the ratio of the total flux and the magnetic flux unit.

Nonzero Electric Field

In the presence of an electric field of magnitude E along the x-direction, the Hamiltonian becomes

$$H = \frac{1}{2m} \left|\mathbf{P} + \frac{e}{c}\mathbf{A}\right|^2 - eEx$$

$$H = \frac{\mathbf{P}_x^2}{2m} + \frac{1}{2}m\omega_C^2 \left(x - x_0(E, B)\right)^2 + \frac{\hbar k_y cE}{B} - \frac{1}{2}mc^2\frac{E^2}{B^2}$$

$$x_0(E, B) = \frac{\mathbf{P}_y c}{eB} - \frac{eE}{m\omega_C^2}. \tag{26.12}$$

The energy eigenvalues now become

$$E_n(k_y) = \left(n + \frac{1}{2}\right) \hbar \omega_C + \frac{\hbar k_y cE}{B} - \frac{1}{2}mc^2\frac{E^2}{B^2} \tag{26.13}$$

and we see that the degeneracy of the Landau levels is lifted.

We may now determine, from the quantum-mechanical treatment just exposed, the average velocity in the y-direction, and from it obtain the transverse current density $\mathbf{j}_y$. The average velocity corresponds to the group velocity

$$v_y = \frac{1}{\hbar} \frac{\partial E_n(k_y)}{\partial k_y} = \frac{cE}{B}. \tag{26.14}$$

From this we obtain

$$\mathbf{j}_y = ne v_y = \frac{nec}{B} \mathbf{E}_x = \sigma_{xy} \mathbf{E}_x. \tag{26.15}$$

Considering that $n = \frac{N_e}{A}$, where N_e is the number of electrons and A is the area, we can write the Hall conductivity as

$$\sigma_H = \sigma_{xy} = \frac{N_e ec}{BA} = \frac{N_e}{\frac{BA}{hc/e}} \frac{e^2}{h}$$

$$\sigma_H = \frac{N_e}{N_\Phi} \frac{e^2}{h}. \tag{26.16}$$

Now, we have just seen that N_Φ, the number of magnetic flux units, is the degeneracy of the Landau levels, hence we conclude that the Hall conductivity is given by

$$\sigma_H = \nu \frac{e^2}{h}, \tag{26.17}$$

where $\nu = N_e/N_\Phi$ is the occupation number of the Landau levels. For a fixed electron density, the number of occupied Landau levels will vary as we change the magnetic field, since this will change the maximum occupation capacity of each level.

We see that an integer multiple of e^2/h is the value obtained for the transverse conductivity whenever a Landau level is completely filled. It is likely, therefore, that the conductivity values observed at the plateaus are related to the complete filling of successive Landau levels. Basic questions, however, still remain unanswered: why there are plateaus? The set of values $\nu = 1, 2, 3, \ldots$ are just points in a curve conductivity $\times$ magnetic field. Why does the conductivity remain constant while the magnetic field is varied within a certain finite range, thus forming the plateaus? There is not a simple explanation. As we will see below, it involves a peculiar combination of disorder and topology.

26.2.2 The Effect of Disorder

Let us remember how the presence of disorder affects the spectrum of a system, or, more specifically, a non-interacting system such as the two-dimensional electron gas we have been considering.

According to (13.36), for a system of non-interacting electrons possessing one-particle states with discrete energies $E_n = \hbar\omega_n$, the Landau levels, the energy density of states is given by

$$N(E) = \sum_n \delta(E - E_n).\tag{26.18}$$

This can also be written as

$$N(E) = -\frac{1}{\pi}\text{Im}\sum_n \frac{1}{E - E_n + i\epsilon}.\tag{26.19}$$

In the presence of disorder, we replace this expression with

$$N(E) = -\frac{1}{\pi}\text{Im}\sum_n \frac{1}{E - E_n + i\Delta + i\epsilon},\tag{26.20}$$

where Δ conveys the information about the amount of disorder in the system, typically $\Delta \ll \hbar\omega_C$.

Now we can take the limit $\epsilon \to 0$, obtaining

$$N(E) = \frac{1}{\pi}\sum_n \frac{\Delta}{(E - E_n)^2 + \Delta^2}.\tag{26.21}$$

Disorder broadens the spectral weight, replacing the deltas with Lorentzians. Degeneracy of the Landau levels is lifted, but still the total number of available states in each band centered in E_n remains $G(n)$, given by (26.11).

Broadening of the spectral weight, however, is not the only effect produced by disorder. We have seen in Chapter 13 that disorder also produces the emergence of a mobility edge, such that only the states in the middle of the band would be extended, thus being able to conduct electrons. The states located away from the center, beyond the mobility edge would be localized, and therefore unable to transport electric charge.

Can the effects of disorder explain the QHE plateaus? Well, as we tune the magnetic field, say by decreasing it, we will decrease the number of available states in each band, thereby populating a larger number of bands to accommodate the N_e electrons. As we start filling a new band, however, the edge states would not conduct, hence the conductivity will remain fixed for a certain interval of values of the magnetic field: a plateau. As soon as we cross the mobility edge, the available states start to conduct and, consequently, the conductivity starts to increase until we find the other mobility edge and a new plateau forms.

This seems to be a promising explanation of the QHE plateaus, yet this cannot be the whole story. We have seen in the previous subsection that the integer ν appearing in the expression of the Hall conductivity is the ratio between the number of electrons and the number of available electron states in each band. For instance, for just one filled band, $\nu = 1$, meaning the number of states equals the number of electrons and $\sigma_H = e^2/h$. However, how can this be possible, considering that not all available states are conducting, only the fraction of states that are located in the center of the band?

In order to provide the answer, we must resort to topological properties of quantum mechanics, which reflect deep theorems of pure mathematics.

26.2.3 The Berry Phase

The Adiabatic Theorem

The adiabatic theorem [229] concerns the determination of the state vector in a quantum-mechanical system described by a Hamiltonian, which depends on a set of external parameters $\{\lambda_i(t)\}$, which may vary as a function of time. Then, the energy eigenvectors and eigenvalues will depend on time as well, namely,

$$H[\lambda_i(t)]|\psi_n[\lambda_i(t)]\rangle = E_n[\lambda_i(t)]|\psi_n[\lambda_i(t)]\rangle \qquad (26.22)$$

with

$$\langle \psi_n[\lambda_i(t)]|\psi_m[\lambda_i(t)]\rangle = \delta_{nm}$$
$$\sum_n |\psi_n[\lambda_i(t)]\rangle\langle\psi_n[\lambda_i(t)]| = \mathbb{I}. \qquad (26.23)$$

Assuming the variation is slow, the spectrum is discrete and there is no degeneracy among the energy eigenstates, then the theorem states that if the initial state of the system is one of the energy eigenvectors, as time evolves it will remain in the same state, except for a phase $\theta_n(t)$:

$$|\Psi(0)\rangle = |\psi_n[\lambda_i(t=0)]\rangle$$
$$|\Psi(t)\rangle = e^{i\theta_n(t)}|\psi_n[\lambda_i(t)]\rangle. \qquad (26.24)$$

The form of the phases has a physical meaning, because, assuming the initial state would be a linear combination of energy eigenstates, then the different phases would interfere, producing measurable effects.

In order to prove this result, let us use the completeness relation and expand the state vector as

$$|\Psi(t)\rangle = \sum_m c_m(t)|\psi_m[\lambda_i(t)]\rangle. \qquad (26.25)$$

Then, using the time evolution equation and applying $\langle \psi_n[\lambda_i(t)]|$, we get

$$\frac{dc_n(t)}{dt} = \left[-\frac{i}{\hbar}E_n(t) - \langle\psi_n|\frac{d\psi_n}{dt}\rangle\right]c_n(t) - \sum_{m\neq n}\langle\psi_n|\frac{d\psi_m}{dt}\rangle c_m(t), \qquad (26.26)$$

where we have dropped the λ_i dependence in order to simplify the notation.

Now, differentiating (26.22) with respect to time and applying $\langle \psi_m |; m \neq n$, we obtain

$$\langle \psi_m | \frac{d}{dt} | \psi_n \rangle (E_m - E_n) = \langle \psi_n | \frac{dH}{dt} | \psi_m \rangle, \qquad (26.27)$$

whereby we can write the last term in (26.26) as

$$\sum_{m \neq n} \frac{\langle \psi_n | \frac{dH}{dt} | \psi_m \rangle}{E_m - E_n} c_m(t). \qquad (26.28)$$

The result, so far, is exact. The adiabatic approximation consists in neglecting the last term in (26.26). Then, dividing by $c_n(t)$ and integrating in time, we obtain

$$c_n(t) = c_n(0) \exp \left\{ \frac{i}{\hbar} \int_0^t dt' E_n(t') + i\gamma_n \right\}, \qquad (26.29)$$

where

$$\gamma_n(t) = i \int_0^t dt' \langle \psi_n | \frac{d}{dt} | \psi_n \rangle \qquad (26.30)$$

is the so-called geometrical phase.

Result (26.29) is the proof of the adiabatic theorem. Inserting in (26.25), we see that it will have observable physical consequences.

The Berry Phase

In 1984, M. Berry noticed [230] that the geometrical phase could be written as a line integral going from $\{\lambda_i(0)\}$ to $\{\lambda_i(t)\}$, along a path in the parameter space:

$$\gamma_n(t) = i \int_0^t dt' \langle \psi_n | \frac{\partial}{\partial \lambda_i} | \psi_n \rangle \frac{d\lambda_i}{dt}$$

$$\gamma_n(t) = i \int_{C(t)} d\lambda_i \langle \psi_n | \frac{\partial}{\partial \lambda_i} | \psi_n \rangle. \qquad (26.31)$$

The geometrical phase, since then known as Berry phase, actually does not depend on the time, but, rather, on the path length in parameter space.

Defining the so-called Berry connection as

$$\mathcal{A}_i^n = i \langle \psi_n | \frac{\partial}{\partial \lambda_i} | \psi_n \rangle, \qquad (26.32)$$

we can express the Berry phase as

$$\gamma_n(C) = \int_C d\lambda_i \mathcal{A}_i^n. \qquad (26.33)$$

Particularly interesting is the case when C is a closed path in parameter space. Using Stokes' theorem, in this case, we have

$$\gamma_n(C) = \oint_C d\lambda \cdot A^n = \int_{S(C)} dS \cdot B^n \tag{26.34}$$

and we see that the Berry phase is the flux of the "magnetic field,"

$$B^n = \nabla \times A^n, \tag{26.35}$$

associated to the Berry connection through the surface $S(C)$ enclosed by the curve C.

It is convenient to introduce the Berry curvature tensor as

$$\mathcal{F}_{ij}^n = \partial_{\lambda_i} A_j^n - \partial_{\lambda_j} A_i^n. \tag{26.36}$$

In terms of this, the Berry phase becomes

$$\gamma_n(C) = \int_{S(C)} dS^{ij} \mathcal{F}_{ij}^n, \tag{26.37}$$

where

$$dS^{ij} = \frac{1}{2} dS^k \epsilon^{ijk}.$$

Inserting (26.36) and (26.32) in (26.37), we obtain

$$\gamma_n(C) = i \int_{S(C)} dS^k \epsilon^{ijk} \langle \frac{\partial}{\partial \lambda_i} \psi_n | \frac{\partial}{\partial \lambda_j} | \psi_n \rangle \tag{26.38}$$

or, introducing the complete set of energy eigenvectors,

$$\gamma_n(C) = i \int_{S(C)} dS^k \epsilon^{ijk} \sum_{m \neq n} \langle \frac{\partial}{\partial \lambda_i} \psi_n | \psi_m \rangle \langle \psi_m | \frac{\partial}{\partial \lambda_j} | \psi_n \rangle, \tag{26.39}$$

where the $n = m$ term vanishes. For the $n \neq m$ terms, differentiating (26.22) with respect to λ_i and applying $\langle \psi_m |$; $m \neq n$, we get

$$\gamma_n(C) = i \int_{S(C)} dS^k \epsilon^{ijk} \sum_{m \neq n} \frac{\langle \psi_n | \frac{\partial H}{\partial \lambda_i} | \psi_m \rangle \langle \psi_m | \frac{\partial H}{\partial \lambda_j} | \psi_n \rangle}{(E_m - E_n)^2}. \tag{26.40}$$

The First Chern Number

It is instructive to evaluate the Berry phase for a closed curve C belonging to a closed surface. Assume $S(C)$ in (26.34) is a sphere S^2 and the "magnetic field" is that of a monopole with magnetic charge q_M centered at the center of the sphere, namely

$$B^n = q_M \frac{\hat{r}}{r^2}.$$

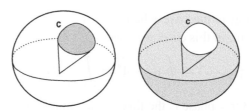

Figure 26.4 The closed curve C determines two complementary surfaces, the union of which is the spherical surface. The corresponding solid angles' sum is 4π.

Then $\gamma_n(C) = q_M\,\Omega(C)$, where $\Omega(C)$ is the solid angle corresponding to the curve C. Nevertheless, there are two complementary surfaces associated to the curve C. By choosing the other surface, we will find instead $\gamma_n(C) = -q_M(4\pi - \Omega(C))$. The consistency condition on $e^{i\gamma_n}$, then, imposes $4\pi q_M = 2\pi N$, with $N \in \mathbb{Z}$, or $2q_M = N$.

This observation has a strong consequence on the possible values of the total flux of the Berry magnetic field through the total sphere S^2:

$$\Phi(S^2) = \int_{S^2} d\mathbf{S} \cdot \mathcal{B}^n = 2\pi\mathcal{C} \;\; ; \;\; \mathcal{C} \in \mathbb{Z}. \tag{26.41}$$

This result actually holds in general for the integral of the Berry curvature tensor along any closed surface S, namely

$$\int_S \frac{d\mathbf{S}^{ij}}{2\pi} \mathcal{F}^n_{ij} = \mathcal{C} \;\; ; \;\; \mathcal{C} \in \mathbb{Z}, \tag{26.42}$$

where the integer $\mathcal{C} \in \mathbb{Z}$ is called the First Chern Number.

This result, obtained by Chern, is a generalization of the Gauss–Bonnet theorem. This is a deep theorem, which states that if we replace the Berry curvature by the intrinsic geometric curvature of the surface S in (26.42), we obtain $2(1 - g)$ as a result of the integral, where g is the "genus" of the surface S, which expresses the number of handles it has. The Gauss–Bonnet theorem is remarkable, among other things, for establishing a deep connection between two of the great areas of mathematics, namely, geometry and topology.

The Aharonov–Bohm effect

This beautiful effect [238] concerns a phase that the wave-function of a charged particle acquires due to the presence of a magnetic flux even if this vanishes where the particle is, provided the vector potential associated to the magnetic flux is nonzero at the particle's position. For a particle of charge q taken around a magnetic flux Φ, the wave-function will pick up a phase

$$\varphi = \frac{q\Phi}{\hbar c}. \tag{26.43}$$

One can show that this is precisely the Berry phase for the wave-function transported around the magnetic flux, so the Aharonov–Bohm effect is another manifestation of the Berry phase.

Generalized Statistics

We have seen in Chapter 10 that a charge with an attached magnetic flux will have its statistics changed by Δs, where

$$2\pi \Delta s = \frac{q\Phi}{\hbar c} \tag{26.44}$$

(in units where $\hbar, c \neq 1$). We now see that this follows from the Berry phase/Aharonov–Bohm picture.

Indeed, exchanging the positions of two charge-flux sets is equivalent to rotating one around the other by 180°. This, by its turn, is equivalent to rotating *one* charge around *one* flux by 360°. But this is the Aharonov–Bohm effect. That is why $2\pi \Delta s = \varphi$.

26.2.4 Topology and the QHE

Here we consider the last ingredient required for the explanation of the integer QHE. A crucial step is the realization that for a completely filled band, the states in the first Brillouin zone form a closed surface, namely, a torus. Then, as we show below, the transverse conductivity, expressed by the Kubo formula, can be written as an integral of the Berry curvature on this torus, being consequently proportional to the Chern First Number, which is an integer.

The Kubo Formula

According to the Kubo formula (4.61), the DC transverse conductivity of a state $|\psi\rangle_n$ is given by

$$\sigma_{xy} = \frac{i}{\omega} \left[\langle \psi_n | \mathbf{J}_x \mathbf{J}_y | \psi_n \rangle - \langle \psi_n | \mathbf{J}_x | \psi_n \rangle \langle \psi_n | \mathbf{J}_y | \psi_n \rangle \right]$$

$$\sigma_{xy} = \frac{i}{\omega} \sum_{m \neq n} \langle \psi_n | \mathbf{J}_x | \psi_m \rangle \langle \psi_m | \mathbf{J}_y | \psi_n \rangle. \tag{26.45}$$

or, equivalently,

$$\sigma_{xy} = \frac{i}{\hbar\omega} \sum_{m \neq n} \int_0^\infty dt\, e^{i[\omega + i\epsilon]t} \langle \psi_n | \mathbf{J}_x(t) | \psi_m \rangle \langle \psi_m | \mathbf{J}_y(0) | \psi_n \rangle, \tag{26.46}$$

where the limits $\mathbf{k}, \omega \to 0$ are implicitly understood.

This can be written as

$$\sigma_{xy} = \frac{i}{\hbar\omega} \sum_{m \neq n} \int_0^\infty dt\, e^{i[\omega + i\epsilon - \frac{E_m - E_n}{\hbar}]t} \langle \psi_n | \mathbf{J}_x(0) | \psi_m \rangle \langle \psi_m | \mathbf{J}_y(0) | \psi_n \rangle,$$

$$\sigma_{xy} = \frac{i}{\omega} \sum_{m \neq n} \frac{\langle \psi_n | \mathbf{J}_x | \psi_m \rangle \langle \psi_m | \mathbf{J}_y | \psi_n \rangle}{\hbar\omega + i\epsilon + E_m - E_n}. \tag{26.47}$$

In the limit $\omega \to 0$,

$$\left(\frac{i}{\omega} \right) \frac{1}{\hbar\omega + i\epsilon + E_m - E_n} \overset{\omega \to 0}{\to} \left(\frac{i}{\omega} \right) \frac{1}{E_m - E_n} + \frac{\hbar}{(E_m - E_n)^2}. \tag{26.48}$$

Inserting this in (26.47), we can see that the first term vanishes by $n \leftrightarrow m$ symmetry, whereas the second gives

$$\sigma_{xy} = i\hbar \sum_{m \neq n} \frac{\langle \psi_n | \mathbf{J}_x | \psi_m \rangle \langle \psi_m | \mathbf{J}_y | \psi_n \rangle}{(E_m - E_n)^2}. \tag{26.49}$$

This form of the Kubo formula will be convenient for what follows.

The TKNN Result

Let us derive here the connection between the Kubo formula for the transverse conductivity and the First Chern Number topological invariant. This important result was obtained by Thouless, Kohmoto, Nightingale and den Nijs in 1982 [231].

We start by observing that the current density operator differential element can be expressed as

$$d\mathbf{J}_i = \frac{e}{\hbar} \frac{d^2 k}{(2\pi)^2} \nabla_{k_i} H. \tag{26.50}$$

This follows from the fact that the group velocity can be expressed as $\mathbf{v} = \frac{1}{\hbar}\nabla H$, where H is the Hamiltonian, and also by observing that the density of states is given by $d\rho = \frac{1}{A} = \frac{d^2 k}{(2\pi)^2}$. The current density element then reads $d\mathbf{J}_i e d\rho v_i$. Choosing the states $|\psi_m\rangle$ as the non-interacting one-particle states associated to the nth Landau level, which completely fill the first Brillouin zone, and using (26.27), we have

$$\sigma_{xy} = i \frac{e^2}{\hbar} \int \frac{d^2 k}{(2\pi)^2} \int \frac{d^2 k'}{(2\pi)^2} \sum_{m \neq n} \langle \nabla_{k_x} \psi_n(\mathbf{k}) | \psi_m(\mathbf{k}') \rangle \langle \psi_m(\mathbf{k}') | \nabla_{k_y} \psi_n(\mathbf{k}) \rangle, \tag{26.51}$$

which yields

$$\sigma_{xy} = i \frac{e^2}{\hbar} \sum_n \int \frac{d^2k}{(2\pi)^2} \langle \nabla_{k_x} \psi_n(\mathbf{k}) | \nabla_{k_y} \psi_n(\mathbf{k}) \rangle. \tag{26.52}$$

This, however, is

$$\sigma_{xy} = \frac{e^2}{2\pi \hbar} \sum_n \int_{T^2} \frac{d^2k}{2\pi} \mathcal{F}_{xy}^n, \tag{26.53}$$

which is just the integral of the Berry curvature over the torus corresponding to a completely filled first Brillouin zone for the band associated by the nth Landau level. This is precisely the First Chern Number, hence we finally get

$$\sigma_{xy} = \frac{e^2}{h} \sum_n C_n. \tag{26.54}$$

This explains why the Hall conductivity is an integer multiple of e^2/h. The result above shows that the completely filled band with the mobility edge separating conducting from non-conducting states is in the same topological class as the band without disorder, where all the states are conducting. This is the key for understanding the effect.

26.3 The Fractional Quantum Hall Effect

The understanding of the integer QHE was achieved without considering the interactions that unavoidably exist among the electrons. The situation is somehow similar to the one found in the band theory of crystalline solids, in which the solutions of Schrödinger's equation for non-interacting electrons in the presence of a periodic potential lead to the Bloch theorem, the consequences of which are in excellent agreement with the experiments. Actually, the success of the non-interacting picture of electrons in solids was so good that it formed the prevailing paradigm of condensed matter physics for a long while. By the way, we have just used precisely this picture for the explanation of the integer QHE.

For describing the fractional QHE, however, we can no longer avoid the inclusion of interactions in our model. Instead of the single-particle Hamiltonian (26.9) in the presence of a uniform magnetic field, which is appropriate for non-interacting electrons, we must have

$$H = \frac{1}{2m} \sum_{i=1}^{N_e} |\mathbf{P}_i + \frac{e}{c}\mathbf{A}|^2 + \sum_{i<j} \frac{e^2}{|\mathbf{r}_i - \mathbf{r}_j|}, \tag{26.55}$$

where we included the static Coulomb potential among the electrons. Despite the fact that the electrons are far from being static, this may be justified by the fact that the electrons' speed is much less than the speed of light.

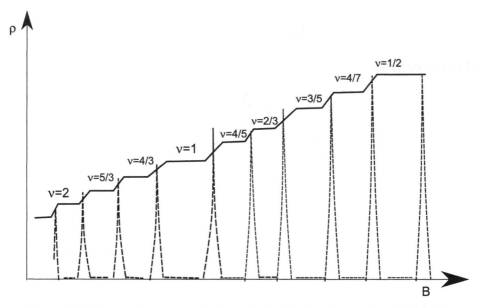

Figure 26.5 Schematic representation of the fractional quantum Hall effect [227]. Solid line represents the transverse (Hall) resistivity and dashed line the longitudinal resistivity, both as a function of the magnetic field.

26.3.1 The Laughlin Wave Function

In 1983, Laughlin proposed a wave-function for the ground state of the above Hamiltonian [232]. Using a complex notation $z = x + iy$ for the electron position in the plane, this wave-function reads

$$\Psi(z_1, \ldots, z_{N_e}) = C \prod_{i<j} (z_i - z_j)^m \exp\left\{ -\sum_i^{N_e} \frac{|z_i|^2}{4l_b^2} \right\}, \qquad (26.56)$$

where l_b is the magnetic length, introduced before and $m = 1, 3, 5, \ldots$ As we shall see, this wave-function describes the essential features of the FQHE.

Notice that Laughlin's ansatz is inspired on the eigenfunctions of the non-interacting case, which are given by sums of terms of the form

$$\Psi(z_1, \ldots, z_{N_e}) = C \prod_i z_i^n \exp\left\{ -\sum_i^{N_e} \frac{|z_i|^2}{4l_b^2} \right\}, \qquad (26.57)$$

and it takes into account the anti-symmetry of the many-electron wave-function, which imposes m as a positive odd integer.

The Coulomb Liquid

Consider the probability density corresponding to the wave-function (26.57), namely

$$|\Psi(z_1, \ldots, z_{N_e})|^2 = \prod_{i<j} \left(\frac{|z_i - z_j|^2}{l_b^2} \right)^m \exp\left\{ -\sum_i^{N_e} \frac{|z_i|^2}{2l_b^2} \right\}, \qquad (26.58)$$

where the normalization factor is chosen as $C = l_b^{-m \frac{N_e(N_e+1)}{2}}$. The quantum average of a quantity $A(z_1, \ldots, z_{N_e})$ will be, then, given by

$$\langle A \rangle = \int dz_1 \ldots dz_{N_e} A(z_1, \ldots, z_{N_e}) |\Psi(z_1, \ldots, z_{N_e})|^2. \qquad (26.59)$$

We can write the probability density as

$$|\Psi(z_1, \ldots, z_{N_e})|^2 = \exp\left\{ 2m \sum_{i<j} \ln \frac{|z_i - z_j|}{l_b} - \sum_i^{N_e} \frac{|z_i|^2}{2l_b^2} \right\}$$

$$\equiv e^{-\beta U(z_1, \ldots, z_{N_e})}, \qquad (26.60)$$

where $U(z_1, \ldots, z_{N_e})$ is formally identical to the energy of a classical gas of point charges, located at the positions z_i in a uniformly charged background of the opposite sign. Then, $\langle A \rangle$ becomes a classical average in the canonical ensemble at a temperature $T = 1/k_B\beta$.

Choosing $\beta \equiv 2/m$, we have

$$U(z_1, \ldots, z_{N_e}) = -m^2 \sum_{i<j} \ln \frac{|z_i - z_j|}{l_b} + m \sum_i^{N_e} \frac{|z_i|^2}{4l_b^2}. \qquad (26.61)$$

This is the electrostatic energy of a fictitious system of N_e point particles of charge $q = -m$ on a uniformly charged background of total charge $+\overline{\rho}A$.

The charge density of this system is

$$\rho(z; z_1, \ldots, z_{N_e}) = -m \sum_i^{N_e} \delta(z - z_i) + \overline{\rho}. \qquad (26.62)$$

Equilibrium requires the neutrality of the associated system. This imposes $\overline{\rho} = +\frac{N_e m}{A}$.

Now, using the Poisson equation, relating the scalar potential to the charge density, namely

$$-\nabla^2 \varphi(z) = 2\pi \rho(z), \qquad (26.63)$$

it follows that

$$2\pi \overline{\rho} = -\nabla^2 \left(-\frac{|z_i|^2}{4l_b^2}\right) = \frac{1}{l_b^2}$$

$$2\pi \frac{N_e m}{A} = \frac{eB}{\hbar c} \implies \nu \equiv \frac{N_e}{BA/\Phi_0} = \frac{1}{m}. \tag{26.64}$$

This explains the series of Hall plateaus: $\nu = \frac{1}{3}, \frac{1}{5}, \frac{1}{7}, \dots$

Composite Fermions

The statistics of an electron does not change if we attach to it an even number of flux quanta. According to Laughlin's wave-function, if we transport an electron along a closed path, back to its original position, it will pick up a phase $2\pi m$, namely, an amount of m flux quanta.

We can write Laughlin's wave function in the form

$$\Psi(z_1, \dots, z_{N_e}) = C \prod_{i<j} (z_i - z_j)^{m-1} \prod_{i<j} (z_i - z_j) \exp\left\{-\sum_i^{N_e} \frac{|z_i|^2}{4l_b^2}\right\}. \tag{26.65}$$

This can be interpreted as the wave-function of N_e non-interacting composite fermions, each one constituted by an electron with $m-1$ attached units of magnetic flux [233]. Within this picture, the effective magnetic field felt by the composite fermions becomes

$$B^* = B \pm N_e(m-1)\frac{\Phi_0}{A}, \tag{26.66}$$

where $\Phi_0 = hc/e$ is the flux quantum, A is the area of the system and the $\pm$ signs correspond, respectively, to a situation where the attached magnetic fluxes are parallel or anti-parallel to the applied magnetic field.

Remembering that the filling fraction is $\nu = N_e/(BA/\Phi_0)$, we can cast (26.66) in the form

$$\nu = \frac{\nu^*}{1 + (m-1)\nu^*} \quad ; \quad \nu = \frac{\nu^*}{(m-1)\nu^* - 1}, \tag{26.67}$$

where $\nu^* = N_e/(B^*A/\Phi_0)$ is the filling factor of the composite fermions: $\nu^* = 1, 2, 3, \dots$ From this, we obtain, among others, the following Hall plateaus: $\nu = \frac{1}{3}, \frac{1}{5}, \frac{1}{7} \dots, \nu = \frac{2}{3}, \frac{2}{5}, \frac{2}{7} \dots; \nu = \frac{3}{5}, \frac{3}{7}, \frac{3}{9} \dots$ These are the ones observed in Fig. 1.8 in the FQHE.

26.3.2 Excitations with Fractional Charge and Statistics

Interacting systems usually exhibit surprising features in their spectrum of quantum excitations. It frequently happens that these excitations keep no trace of the original degrees of freedom, in terms of which the corresponding Hamiltonian is formulated. Well-known examples are QCD, where one starts formulating a theory in terms of colored quarks and gluons and ends up with a spectrum containing colorless hadrons; QED in 1+1D, which is formulated in terms of charged fermions and "photons," whereas the spectrum contains just a neutral massive scalar boson; the Tomonaga–Luttinger and Hubbard models, where one starts with fermions with charge and spin, as we saw in Chapter 16, and at the end, obtains a spectrum of bosonic excitations with separated charge and spin degrees of freedom; and polyacetylene, where one starts with a Hamiltonian with gapless electrons and phonons and ultimately obtains a spectrum containing gapped electrons and spinless charged soliton excitations.

A similar phenomenon occurs in the system undergoing the fractional QHE. Starting from an electron gas confined on a quasi-two-dimensional region, subject to the mutual Coulomb repulsion and to a strong perpendicular magnetic field, we reached a quantum-mechanical description in terms of the Laughlin wave-function, which accounts for the observed features. Let us now examine what are the elementary excitations obtained out of it. One can verify that, out of Laughlin's state one can construct quasi-particle an quasi-hole elementary excitations, as well as collective excitations. Here, we are going to concentrate on the quasi-holes.

Quasi-Holes with Fractional Charge

Consider the wave-function

$$\Psi_H(z_i; \omega) = C \prod_{i=1}^{N_e} (z_i - \omega) \prod_{i<j} (z_i - z_j)^m \exp\left\{-\sum_i^{N_e} \frac{|z_i|^2}{4l_b^2}\right\} \tag{26.68}$$

obtained from Laughlin's wave-function by introducing the first multiplicative factor. It vanishes at the position corresponding to the complex parameter ω. Since this is not an electron's position, such wave-function describes a state where the electron density is zero, at $z_i = \omega$. This is what we call a hole. It is naturally identified as an excited state, not only because the Laughlin state is, by hypothesis, the ground state, but also because it has an additional node.

Suppose now we introduce m of such holes at the positions ω_k, $k = 1, 2, \ldots, m$, so that when we bring the m holes together at the point ω, the resulting wave-function becomes

$$\Psi_{mH}(z_i; \omega_1, \ldots, \omega_m) \to C \prod_{i=1}^{N_e} (z_i - \omega)^m \prod_{i<j} (z_i - z_j)^m \exp\left\{-\sum_i^{N_e} \frac{|z_i|^2}{4l_b^2}\right\}. \tag{26.69}$$

We see it would be identical to Laughlin's wave function of $N_e + 1$ electrons if ω were the additional electron's position. Since it is not, it represents a (multiple) hole associated with the deficit of this electron. The charge of the above state, therefore, must be the opposite of one electron's charge, namely $+e$. Now, since Ψ_{mH} results from the combination of m wave-functions $\Psi_H(z_i; \omega)$, it follows immediately that the elementary excitation described by the wave-function (26.68) possesses a fractional charge $+e/m$. This amazing prediction about systems undergoing the FQHE was experimentally confirmed in 1997 [235].

Quasi-Holes with Fractional Statistics

It was Halperin who first pointed out that the excited quasi-holes with fractional charge, obtained out of Laughlin's wave-function, should have fractional statistics as well [234]. This was confirmed by a calculation based on the Berry phase/Aharonov–Bohm effect [236]. Yet, the fact that quasi-holes have generalized statistics can be inferred by a simple reasoning, which follows below.

We have just seen that the quasi-holes possess a fractional charge $q = e/m$. From the wave-function Ψ_{mH}, (26.69), describing m holes we infer that m magnetic fluxes are attached to each of the points $\omega, z_1, \ldots, z_{N_e}$. Now, considering that in the point ω there are m quasi-holes, it follows that each quasi-hole bears one unit of magnetic flux: $\Phi = \Phi_0 = \frac{2\pi\hbar c}{e}$, apart from a charge $q = e/m$.

It follows then, from (26.44), that the quasi-holes have statistics

$$s = \frac{1}{m}, \tag{26.70}$$

being therefore an example of excitations of generalized statistics occurring in a realistic system.

Quasi-holes with a fractional charge $e/3$ and fractional statistics $s = 1/3$ were experimentally observed in 2005 [237] in the $\nu = 1/3$ plateau of a material system undergoing the FQHE.

26.4 The Zhang–Hansson–Kivelson Theory

Even though BCS theory provides a complete microscopic description of superconductivity, still the phenomenological Landau–Ginzburg theory frequently offers a picture of this phenomenon that is at least as useful. Similarly, it would be very convenient to count on a Landau–Ginzburg-type theory of the QHE. This is achieved by the Zhang–Hansson–Kivelson (ZHK) theory of the QHE [239].

The basic idea is to use the mechanism of statistical transmutation presented in Chapter 11 in order to describe the electrons in the d = 2 gas undergoing the QHE as a boson field ϕ with an odd number of attached magnetic flux quanta provided by a Chern–Simons field $\mathcal{A}_\mu$.

According to the results of subsection 11.2.2, by coupling a matter field to a Chern-Simons term with a coefficient $\theta/2$, a magnetic flux $\Phi = \frac{e}{\theta}$ is attached to each charged particle associated to this field. The spin/statistics of such a field, accordingly, gets changed by an amount $\Delta s = \frac{e^2}{4\pi\theta}$, as we can infer from (11.13).

Considering that the spin/statistics parameter s should change by a factor $\Delta s = (2k + 1)\frac{1}{2}$; $k \in \mathbb{N}$, in order to comply with the fact that the electron is a fermion, we must have $1/\theta = \frac{2\pi}{e^2}(2k + 1)$. The total magnetic flux thereby attached to each bosonic particle is, therefore, $\Phi = (2k + 1)\frac{2\pi}{e}$ or, reinstating the physical units, $\Phi = (2k + 1)\frac{hc}{e}$. We see that the electron may be associated to a bosonic scalar field plus an odd number of magnetic flux units attached to it. Again, we are going to neglect the electronic spin degrees of freedom, which are supposedly frozen by the strong external magnetic field.

The action that will be used to describe this system can be written as [239]

$$S = \int d^3x \left\{ i\phi^* D_0\phi + \frac{|\mathbf{D}\phi|^2}{2m} + \frac{\theta}{2}\epsilon^{\mu\alpha\beta}\mathcal{A}_\mu\partial_\alpha\mathcal{A}_\beta + \mu|\phi|^2 - \frac{\lambda}{2}|\phi|^4 \right\}$$
$$S = S_\phi\left[\phi, A_\mu + \mathcal{A}_\mu\right] + S_{CS}\left[\mathcal{A}_\mu\right], \tag{26.71}$$

where the last term is the Chern–Simons term.

In the above expression $D_\mu = \partial_\mu + i\frac{e}{c}\left(A_\mu + \mathcal{A}_\mu\right)$, where $\mathcal{A}_\mu$ is the statistical gauge field and A_μ is the external applied electromagnetic field. ϕ is the bosonic field, which, when appended $2k+1$ magnetic fluxes of the statistical field, describes the electron (with frozen spin). μ is the chemical potential, and the last term is an approximation for the interaction

$$\frac{1}{2}\int d^3x d^3y |\phi(x)|^2 V(x-y) |\phi(y)|^2$$
$$V(x-y) \simeq \lambda\delta^3(x-y). \tag{26.72}$$

The mean-field solution to the ZHK theory corresponds to the stationary conditions

$$\frac{\delta S}{\delta\phi^*} = \frac{\delta S}{\delta\phi} = \frac{\delta S}{\delta\mathcal{A}_\mu} = 0, \tag{26.73}$$

which imply

$$i D_0\phi + \frac{\mathbf{D}^2\phi}{2m} = -\phi\left[\mu - \lambda|\phi|^2\right] \tag{26.74}$$

and

$$\frac{\delta S_\phi}{\delta\mathcal{A}_\mu} = -\theta\epsilon^{\mu\alpha\beta}\partial_\alpha\mathcal{A}_\beta. \tag{26.75}$$

The first term above is just the charge current j_μ, hence, we have

$$j^\mu = -\theta\epsilon^{\mu\alpha\beta}\partial_\alpha\mathcal{A}_\beta. \tag{26.76}$$

Eq. (26.74) admits a constant solution: $\partial_0\phi = \nabla\phi = 0$, with

$$\mathcal{A}_\mu = -A_\mu$$
$$\phi\left[\mu - \lambda|\phi|^2\right] = 0 \quad ; \quad |\phi|^2 = \frac{\mu}{\lambda}, \tag{26.77}$$

which characterizes an incompressible fluid, in view of the constant density.

Inserting the first part of (26.77) in (26.76), then, gives

$$j^\mu = \theta \epsilon^{\mu\alpha\beta} \partial_\alpha A_\beta. \tag{26.78}$$

The time and spatial components of this read

$$j^0 = \theta B$$
$$j^i = \theta \epsilon^{ij} E^j, \tag{26.79}$$

where E^i and B are the applied electric and magnetic fields. We conclude that the Chern–Simons parameter θ is just the Hall conductivity. As we have seen, in physical units this is given by $\theta = \frac{e^2}{h}\frac{1}{2k+1}$ to ensure that electrons are fermions, hence we reproduce in the ZHK theory the expression for the Hall conductivity derived in Laughlin's theory.

We see that the applied external field cancels the internal Chern–Simons statistical field, producing an effective "Meissner effect," effectively fixing $\mathcal{B} = -B$. Since the first of the previous equations relates this to the matter density, we conclude that the ZHK theory describes an incompressible fluid, precisely in the same way as the Laughlin theory.

By changing the applied field B, we will change the CS field, $\mathcal{B}$, accordingly. This corresponds to a number $2k + 1$ of magnetic flux lines piercing the plane. As we increase $\mathcal{B}$, assuming the behavior of a type-II superconductor, the bulk properties of the fluid would remain unchanged until we reach a critical value of the field, at which a new flux unit will pierce the plane. This explains the formation of plateaus.

The theory admits vortex excitations, which at the classical level correspond to fields with the large-distance behavior

$$\phi(\mathbf{r}) \xrightarrow{\mathbf{r} \sim \infty} \rho_0 e^{i\,\arg(\mathbf{r})}$$
$$\mathcal{A}_i(\mathbf{r}) \xrightarrow{\mathbf{r} \sim \infty} \nabla_i \arg(\mathbf{r})$$
$$\mathcal{A}_0(\mathbf{r}) = 0. \tag{26.80}$$

These vortex excitations carry magnetic flux and charge given, respectively by

$$\Phi = \oint \mathcal{A} \cdot d\mathbf{l} = 2\pi$$
$$Q = \int d^2x\, j^0 = -\theta \int d^2x \mathcal{B} = -\theta\Phi = -2\pi\theta. \tag{26.81}$$

Considering that $\theta = \frac{1}{2\pi(2k+1)}$ in natural units, we see that the vortex excitations will carry a charge $Q = \frac{1}{(2k+1)}$. The minus sign indicates the vortex charge has opposite sign of the electron charge, therefore characterizing these excitations as holes, in analogy to the related excitations of Laughlin's wave-function. These vortices will possess spin/statistics $s = Q\Phi/2\pi = \frac{1}{(2k+1)}$. Observe, consequently, that the vortex excitations of the ZHK theory possess the same quantum numbers as the hole excitations of Laughlin's theory, in particular the charge and spin. The whole hierarchy of quantum Hall plateaus may be derived from the ZHK theory as well, by following the same steps as in the framework of Laughlin's theory or in the composite fermion picture [240]. Consider, for instance, the piece of the action

$$S_0 = \int d^3x \left[j_\mu(A_\mu + \mathcal{A}_\mu) + \frac{e^2}{2h}\frac{1}{2k+1}\epsilon^{\mu\alpha\beta}\mathcal{A}_\mu\partial_\alpha\mathcal{A}_\beta \right], \tag{26.82}$$

which imparts an odd number of magnetic fluxes on the particles associated to the current j_μ, leading to a Hall conductivity

$$\frac{1}{2k+1}\frac{e^2}{h}.$$

We may re-write this as the following integral on $\tilde{\mathcal{A}}_\mu$,

$$e^{\frac{i}{\hbar}S_0} = \int D\tilde{\mathcal{A}}_\mu \exp\left\{ \frac{i}{\hbar}\int d^3x \left[j_\mu(A_\mu + \mathcal{A}_\mu) + \frac{e^2}{2h}(2k+1)\epsilon^{\mu\alpha\beta}\tilde{\mathcal{A}}_\mu\partial_\alpha\tilde{\mathcal{A}}_\beta \right.\right.$$
$$\left.\left. + \epsilon^{\mu\alpha\beta}\mathcal{A}_\mu\partial_\alpha\tilde{\mathcal{A}}_\beta \right]\right\}. \tag{26.83}$$

Going one step further, we can impart an even number of magnetic flux units to the particles associated to the current j_μ by integrating over the field $\tilde{\tilde{\mathcal{A}}}_\mu$

$$\int D\tilde{\mathcal{A}}_\mu D\tilde{\tilde{\mathcal{A}}}_\mu \exp\left\{ \frac{i}{\hbar}\int d^3x \left[j_\mu(A_\mu + \mathcal{A}_\mu) + \frac{e^2}{2h}(2k+1)\epsilon^{\mu\alpha\beta}\tilde{\mathcal{A}}_\mu\partial_\alpha\tilde{\mathcal{A}}_\beta \right.\right.$$
$$\left.\left. + \epsilon^{\mu\alpha\beta}\mathcal{A}_\mu\partial_\alpha\tilde{\mathcal{A}}_\beta \pm \frac{e^2}{2h}(2l)\epsilon^{\mu\alpha\beta}\tilde{\tilde{\mathcal{A}}}_\mu\partial_\alpha\tilde{\tilde{\mathcal{A}}}_\beta + \epsilon^{\mu\alpha\beta}\tilde{\mathcal{A}}_\mu\partial_\alpha\tilde{\tilde{\mathcal{A}}}_\beta \right]\right\}. \tag{26.84}$$

Integrating in $\tilde{\mathcal{A}}_\mu$ and in $\tilde{\tilde{\mathcal{A}}}_\mu$, we see that the coefficient in (26.83) becomes

$$\frac{e^2}{2h}(2k+1) \longrightarrow \frac{e^2}{2h}\left[(2k+1) \pm \frac{1}{2l} \right] \tag{26.85}$$

in such a way that the effective filling fraction becomes

$$\nu = \frac{1}{(2k+1) \pm \frac{1}{2l}}. \tag{26.86}$$

Repeating this procedure, we eventually obtain the whole hierarchy

$$\nu = \frac{1}{m \pm \dfrac{1}{m_1 \pm \frac{1}{m_2 \pm \cdots}}} \tag{26.87}$$

where m_i are even integers. This can be shown to be equivalent to the one obtained within the composite fermion picture.

26.5 The Edges

The fact that systems undergoing the QHE are finite implies there are boundaries within which the electrons are confined. The confining potential bends the Landau levels near the boundaries, in such a way that the gap separating these energy levels, which exists in the bulk, disappears in the edges, leading consequently to conducting edge states.

Already within a classical picture we may infer that there will be edge currents. Indeed, applying a magnetic field to a system of electrons moving on the plane will produce a stable state where the electrons move in circles oriented in the same direction. This will naturally produce a net edge current in the same sense, while the net current in the bulk would be zero.

A more detailed description can be achieved by calling x and y, respectively, the coordinates of the quantum Hall fluid along the edge and the one perpendicular to it. Then, the differential element of charge at a certain point of the edge will be

$$dQ_E = enydx \equiv \rho_E dx, \qquad (26.88)$$

where $n = \frac{N_e}{A}$ is the electronic surface density and $\rho_E = eny$ is the linear charge density on the edge. The current along the edge is

$$I_E = eny\frac{dx}{dt} = enyv_E = \rho_E v_E. \qquad (26.89)$$

Since in $d = 1$ current coincides with current density, we have $j_E = \rho_E v_E$, which, along with the linear density, satisfies the continuity equation

$$\frac{\partial \rho_E}{\partial t} + \frac{\partial j_E}{\partial x} = 0. \qquad (26.90)$$

The confining potential at the edge produces an electric field of magnitude $E_E = \frac{v_E}{c}B$, perpendicular to the edge. We can, therefore, write the edge current density as $j_E = eny\frac{cE_E}{B}$.

Now, considering that $\frac{nc}{B} = \nu\frac{e}{h}$, we find that the linear current density is

$$j_E = \nu y\frac{e^2}{h}E_E. \qquad (26.91)$$

Since the $d = 2$ current density is given in terms of the edge linear current as $j_{2d} = j_E/y$, we have $j_{2d} = \nu\frac{e^2}{h}E_E$, which is precisely the Hall current.

We find that a one-to-one correspondence exists between the filled bulk states of the degenerate, gapped, Landau levels and the conducting gapless edge states.

The modulus of the Chern number, therefore, expresses the number of such gapless edge states:

$$|\mathcal{C}| = \#\text{of gapless edge states.} \qquad (26.92)$$

This double character; bulk insulator, edge metal, first observed in systems exhibiting a quantum Hall phase, is the general feature of a new class of condensed matter systems, namely the topological insulators, which we will study in Chapter 29.

Now, from (26.88) to (26.90), it follows that the edge charge and current densities may be written in terms of a bosonic field as

$$\rho_E = \frac{1}{\sqrt{\pi}}\partial_x\phi(x - v_E t)$$

$$j_E = -\frac{1}{\sqrt{\pi}}\partial_t\phi(x - v_E t). \qquad (26.93)$$

Because of its peculiar space and time dependence, the ϕ excitations move in a unique sense along the edges. Consequently, the edge current will also have a unique orientation, hence being a chiral current. The ϕ-field, accordingly, is a chiral boson.

The electric potential corresponding to the field E_E is $V_E = eE_E y$. It follows that the interaction energy of the linear edge charge density will be

$$H_I = \int dx e^2 n E_E y^2(x) = \int dx e^2 n v_E B y^2(x)$$

$$H_I = \frac{hc}{e}\frac{v_E}{v}\int dx \rho_L^2 = \frac{hc}{e}\frac{v_E}{\pi v}\int dx \partial_x\phi\partial_x\phi = \frac{hc}{e}\frac{v_E}{4\pi v}\int dx \partial_-\phi\partial_-\phi$$

$$\longrightarrow \frac{v_E}{2v}\int dx \partial_-\phi\partial_-\phi, \qquad (26.94)$$

where $\partial_\pm = \partial_x \pm \frac{1}{v_E}\partial_t$ and we returned to the natural units system in the last step. We see that the only excitations are the bosonic, Tomonaga–Luttinger charged collective gapless excitations $\phi(x_-)$, in terms of which the current density is given by (26.93).

26.6 Even Denominators

Even though most of the plateaus observed in systems undergoing the FQHE correspond to fractions with odd denominators, there are clear exceptions, such as the states with $v = 5/2, 7/2 \ldots$ A wave-function describing these states has been proposed by Moore and Read [241]. It is a generalization of Laughlin's wave-function, given by

$$\Psi(z_1, \ldots, z_{N_e})_{MR} = C \operatorname{Pf}\left(z_i - z_j\right) \prod_{i<j}\left(z_i - z_j\right)^m \exp\left\{-\sum_i^{N_e} \frac{|z_i|^2}{4l_b^2}\right\}, \quad (26.95)$$

where l_b is the magnetic length, introduced before and $m = 2, 4, 6, \ldots \operatorname{Pf}\left(z_i - z_j\right)$ is the Pfaffian or completely anti-symmetrized product of pairs $(z_i - z_j)$. The wave-function, as a whole, is completely anti-symmetric since, for m even, the second factor is completely symmetric. It is, consequently, apt to describe electrons. It corresponds, indeed, to a fermionic state with occupation fraction $\nu = 1/m$. The plateau $5/2 = 2 + 1/2$, for instance, would correspond to two full Landau levels plus a half-filled one with a wave-function with $m = 2$. One can obtain the excitations above the Moore–Read state proceeding in a similar way as we did in the case of Laughlin's wave-function. As it turns out, these are quasi-holes with non-abelian statistics. This is a completely novel feature that has produced an enormous amount of new results, actually opening new areas of research such as topological quantum computation. We will consider this with more detail in Chapter 30.

It has been shown by Fradkin, Nayak, Tsvelik and Wilczek [242] that, in the same way that the ZHK provides a QFT description of the state associated with the Laughlin wave-function, the QFT that describes Moore–Read state, which is associated to the Pfaffian wave-function, is the non-abelian SU(2) Chern–Simons theory of level 2 ($k = 2$). This theory possesses vortex excitations obeying non-abelian statistics, which are the QFT analogs of the Moore–Read quasi-hole excitations. The Majorana states constructed out of these non-abelian anyons form qubits that can be used in quantum computation, and are robust against loss of quantum coherence. We will come to this point in Chapter 30.

27

Graphene

Graphene is one of the most remarkable materials ever found. Also one of the most studied. It presents a number of unique features that have attracted the attention of both theoreticians and experimentalists. Investigation of its properties has led to breakthroughs, not only from the perspective of fundamental research but also from the point of view of applied science. Theoretically conjectured long ago, it was concretely obtained in 2004 by Geim and Novoselov [243]. Graphene properties include an outstanding mechanical robustness, being orders of magnitude stronger than steel; high electric and thermal conductivities, despite the absence of a Fermi surface; finite resistivity even without impurities, despite the absence of a gap; and relativistic dispersion relation for the active electrons, implying their kinematics are described by the Dirac equation and not by the Schrödinger equation, among others. This last property makes of graphene a concrete realization of the Dirac sea, a concept that in spite of not manifesting itself in nature, in the absence of matter has enabled Dirac to predict the existence of antimatter. The observation of antimatter *in vacuo* and the subsequent concrete realization of the Dirac sea in a material system such as graphene is an outstanding example of the great unity that exists in physics. Other properties of Dirac particles such as the Klein tunneling have been observed as well in graphene. As a consequence of charge conjugation symmetry, both electrons and holes possess the same mobility in graphene, a feature that is not found, for instance, in regular (*Si* or *Ge* based) semiconductors. Further properties of this extraordinary material include, for instance, the occurrence of the integer and fractional quantum Hall effects, in the presence of an external perpendicular magnetic field and the Zitterbewegung.

27.1 Crystal Structure and Tight-Binding Approach

Graphene is a one-atom-wide sheet of carbon with a sp^2 hybridization, assembled in a honeycomb crystal structure, consisting of a Bravais triangular lattice with spacing a and a base of two atoms, respectively placed at $(0, 0)$ and $(0, h)$, with respect to the Bravais lattice sites. We have $h = a/\sqrt{3} = 0.142nm$.

The primitive vectors of this can be taken as

$$\mathbf{a}_1 = a\hat{x} \qquad \mathbf{a}_2 = a\left(\frac{1}{2}\hat{x} + \frac{\sqrt{3}}{2}\hat{y}\right). \tag{27.1}$$

The corresponding vectors of the reciprocal lattice are

$$\mathbf{b}_1 = \frac{4\pi}{a}\left(\frac{1}{2}\hat{x} - \frac{\sqrt{3}}{2}\hat{y}\right) \qquad \mathbf{b}_2 = \frac{4\pi}{a\sqrt{3}}\hat{y}. \tag{27.2}$$

The first Brillouin zone is a hexagon, centered at the origin, with a side $l = 4\pi/3a$.

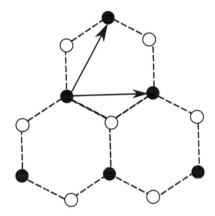

Figure 27.1 The crystal structure of graphene with the primitive vectors $\mathbf{a}_1$ and $\mathbf{a}_2$

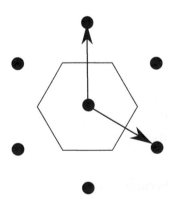

Figure 27.2 The reciprocal lattice of graphene with the corresponding primitive vectors $\mathbf{b}_1$ and $\mathbf{b}_2$ and the first Brillouin zone: an hexagon with side $l = 4\pi/3a$

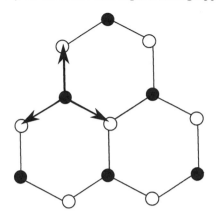

Figure 27.3 The crystal structure of graphene with the nearest neighbor vectors, $\mathbf{d}_i$, $i = 1, 2, 3$

Let us consider now the tight-binding approach to graphene. In this, we will be concerned with the electron occupying the non-hybridized, p_z orbitals of carbon. This is the dynamical degree of freedom, which is relevant for most of the physical properties of graphene, especially the electronic and transport properties.

For this purpose, notice that the nearest neighbors belong to different sublattices, which we denote A and B. Given a point in one certain sublattice, it will have three nearest neighbors in the opposite sublattice, located at the points corresponding to the three vectors,

$$\mathbf{d}_1 = \frac{a}{\sqrt{3}}\hat{y} \quad \mathbf{d}_2 = -\frac{a}{\sqrt{3}}\left(\frac{\sqrt{3}}{2}\hat{x} + \frac{1}{2}\hat{y}\right) \quad \mathbf{d}_3 = \frac{a}{\sqrt{3}}\left(\frac{\sqrt{3}}{2}\hat{x} - \frac{1}{2}\hat{y}\right). \quad (27.3)$$

The corresponding tight-binding Hamiltonian will be

$$H_{TB} = -t \sum_{\mathbf{R}, i=1,2,3, \sigma=\uparrow,\downarrow} \left[c_B^\dagger(\mathbf{R} + \mathbf{d}_i, \sigma)c_A(\mathbf{R}, \sigma) + H.C.\right], \quad (27.4)$$

where $c_{A,B}^\dagger(\mathbf{r}, \sigma)$ is the creation operator of an electron, with spin σ, in the orbital p_z of the carbon atom located at the position $\mathbf{r}$ in sublattice: A, B.

Introducing

$$c_{A,B}(\mathbf{r}, \sigma) = \frac{1}{\sqrt{N}} \sum_{\mathbf{k}} e^{i\mathbf{k}\cdot\mathbf{r}} c_{A,B}(\mathbf{k}, \sigma), \quad (27.5)$$

we may re-write the Hamiltonian as

$$H_{TB} = \sum_{\mathbf{k},\sigma} c_A^\dagger(\mathbf{k}, \sigma)c_B(\mathbf{k}, \sigma)\left[-t \sum_{i=1,2,3} e^{i\mathbf{k}\cdot\mathbf{d}_i}\right] + H.C. \quad (27.6)$$

Introducing $\Psi^\dagger(\mathbf{k}, \sigma) = (c_A^\dagger(\mathbf{k}, \sigma) c_B^\dagger(\mathbf{k}, \sigma))$, we may express this as

$$H_{TB} = \sum_{\mathbf{k},\sigma} \Psi^\dagger(\mathbf{k}, \sigma) \begin{pmatrix} 0 & \phi \\ \phi^* & 0 \end{pmatrix} \Psi(\mathbf{k}, \sigma), \qquad (27.7)$$

where

$$\phi(\mathbf{k}) = -t \sum_{i=1,2,3} e^{i\mathbf{k}\cdot\mathbf{d}_i}. \qquad (27.8)$$

It follows, immediately, that the energy eigenvalues are

$$E(\mathbf{k}) = \pm|\phi(\mathbf{k})| = \pm t \sqrt{\sum_{i,j=1,2,3} e^{i\mathbf{k}\cdot(\mathbf{d}_i - \mathbf{d}_j)}}. \qquad (27.9)$$

This expression gives the two bands of graphene, first obtained by Wallace in 1947 [303]. The two bands touch at the points $\mathbf{K}$, where $\phi(\mathbf{K}) = 0$. It is not difficult to see that for

$$\mathbf{K} = \frac{4\pi}{3a}\hat{x} \quad ; \quad \mathbf{K}' = -\frac{4\pi}{3a}\hat{x} \qquad (27.10)$$

we have

$$\phi(\mathbf{K}) = \phi(\mathbf{K}') = 1 + e^{i\frac{2\pi}{3}} + e^{-i\frac{2\pi}{3}} = 0. \qquad (27.11)$$

These points are located at the vertices of the hexagon delimiting the first Brillouin zone, consisting in $\mathbf{K}$, $\mathbf{K}'$ plus the four vertices obtained by multiple rotations of $60°$, two of which are connected to $\mathbf{K}$ and the other two with $\mathbf{K}'$ by reciprocal lattice vectors being, therefore, essentially the same point. $\mathbf{K}$ and $\mathbf{K}'$ are, consequently, the only two inequivalent points where the valence and conduction bands of graphene touch. Since carbon in the sp^2 hybridization provides one dynamical electron per atom, the lower band will be completely filled. The Fermi surface, consequently, reduces to two inequivalent Fermi points.

27.2 A Concrete Realization of the Dirac Sea

Almost 40 years after the tight-binding study of graphene was first made, Di Vincenzo and Mele [249] and Semenoff [248] independently demonstrated that the low-energy excitations of graphene, which occur precisely around the Fermi points $\mathbf{K}$ and $\mathbf{K}'$, possess a dispersion relation, which would be appropriate to a relativistic particle and, hence, are governed by the Dirac equation.

In order to see that, let us re-write the Hamiltonian (27.7) as

$$H_{TB} = \sum_{\mathbf{k},\sigma} \Psi^\dagger(\mathbf{k}, \sigma) h_{TB} \Psi(\mathbf{k}, \sigma)$$

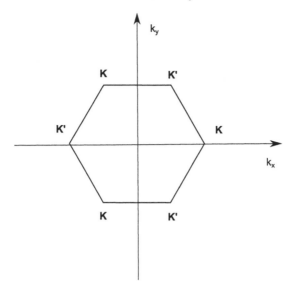

Figure 27.4 The first Brillouin zone of graphene. The three **K** and three **K'** Dirac points are connected by reciprocal lattice vectors, being, therefore, equivalent. There are two inequivalent Dirac points: **K** and **K'**. The hexagon has a side $l = \frac{4\pi}{3a}$.

$$h_{TB} = -t \left[\sum_i \cos(\mathbf{k} \cdot \mathbf{d}_i)\sigma_x + \sum_i \sin(\mathbf{k} \cdot \mathbf{d}_i)\sigma_y \right], \qquad (27.12)$$

where the σs are Pauli matrices.

Let us expand now the above Hamiltonian around the points **K** and **K'**. Making, respectively, $\mathbf{k} = \mathbf{K} + \mathbf{p}$ and $\mathbf{k} = \mathbf{K'} + \mathbf{p}$ we get, up to the first order in $\mathbf{p}$,

$$h_K = -\frac{ta}{\sqrt{3}} \left[-\sum_i \mathbf{p} \cdot \mathbf{d}_i \sin(\mathbf{K} \cdot \mathbf{d}_i)\sigma_x + \sum_i \mathbf{p} \cdot \mathbf{d}_i \cos(\mathbf{K} \cdot \mathbf{d}_i)\sigma_y \right] \qquad (27.13)$$

and

$$h_{K'} = -\frac{ta}{\sqrt{3}} \left[-\sum_i \mathbf{p} \cdot \mathbf{d}_i \sin(\mathbf{K'} \cdot \mathbf{d}_i)\sigma_x + \sum_i \mathbf{p} \cdot \mathbf{d}_i \cos(\mathbf{K'} \cdot \mathbf{d}_i)\sigma_y \right], \qquad (27.14)$$

or, equivalently

$$h_K = \frac{\sqrt{3}ta}{2} \left[p_x\sigma_x + p_y\sigma_y \right] + O(p^2)$$

$$h_{K'} = \frac{\sqrt{3}ta}{2} \left[-p_x\sigma_x + p_y\sigma_y \right] + O(p^2). \qquad (27.15)$$

We now identify h_K as the Dirac Hamiltonian of a massless particle with velocity $v_F = \frac{\sqrt{3}ta}{2}$, instead of c, which would be the speed of a massless particle in a relativistic theory. Actually $v_F \simeq 10^6 \ m/s \simeq c/300$ and $t \simeq 3 \ eV$. As for $h_{K'}$, we can put it on the standard Dirac form by performing a canonical transformation

$$\Psi_{K'}(\mathbf{k}, \sigma) \longrightarrow \exp\left\{i\frac{\pi}{2}\sigma_z\right\}\sigma_x \Psi_{K'}(\mathbf{k}, \sigma)$$

$$\Psi_{K'}^{\dagger}(\mathbf{k}, \sigma) \longrightarrow \Psi_{K'}^{\dagger}(\mathbf{k}, \sigma)\sigma_x \exp\left\{-i\frac{\pi}{2}\sigma_z\right\}, \qquad (27.16)$$

whereupon

$$H_K = \sum_{\mathbf{k},\sigma} \Psi_K^{\dagger}(\mathbf{k}, \sigma)h_K \Psi_K(\mathbf{k}, \sigma)$$

$$H_{K'} = \sum_{\mathbf{k},\sigma} \Psi_{K'}^{\dagger}(\mathbf{k}, \sigma)h_{K'} \Psi_{K'}(\mathbf{k}, \sigma). \qquad (27.17)$$

Here

$$h_K = h_{K'} = v_F\left[p_x\sigma_x + p_y\sigma_y\right] \qquad (27.18)$$

The vector Dirac matrices are, then, $\alpha_i = \sigma_i$, $i = x, y$ and $\beta = \sigma_z$, hence the covariant Dirac matrices are given by

$$\gamma^0 = \sigma_z \ ; \ \gamma^1 = i\sigma_y \ ; \ \gamma^2 = -i\sigma_x. \qquad (27.19)$$

The energy eigenvalues $E(\mathbf{p})$, with respect to the origin, can now be easily found by noting that $h_{K,K'}^2 = E^2(\mathbf{p})\mathbb{I} = v_F^2|\mathbf{p}|^2\mathbb{I}$, hence, for both K and K',

$$E(\mathbf{p}) = \pm v_F|\mathbf{p}| \ ; \ v_F = \frac{\sqrt{3}ta}{2}. \qquad (27.20)$$

We see that the energy eigenvalues are such that two opposite cones are formed at each point K, K', where the valence and conduction bands touch. Each set of two opposite Dirac cones is called a "valley." Notice the fact that $E(\mathbf{p})$ is equal for the two valleys, thus expressing the invariance of the system under the time-reversal symmetry, which connects the two valleys.

Moving to the coordinates representation, we introduce the Dirac field

$$\psi_{K\sigma} = \begin{pmatrix} \Psi_A \\ \Psi_B \end{pmatrix}_{K\sigma}, \qquad (27.21)$$

the components of which correspond to the two different sublattices A and B, whereas $K, K', \uparrow, \downarrow$ are four different flavors, specifying to which valley (Fermi point) the electron belongs, as well as its spin component orientation. In terms of this we can express the Hamiltonian density as

$$\mathcal{H} = -i\hbar v_F \sum_{A=K,K';\sigma-\uparrow,\downarrow} \psi_{A\sigma}^{\dagger}\vec{\Sigma} \cdot \nabla\psi_{A\sigma} \qquad (27.22)$$

where

$$\vec{\Sigma} = \left(\sigma_x, \sigma_y \right).$$ (27.23)

The corresponding Lagrangean density is

$$\mathcal{L} = -i\hbar v_F \sum_{A=K,K';\sigma=\uparrow,\downarrow} \overline{\psi}_{A\sigma} \gamma^\mu \partial_\mu \psi_{A\sigma},$$ (27.24)

where $\overline{\psi} = \psi^\dagger \gamma^0 = \psi^\dagger \sigma^z$, the sum over flavors in the set: $K \uparrow$, $K \downarrow$, $K' \uparrow$, $K' \downarrow$ is understood.

We see that graphene provides a concrete material realization of the Dirac sea with holes, associated to vacancies in the valence band, corresponding to positrons. The fact that the carbon p-electrons in graphene are governed by a massless relativistic Dirac equation has a deep impact on the transport properties of graphene. Different effects observed in relativistic quantum mechanics, which can be understood on the basis of the Dirac sea, have been predicted and/or experimentally observed in graphene [245, 246]. We describe some of these below.

27.3 Pseudo-Chirality, Klein Phenomena and Zitterbewegung

Pseudo-Chirality

The Hamiltonian (27.22) decouples into two-independent components, K and K', as we can see in (28.6). This is in complete analogy to what happens with the massless Dirac field in $D = 3 + 1$ when we use the Weyl representation of the Dirac matrices, as we saw in Section 10.6. The two valleys, therefore, play a role analogous to the left and right components found in a massless Dirac field in three-dimensional space and, consequently, could be called pseudo-chiralities.

Furthermore, from (27.22) we infer that the quantity

$$h = \vec{\Sigma} \cdot \mathbf{p},$$ (27.25)

which may be called pseudo-helicity, commutes with the Hamiltonian, being therefore conserved. This conservation is a direct consequence of the masslessness of the Dirac quasi-particles, and severely restricts the backscattering of electrons in graphene. Indeed, since backscattering takes $\mathbf{k}$ into $-\mathbf{k}$, then, for $\mathbf{k}$ near the point K, we can see from the positions of K and K' in the first Brillouin zone that such a process would require, at the same time, that $K \rightarrow K'$.

This can be understood once we realize that, for $\mathbf{p}$ near K, it follows that $-\mathbf{p}$ is near K'. Hence, because of their effective masslessness, electrons belonging to valley K must be backscattered into electrons of the opposite valley, K'. Consequently, admitting that intervalley scattering is suppressed, we may conclude that backscattering will be suppressed as well. Now, since backscattering is a major source of resistance, it follows that conductivity will be ultimately enhanced through this mechanism.

It is remarkable that pure graphene, even though exhibiting a zero density of states at the Fermi points, presents a minimum conductivity, as we will see below. This fact can be ascribed to the effective masslessness of the p-electrons.

Klein Paradox and Klein Tunneling

These are two phenomena occur when we solve the Dirac equation in the presence of a step potential and of a rectangular barrier potential, respectively [252]. In the first case, we get a reflection coefficient larger than one and a nonzero transmission coefficient into a classically forbidden region, thus justifying the name "paradox" given to this process.

The Klein tunneling, conversely, occurs when electrons incide on a rectangular barrier and, as can be inferred from the Dirac equation, are transmitted with probability one in certain directions [250, 253, 245].

Already at the classical level we can have a hint on how to understand these phenomena. In the presence of a potential step or barrier of height U_0 the electrons in graphene would have their energy given by

$$E(\mathbf{p}) = v_F \sqrt{|\mathbf{p}|^2} + U_0. \tag{27.26}$$

The group velocity, then, is given by

$$\mathbf{v} = \nabla_{\mathbf{p}} E(\mathbf{p}) = v_F \frac{\mathbf{p}}{\sqrt{|\mathbf{p}|^2}} = v_F^2 \frac{\mathbf{p}}{E - U_0}. \tag{27.27}$$

We see that, outside the barrier, where $E > U_0$, $\mathbf{v}$ and $\mathbf{p}$ are parallel, whereas under the barrier, where $E < U_0$, they are anti-parallel. The momentum, therefore, can be reverted in the scattering process without reversing the sign of the group velocity, as the particle goes through the potential! That is why we can have transmission without any reflection, in the case of a rectangular barrier. On a quantum-mechanical basis, we can understand the transmission coefficient being equal to one as a consequence of the absence of intervalley scattering, which would preclude backscattering.

The fact that the reflection coefficient is larger than one, for a step potential barrier, stems from the fact that the applied potential energy will create electron-hole pairs and such holes, in the presence of the corresponding field, will propagate leftward, thus making the reflected wave more intense than the incident one. It is a direct consequence of the fact that Dirac particles are always in the presence of a Dirac sea.

Zitterbewegung

Usually, when we form a wave-packet for describing the quantum state of a particle, the average of a given observable for this wave-function behaves precisely as the corresponding classical quantity. Hence the average velocity for the wave packet of a free particle is expected to be a time-independent quantity. One verifies, nevertheless, for particles governed by the Dirac equation, that if we try to compress the wave-packet to a size smaller than the particle's Compton wavelength, the average velocity exhibits an oscillatory motion with an extremely high frequency, which has been called Zitterbewegung. The name, which means "trembling motion," was coined by Schrödinger, who first realized its existence. How to interpret this result on the basis of the principles of quantum mechanics?

It happens that strictly speaking, in the presence of a Dirac sea, Dirac's theory cannot be considered as a one-particle quantum mechanics. The comprehension of this phenomenon requires the Dirac field to be considered as a fully quantized operator. In this framework, the Zitterbewegung can be ascribed to the fluctuations produced by the contributions of the Dirac sea to the wave-packet of a one-particle state and shall be observed in graphene, as described in [251, 245].

27.4 Quantum Hall Effect in Graphene

We have seen in the previous chapter that a quantum regime of the Hall effect sets in when a sufficiently strong magnetic field is applied at a low enough temperature in a two-dimensional electron gas. We should expect, therefore, that graphene, being maybe the best example of a $d = 2$ electron gas, should exhibit the same effect under similar conditions.

In a broad sense, we can say that the starting point for the description of the QHE is the study of the Landau levels generated by the action of an external magnetic field. Let us investigate, then, the structure of such Landau levels in graphene.

A constant, uniform external magnetic field B, associated to a vector potential $\mathbf{A} = (0, -Bx)$, is introduced by the minimal coupling: $\mathbf{p} \to \mathbf{p} + \frac{e}{c}\mathbf{A}$.

From (27.15), we see therefore that the eigenvalue equations for the one-particle Hamiltonians of the electrons of graphene, belonging, respectively, to valleys K and K', in the presence of an external magnetic field, are given by

$$\det \begin{pmatrix} -\frac{E}{v_F} & P_x - iP_y + i\frac{e}{c}BX \\ P_x + iP_y - i\frac{e}{c}BX & -\frac{E}{v_F} \end{pmatrix} = 0 \tag{27.28}$$

and

$$\det \begin{pmatrix} -\frac{E}{v_F} & P_x - iP_y + i\frac{e}{c}BX \\ -P_x + iP_y - i\frac{e}{c}BX & -\frac{E}{v_F} \end{pmatrix} = 0. \tag{27.29}$$

Notice that $[P_y, h_K] = [P_y, h_{K'}] = 0$, hence, being a constant of motion, P_y can be discarded. Then, using the commutator $[X, P_x] = i\hbar$, we get

$$\frac{E_{K,K'}^2}{v_F^2} = P_x^2 + \frac{e^2 B^2}{c^2} X^2 \mp \hbar \frac{eB}{c}, \qquad (27.30)$$

where the $\mp$ signs corresponds, respectively, to the valleys K and K'.

One recognizes in the first term on the right-hand side a one-dimensional harmonic oscillator Hamiltonian with mass $m = 1/2$ and frequency $\omega = 2\frac{eB}{c}$, which has eigenvalues $\epsilon_n = (2n + 1)\hbar\frac{eB}{c}$.

It follows that the energy eigenvalues for the electrons of graphene will be, for $n \in \mathbb{N}$,

$$E_K = \pm v_F \sqrt{2n\hbar \frac{eB}{c}}$$

$$E_{K'} = \pm v_F \sqrt{2(n + 1)\hbar \frac{eB}{c}}, \qquad (27.31)$$

respectively, for the valleys K and K'. We see that the presence of a magnetic field breaks the symmetry between the two valleys, by making $E_K \neq E_{K'}$.

Keeping in mind that each Landau level yields an edge state, which by its turn contributes one conductivity quantum, $\frac{e^2}{h}$, to the Hall conductivity, we see that the two valleys will contribute $2 \times (2n + 1)$ edge states, the factor 2 corresponding to the two spin orientations. Through this reasoning, therefore, we infer the Hall conductivity of graphene is given by

$$\sigma_H = 4\left(n + \frac{1}{2}\right)\frac{e^2}{h} \quad ; \quad \nu = \pm 2, \pm 6, \pm 10, \ldots \qquad (27.32)$$

This Hall conductivity was measured in graphene [244, 254] at $T = 4K$ under a magnetic field of $15\ T$. The fractional quantum Hall effect has also been experimentally observed, more recently, in graphene [255, 256].

We have so far neglected the electronic interactions in graphene; nevertheless, the observation of the fractional quantum Hall effect in this material is a strong indication that these must be important. In the rest of this chapter we include the electromagnetic electronic interactions in our description of graphene.

27.5 Electronic Interactions in Graphene

Being charged particles, electrons interact by means of the electromagnetic interaction. This is certainly the case of the carbon p-orbital electrons of graphene. Nevertheless, despite the fact that these are confined to a plane, the electromagnetic fields through which they interact are clearly not. Also, within a quantum-mechanical description, the photons, which such electrons exchange when they interact, are not restricted to the plane delimited by the honeycomb structure of graphene, where the electrons move. One then faces the possibility of formulating a theory for describing the electronic interactions in graphene, in which all matter lives in $d = 2$ while the fields that intermediate its interaction live in $d = 3$. This is not convenient, for practical, calculational and esthetical reasons. Hence, by all standards, it would be preferable to have available an effective field theory, which albeit strictly formulated in a $d = 2$ space, yet would describe the same electromagnetic interaction as QED_4 would do. This theory is not QED_3, as we may obviously infer, for instance, from the fact that it produces a logarithmic Coulomb potential between static charges, whereas the correct result would be the familiar $1/r$ potential.

In Chapter 12, we described in detail such an effective theory, called pseudo quantum electrodynamics (PQED), which reproduces the properties of QED_4, despite being completely formulated on the plane [89]. The electromagnetic interaction of the carbon p-electrons in graphene, therefore, can be conveniently described by minimally coupling the Dirac field introduced previously in this chapter to the pseudo electromagnetic field, governed by PQED. We thereby combine the results obtained from the tight-binding approach and introduce the following interacting theory for graphene [257],

$$\mathcal{L} = -\frac{1}{4} F_{\mu\nu} \left[\frac{2}{\sqrt{\Box}} \right] F^{\mu\nu} - i\hbar v_F \overline{\psi}_{K\sigma} \gamma^\mu D_\mu \psi_{K\sigma}, \qquad (27.33)$$

where $D_\mu = \partial_\mu + ieA_\mu$ and $F_{\mu\nu} = \partial_\mu A_\nu - \partial_\nu A_\mu$ is the field intensity tensor of the pseudo-electromagnetic field.

It could be argued that, since the characteristic speed of the electrons in graphene, namely, v_F, is of the order of 300 times less than the speed of light, which is the speed of the photons, then it would suffice to use just the static Coulomb potential for describing the electronic interactions in graphene. Albeit this is true to a large extent, yet there are some subtle effects that are missed if one uses the static interaction right from the beginning.

A first example of such effects consists of anomalies, such as the chiral or parity anomalies, which will only occur in the presence of the full trilinear interaction vertex generated by the minimal coupling with a dynamical electromagnetic field. Another example occurs when using Kubo's formula, for instance, for determining the DC-conductivity. In order to obtain the correct result one must take the limit $\omega \to 0$ at the very end, after every other limit has been taken. By using just the static potential, we are in a way actually imposing $\omega \to 0$, from the very beginning, thus violating that prescription.

Notice that the description of the electronic interactions in graphene by the full electromagnetic interaction contains the static Coulomb interaction as a particular case; hence, it is anyway more complete.

In practical applications, the use of PQED amounts to considering the trilinear vertex, coupling the electron current to the pseudo-electromagnetic field A_μ, and using the propagator $G_{\mu\nu}$ of this field, given by (12.13). In what follows, we will explore different consequences of the electronic interactions in graphene.

27.6 Velocity Renormalization

A first consequence of the electronic interactions in graphene is the renormalization of the Fermi velocity v_F. We address this issue by using a renormalization group method in the framework of PQED. The results shown in this section are a nice example of how the presence of interactions strongly influences the behavior of physical quantities in a system described by a certain quantum field theory.

Let us consider the vertex function $\Gamma^\mu(p_1, p_2; \alpha; v_F)$, which is given in one loop by the graph of Fig. 27.5, and where $\alpha = e^2/4\pi\epsilon_0\hbar c$ is the fine-structure constant. The (pseudo) photon propagator in the triangle graph is given by (12.13).

As we saw in Section 6.5, such a vertex function must obey the renormalization group equation

$$\left[\mu^2 \frac{\partial}{\partial\mu^2} + \beta_v(v_R, e^2)\frac{\partial}{\partial v_R} + \beta_e(v_R, e^2)\frac{\partial}{\partial e} - \gamma_{A_\mu} - 2\gamma_\psi\right]\Gamma^\mu_R(p_1, p_2; \alpha; v_R) = 0,$$
(27.34)

where v_R is the renormalized Fermi velocity and

$$\beta_v(v_R, e^2) = \mu^2\frac{\partial v_R}{\partial\mu^2}.$$
(27.35)

Figure 27.5 The vertex function

From the last equation, we obtain

$$\int_{v_R(\mu_0)}^{v_R(\mu)} \frac{dv_R}{\beta_v(v_R)} = \int_{\mu_0^2}^{\mu^2} \frac{d\mu^2}{\mu^2} = \ln\frac{\mu^2}{\mu_0^2}. \tag{27.36}$$

Knowledge of the beta function $\beta_v(v_R, e^2)$, consequently, allows us to determine the renormalized Fermi velocity in graphene.

In order to determine the $\beta_v(v_R, e^2)$ function, one evaluates the above vertex function, within a certain approximation; then, one may extract its explicit form from (27.36). Through this method, one obtains, in one loop, [267] $\beta_e = 0$ and

$$\beta_v(v_R, e^2) = \frac{e^2}{8\pi^2} \int_0^1 dx \frac{\sqrt{1-x}}{x(1-\beta^2)-1} \left[1 - 2\beta^2 - \frac{1}{x(1-\beta^2)-1}\right], \tag{27.37}$$

where $\beta = v_R/c$. Notice that the $\beta_v(v_R, e^2)$ function vanishes at $\beta = 1$, hence $v_R = c$ is a fixed point for the renormalized velocity [269, 274].

We shall analyze the $\beta_v(v_R, e^2)$ function in two different regimes, namely, $\beta \ll 1$ and $\beta \simeq 1$. From (27.37), we get

$$\beta_v(v_R, e^2) = -\frac{e^2}{16\pi}\left(1 - \frac{\beta^2}{2}\right) \quad ; \quad \beta \ll 1$$

$$\beta_v(v_R, e^2) = -\frac{2e^2}{5\pi^2}(1-\beta) \quad ; \quad \beta \simeq 1. \tag{27.38}$$

Inserting each one of the above expressions in (27.36) and performing the integration in v_R, we obtain, respectively, in the two regimes [269, 267],

$$v_R(\mu) = v_R(\mu_0)\left[1 - \frac{e^2}{8\pi v_R(\mu_0)} \ln\left(\frac{\mu}{\mu_0}\right)\right] \quad ; \quad \beta \ll 1$$

$$v_R(\mu) = c\left[1 - \left(1 - \frac{v_R(\mu_0)}{c}\right)\left(\frac{\mu}{\mu_0}\right)^{2\gamma}\right] \quad ; \quad \beta \simeq 1, \tag{27.39}$$

where $\gamma = \frac{e^2}{5\pi^2 c}$.

We may exchange the above dependence of the renormalized velocity on the energy scale by the density of electrons. Indeed, by adding or removing p-electrons from the carbon atoms of graphene, we can create a Fermi circle of radius p_F. The density of electrons, then, can be related to p_F as

$$N_e = \frac{\pi p_F^2}{\frac{(2\pi)^2}{A}} \times 4 \quad ; \quad p_F = \sqrt{\pi n}, \tag{27.40}$$

where we used the fact that the number of available states is the area of the Fermi circle divided by the area occupied by each state. Each of these states can accommodate four electrons, corresponding to the two spins and two valleys.

Using the above relation, we may trade the energy scale set by p_F by the electronic density n:

$$v_R(n) = v_R(n_0)\left[1 - \frac{\alpha_0}{4}\ln\left(\frac{n}{n_0}\right)\right] \quad ; \quad \beta \ll 1$$

$$v_R(n) = c\left[1 - \left(1 - \frac{v_R(n_0)}{c}\right)\left(\frac{n}{n_0}\right)^\gamma\right] \quad ; \quad \beta \simeq 1, \qquad (27.41)$$

where $\alpha_0 = \frac{e^2}{4\pi v_R(n_0)} = \frac{c}{v_R(n_0)} \times \frac{1}{137}$ is the graphene fine structure constant corresponding to $v_R(n_0)$.

It is interesting to note that v_F renormalizes to the speed of light exactly at half-filling, $n = 0$, which is the infrared fixed point [268, 269, 274]. This regime is not likely to be attained in real samples, because of the presence of impurities and as a consequence of the finite size of the graphene sheet. Nevertheless, a clear enhancement of the effective velocity has been experimentally observed as the electronic density is reduced [270]. Observe, however, that in the low-density regime, we must use the second expression above, because in this regime one violates the $\beta \ll 1$ condition.

27.7 DC-Conductivity

In the previous section, we have just determined how the interactions produce a renormalization of the Fermi velocity in graphene. DC-conductivity is another physical quantity that is influenced by such electronic interactions.

DC-conductivity is given by Kubo's formula,

$$\sigma^{ij} = \lim_{\omega \to 0} \frac{i}{\omega}\langle j^i j^j\rangle(\omega, \mathbf{p} = 0). \qquad (27.42)$$

As we have seen in Section 6.5, the current correlator is given by the vacuum polarization tensor, namely

$$\langle j^i j^j\rangle(\omega, \mathbf{p}) = \Pi^{ij}(\omega, \mathbf{p}). \qquad (27.43)$$

The vacuum polarization tensor of PQED has been calculated up to two loops. The one-loop result contains the trace

$$\text{tr}\left[\gamma^\mu \frac{(\gamma^\mu k_\mu \pm m)}{k^2 + m^2}\gamma^\nu \frac{(\gamma^\mu k'_\mu \pm m)}{k'^2 + m^2}\right], \qquad (27.44)$$

where m is an electron mass that will be set to zero. We will see in the next chapter that the mass signs $\pm$ correspond, respectively, to each of the two valleys K and K'. We then conclude that terms corresponding to the trace of an even number of

γ-matrices will be valley-insensitive, even functions of m. Those corresponding to the trace of an odd number of such matrices will be proportional to

$$\pm m \text{tr}\left[\gamma^\mu \gamma^\alpha \gamma^\beta\right] = \pm 2m\epsilon^{\mu\alpha\beta}.$$

Combining these observations with the one-loop calculation for one single flavor, performed in [271], one obtains, in the $m = 0$ limit [257],

$$i\Pi^{ij}_{(1)}(\omega, \mathbf{p}) = \frac{1}{16}\left[\frac{\delta^{ij}(\omega^2 - v_F^2|\mathbf{p}|^2) - p^i p^j}{\sqrt{\omega^2 - v_F^2|\mathbf{p}|^2}}\right] \pm \frac{1}{2\pi}\left(n + \frac{1}{2}\right)\epsilon^{ij}\omega, \quad (27.45)$$

where $n \in \mathbb{Z}$ is a topological invariant [271] and the last term corresponds to the trace of three γ-matrices. The two signs correspond, as announced, to the contributions of the two valleys K and K'.

The two-loops result has been obtained in [272, 273, 274, 275]. It is given by the first factor of $\Pi^{ij}_{(1)}(\omega, \mathbf{p})$ multiplied by a constant $C\alpha_0$

$$i\Pi^{ij}_{(2)}(\omega, \mathbf{p}) = \frac{1}{16}\left[\frac{\delta^{ij}(\omega^2 - v_F^2|\mathbf{p}|^2) - p^i p^j}{\sqrt{\omega^2 - v_F^2|\mathbf{p}|^2}}\right] C\alpha_0, \quad (27.46)$$

where C depends on the Fermi velocity beta function (β_{v_R}) regime: $v_R \ll c$ or $v_R \simeq c$. $\alpha_0 \simeq \frac{300}{137} \simeq 2.19$ is the fine structure constant corresponding to the velocity v_F of graphene.

Observe that the transverse term of the vacuum polarization tensor does not receive higher-order radiative corrections, in compliance with the Coleman–Hill theorem [278].

Now, since we are interested in the response to a constant electric field, we must take the limit $\mathbf{p} \to 0$ in the previous expressions. Then, all the v_F-dependence is washed out, and the system behaves as if $v_F = c$. Hence, for determining the constant C, for the sake of consistency, we choose the ultra-relativistic prescription leading to the value [275],

$$C = \frac{92 - 9\pi^2}{18\pi} \simeq 0.056. \quad (27.47)$$

Notice that the two-loops correction is much smaller than the one-loop, thus justifying the use of perturbation theory in this case.

After taking the limit $\mathbf{p} \to 0$, the static limit $\omega \to 0$ can be safely and consistently taken. Inserting in Kubo's formula, we get, for each valley and spin component, the conductivity

$$\sigma^{ij} = \frac{1}{16}\frac{e^2}{\hbar}\left[1 + \frac{92 - 9\pi^2}{18\pi}\alpha_0\right]\delta^{ij} \pm \frac{1}{2\pi}\left(n + \frac{1}{2}\right)\frac{e^2}{\hbar}\epsilon^{ij}, \quad (27.48)$$

where the two signs correspond to the valleys K and K', respectively, and we have reinstated the physical units of $\frac{e^2}{\hbar}$.

Summing the contributions of the four flavors in the set: $K \uparrow, K \downarrow, K' \uparrow, K' \downarrow$, we get the total conductivity [257]

$$\sigma_T^{ij} = 4 \times \frac{\pi}{8} \frac{e^2}{h} \left[1 + \frac{92 - 9\pi^2}{18\pi} \alpha_0 \right] \delta^{ij}$$

$$\sigma_T^{ij} = \frac{\pi}{2} [1 + 0.056 \, \alpha_0] \frac{e^2}{h} \delta^{ij} \simeq 1.76 \frac{e^2}{h} \delta^{ij}. \tag{27.49}$$

Observe that the electrons of each valley give opposite contributions to the transverse conductivity and, consequently, this vanishes. We have, in summary,

$$\sigma_T^{xx} = \frac{\pi}{2} [1 + 0.056 \, \alpha_0] \frac{e^2}{h} \simeq 1.76 \frac{e^2}{h}$$

$$\sigma_T^{xy} = 0. \tag{27.50}$$

Even though this is still not so close to the experimental result [276]

$$\sigma_{T,exp}^{ij} \simeq 2.16 \frac{e^2}{h} \delta^{ij},$$

yet it represents an improvement with respect to the non-interacting minimal conductivity

$$\sigma_{0,T}^{ij} = \frac{\pi}{2} \frac{e^2}{h} \delta^{ij} \simeq 1.57 \frac{e^2}{h} \delta^{ij}$$

and is the closest to the experimental value [277].

27.8 The Quantum Valley Hall Effect

In the previous section, we found that graphene exhibits a longitudinal conductivity, which is corrected by the electromagnetic interactions. No transverse conductivity, however was obtained in the absence of an external magnetic field.

We now introduce the concept of a "valley conductivity," which is naturally defined in terms of the *difference* between the conductivities of each valley, namely

$$\sigma_V^{ij} = \sum_{\sigma=\uparrow,\downarrow} \left[\sigma_{K,\sigma}^{ij} - \sigma_{K',\sigma}^{ij} \right]. \tag{27.51}$$

The valley conductivity, therefore, relates an applied electric field to the *relative* current between the two valleys.

From (27.48), we infer that

$$\sigma_V^{ij} = 4\left(n + \frac{1}{2}\right)\frac{e^2}{h}\epsilon^{ij},$$ (27.52)

or

$$\sigma_V^{xx} = 0$$
$$\sigma_V^{xy} = 4\left(n + \frac{1}{2}\right)\frac{e^2}{h}.$$ (27.53)

We see now that the longitudinal valley conductivity vanishes, whereas the transverse one is nonzero. Also, interestingly, it is identical to the one obtained for the quantum Hall effect in graphene in the presence of an applied external magnetic field. This is an exact result, by virtue of the Coleman–Hill theorem that rules out the correction of the transverse part of the vacuum polarization tensor by higher-order terms.

The fact that these two conductivities are identical is by no means a coincidence. It stems from the fact that both conductivities can be expressed in terms of topological invariants. These, on the other hand, are closely related to conducting edge states, which, by their turn, correspond to a discrete set of gapped bulk states. In the case of the QHE produced by an external magnetic field in graphene, these bulk states are the Landau levels studied in Section 27.4. In the present case of the spontaneous generation of a quantum valley Hall effect (QVHE) in graphene, we will demonstrate the existence of the corresponding gapped bulk states in the next section.

The emergence of the QVHE in graphene is the consequence of the electromagnetic interactions among the p-electrons of carbon described by PQED [257]. For this, we unavoidably need the full dynamical interaction; should we use just the static Coulomb interaction, even though it could be justified by the fact that $v_F \ll c$, we would completely miss this effect. This is a striking example of the successful use of PQED in a condensed matter system.

27.9 The Electronic Spectrum of Graphene

Let us study here the energy spectrum of graphene in the absence of external fields [257]. We saw in Chapter 13 that energy eigenstates appear as poles of the Green function. We had shown before, in Chapter 6, that the value of these eigen-energies at zero momentum (masses) can be obtained from the self-energy, according to (6.59). For this purpose, therefore, we shall determine the electron self-energy function of graphene, which is given by the graph in Fig. 27.6. In

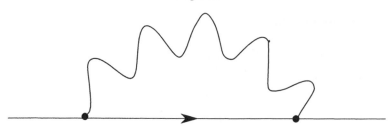

Figure 27.6 The electron self-energy

this, the curly line is the gauge field propagator of PQED, whereas the straight line is the full electron propagator $S(p)$. This is given by the Schwinger–Dyson equation

$$S^{-1}(p) = S_0^{-1}(p) - \Sigma(p)$$
$$S(p) = \frac{1}{\not{p} - \Sigma(p)}, \tag{27.54}$$

where S_0 is the electron free propagator and Σ is the self-energy itself.

Since the electrons in graphene are massless Dirac fermions, the sought energy eigenstates, ϵ, are a solution of the following version of (6.59):

$$\mathrm{Re}\ \Sigma(p = \epsilon) = \epsilon, \tag{27.55}$$

where we took into account the fact that electrons in graphene are massless Dirac fermions.

The imaginary part of the self-energy is related to the lifetimes, τ, of these states, namely,

$$\mathrm{Im}\ \Sigma(p = \epsilon) = \frac{\hbar}{\tau}. \tag{27.56}$$

Our task, therefore, is to determine the self-energy, such that the energy eigenvalues could be extracted from the equation above. Since the self-energy graph is an integral that also contains the self-energy as part of the integrand, via the electron propagator, it actually is an integral equation for the self-energy. Going through a standard procedure [257, 258], one can transform such into a differential equation for the (scalar) trace of the self-energy

$$\mathrm{tr}\Sigma \equiv 2\Sigma_1, \tag{27.57}$$

namely

$$p^2 \frac{d^2\Sigma_1}{dp^2} + 2p \frac{d\Sigma_1}{dp} + \frac{\alpha}{4\alpha_c}\Sigma_1 = 0, \tag{27.58}$$

where α is the fine structure constant of graphene and $\alpha_c \simeq 1.02$ is a critical coupling.

The Schwinger–Dyson method has been used in several interesting studies of graphene found in the literature [259, 260, 261, 262, 265, 263, 264, 266]. Most of the time, however, the static Coulomb potential is used in these approaches, instead of the full electrodynamical interaction described by PQED (see Chapter 12). The differential equation above is Euler's equation, which admits the solutions [257]

$$\Sigma_1 = \frac{A}{\sqrt{p}} e^{-i[\gamma \ln \frac{p}{\Lambda} + \varphi]}, \tag{27.59}$$

where A and φ are arbitrary real constants and Λ is an ultraviolet energy-momentum cutoff. The parameter γ is given by

$$\gamma = \frac{2}{\sqrt{\pi}} \sqrt{\left(\frac{\alpha}{2 + \pi\alpha}\right) - \left(\frac{\alpha_c}{2 + \pi\alpha_c}\right)}. \tag{27.60}$$

By taking the trace of (27.55), we see that

$$\Sigma_1(p = \epsilon) = \epsilon. \tag{27.61}$$

Then, choosing $A = \Lambda^{3/2}$ and $\varphi = 0$, we get

$$\text{Re } \Sigma_1 = \frac{\Lambda^{3/2}}{\sqrt{\epsilon}} \cos\left[\gamma \ln \frac{\epsilon}{\Lambda}\right] = \epsilon, \tag{27.62}$$

and

$$\text{Im } \Sigma_1 = \frac{\Lambda^{3/2}}{\sqrt{\epsilon}} \sin\left[\gamma \ln \frac{\epsilon}{\Lambda^2}\right]. \tag{27.63}$$

Introducing z, in such a way that

$$\epsilon = \Lambda e^{-\frac{z}{\gamma}}, \tag{27.64}$$

we obtain [257] the following condition on the zs:

$$e^{-\frac{3z}{2\gamma}} = \cos z. \tag{27.65}$$

Calling Z_n, $n \in \mathbb{N}$, the solutions of this transcendental equation, we find an infinite number of eigenenergies given by [257]

$$\epsilon_n = \Lambda e^{-\frac{Z_n}{\gamma}} \quad n \in \mathbb{N}. \tag{27.66}$$

In (27.61), we have $p = \sqrt{E^2 - |\mathbf{p}|^2}$, and consequently the electron propagator will have poles at $p^2 = \epsilon_n^2$, corresponding to energy eigenstates with eigenvalues

$$E_n = \pm\sqrt{|\mathbf{p}|^2 + |\epsilon_n|^2}. \tag{27.67}$$

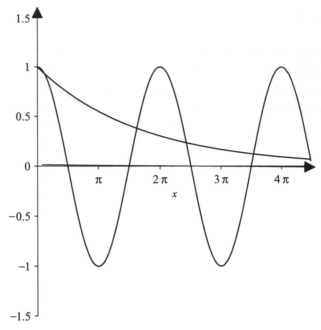

Figure 27.7 The solutions of the transcendent equation, Eq. (27.65). Observe they quickly tend to the zeros of the cosine function: $(n + \frac{1}{2})\pi$.

We conclude, therefore, that the set of dynamically generated energy states is symmetric about the Fermi level, which is located at $E = 0$ for undoped graphene.

The lifetime of these states is given by the imaginary part of the self-energy. We, therefore find, for the nth state,

$$\tau_n = \frac{\hbar}{\Lambda} \frac{\left| \cos Z_n \right|^{1/3}}{\left| \sin Z_n \right|}$$

$$\tau_n = \frac{\hbar}{\Lambda} \frac{e^{-Z_n/\gamma}}{\sqrt{1 - e^{-3Z_n/\gamma}}}. \tag{27.68}$$

We see that the lifetimes decay exponentially with Z_n and, consequently, only the first few states could be observed. Choosing the cutoff as $\Lambda \simeq 1.0 \ eV$, and considering that $\hbar = 6.58 \times 10^{-16} eV \cdot s$, we conclude that the first states will have a lifetime of the order $\tau \simeq 1.0 \ fs$. There is, however, a special state, corresponding to the solution $Z_0 = 0$, that has an infinite lifetime and is, consequently, stable. It has a rest energy $\epsilon_0 = \Lambda$, and therefore opens a gap 2Λ in the electronic spectrum.

These are the announced gapped bulk quantum states that are spontaneously generated by the electronic interaction, described by PQED. To each of them there will correspond a gapless edge state which will correspond to the transverse valley conductivity found in the previous section. The existence of states of opposite signs reflects in the existence of equal but opposite currents that correspond to a nontrivial valley conductivity, despite the fact that the total conductivity vanishes. There is a characteristic temperature $T^* \simeq 2K$ above which the dynamically generated gapped states would be washed out by thermal activation. The whole effect should be unobservable above such temperature [257]. Below such temperatures, and moreover, close to $0\,K$, there will be a gap $E_g = 2\Lambda$ separating the valence and conduction bands. The DC-conductivity, consequently, will vanish at temperatures in the range $T \ll T^*$, returning to the value determined above as we go to the range $T > T^*$.

28

Silicene and Transition Metal Dichalcogenides

Graphene is an atom-wide monolayer gapless material, which is classified as a semi-metal and exhibits outstanding transport, electronic and structural properties. When carbon is replaced by silicon in the underlying honeycomb crystal structure, a new phenomenon occurs. Due to the larger atomic size, effects such as the core's ionic repulsion make the system stabilize in such a way that the triangular sublattices A and B are no longer in the same plane, but become separated by a distance of approximately 0.046 nm [279, 280]. This produces a spin-orbit coupling that presents a sublattice (A or B) dependent sign [280]. This staggered energy generates an asymmetry between the two sublattices, which has the effect of producing a gap in the energy spectrum. As a consequence, silicene, as this material is known, is a semiconductor, rather than a semi-metal. Silicene was synthesized in 2010 [283] and since then has been attracting the attention of the condensed matter physics community. Related materials made of germanium and tin [284] have also been synthesized since then [285, 286].

Transition metal dichalcogenides (TMD) such as WSe_2, WS_2, $MoSe_2$ and MoS_2 are another class of extremely interesting materials, which present a crystal structure similar to silicene, but with the sites belonging to sublattices A and B being occupied, respectively, by transition metal and chalcogen atoms. The net effect is, again, a breakdown of the symmetry between the A and B sublattices of the honeycomb structure, which existed in graphene. This, again, generates a staggered local energy that, ultimately, produces a gap in the energy spectrum. Silicene and TMDs present peculiar physical properties such as strong electron-hole interaction, leading to exciton states with a binding energy much larger than that of conventional semiconductors. Silicene, germanene and stanene, along with the transition metal dichalcogenides, form a class of materials that can be described by the massive Dirac equation.

28.1 The Gapped Dirac Hamiltonian

We can model silicene and TMD by adding to (27.7) a staggered sublattice-dependent M term, as follows,

$$H_{TB} = \sum_{\mathbf{k},\sigma} \Psi^\dagger(\mathbf{k},\sigma) \begin{pmatrix} M & \phi \\ \phi^* & -M \end{pmatrix} \Psi(\mathbf{k},\sigma), \qquad (28.1)$$

where $\phi(\mathbf{k})$ is given by (27.8). Notice that the M-term, being proportional to a σ_z matrix in the (A,B) space, describes the staggered energy, provenient either from the spin-orbit coupling or from the local atomic asymmetry between the two sublattices that occur in TMDs.

Diagonalizing the Hamiltonian, we readily find that the energy eigenvalues are now

$$E(\mathbf{k}) = \pm\sqrt{|\phi(\mathbf{k})|^2 + M^2} = \pm\sqrt{t^2 \sum_{i,j=1,2,3} e^{i\mathbf{k}\cdot(\mathbf{d}_i - \mathbf{d}_j)} + M^2}, \qquad (28.2)$$

which clearly show the presence of a gap $2M$.

We may re-write the Hamiltonian as

$$H_{TB,M} = \sum_{\mathbf{k},\sigma} \Psi^\dagger(\mathbf{k},\sigma) h_{TB,M} \Psi(\mathbf{k},\sigma)$$

$$h_{TB,M} = -t\left[\sum_i \cos(\mathbf{k}\cdot\mathbf{d}_i)\sigma_x + \sum_i \sin(\mathbf{k}\cdot\mathbf{d}_i)\sigma_y\right] + M\sigma_z, \qquad (28.3)$$

where the σs are Pauli matrices with entries in the (A,B) space.

Now, expanding around the points K and K', we obtain, respectively,

$$h_{K,M} = \frac{\sqrt{3}ta}{2}\left[p_x\sigma_x + p_y\sigma_y + M\sigma_z\right] + O(p^2)$$

$$h_{K',M} = \frac{\sqrt{3}ta}{2}\left[-p_x\sigma_x + p_y\sigma_y + M\sigma_z\right] + O(p^2). \qquad (28.4)$$

Then, performing the same canonical transformation (27.16), in $\Psi_{K',\sigma}(\mathbf{k})$, we see that the mass term transforms as

$$M\Psi^\dagger_{K',\sigma}\sigma_z\Psi_{K',\sigma} \longrightarrow M\Psi^\dagger_{K',\sigma}\sigma_x \exp\left\{-i\frac{\pi}{2}\sigma_z\right\}\sigma_z \exp\left\{i\frac{\pi}{2}\sigma_z\right\}\sigma_x\Psi_{K',\sigma}$$

$$= -M\Psi^\dagger_{K',\sigma}\sigma_z\Psi_{K',\sigma}. \qquad (28.5)$$

The terms that are linear in the momentum transform as before, hence we obtain, in the massive case,

$$H_K = \sum_{\mathbf{k},\sigma} \Psi_K^\dagger(\mathbf{k},\sigma) h_K \Psi_K(\mathbf{k},\sigma)$$

$$H_{K'} = \sum_{\mathbf{k},\sigma} \Psi_{K'}^\dagger(\mathbf{k},\sigma) h_{K'} \Psi_{K'}(\mathbf{k},\sigma), \tag{28.6}$$

where

$$h_K = v_F \left[p_x \sigma_x + p_y \sigma_y + M \sigma_z \right] \tag{28.7}$$

and

$$h_{K'} = v_F \left[p_x \sigma_x + p_y \sigma_y - M \sigma_z \right]. \tag{28.8}$$

Notice that the mass at valley K' has an opposite sign to that at valley K.

Using the Dirac matrices (27.19) we can obtain the Hamiltonian and Lagrangean densities corresponding to the above one-particle Hamiltonians,

$$\mathcal{H} = \psi_{K\sigma}^\dagger \left[-i\hbar v_F \vec{\Sigma} \cdot \nabla + M \tau_{KK'}^z \beta \right] \psi_{K'\sigma'} \tag{28.9}$$

and

$$\mathcal{L} = \overline{\psi}_{K\sigma} \left[-i\hbar v_F \gamma^\mu \partial_\mu - M \tau_{KK'}^z \right] \psi_{K'\sigma'}, \tag{28.10}$$

where, as before, $\overline{\psi} = \psi^\dagger \gamma^0 = \psi^\dagger \sigma^z$, the sum over flavors in the set: $K\uparrow$, $K\downarrow$, $K'\uparrow$, $K'\downarrow$ is understood, Σ is given by (27.23). The 2×2 Dirac matrices have entries in the (A,B) sublattice subspace.

A primary mechanism for the existence of a mass term in the systems being examined here is any asymmetry between the two sublattices A and B. In the case of TMDs, the fact that different atoms occupy each of the two sublattices naturally generates a σ_z term in the Hamiltonian, which expresses such asymmetry. Supposing the local difference in chemical potential between the atoms in the two sublattices is 2Δ, then $M = \Delta$.

In the case of silicene and related monoatomic compounds, conversely, we may simulate the staggered local energy difference by applying a perpendicular electric field, E_z. Supposing the sublattices are a distance l apart, then $M = \Delta = \frac{l}{2} E_z$ in this case. Assuming the electric field is applied from the lower (B) to the upper (A) sublattice, it follows that in that case we will have $\Delta < 0$. Another mechanism for mass generation in the systems under consideration is the spin-orbit interaction. In this case, the mass M may be written as [280]

$$M = \xi \xi_s \Delta_{SO}, \tag{28.11}$$

where Δ_{SO} is the magnitude of the spin-orbit interaction energy. In the case of silicene this is $\Delta_{SO} \simeq 1.55 \; meV$, whereas for germanene, $\Delta_{SO} \simeq 23.9 \; meV$. Both are much larger than the corresponding values for graphene. Then, ξ and ξ_s are flavor-dependent signs: $\xi = \pm 1$, corresponding respectively, to the valleys K

and K', whereas $\xi_s = \pm 1$, by its turn, corresponds, respectively, to the two spin orientations $\sigma = \uparrow$ and $\sigma = \downarrow$.

We conclude that silicene and the related germanium and tin compounds, as well as the TMDs, can all be described, in general terms, by a massive Dirac theory with the appropriate inclusion of the flavor matrix τ^z.

The electromagnetic interaction, as in the case of graphene, can be described by minimally coupling the Dirac field to the vector gauge field A_μ of pseudo quantum electrodynamics. In the next subsections, we explore physical properties of silicene and related materials by making use of such field theoretic description.

28.2 Time Reversal Symmetry

A mass term in the Dirac equation breaks the time-reversal (TR) symmetry. Indeed, under the TR operation, the Dirac field transforms as [163]

$$\psi(\mathbf{r}, t) \longrightarrow -i\sigma_y \psi(-\mathbf{r}, t)$$

in such a way that

$$\int d^2r\, \overline{\psi}(\mathbf{r}, t)\psi(\mathbf{r}, t) = \int d^2r\, \psi^\dagger(\mathbf{r}, t)\sigma_z\psi(\mathbf{r}, t) \longrightarrow$$
$$-\int d^2r\, \overline{\psi}(-\mathbf{r}, t)\psi(-\mathbf{r}, t) = -\int d^2r\, \overline{\psi}(\mathbf{r}, t)\psi(\mathbf{r}, t). \tag{28.12}$$

A TR invariant massive theory, nevertheless, may be obtained by ascribing opposite masses to the TR connected K and K' valley flavors. In this case, clearly, the sum of mass terms becomes TR invariant:

$$M \int d^2r\, \overline{\psi}_K \psi_K - M \int d^2r\, \overline{\psi}_{K'} \psi_{K'} \longrightarrow$$
$$-M \int d^2r\, \overline{\psi}_{K'} \psi_{K'} + M \int d^2r\, \overline{\psi}_K \psi_K. \tag{28.13}$$

We conclude that the gapped Hamiltonian, used for describing TMDs, silicene, etc., is TR invariant.

28.3 Parity Anomaly: Total and Valley Conductivities

In order to determine the total and valley conductivities, one must determine firstly the contribution of each one of the four flavors K, K', $\uparrow$, $\downarrow$ to the DC-conductivity. For this purpose, one should follow the same procedure as the one we followed in Section 27.7, using the fact that the current-current correlation function in the Kubo formula is given by the vacuum polarization tensor, with the only difference being that this now is evaluated with a massive electron propagator [287, 271].

As we have seen in the case of graphene, the one-loop contribution to the vacuum polarization tensor possesses a transverse component [271], which dynamically violates both TR and inversion symmetries. In the zero fermion mass case, which applies to graphene, such an anomalous term generates contributions to the transverse conductivity for each of the valleys K and K', which are equal in magnitude but have opposite signs [257]. When we sum the contributions of the two valleys, consequently, we obtain a zero total Hall conductivity, but a nonzero valley conductivity, as is shown in (27.50) and (27.53).

Now, when the fermion mass is different from zero, one can infer from the results in [271] that masses with a different sign produce contributions of different magnitudes to the Hall conductivity. Therefore, when we choose masses with different signs for each valley K and K', it means each valley will generate transverse conductivities of different *magnitudes*. Hence, even though they have opposite signs, when we sum the contributions from each valley to the transverse conductivity, we obtain a nonzero total Hall conductivity and a nonzero valley transverse conductivity [287].

The total transverse conductivity, obtained by summing the contributions from each of the four flavors, yields [287]

$$\sigma_T^{ij} = 2\,\frac{e^2}{h}\epsilon^{ij},$$

whereas the valley conductivity, defined in (27.51) gives [287]

$$\sigma_V^{ij} = 4\left(n + \frac{1}{2}\right)\frac{e^2}{h}\epsilon^{ij}. \tag{28.14}$$

Similarly to what happened in graphene, the onset of the emergent Hall conductivity and transverse valley conductivity will be accompanied by the dynamical generation of an infinite set of midgap states [287] analogous to the ones in (27.66) and (27.67).

28.4 Overview

We have seen that the low-energy excitations of a system with one electron per site in a honeycomb crystal structure have their kinematics described by a massless Dirac equation. Such is the case of graphene and any other system exhibiting a symmetry between the two sublattices. When such symmetry is broken, the corresponding excitations are described by a massive Dirac equation. A primary way of breaking the sublattice symmetry is by having different atoms in each of them. Such is the case of the TMDs. Silicene and other monoatomic materials of the carbon group, conversely, because of their size, stabilize in a buckled structure, where the sublattices are spatially separated. It follows that under the action of a perpendicular electric field, the two sublattices will be at different chemical potentials, thereby generating the same effect.

We have seen that because of the parity anomaly, a spontaneous transverse conductivity is generated in such types of systems. In the case of massless systems such as graphene, by virtue of the sublattice symmetry, the two valleys generate identical transverse currents. Since they counter-propagate, the net transverse current vanishes, yielding a zero Hall conductivity, but a nonzero transverse valley conductivity. In the case of massive systems, however, because of the sublattice asymmetry, the two valleys yield different transverse currents, thereby leading to a net transverse conductivity, which characterizes an emergent quantum Hall effect in the absence of any external agent [257, 287].

These effects are accompanied by the corresponding generation of midgap states, which occur in equal numbers for each valley in the massless case, whereas in the massive case the number of states from each valley are different. Such interesting effects are produced whenever the full electromagnetic interaction is taken into account and, therefore, the full trilinear interaction vertex is present. Then, the one loop transverse contribution to the vacuum polarization tensor, which is a topological quantity, yields a nontrivial result [257, 287].

29

Topological Insulators

The discovery of systems undergoing the quantum Hall effect has revealed the existence of materials that exhibit a vanishing bulk conductivity despite the presence of a nonzero conductivity on the edges. Moreover such an edge conductivity was shown to be proportional to a nonzero topological invariant, which was responsible for the stability of the edge current. Materials presenting these properties, namely (a) gapped bulk excitations with an associated zero bulk conductivity, (b) gapless edge excitations with an associated nonzero edge conductivity, and (c) a nontrivial topological invariant protecting the robustness of the edge currents, have been called "topological insulators" [311, 312, 313]. The number of materials found to belong to this new class has been ever growing since the discovery and comprehension of the quantum Hall effect. Among these, we find quantum wells of $HgTe$ sandwiched between $Hg_{1-x}Cd_xTe$, which are essentially two-dimensional. Three-dimensional topological insulators have also been found, usually structured in the form of stacked $d = 2$ layers. Examples are Bi_2Te_3 and Bi_2Se_3.

Topological insulators are characterized by a new kind of "order," in which the non-vanishing element is a topological invariant rather than an order parameter that is usually related to the symmetry of the system. Topological ordering, hence, differs from the Landau–Ginzburg type of order, which has been a paradigm for decades.

Band insulators exhibit a completely filled valence band, which is separated by an energy gap from the conduction band. In two-dimensional space, however, for a rectangular (or square) lattice, a full valence band, namely, the whole first Brillouin zone, is topologically equivalent to a torus, as we can infer by connecting opposite boundaries by reciprocal lattice vectors. Consider a gapped relativistic system such as the ones found in the preceding chapter. The energy of such a system turns out to be a function $E = E(|\mathbf{p}|^2 + M^2)$, hence constant energy states generate a sphere $|\mathbf{p}| = C$. Since $\mathbf{p}$ is in the first BZ, we see that such states produce a mapping between a torus T_2 and a sphere S_2, which has nontrivial topological classes. The topological invariant that classifies the inequivalent classes for this mapping is the Chern number. Following the above reasoning, we understand why there are no topological metals, but just insulators: an incomplete valence band is not equivalent to a torus, but only to a part of it.

In this chapter, we describe topological insulators associated either to the Chern number or to the so-called $\mathbb{Z}_2$ topological invariant, in two spatial dimensions. We just make a brief comment about topological insulators in three spatial dimensions, which has become a rather fertile field of research recently.

29.1 Chern Topological Insulators

Systems exhibiting the quantum Hall effect were the first topological insulators. The gapped bulk states associated with a zero longitudinal conductivity are the Landau levels. The topological invariant is the Chern number that Thouless *et al.* [231] have shown to be proportional to the transverse conductivity, which by its turn is proportional to the edge conductivity. These topological insulators were studied in detail in Chapter 26. In this section we study other topological insulators, which are also characterized by a nonzero Chern number.

29.1.1 The Two-Bands Topological Insulator

Consider a system described by the following Hamiltonian, in momentum space

$$H = n_0(\mathbf{k})\mathbb{I} + \mathbf{n}(\mathbf{k}) \cdot \vec{\sigma}, \tag{29.1}$$

where $\mathbf{n}(\mathbf{k})$ is a vector and the σs are Pauli matrices. Notice that, if we consider the two entries of the above matrices as corresponding to the two sublattices of a bipartite lattice, this Hamiltonian has the general form of the Hamiltonians we found in graphene, silicene and dichalcogenides. In these cases, we had $\mathbf{n} = (v_F\mathbf{k}, M)$. The mass term in particular is given by the σ^z term and, therefore, implies an energy asymmetry between the two sublattices.

It is easy to show that the energy eigenvalues are given by

$$E_\pm = n_0(\mathbf{k}) \pm R(\mathbf{k})$$
$$R(\mathbf{k}) = \sqrt{\mathbf{n}(\mathbf{k}) \cdot \mathbf{n}(\mathbf{k})}, \tag{29.2}$$

with the corresponding eigenvectors

$$u_+(\mathbf{k}) = \begin{pmatrix} e^{-i\varphi} \cos\frac{\theta}{2} \\ \sin\frac{\theta}{2} \end{pmatrix} \quad ; \quad u_-(\mathbf{k}) = \begin{pmatrix} -e^{-i\varphi} \sin\frac{\theta}{2} \\ \cos\frac{\theta}{2} \end{pmatrix}, \tag{29.3}$$

where

$$\mathbf{n}(\mathbf{k}) = R(\mathbf{k}) \begin{pmatrix} \sin\theta\cos\varphi \\ \sin\theta\sin\varphi \\ \cos\theta \end{pmatrix}. \tag{29.4}$$

From (29.2), we see that the system possesses a gap $\Delta = 2R(\mathbf{k}=0)$ separating two energy bands placed symmetrically with respect to n_0. Let us compute now the Berry phase corresponding to the states $u_\pm(\theta,\varphi)$. According to (26.32), the Berry connection is given by

$$\mathcal{A}_\theta^\pm = i\langle u_\pm| \frac{\partial}{\partial\theta}|u_\pm\rangle = 0$$

$$\mathcal{A}_\varphi^+ = i\langle u_+| \frac{\partial}{\partial\varphi}|u_+\rangle = \cos^2\frac{\theta}{2}$$

$$\mathcal{A}_\varphi^- = i\langle u_-| \frac{\partial}{\partial\varphi}|u_-\rangle = -\sin^2\frac{\theta}{2}, \tag{29.5}$$

or, in vector form,

$$\vec{\mathcal{A}}^+ = \frac{\cos^2\frac{\theta}{2}}{R\sin\theta}\hat{\varphi} \quad ; \quad \vec{\mathcal{A}}^- = -\frac{\sin^2\frac{\theta}{2}}{R\sin\theta}\hat{\varphi}. \tag{29.6}$$

It follows that the "magnetic" field associated with the Berry vector connection is

$$\vec{\mathcal{B}}^\pm = \nabla\times\vec{\mathcal{A}}^\pm = \frac{\mathbf{R}}{2R^3}. \tag{29.7}$$

This is a "magnetic monopole" in the space where the vector $\mathbf{n}$ is defined. We can express the First Chern Number, $\mathcal{C}$, in terms of the total flux of such field through a closed surface containing the origin:

$$2\pi\mathcal{C} = \int_S d^2\mathbf{S}\cdot\vec{\mathcal{B}}^\pm = \frac{1}{2}4\pi = 2\pi. \tag{29.8}$$

We see that the Chern number is $\mathcal{C} = 1$ for the energy eigenstates of the two-band topological insulator.

29.1.2 *The Chern Number as a Mapping $T_2 \mapsto S_2$*

Let us assume $\mathbf{n} = (n^1, n^2, n^3)$, with $|\mathbf{n}| = R$, so $\mathbf{n}$ is in a sphere of radius R. In this case, we can get an insight on the nature of the mapping $\mathbf{n}(\mathbf{k})$.

Starting from (29.8), we can write the Chern number in terms of the solid angle: $\mathcal{C} = \Omega/4\pi$, namely

$$\mathcal{C} = \frac{1}{4\pi}\int_{S_2} d^2\mathbf{S}^i\frac{\hat{\mathbf{n}}^i}{|\mathbf{n}|^2}. \tag{29.9}$$

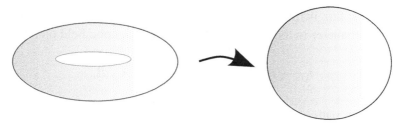

Figure 29.1 The mapping of the torus on a sphere has inequivalent classes classified by the First Chern Number.

Now, considering that $\mathbf{n} = \mathbf{n}(k_x, k_y)$, where (k_x, k_y) is in the first Brillouin torus, we have

$$d^2\mathbf{S}^i = (d\mathbf{n}^2 d\mathbf{n}^3, d\mathbf{n}^3 d\mathbf{n}^1, d\mathbf{n}^1 d\mathbf{n}^2).$$

Then, using the Jacobian for the change of variables $(\mathbf{n}^i, \mathbf{n}^j) \Leftrightarrow (k_x, k_y)$, namely

$$d\mathbf{n}^j d\mathbf{n}^k = \left[\frac{\partial \mathbf{n}^j}{\partial k_x} \frac{\partial \mathbf{n}^k}{\partial k_y} - \frac{\partial \mathbf{n}^j}{\partial k_y} \frac{\partial \mathbf{n}^k}{\partial k_x} \right] dk_x dk_y,$$

we obtain, from (29.9)

$$d^2\mathbf{S}^i \mathbf{n}^i = \epsilon^{ijk} \mathbf{n}^i \frac{\partial \mathbf{n}^j}{\partial k_x} \frac{\partial \mathbf{n}^k}{\partial k_y} dk_x dk_y. \tag{29.10}$$

Inserting this in (29.9) and choosing a unit sphere, $|\mathbf{n}| = 1$, we obtain

$$\mathcal{C} = \frac{1}{4\pi} \int_{T_2} d^2 k \; \mathbf{n} \cdot \frac{\partial \mathbf{n}}{\partial k_x} \times \frac{\partial \mathbf{n}}{\partial k_y}. \tag{29.11}$$

This expression gives the area of the unit sphere covered by $\mathbf{n}$ as we sweep the torus in (k_x, k_y), thus making very clear the meaning of the Chern number in this case: the number of times the sphere is covered in this map.

29.2 The Haldane Topological Insulator

Haldane's topological insulator is formulated on a honeycomb lattice structure identical to that of graphene [314]. It includes, though, three new features with respect to graphene: a second nearest neighbors hopping, t_2; a term breaking the symmetry between the two sublattices; and a local magnetic flux ϕ, such that for a plaquette of nearest neighbors the net flux is zero, while for a plaquette of second nearest neighbors, which contains three points, there is a nonvanishing net flux.

The magnetic flux is introduced through the standard prescription of modifying the hopping parameter in the presence of an electromagnetic field on a lattice

$$t_{ij} \rightarrow t_{ij} e^{i\Phi_{ij}} \qquad \Phi_{ij} = \frac{ec}{\hbar} \int_{L_{ij}} \mathbf{A} \cdot d\mathbf{l}, \tag{29.12}$$

where L_{ij} is the path along the line connecting sites i and j. This is the lattice version of the minimal coupling.

Since the magnetic flux vanishes identically for plaquettes formed by nearest neighbors, we shall only use the minimal coupling prescription for the second nearest neighbor plaquettes. The basic plaquettes of second nearest neighbors of sublattices A and B, however, possess opposite signs. The total magnetic flux, therefore, is zero.

The Haldane Hamiltonian is given by

$$H_H = \sum_{\mathbf{k}} c_A^\dagger(\mathbf{k}) c_B(\mathbf{k}) \left[-t \sum_{i=1,2,3} e^{i\mathbf{k}\cdot\mathbf{d}_i} \right] + H.C.$$

$$\sum_{\mathbf{k}} c_A^\dagger(\mathbf{k}) c_A(\mathbf{k}) \left[M - t_2 \sum_{i=1,2,3} e^{i\mathbf{k}\cdot\mathbf{e}_i + i\Phi} \right] + H.C.$$

$$\sum_{\mathbf{k}} c_B^\dagger(\mathbf{k}) c_B(\mathbf{k}) \left[-M - t_2 \sum_{i=1,2,3} e^{i\mathbf{k}\cdot\mathbf{e}_i + i\Phi} \right] + H.C. \tag{29.13}$$

Then, this can be written in terms of the field $\Psi^\dagger(\mathbf{k}) = (c_A^\dagger(\mathbf{k}) c_B^\dagger(\mathbf{k}))$ as

$$H_H =$$
$$\sum_{\mathbf{k}} \Psi^\dagger(\mathbf{k}) \begin{pmatrix} M + 2t_2 \sum_i \cos[\mathbf{k}\cdot\mathbf{e}_i + \Phi] & \phi \\ \phi^* & -M + 2t_2 \sum_i \cos[\mathbf{k}\cdot\mathbf{e}_i - \Phi] \end{pmatrix} \Psi(\mathbf{k}), \tag{29.14}$$

where

$$\phi(\mathbf{k}) = -t \sum_{i=1,2,3} e^{i\mathbf{k}\cdot\mathbf{d}_i} \tag{29.15}$$

and the vectors $\mathbf{e}_i$ are the ones separating the first neighbors within the triangular Bravais sublattices A and B, namely

$$\mathbf{e}_1 = \mathbf{d}_2 - \mathbf{d}_3 \; ; \quad \mathbf{e}_2 = \mathbf{d}_3 - \mathbf{d}_1 \; ; \quad \mathbf{e}_3 = \mathbf{d}_1 - \mathbf{d}_2$$

where the $\mathbf{d}_i$s are defined in (27.3).

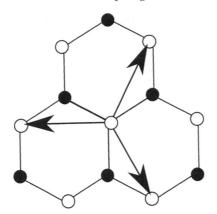

Figure 29.2 The vectors $\mathbf{e}_i$, $i = 1, 2, 3$, which along with $-\mathbf{e}_i$ connect the nearest neighbors within the Bravais triangular sublattice of the honeycomb crystal structure. These are second nearest neighbors in the honeycomb structure as a whole.

Haldane's one-particle Hamiltonian may be written as

$$
h_H = 2t_2 \cos \Phi \sum_i \cos(\mathbf{k} \cdot \mathbf{e}_i)\mathbb{I} + \left[\sum_i \cos(\mathbf{k} \cdot \mathbf{d}_i)\sigma_x + \sum_i \sin(\mathbf{k} \cdot \mathbf{d}_i)\sigma_y \right]
$$

$$
+ \left[M - 2t_2 \sin \Phi \sum_i \sin(\mathbf{k} \cdot \mathbf{e}_i) \right] \sigma_z, \tag{29.16}
$$

where the σs are Pauli matrices.

The energy eigenvalues are

$$
E_\pm = 2t_2 \cos \Phi \sum_i \cos(\mathbf{k} \cdot \mathbf{e}_i) \pm \sqrt{|\phi|^2 + \left(M - 2t_2 \sin \Phi \sum_i \sin(\mathbf{k} \cdot \mathbf{e}_i) \right)^2}.
$$

$$
\tag{29.17}
$$

Expanding the Hamiltonian around the Dirac points K and K', we obtain, respectively,

$$
h_{H,K} = \frac{\sqrt{3}ta}{2} \left[p_x \sigma_x + p_y \sigma_y \right] + M_K \sigma_z
$$

$$
h_{H,K'} = \frac{\sqrt{3}ta}{2} \left[-p_x \sigma_x + p_y \sigma_y \right] + M_{K'}\sigma_z, \tag{29.18}
$$

where

$$
M_K = M - 3\sqrt{3}t_2 \sin \Phi \quad ; \quad M_{K'} = M + 3\sqrt{3}t_2 \sin \Phi. \tag{29.19}
$$

Performing the canonical transformation (27.16), we can identify $h_{H,K}$ and $h_{H,K'}$, respectively, as the Dirac Hamiltonian

$$h_{H,M_a} = \frac{\sqrt{3}ta}{2}\left[p_x\sigma_x + p_y\sigma_y\right] + M_a\sigma_z \tag{29.20}$$

of particles with masses

$$M_{a=K} = M - 3\sqrt{3}t_2\sin\Phi \quad ; \quad M_{a=K'} = -M - 3\sqrt{3}t_2\sin\Phi. \tag{29.21}$$

Now for such a Dirac system, we have the unit vector

$$\mathbf{n}(\mathbf{k}) = \frac{(v_F k_x, v_F k_y, M_a)}{\sqrt{v_F^2|\mathbf{k}|^2 + M_a^2}} = (\sin\theta\cos\varphi, \sin\theta\sin\varphi, \cos\theta), \tag{29.22}$$

where $a = K, K'$.

The Chern number may be expressed, in terms of this, as

$$\mathcal{C} = \frac{1}{4\pi}\int_0^{2\pi}d\varphi\int_0^{\pi}d\theta\,\epsilon^{ijk}\mathbf{n}^i\partial_\theta\mathbf{n}^j\partial_\varphi\mathbf{n}^k. \tag{29.23}$$

Now, observe that the n^3-component in (29.22) has a fixed sign determined by the sign of the mass M_a. Hence, for each Dirac point we will have either $\theta \in [0, \pi/2]$ or $\theta \in [\pi/2, \pi]$, implying the θ-integral above will cover just a hemisphere for each Dirac point and each will, consequently, contribute a half for the Chern number. Indeed, evaluating the integral above, we find, for $a = K, K'$, respectively,

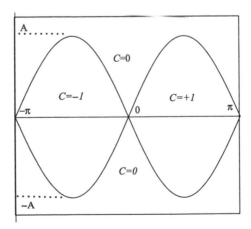

Figure 29.3 The phase diagram of Haldane's topological insulator, showing the different topological phases, with the corresponding value of the First Chern Number, $\mathcal{C}$. The vertical axis is M/t_2, while the horizontal is the magnetic flux Φ. The constant A is $3\sqrt{3}$.

$$C_K = \frac{1}{2}\text{sign}(M_K) \quad ; \quad C_{K'} = \frac{1}{2}\text{sign}(M_{K'}), \tag{29.24}$$

whereby the Chern number of the Haldane model is given by

$$C = \frac{1}{2}\left[\text{sign}(M - 3\sqrt{3}t_2 \sin \Phi) - \text{sign}(M + 3\sqrt{3}t_2 \sin \Phi)\right]. \tag{29.25}$$

We see that $C = 0, \pm 1$, according to the phase diagram depicted in Fig. 29.3.

29.3 Chiral Edge States in the Haldane Model

One of the most basic features of a quantum insulator is the presence of topologically stabilized edge currents. We may consider as an "edge" the interface between two regions, which are, respectively, trivial and nontrivial from the topological point of view. For this purpose, let us create such an edge by making one of the two Dirac masses depend on the spatial coordinate y in such a way that

$$M_1 = M(y) \quad ; \quad \begin{cases} M(y) < 0 & y < 0 \\ M(0) = 0 & \\ M(y) > 0 & y > 0 \end{cases}, \tag{29.26}$$

while the other we keep constant

$$M_2 = M = \lim_{y \to \infty} M(y). \tag{29.27}$$

According to (29.25), this way we create an interface at $y = 0$, separating a region with zero Chern number from another with a nonzero Chern number, namely an interface separating a topologically trivial region from a nontrivial one.

Inserting in the Dirac equation corresponding to the one-particle Dirac Hamiltonians (29.18), we find the exact solution

$$\psi(x, y) = e^{ip_x x} \exp\left\{-\int_0^y dy' M(y')\right\} \begin{pmatrix} 1 \\ 1 \end{pmatrix}. \tag{29.28}$$

This is localized along y and propagates along the edge (x) in the positive direction, being therefore chiral. It cuts the Fermi level at $E = 0$, with a positive group velocity. Notice that the existence of such a solution is guaranteed by the fact that a non-vanishing topological number exists in one of the two regions separated by the interface. This ensures its robustness.

A nontrivial topology on the bulk gapped band implies the existence of gapless edge modes, according to the relation

$$\Delta C = N_R - N_L, \tag{29.29}$$

where $N_{R,L}$ are the numbers of chiral edge modes and ΔC is the variation of the topological number across the interface. The different number of chiral edge modes is a manifestation of the time reversal symmetry breakdown occurring in Haldane's model. Such effect is called the quantum anomalous Hall effect. An experimental realization of Haldane's topological insulator was found in thin films of chromium-doped $(Bi,Sb)_2Te_3$ [315].

Systems that generalize Haldane's model of topological insulators, possessing a higher Chern number, can be obtained by including the interaction with a next hierarchy of neighbors [316].

29.4 $\mathbb{Z}_2$ Topological Insulators

A new class of topological insulators characterized by a topological invariant different than the Chern number was proposed in 2005 [317]. Two relevant features of these systems are the fundamental role played by the spin degrees of freedom and the assumption of time reversal symmetry. A new topological invariant presenting just two inequivalent classes uprises, namely the $\mathbb{Z}_2$ topological invariant, which classifies the topological phases of the new topological insulators. Again, gapped bulk states give place to gapless edge states, which are stabilized by the nontrivial topology expressed by this new index.

29.4.1 Time Reversal Symmetry

Time reversal is an operation taking $t \to -t$. Observables such as momentum, angular momentum and spin, accordingly, will change sign under time reversal, while energy, for instance, will remain invariant. At a quantum-mechanical level, according to Wigner's theorem, the time reversal operation must be implemented either by a unitary or anti-unitary operator: Θ. In both cases, $|\langle \Theta\psi|\Theta\phi\rangle| = |\langle\psi|\phi\rangle|$. Unitary operators are linear, however anti-unitary operators are antilinear, namely

$$\Theta(\alpha_1|\psi_1\rangle + \alpha_2|\psi_2\rangle) = \alpha_1^*\Theta|\psi_1\rangle + \alpha_2^*\Theta|\psi_2\rangle,$$

where $\alpha_1, \alpha_2 \in \mathbb{C}$.

For an anti-unitary operator $\langle A\psi|A\phi\rangle = \langle\phi|\psi\rangle$, while for a unitary one $\langle U\psi|U\phi\rangle = \langle\psi|\phi\rangle$.

The following argument shows Θ must be anti-unitary. Consider an arbitrary, time-evolving state-vector $|\Psi(t)\rangle$. Under the time-reversal operation, it transforms to

$$|\Psi(t)\rangle_\Theta = \Theta|\Psi(t)\rangle = \Theta e^{-\frac{i}{\hbar}Ht}|\Psi(0)\rangle. \tag{29.30}$$

If the system is invariant under time-reversal symmetry, then $[\Theta, H] = 0$. An antilinear Θ operator is therefore required for the consistency of time-reversed evolution:

$$|\Psi(t)\rangle_\Theta = e^{-\frac{i}{\hbar}H(-t)}\Theta|\Psi(0)\rangle = e^{-\frac{i}{\hbar}H(-t)}|\Psi(0)\rangle_\Theta$$

$$|\Psi(t)\rangle_\Theta = |\Psi(-t)\rangle_\Theta, \tag{29.31}$$

The time-reversal operator has a peculiar property for spin $s = 1/2$ particles, which has a profound impact in the energy spectrum. Indeed, in this case it can be shown that $\Theta^2 = -1$. Consider a time-reversal symmetric spin $1/2$ system for which $[\Theta, H] = 0$. Then, given an energy eigenstate $|\psi\rangle$, it follows that $\Theta|\psi\rangle$ is also an eigenstate with the same energy. Now,

$$\langle\psi|\Theta\psi\rangle = \langle\Theta^2\psi|\Theta\psi\rangle = -\langle\psi|\Theta|\psi\rangle$$

$$\langle\psi|\Theta|\psi\rangle = 0. \tag{29.32}$$

This implies $|\psi\rangle$ and $\Theta|\psi\rangle$ are orthogonal and therefore genuine degenerate energy eigenstates, in such a way that the energy spectrum is doubly degenerate. This is known as the Kramers degeneracy.

29.5 The $\mathbb{Z}_2$ Topological Invariant

Consider the matrix $W_{ij} = \langle\psi_i|\Theta|\psi_j\rangle$, representing the time reversal operator in a certain base $\{|\psi_1\rangle, \ldots, |\psi_{2N}\rangle\}$ containing an even number of elements. We can see that this matrix is anti-symmetric. Indeed

$$W_{ij} = \langle\psi_i|\Theta|\psi_j\rangle = \langle\Theta^2\psi_j|\Theta|\psi_i\rangle = -\langle\psi_j|\Theta|\psi_i\rangle = -W_{ji}. \tag{29.33}$$

Being an anti-symmetric matrix of even dimension, as we saw in Chapter 5, it follows that the determinant of W_{ij} is expressed in terms of the Pfaffian as

$$\det[W_{ij}] = \text{Pf}^2[W_{ij}]. \tag{29.34}$$

Notice that if we change the sign of the matrix W_{ij} elements, the Pfaffian will change its sign, while the determinant will remain unchanged. This observation is at the root of the construction of the $\mathbb{Z}_2$ topological invariant [317].

As an example, consider the matrices

$$A = \begin{pmatrix} 0 & a \\ -a & 0 \end{pmatrix} \quad ; \quad B = \begin{pmatrix} 0 & -a \\ a & 0 \end{pmatrix}. \tag{29.35}$$

Both have the same determinant $\det[A] = \det[B] = a^2$, while the corresponding Pfaffians are different, namely $\text{Pf}[A] = a$ and $\text{Pf}[B] = -a$.

Let us now consider more specifically the setup for the definition of the new topological invariant. Remember that Bloch energy eigenstates, corresponding to a periodic potential, can be written as abstract vectors in the Hilbert space as

$$|\psi_n(\mathbf{k})\rangle = e^{i\mathbf{k}\cdot\mathbf{r}}|u_n(\mathbf{k})\rangle, \tag{29.36}$$

where $\mathbf{k}$ is in the first Brillouin zone.

It follows that the $|u_n(\mathbf{k})\rangle$ states are eigenvectors of the Bloch Hamiltonian

$$\mathcal{H}(\mathbf{k})|u_n(\mathbf{k})\rangle = E_n|u_n(\mathbf{k})\rangle, \tag{29.37}$$

which is defined as

$$\mathcal{H}(\mathbf{k}) = e^{-i\mathbf{k}\cdot\mathbf{r}}He^{i\mathbf{k}\cdot\mathbf{r}}. \tag{29.38}$$

We can take the momentum $\mathbf{k}$ as a parameter on which the Bloch Hamiltonian depends, in the spirit of the adiabatic theorem. Then the Hamiltonian can be continuously deformed without changing the eigenstate, provided there is no degeneracy. In other words, the system remains in the same topological class in the absence of degeneracy.

Notice that, for a time-reversal symmetric system, even though $[H, \Theta] = 0$, we have

$$\Theta\mathcal{H}(\mathbf{k})\Theta^{-1} = \mathcal{H}(-\mathbf{k}), \tag{29.39}$$

which means the Bloch Hamiltonian will not be invariant, in general.

There are, however, special points in the first Brillouin zone for which $-\mathbf{k} = \mathbf{k} + \mathbf{G}$, where this last vector belongs to the reciprocal lattice. For these points the vectors $\mathbf{k}$ and $-\mathbf{k}$ are equivalent, and therefore constitute fixed points for the time-reversal operation. There are four of these points, in $d = 2$, which we denote by Λ_i, $i = 1, \ldots, 4$. For these points, $[\mathcal{H}, \Theta] = 0$, and we can have the Kramers degeneracy for the $|u_n(\mathbf{k})\rangle$ states. This means level crossing may occur and, consequently, a change of topological class.

In order to construct a mathematical object that captures such a change, consider the matrix W_{ij} defined as

$$W_{ij}(\mathbf{k}) = \langle u_i(-\mathbf{k})|\Theta|u_j(\mathbf{k})\rangle. \tag{29.40}$$

This in general will not be an anti-symmetric matrix, because $W_{ji}(\mathbf{k}) = -W_{ij}(-\mathbf{k})$. For the special points Λ_i of the first Brillouin zone, however,

$$W_{ij}(\Lambda_i) = -W_{ji}(\Lambda_i). \tag{29.41}$$

Then the $\mathbb{Z}_2$ topological invariant is defined as

$$(-1)^\nu = \prod_{i=1}^{4} \delta_i, \tag{29.42}$$

in terms of $\delta_i\ i = 1, \ldots, 4$, which are given by

$$\delta_i = \frac{\mathrm{Pf}[W_{ij}(\Lambda_i)]}{\sqrt{\det[W_{ij}(\Lambda_i)]}} = \pm 1 \tag{29.43}$$

in each of the four time-reversal fixed points Λ_i.

The index ν equals 0 or 1, $\nu = 0$ corresponding to the trivial topological class, while $\nu = 1$ corresponds to the nontrivial one.

29.6 The Kane–Mele Topological Insulator

The first example of a topological insulator based on the $\mathbb{Z}_2$ topological invariant was proposed by Kane and Mele in 2005 [317]. The underlying base includes both the sublattice and spin degrees of freedom

$$(A, B) \otimes (\uparrow, \downarrow) = (A \uparrow, A \downarrow, B \uparrow, B \downarrow),$$

which will correspond to a field

$$\Phi^\dagger(\mathbf{k}) = (c_{A\uparrow}^\dagger(\mathbf{k}), c_{A\downarrow}^\dagger(\mathbf{k}), c_{B\uparrow}^\dagger(\mathbf{k}), c_{B\downarrow}^\dagger(\mathbf{k})).$$

The Kane–Mele Hamiltonian is written in terms of this as

$$H_{KM} = \sum_{\mathbf{k}} \Phi^\dagger(\mathbf{k}) h_{KM} \Phi(\mathbf{k}), \tag{29.44}$$

where h_{KM} is the 4×4, one-particle Hamiltonian

$$h_{KM}(\mathbf{k}) = d_0(\mathbf{k})\mathbb{I} + \sum_{i=1}^{5} d_i(\mathbf{k})\Gamma_i. \tag{29.45}$$

Here the Γ_i, $i = 1, \ldots, 5$ are the following 4×4 matrices:

$$\Gamma_1 = \sigma_x \otimes \mathbb{I} \quad \Gamma_2 = \sigma_y \otimes \mathbb{I} \quad \Gamma_3 = \sigma_z \otimes s_x \quad \Gamma_4 = \sigma_z \otimes s_y \quad \Gamma_5 = \sigma_z \otimes s_z. \tag{29.46}$$

The first factor in the direct products above acts on the sublattice (A, B) subspace, while the second factor acts on the spin $(\uparrow, \downarrow)$ subspace (s_i, $i = x, y, z$ are Pauli matrices in the spin subspace).

The energy eigenvalues of the Kane–Mele model are

$$E_\pm = d_0(\mathbf{k}) \pm \sqrt{\sum_{i=1}^{5} |d_i(\mathbf{k})|^2}. \tag{29.47}$$

Considering now that, in the above base, the time-reversal operator may be represented as $\Theta = i\mathbb{I} \otimes \sigma_y \mathcal{K}$, where $\mathcal{K}$ is the complex conjugation operator, it follows from imposing time-reversal symmetry on the Kane–Mele Hamiltonian, that the $d_i(\mathbf{k})$, $i = 2, 3, 4, 5$ should all be odd under the time-reversal operation, while $d_1(\mathbf{k})$ should be even. Consequently, we must have at the time-reversal fixed points

$$d_i(\Lambda_a) = 0, \quad i = 2, 3, 4, 5. \tag{29.48}$$

The gap at these points, therefore, is given by

$$\Delta(\Lambda_a) = 2d_1(\Lambda_a). \tag{29.49}$$

29.6.1 Inversion Symmetry

It follows from the choice of the base in which the Kane–Mele Hamiltonian is written that the spatial inversion operator is represented by $\mathcal{P} = \sigma_x \otimes \mathbb{I} \equiv \Gamma_1$. Then imposing spatial inversion symmetry, apart from the time-reversal one, forces $d_1(\mathbf{k})$ to be even and $d_i(\mathbf{k})$ $i = 2, 3, 4, 5$ to be odd under that symmetry operation.

It was shown in a very simple form as a function of the parity eigenvalues of the filled bands at the time-reversal invariant fixed points; namely, [318] that in this case, the $\mathbb{Z}_2$ topological invariant can be expressed as

$$(-1)^\nu = \prod_{a=1}^{4} \mathrm{sign}[d_1(\Lambda_a)], \tag{29.50}$$

where Λ_a are the special points of the Brillouin zone, which are invariant under the time-reversal symmetry. It becomes clear, therefore, that a nontrivial topology, characterized by a value $\nu = 1$, necessarily implies a change in the sign of $d_1(\Lambda_a)$, which will be associated to the closing of the gap, according to (29.49).

29.7 Edge States and the Quantum Spin Hall Effect

We can study the edge states in the Kane–Mele model of a topological insulator by using the same method as in the case of the Haldane model. Expanding and linearizing around one of the points Λ_a and making $d_1(\Lambda_a) = M(y)$, it can be shown [319] that, in the base $(A \uparrow, A \downarrow, B \uparrow, B \downarrow)$, we obtain two solutions, namely

$$\psi_{L\uparrow}(x, y) = e^{-ip_x x} \exp\left\{-\int_0^y dy' M(y')\right\} \begin{pmatrix} 0 \\ 1 \\ 0 \\ 0 \end{pmatrix}$$

$$\psi_{R\downarrow}(x, y) = e^{+ip_x x} \exp\left\{-\int_0^y dy' M(y')\right\} \begin{pmatrix} 0 \\ 0 \\ 0 \\ 1 \end{pmatrix}. \qquad (29.51)$$

These two solutions form a Kramers pair that propagates along the interface between a topologically nontrivial phase and a trivial one. Time-reversal invariance forces the solutions to occur in pairs with opposite spins counter-propagating. There is, consequently, no net current on the edges and absence of the anomalous quantum Hall effect. Conversely, a separation of the spin components does occur, since opposite spins propagate in opposite directions. This has been called the quantum spin Hall effect (QSHE). A model similar to the Kane–Mele model was proposed by Bernevig, Hughes and Zhang [320], and soon an experimental realization of the (QSHE) was obtained using $HgTe$ quantum wells [321].

The Kane–Mele model of topological insulators, which is characterized by a $\mathbb{Z}_2$ topological invariant, was, then, generalized to three-dimensional materials [322, 323, 324] and from then on the area of three-dimensional topological insulators has been continuously growing.

30

Non-Abelian Statistics and Quantum Computation

Classical computation is based on algorithms that manipulate binary variables (bits), which convey the information being processed. Such algorithms are formed by logic units (logic gates), which are basic functions of binary variables yielding an output that will be used in the next step of the algorithm, and so on. Algorithms of classical computation make extensive use of irreversible gates, which can only be used in one direction and, consequently, loose information. Quantum computation, conversely, uses the quantum states of a given Hilbert space (qubits) to store the information [290, 291, 292]. Then, it takes advantage of the fact that these states have a unitary evolution to build data-processing algorithms, which only possess reversible, unitary logic gates. An important point in this scheme is that coherence of the quantum states storing the information is a crucial requirement; otherwise, the information is lost. Nevertheless, we know that quantum systems rapidly loose their coherence as we increase their size. A clever method of coherence protection, therefore, must be devised if we want to have a quantum computer operating on a human scale. For this reason, quantum computation, since its inception, has been always related to the principles of quantum mechanics.

The main method of coherence protection involves excitations having the so-called non-abelian statistics. Such excitations were shown to occur in the $\nu = 5/2$ plateaus of systems presenting the quantum Hall effect, which is described by the Moore–Read Pfaffian wave-function. Such a state is associated to a Landau–Ginzburg type field theory consisting in the level 2 non-abelian SU(2) Chern–Simons theory in the same way the odd denominator plateaus described by the Laughlin wave-function are associated to the Zhang–Hansson–Kivelson abelian Chern–Simons–Landau–Ginzburg theory. We have shown that the vortex excitations of this theory correspond to the (abelian) anyon excitations of Laughlin's wave-function. In an analogous way, the vortex excitations of the non-abelian Chern–Simons theory are associated to the Pfaffian wave-function excitations, which present non-abelian statistics. Such gapped bulk excitations produce gapless

478

Dirac excitations on the edges, which may combine in specific ways to reproduce the non-abelian statistics exhibited by the bulk vortex states.

Both the bulk and edge versions of the excitations with non-abelian statistics are such that, for the specific value of the spin $s = 1/4$, the braiding matrices expressing the non-abelian character of the statistics become the logic gates NOT and CNOT.

30.1 Bits, Qubits and Logic Gates

In a classical computer, information is stored in binary units 0 or 1 known as bits. In quantum computation, conversely, one uses the quantum states of a two-dimensional Hilbert space with a base $\{|0\rangle, |1\rangle\}$, namely

$$\alpha|0\rangle + \beta|1\rangle \; ; \; \alpha, \beta \in \mathbb{C} \; ; \; |\alpha|^2 + |\beta|^2 = 1. \tag{30.1}$$

The information is encoded in the pair of complex numbers (α, β), the so-called qubit, or "quantum bit." The logic algorithms of quantum computation make use of reversible logic units, or logic gates, among which the most basic are the NOT and CNOT (controlled NOT) gates. These are represented, respectively, by the matrices

$$\text{NOT} = \begin{pmatrix} 0 & 1 \\ 1 & 0 \end{pmatrix} \tag{30.2}$$

and

$$\text{CNOT} = \begin{pmatrix} 0 & 1 & 0 & 0 \\ 1 & 0 & 0 & 0 \\ 0 & 0 & 1 & 0 \\ 0 & 0 & 0 & 1 \end{pmatrix}.$$

The NOT gate, when fed the input (α, β), produces the output (β, α). The CNOT gate, conversely, operates with two qubits

$$\begin{pmatrix} \alpha_1 \\ \beta_1 \end{pmatrix} \otimes \begin{pmatrix} \alpha_2 \\ \beta_2 \end{pmatrix} = \begin{pmatrix} \alpha_1\alpha_2 \\ \alpha_1\beta_2 \\ \beta_1\alpha_2 \\ \beta_1\beta_2 \end{pmatrix} \overset{\text{CNOT}}{\Longrightarrow} \begin{pmatrix} \alpha_1\beta_2 \\ \alpha_1\alpha_2 \\ \beta_1\alpha_2 \\ \beta_1\beta_2 \end{pmatrix} \tag{30.3}$$

This means the output of the second qubit depends on the result of the measurement of the first one.

30.2 Non-Abelian Statistics

For a system of N identical particles, described by a state-vector $|\varphi_{n_1}(1) \ldots \varphi_{n_N}(N)\rangle$, we must have

$$||\varphi_{n_1}(1)\dots\varphi_{n_i}(i)\dots\varphi_{n_j}(j)\dots\varphi_{n_N}(N)\rangle|^2$$
$$= ||\varphi_{n_1}(1)\dots\varphi_{n_i}(j)\dots\varphi_{n_j}(i)\dots\varphi_{n_N}(N)\rangle|^2. \tag{30.4}$$

We have seen in Chapter 10 that, in order to comply with that, we have in general

$$|\varphi_{n_1}(1)\dots\varphi_{n_i}(i)\dots\varphi_{n_j}(j)\dots\varphi_{n_N}(N)\rangle$$
$$= e^{i\theta}|\varphi_{n_1}(1)\dots\varphi_{n_i}(j)\dots\varphi_{n_j}(i)\dots\varphi_{n_N}(N)\rangle. \tag{30.5}$$

In $d = 3$ only two possibilities are allowed, in agreement with the symmetrization postulate of quantum mechanics, namely $\theta = 0, \pi$. The corresponding particles are, respectively, bosons or fermions. In $d = 2, 1$, however, we can have $\theta \in \mathbb{R}$ and the associated particles for which $\theta \neq 0, \pi$ are the anyons, which violate the symmetrization postulate of quantum mechanics. For all above cases, if two transpositions $i \leftrightarrow j$ are made in a different order, the result is precisely the same. Such particles, therefore, are said to obey an abelian statistics with the spin-statistics parameter given by $s = \frac{\theta}{2\pi}$.

There is, nevertheless, an even more radical way in which the symmetrization postulate is violated. This happens when the N-particle state-vectors are G-fold degenerate, namely, belong to a set

$$\{|\Psi_1(N)\rangle, \dots, |\Psi_G(N)\rangle\} \tag{30.6}$$

in such a way that, under the exchange of particles $i \leftrightarrow j$,

$$|\Psi_A(N)\rangle \xrightarrow{i\leftrightarrow j} \sum_{B=1}^{G} \mathcal{M}_{AB}(ij)|\Psi_B(N)\rangle, \tag{30.7}$$

where $\mathcal{M}_{AB}(ij)$ is the so-called monodromy matrix, which, being unitary, guarantees that $|\Psi_A(N)\rangle$ will satisfy (30.4). Two transpositions, or braiding operations, as the exchange of identical particles is usually called, will correspond, respectively, to different monodromy matrices that in general do not commute. Hence the name "non-abelian statistics."

States with non-abelian statistics have been reported in the literature, as for instance the superconductors with a $p_x + ip_y$-symmetric gap [294, 295, 296, 290] and Kitaev's honeycomb lattice model [297]. Both systems present non-abelian vortices and Majorana states. Non-abelian statistics has also been obtained for certain quasi-particle, anyon excitations of the quantum Hall liquid [298] as well for Ising anyons [301]. Non-abelian anyons have therefore inspired the proposition of several interesting models, such as the one in [300].

30.3 The Non-Abelian Chern–Simons Theory

30.3.1 The Theory

We introduce here the SU(2) non-abelian Chern–Simons theory coupled to a Higgs field in the adjoint representation. This was shown to correspond to the Pfaffian Moore–Read wave-function of the $\nu = 5/2$ plateau [242] in an analogous way as the ZHK abelian Chern–Simons–Landau–Ginzburg theory corresponds to the Laughlin states associated to odd-denominator FQHE.

The theory is defined by the action

$$S_{CS}[A] = \frac{k}{4\pi} \int d^3z \left\{ \epsilon^{\mu\nu\rho} \left[A_\mu^a \partial_\nu A_\rho^a + \frac{2}{3} \epsilon^{abc} A_\mu^a A_\nu^b A_\rho^c \right] \right\}. \tag{30.8}$$

We will be interested in the vortex excitations that this theory might possess. In order to stabilize them, we need to couple a Higgs field $\Phi = \Phi^a T^a$ where T^a are the SU(2) group generators. This is done by adding to the Chern–Simons Lagrangean above, the term [293]

$$\text{Tr} D_\mu \Phi D^\mu \Phi - V(|\Phi|, \eta), \tag{30.9}$$

where the covariant derivative is given by

$$D_\mu^{ij} = \partial_\mu \delta^{ij} - i(T^a)^{ij} A_\mu^a$$

and the Higgs field potential, by

$$V = \lambda \text{Tr} \Phi^2 (\eta^2 + \Phi^2)^2. \tag{30.10}$$

The Euler–Lagrange equation corresponding to this Lagrangean is

$$\frac{k}{2\pi} \epsilon^{\mu\nu\rho} D_\nu^{ac} A_\rho^c = J^{\mu a}, \tag{30.11}$$

where $J^{\mu a} = -2[\Phi, D^\mu \Phi]^a$.

The theory will be in an ordered or disordered phase, depending on η^2 being positive or negative. For $\eta^2 < 0$ we will have $\langle \Phi \rangle \neq 0$, which corresponds to an ordered phase, while for $\eta^2 > 0$, we would have $\langle \Phi \rangle = 0$, characterizing a disordered phase.

Classical vortex solutions of this theory have been studied in [293], and resemble the classical vortices described in Chapter 8. Here we are going to develop a fully quantized approach to these vortices.

30.3.2 *Charge and Magnetic Flux Carrying Operators*

We now introduce the charge and magnetic flux operators. This is made by means of an abelian projection on n^a, a unit vector transforming under the adjoint representation of the group, such as $n^a = \frac{\phi^a}{|\phi|}$ where $\phi^a \equiv \langle \Phi^a \rangle$.

The charge operator, then, is given by

$$Q = \int d^2x \, J^{0a} n^a \tag{30.12}$$

in terms of the current density

$$J^{0a} = \frac{k}{2\pi} \epsilon^{ij} D_i^{ac} A_j^c(x), \tag{30.13}$$

whereas the magnetic flux operator, accordingly, is defined by

$$\Phi_M = \int d^2x \, B^a n^a \tag{30.14}$$

where B^a is the non-abelian magnetic field, given by

$$B^a(x) = \frac{1}{2} \epsilon^{ij} \, F_{ij}^a. \tag{30.15}$$

We now introduce the operators that create states carrying, respectively, charge and magnetic flux, namely, eigenstates of Q and Φ_M, respectively denoted σ and μ. The latter will be the vortex creation operator.

We have seen in Chapter 10 that there exists a duality relation between charge and topological charge bearing operators. Since magnetic flux is precisely the topological charge in the present system, it follows that the σ and μ operators will have the general form of the dual operators studied before.

Indeed, we have [299]

$$\mu(\mathbf{x}, t) = \exp \left\{ i \frac{a}{2\pi} n^a \int_{\mathbf{x}}^{\infty} d\xi^i A_i^a(\mathbf{x}, t) \right\} \tag{30.16}$$

and

$$\sigma(\mathbf{x}, t) = \exp \left\{ \frac{b}{2\pi} n^a \int_{S_x} d^2\xi^\lambda \epsilon^{\lambda\mu\nu} \partial_\mu A_\nu^a(\vec{\xi}, t) \arg(\vec{\xi} - \mathbf{x}) \right\}. \tag{30.17}$$

Using the canonical commutation relation corresponding to (30.8), namely

$$[A_i^a(\mathbf{x}, t), A_j^b(\mathbf{y}, t)] = \frac{2\pi}{k} \epsilon_{ij} \delta^{ab} \delta^2(\mathbf{x} - \mathbf{y}), \tag{30.18}$$

we demonstrate in Section 30.11 the following relations involving the σ and μ operators as well as the charge and magnetic flux operators.

Firstly there are the relations, which imply the σ and μ operators create states bearing, respectively charge and magnetic flux:

$$[\sigma(\mathbf{x}, t), Q] = b\sigma(\mathbf{x}, t) \tag{30.19}$$

$$[\mu(\mathbf{x}, t), \Phi_M] = \frac{a}{k}\mu(\mathbf{x}, t). \tag{30.20}$$

Then there is the order/disorder dual algebra

$$\mu(\mathbf{y}, t)\sigma(\mathbf{x}, t) = \sigma(\mathbf{x}, t)\mu(\mathbf{y}, t)e^{i\frac{ab}{2\pi k}\arg(\mathbf{y}-\mathbf{x})}. \tag{30.21}$$

This implies that the composite operator

$$\Psi(x) = \lim_{x^a, x^b \to x} \sigma(x^a)\mu(x^b)f(x^a - x^b), \tag{30.22}$$

defined in terms of an arbitrary function $f(x)$, has commutation

$$\Psi(\mathbf{x}, t)\Psi(\mathbf{y}, t) = \Psi(\mathbf{y}, t)\Psi(\mathbf{x}, t) \exp\left\{i\frac{ab}{2\pi k}\left[\arg(\mathbf{y} - \mathbf{x}) - \arg(\mathbf{x} - \mathbf{y})\right]\right\}$$

$$\Psi(\mathbf{x}, t)\Psi(\mathbf{y}, t) = \Psi(\mathbf{y}, t)\Psi(\mathbf{x}, t) \exp\left\{i\frac{ab}{2\pi k}\pi\epsilon\left(\pi - \arg(\mathbf{y} - \mathbf{x})\right)\right\}. \tag{30.23}$$

An excitation carrying magnetic flux, which in D = 2+1 is the topological charge, is by definition a magnetic vortex. We conclude, therefore, that the μ-operator, acting on the vacuum, creates magnetic vortex topological excitations. The composite operators $\Psi = \sigma\mu$, accordingly, create charged vortices, namely vortices that bear both charge and magnetic flux. The commutation relation above implies these composite charged magnetic vortices possess generalized spin-statistics $s = ab/4\pi k$. These are naturally interpreted as the excitations occurring in the $\nu = 5/2$ quantum Hall system. In what follows, we show how to obtain states with non-abelian statistics out of these charged magnetic vortices.

30.4 Quantum Correlation Functions of Charged Vortices

Let us determine here the quantum correlation functions of the vortex excitations. For this purpose, we follow the same procedure developed in Chapter 9 for the obtainment of topological excitation correlation functions. There, we have shown that such correlation functions are given in general by the exponential of the Γ-functional, the generator of proper vertices, calculated at an appropriate external field [39, 40].

The external fields suited for describing the σ and μ correlation functions in the present theory were derived in [299], according to the lines described in Chapter 9. They are given, respectively, by

$$\overline{A}^a_\mu(z; x) = \frac{b}{2\pi} n^a \int_{S_x} d^2\xi^\mu \, \arg(z - x) \, \delta^3(z - \xi) \tag{30.24}$$

and

$$\overline{B}^a_\mu(z; x) = i\frac{a}{2\pi} n^a \int_{x,L}^\infty d\xi_\mu \delta^3(z - \xi). \tag{30.25}$$

Here, S_x is a surface containing the point $(\mathbf{x}, t)$, while L is an arbitrary curve going from $(\mathbf{x}, t)$ to infinity.

We are interested in the large-distance behavior of the composite vortex correlation function. In this case, only the $\Gamma^{(2)}$-component of the functional $\Gamma[\overline{A}^a_\mu, \overline{B}^a_\mu]$ contributes, as we saw in Chapter 9 [39, 40],

$$\langle \sigma(x_1)\mu(x_2)\mu^\dagger(y_2)\sigma^\dagger(y_1)\rangle = \exp\left\{-\frac{1}{2}A^\mu_E\Gamma^{\mu\nu}_{(2)}A^\nu_E\right\}, \tag{30.26}$$

where the external field is

$$A^\mu_E = \overline{A}^a_\mu(z; x_1) - \overline{A}^a_\mu(z; y_1) + \overline{B}^a_\mu(z; x_2) - \overline{B}^a_\mu(z; y_2). \tag{30.27}$$

In the so-called "broken" phase, where $\langle \Phi \rangle = M \neq 0$, we get [299]

$$\langle \sigma(x_1)\mu(x_2)\mu^\dagger(y_2)\sigma^\dagger(y_1)\rangle \overset{|x_i - y_j| \to \infty}{\longrightarrow}$$

$$\exp\left\{-\frac{M^2}{2}\left[\left(\frac{a}{k}\right)^2 |x_1 - y_1| + b^2|x_2 - y_2|\right]\right.$$

$$\left. -i\frac{ab}{4\pi k}\left[\arg(\mathbf{x}_1 - \mathbf{y}_2) + \arg(\mathbf{y}_1 - \mathbf{x}_2) - \arg(\mathbf{x}_1 - \mathbf{x}_2) - \arg(\mathbf{y}_1 - \mathbf{y}_2)\right]\right\}. \tag{30.28}$$

Now we can introduce a composite operator $\Psi(x)$ bearing charge and magnetic flux, through an expression of the form in (30.22):

$$\Psi(x) = \lim_{x_1, x_2 \to x} \sigma(x_1)\mu(x_2) \exp\left\{-i\frac{ab}{4\pi k}\left[\arg(\mathbf{x}_1 - \mathbf{x}_2)\right]\right\}. \tag{30.29}$$

From this and (30.28) we obtain the composite operator correlation function:

$$\langle \Psi(x)\Psi^\dagger(y)\rangle \overset{|x - y| \to \infty}{\longrightarrow} \exp\left\{-\frac{M^2}{2}\left[\left(\frac{a}{k}\right)^2 + b^2\right]|x - y|\right.$$

$$\left. -i\frac{ab}{4\pi k}\left[\arg(\mathbf{x} - \mathbf{y}) + \arg(\mathbf{y} - \mathbf{x})\right]\right\}. \tag{30.30}$$

The spin-statistics of a field may be inferred from the analytic structure of its Euclidean correlation functions [31]. Indeed, multivalued Euclidean functions are such that each sheet corresponds to a different ordering of field operators in the associated Minkowski functions, which, by their turn, correspond to vacuum expectation values of these field operators. Adjacent sheets of a multivalued Euclidean correlator differ by a phase $e^{i2\pi s}$, and it follows that real-time field expectation values will be such that [31]

$$\langle \Psi(x)\Psi^{\dagger}(y)\rangle^{(1)} = e^{i2\pi s}\langle \Psi^{\dagger}(y)\Psi(x)\rangle^{(1)} = e^{i4\pi s}\langle \Psi(x)\Psi^{\dagger}(y)\rangle^{(2)}$$
$$= e^{i6\pi s}\langle \Psi^{\dagger}(y)\Psi(x)\rangle^{(2)} = \ldots \tag{30.31}$$

Observe that for $2s \in \mathbb{N}$ (bosons or fermions), $e^{i2\pi s} = \pm 1$. Then each of the functions $\langle \Psi(x)\Psi^{\dagger}(y)\rangle$ and $\langle \Psi^{\dagger}(y)\Psi(x)\rangle$ is univalued in this case.

Now, whenever $2s \notin \mathbb{N}$, each of the field expectation values is itself multivalued, the particular sheet being indicated by the superscript. The values of the function in two adjacent sheets differ by a factor $e^{i4\pi s}$, implying the field operator vacuum expectation values have a branch cut, the number of sheets being determined by the spin. For $s = 1/N$, $N = 3, 5, 7, \ldots$ for instance, there are N sheets; for $s = 1/4$, there are two sheets; and so on.

Examining the Euclidean vortex correlation function (30.30) in light of the considerations above, we conclude that the vortex excitations possess spin/statistics given by $s = ab/4\pi k$, thus confirming the result derived directly from the commutation relation (30.23).

The vortices carry a magnetic flux $\Phi_M = \frac{a}{k}$ and a charge $Q = b$. Assuming that a is the magnetic flux unit, namely $a = \frac{hc}{Q}$, it follows that the spin of the vortex excitations will be $s = \frac{hc}{2k}$, or $s = 1/2k$ in the natural units.

30.5 Majorana Vortices with Non-Abelian Statistics

We have just seen that the charged, magnetic vortex excitations of the non-abelian Chern–Simons–Higgs model are, in general, anyons, with spin $s = 1/2k$. It has been demonstrated, on the other hand, that this model is associated to the $\nu = 5/2$ Moore–Read, Pfaffian wave-function, provided the "level" is two ($k = 2$). We immediately conclude, following the previous analysis, that the spin/statistics of the charged vortices excitations of the present model is $s = 1/4$. Such vortices are, consequently, abelian anyons. In the remainder of this section we will show how to construct fields with non-abelian statistics out of these $s = 1/4$ abelian anyons. For that purpose, let us define the combined fields

$$\Phi_{\pm} = \frac{1}{2}\left[\Psi \pm \Psi^{\dagger}\right]. \tag{30.32}$$

These are, respectively, self-adjoint and anti-self-adjoint fields, and consequently may be properly called Majorana and anti-Majorana fields. We also have

$$\Psi = \Phi_+ + \Phi_-$$
$$\Psi^\dagger = \Phi_+ - \Phi_-. \tag{30.33}$$

In the next two subsections, we will derive properties of the Φ-field correlation functions that will allow us to infer certain relations among them. Such relations also hold for the corresponding states created by products of Φ-fields acting on the vacuum.

30.5.1 Two-Point Correlation Functions

Let us consider here the two-point correlation function of the fields defined in (30.32).

Using the results $\langle \Psi(x)\Psi(y)\rangle = \langle \Psi^\dagger(x)\Psi^\dagger(y)\rangle = 0$, we obtain

$$\langle \Phi_+(x)\Phi_+(y)\rangle = -\langle \Phi_-(x)\Phi_-(y)\rangle = \frac{1}{4}\left(\langle \Psi(x)\Psi^\dagger(y)\rangle + \langle \Psi^\dagger(x)\Psi(y)\rangle\right) \tag{30.34}$$

and

$$\langle \Phi_-(x)\Phi_+(y)\rangle = -\langle \Phi_+(x)\Phi_-(y)\rangle = \frac{1}{4}\left(\langle \Psi(x)\Psi^\dagger(y)\rangle - \langle \Psi^\dagger(x)\Psi(y)\rangle\right). \tag{30.35}$$

Then, using (30.30), we obtain

$$\langle \Phi_+(x)\Phi_+(y)\rangle = \frac{1}{4}\left[e^{-2is\mathrm{Arg}(x-y)}e^{-is\pi}e^{-\gamma|x-y|} + e^{-2is\mathrm{Arg}(y-x)}e^{-is\pi}e^{-\gamma|x-y|}\right]$$
$$\langle \Phi_-(x)\Phi_+(y)\rangle = \frac{1}{4}\left[e^{-2is\mathrm{Arg}(x-y)}e^{-is\pi}e^{-\gamma|x-y|} - e^{-2is\mathrm{Arg}(y-x)}e^{-is\pi}e^{-\gamma|x-y|}\right] \tag{30.36}$$

where $\gamma = \frac{M^2}{2}\left[\left(\frac{a}{k}\right)^2 + b^2\right]$.

We want to study now how the states associated to the fields $\Phi_\pm$ would transform under braiding operations. Using the properties of Arg (z), we obtain

$$\left[e^{-2is\mathrm{Arg}(x-y)} \pm e^{-2is\mathrm{Arg}(y-x)}\right]e^{-is\pi}e^{-\gamma|x-y|} \; x \overset{\longrightarrow}{\leftrightarrow} y$$
$$\left[e^{-i2\pi s}e^{-2is\mathrm{Arg}(x-y)} \pm e^{i2\pi s}e^{-2is\mathrm{Arg}(y-x)}\right]e^{-is\pi}e^{-\gamma|x-y|}.$$

From this, we can infer that whenever the operator Ψ is bosonic or fermionic the phases generated by braiding the $\Phi_\pm$-particles are identical, namely, $e^{2\pi is} = e^{-2\pi is} = \pm 1$, implying

$$\langle \Phi_\pm(y)\Phi_\pm(x)\rangle = e^{2\pi is}\langle \Phi_\pm(x)\Phi_\pm(y)\rangle = \pm\langle \Phi_\pm(x)\Phi_\pm(y)\rangle.$$

This shows that for bosonic or fermionic vortex fields Ψ, the corresponding Majorana and anti-Majorana combinations $\Phi_\pm$ are also bosonic or fermionic.

Conversely, when the vortex field is an anyon, namely, for $2s \notin \mathbb{N}$, then $e^{2\pi i s} \neq e^{-2\pi i s}$, and the previous braiding operation, consequently, is no longer just the multiplication by a phase factor, but rather a matrix multiplication. This fact is at the root of the non-abelian statistics found in the Majorana and anti-Majorana vortex excitations.

Indeed, under a braiding operation the $\Phi_\pm(x)$ fields behave in the following way:

$$
\begin{aligned}
\langle \Phi_+(y)\Phi_+(x)\rangle &= \frac{1}{4}\left[\alpha^*\langle(\Phi_+(x) + \Phi_-(x))(\Phi_+(y) - \Phi_-(y))\rangle\right.\\
&\quad\left. + \alpha\langle(\Phi_+(x) - \Phi_-(x))(\Phi_+(y) + \Phi_-(y))\rangle\right]\\
&= \frac{1}{2}\left[(\alpha + \alpha^*)\langle\Phi_+(x)\Phi_+(y)\rangle - (\alpha - \alpha^*)\langle\Phi_-(x)\Phi_-(y)\rangle\right]\\
&= \cos\delta\langle\Phi_+(x)\Phi_+(y)\rangle - i\sin\delta\langle\Phi_-(x)\Phi_-(y)\rangle \quad (30.37)
\end{aligned}
$$

and

$$
\begin{aligned}
\langle \Phi_-(y)\Phi_+(x)\rangle &= \frac{1}{4}\left[\alpha^*\langle(\Phi_+(x) + \Phi_-(x))(\Phi_+(y) - \Phi_-(y))\rangle\right.\\
&\quad\left. - \alpha\langle(\Phi_+(x) - \Phi_-(x))(\Phi_+(y) + \Phi_-(y))\rangle\right]\\
&= \frac{1}{2}\left[-(\alpha - \alpha^*)\langle\Phi_+(x)\Phi_+(y)\rangle + (\alpha + \alpha^*)\langle\Phi_-(x)\Phi_-(y)\rangle\right]\\
&= -i\sin\delta\langle\Phi_+(x)\Phi_+(y)\rangle + \cos\delta\langle\Phi_-(x)\Phi_-(y)\rangle, \quad (30.38)
\end{aligned}
$$

where in the above expression $\alpha = e^{i\delta}$ and $\delta = 2\pi s$.

We may conclude, then, that when the composite vortex field Ψ is an anyon, then it follows that the associated Majorana and anti-Majorana field combinations $\Phi_\pm$ will have a non-abelian braiding transformation. This can be written as

$$
\begin{pmatrix} \langle\Phi_+(y)\Phi_+(x)\rangle \\ \langle\Phi_-(y)\Phi_+(x)\rangle \end{pmatrix} = \begin{pmatrix} \cos\delta & -i\sin\delta \\ -i\sin\delta & \cos\delta \end{pmatrix} \begin{pmatrix} \langle\Phi_+(x)\Phi_+(y)\rangle \\ \langle\Phi_-(x)\Phi_+(y)\rangle \end{pmatrix}.
$$

The braiding matrix and its hermitean adjoint

$$
\mathcal{M} = \begin{pmatrix} \cos\delta & -i\sin\delta \\ -i\sin\delta & \cos\delta \end{pmatrix} \qquad \mathcal{M}^\dagger = \begin{pmatrix} \cos\delta & i\sin\delta \\ i\sin\delta & \cos\delta \end{pmatrix}
$$

satisfy $\mathcal{M}^\dagger\mathcal{M} = 1$, being therefore unitary.

Let us finally impose the fact that, as we have seen before, the vortex excitation of the present field theory model associated to the plateaus $\nu = 5/2$ of a system exhibiting the quantum Hall effect are anyons with $s = 1/4$. In this case, we have $\delta = \pi/2$ or, equivalently,

$$\mathcal{M} = -iX, \quad \text{in which} \quad X = \begin{pmatrix} 0 & 1 \\ 1 & 0 \end{pmatrix}. \tag{30.39}$$

We see that a NOT gate is obtained out of the braiding matrix $\mathcal{M}$ (up to an i-factor) [299].

30.5.2 Four-Point Correlation Functions

Other logic gates can be obtained by considering higher correlation functions (more than two-particle quantum states). For states with $2n$ particles, it is easy to show that the dimension of the braiding matrices (or of the space of degenerate states) will be 2^n. Taking four-point functions, for instance, would produce 4×4 matrices.

Let us now turn to the 4-point function of the vortex operator. This is obtained by just inserting in (30.26) two additional external fields similar to the ones in (30.27). The resulting expression is [299]

$$\begin{aligned}
\langle \Psi(x_1)\Psi(x_2)\Psi^\dagger(x_3)\Psi^\dagger(x_4)\rangle &= \langle \Psi^\dagger(x_1)\Psi^\dagger(x_2)\Psi(x_3)\Psi(x_4)\rangle \\
&= \exp\{2is[\text{Arg}(\vec{x}_1 - \vec{x}_2) + \text{Arg}(\vec{x}_3 - \vec{x}_4) \\
&\quad - 2is[\text{Arg}(\vec{x}_1 - \vec{x}_3) + \text{Arg}(\vec{x}_1 - \vec{x}_4) + \text{Arg}(\vec{x}_2 - \vec{x}_3) + \text{Arg}(\vec{x}_2 - \vec{x}_4)] \\
&\quad - \pi a^2 M^2 \left(|x_1 - x_3| + |x_1 - x_4| + |x_2 - x_3| + |x_2 - x_4|\right) \\
&\quad + \pi a^2 M^2 \left(|x_1 - x_2| + |x_3 - x_4|\right)\} \equiv A.
\end{aligned} \tag{30.40}$$

Related correlation functions will be

$$\begin{aligned}
\langle \Psi^\dagger(x_1)\Psi(x_2)\Psi^\dagger(x_3)\Psi(x_4)\rangle &= \langle \Psi(x_1)\Psi^\dagger(x_2)\Psi(x_3)\Psi^\dagger(x_4)\rangle \\
&= \exp\{2is[\text{Arg}(\vec{x}_1 - \vec{x}_3)\text{Arg}(\vec{x}_2 - \vec{x}_4) \\
&\quad - 2is[\text{Arg}(\vec{x}_1 - \vec{x}_2) + \text{Arg}(\vec{x}_1 - \vec{x}_4) + \text{Arg}(\vec{x}_2 - \vec{x}_3) + \text{Arg}(\vec{x}_2 - \vec{x}_4)]+ \\
&\quad - \pi a^2 M^2 \left(|x_4 - x_2| + |x_4 - x_1| + |x_2 - x_3| + |x_2 - x_1|\right) \\
&\quad + \pi a^2 M^2 \left(|x_4 - x_2| + |x_3 - x_1|\right)\} \equiv B
\end{aligned} \tag{30.41}$$

and

$$\begin{aligned}
\langle \Psi^\dagger(x_1)\Psi(x_2)\Psi(x_3)\Psi^\dagger(x_4)\rangle &= \langle \Psi(x_1)\Psi^\dagger(x_2)\Psi^\dagger(x_3)\Psi(x_4)\rangle \\
&= \exp\{2is[\text{Arg}(\vec{x}_1 - \vec{x}_4) + \text{Arg}(\vec{x}_2 - \vec{x}_3) \\
&\quad - 2is[\text{Arg}(\vec{x}_1 - \vec{x}_2) + \text{Arg}(\vec{x}_1 - \vec{x}_3) + \text{Arg}(\vec{x}_2 - \vec{x}_3) + \text{Arg}(\vec{x}_2 - \vec{x}_4)]+ \\
&\quad - \pi a^2 M^2 \left(|x_4 - x_2| + |x_3 - x_1| + |x_2 - x_3| + |x_2 - x_1|\right) \\
&\quad + \pi a^2 M^2 \left(|x_3 - x_2| + |x_4 - x_1|\right)\} \equiv C.
\end{aligned} \tag{30.42}$$

We now express the correlation functions of the new fields given by (30.32) in terms of the three correlation functions above, namely

$$\langle \Phi_+(x_1)\Phi_+(x_2)\Phi_+(x_3)\Phi_+(x_4)\rangle = \langle \Phi_-(x_1)\Phi_-(x_2)\Phi_-(x_3)\Phi_-(x_4)\rangle$$
$$= 2[A + B + C)]$$
$$\langle \Phi_+(x_1)\Phi_+(x_2)\Phi_-(x_3)\Phi_-(x_4)\rangle = \langle \Phi_-(x_1)\Phi_-(x_2)\Phi_+(x_3)\Phi_+(x_4)\rangle$$
$$= 2[A - B - C)]$$
$$\langle \Phi_-(x_1)\Phi_+(x_2)\Phi_-(x_3)\Phi_+(x_4)\rangle = \langle \Phi_+(x_1)\Phi_-(x_2)\Phi_+(x_3)\Phi_-(x_4)\rangle$$
$$= 2[-A + B - C)]$$
$$\langle \Phi_-(x_1)\Phi_+(x_2)\Phi_+(x_3)\Phi_-(x_4)\rangle = \langle \Phi_+(x_1)\Phi_-(x_2)\Phi_-(x_3)\Phi_+(x_4)\rangle$$
$$= 2[-A - B + C)]. \tag{30.43}$$

We now examine how the Majorana and anti-Majorana vortex states behave under the braiding $x_1 \leftrightarrow x_2$. Using the two previous expressions, we obtain

$$\langle \Phi_+(x_1)\Phi_+(x_2)\Phi_+(x_3)\Phi_+(x_4)\rangle \; x_1 \overset{\rightarrow}{\leftrightarrow} x_2 \quad 2\left[e^{2\pi i s}\langle \Psi(x_1)\Psi(x_2)\Psi^\dagger(x_3)\Psi^\dagger(x_4)\rangle \right.$$
$$\left. + e^{-2\pi i s}\langle \Psi^\dagger(x_1)\Psi(x_2)\Psi(x_3)\Psi^\dagger(x_4)\rangle + e^{-2\pi i s}\langle \Psi^\dagger(x_1)\Psi(x_2)\Psi^\dagger(x_3)\Psi(x_4)\rangle\right]$$

$$\langle \Phi_+(x_1)\Phi_+(x_2)\Phi_-(x_3)\Phi_-(x_4)\rangle \; x_1 \overset{\rightarrow}{\leftrightarrow} x_2 \quad 2\left[e^{2\pi i s}\langle \Psi(x_1)\Psi(x_2)\Psi^\dagger(x_3)\Psi^\dagger(x_4)\rangle \right.$$
$$\left. - e^{-2\pi i s}\langle \Psi^\dagger(x_1)\Psi(x_2)\Psi(x_3)\Psi^\dagger(x_4)\rangle - e^{-2\pi i s}\langle \Psi^\dagger(x_1)\Psi(x_2)\Psi^\dagger(x_3)\Psi(x_4)\rangle\right]$$

$$\langle \Phi_-(x_1)\Phi_+(x_2)\Phi_-(x_3)\Phi_+(x_4)\rangle \; x_1 \overset{\rightarrow}{\leftrightarrow} x_2 \quad 2\left[-e^{2\pi i s}\langle \Psi(x_1)\Psi(x_2)\Psi^\dagger(x_3)\Psi^\dagger(x_4)\rangle \right.$$
$$\left. + e^{-2\pi i s}\langle \Psi^\dagger(x_1)\Psi(x_2)\Psi(x_3)\Psi^\dagger(x_4)\rangle - e^{-2\pi i s}\langle \Psi^\dagger(x_1)\Psi(x_2)\Psi^{\dagger}(x_3)\Psi(x_4)\rangle\right]$$

$$\langle \Phi_-(x_1)\Phi_+(x_2)\Phi_+(x_3)\Phi_-(x_4)\rangle \; x_1 \overset{\rightarrow}{\leftrightarrow} x_2 \quad 2\left[-e^{2\pi i s}\langle \Psi(x_1)\Psi(x_2)\Psi^\dagger(x_3)\Psi^\dagger(x_4)\rangle \right.$$
$$\left. - e^{-2\pi i s}\langle \Psi^\dagger(x_1)\Psi(x_2)\Psi^\dagger(x_3)\Psi(x_4)\rangle + e^{-2\pi i s}\langle \Psi^\dagger(x_1)\Psi(x_2)\Psi^\dagger(x_3)\Psi(x_4)\rangle\right]. \tag{30.44}$$

Now with the help of the (30.43) equations, we can write the right-hand side of Eqs. (30.44) in terms of correlators of the new fields Φ_+ and Φ_-, namely

$$\langle \Phi_+(x_1)\Phi_+(x_2)\Phi_+(x_3)\Phi_+(x_4)\rangle \; x_1 \overset{\rightarrow}{\leftrightarrow} x_2 \; \cos\delta\langle \Phi_+(x_1)\Phi_+(x_2)\Phi_+(x_3)\Phi_+(x_4)\rangle$$
$$+ i\sin\delta\langle \Phi_+(x_1)\Phi_+(x_2)\Phi_-(x_3)\Phi_-(x_4)\rangle$$

$$\langle \Phi_+(x_1)\Phi_+(x_2)\Phi_-(x_3)\Phi_-(x_4)\rangle \; x_1 \overset{\rightarrow}{\leftrightarrow} x_2 \; i\sin\delta\langle \Phi_+(x_1)\Phi_+(x_2)\Phi_+(x_3)\Phi_+(x_4)\rangle$$
$$+ \cos\delta\langle \Phi_+(x_1)\Phi_+(x_2)\Phi_-(x_3)\Phi_-(x_4)\rangle$$

$$\langle \Phi_-(x_1)\Phi_+(x_2)\Phi_-(x_3)\Phi_+(x_4)\rangle \; x_1 \overset{\rightarrow}{\leftrightarrow} x_2 \; i\sin\delta\langle \Phi_-(x_1)\Phi_+(x_2)\Phi_-(x_3)\Phi_+(x_4)\rangle$$
$$+ \cos\delta\langle \Phi_-(x_1)\Phi_+(x_2)\Phi_+(x_3)\Phi_-(x_4)\rangle$$

$$\langle \Phi_-(x_1)\Phi_+(x_2)\Phi_+(x_3)\Phi_-(x_4)\rangle \; x_1 \overset{\rightarrow}{\leftrightarrow} x_2 \; \cos\delta\langle \Phi_-(x_1)\Phi_+(x_2)\Phi_-(x_3)\Phi_+(x_4)\rangle$$
$$+ i\sin\delta\langle \Phi_-(x_1)\Phi_+(x_2)\Phi_+(x_3)\Phi_-(x_4)\rangle. \tag{30.45}$$

This can be written in matrix form as

$$
\begin{pmatrix}
\langle \Phi_+(x_1)\Phi_+(x_2)\Phi_+(x_3)\Phi_+(x_4) \rangle \\
\langle \Phi_+(x_1)\Phi_+(x_2)\Phi_-(x_3)\Phi_-(x_4) \rangle \\
\langle \Phi_-(x_1)\Phi_+(x_2)\Phi_-(x_3)\Phi_+(x_4) \rangle \\
\langle \Phi_-(x_1)\Phi_+(x_2)\Phi_+(x_3)\Phi_-(x_4) \rangle
\end{pmatrix}
\xrightarrow{x_1 \leftrightarrow x_2}
\mathcal{M}_{12}
\begin{pmatrix}
\langle \Phi_+(x_1)\Phi_+(x_2)\Phi_+(x_3)\Phi_+(x_4) \rangle \\
\langle \Phi_+(x_1)\Phi_+(x_2)\Phi_-(x_3)\Phi_-(x_4) \rangle \\
\langle \Phi_-(x_1)\Phi_+(x_2)\Phi_-(x_3)\Phi_+(x_4) \rangle \\
\langle \Phi_-(x_1)\Phi_+(x_2)\Phi_+(x_3)\Phi_-(x_4) \rangle
\end{pmatrix},
$$

where $\mathcal{M}_{12}$, the so-called monodromy matrix, or simply braiding matrix, is given by

$$
\mathcal{M}_{12} =
\begin{pmatrix}
\cos \delta & i \sin \delta & 0 & 0 \\
i \sin \delta & \cos \delta & 0 & 0 \\
0 & 0 & i \sin \delta & \cos \delta \\
0 & 0 & \cos \delta & i \sin \delta
\end{pmatrix},
$$

where $\delta = 2\pi s$. We see that it is unitary, namely $\mathcal{M}_{12}^\dagger \mathcal{M}_{12} = 1$.

The monodromy matrices that correspond to other braiding operations, $\mathcal{M}_{13}$, $\mathcal{M}_{14}$, $\mathcal{M}_{23}$, $\mathcal{M}_{24}$ and $\mathcal{M}_{34}$, can be determined analogously. They are given by

$$
\mathcal{M}_{34} = \mathcal{M}_{12}
$$

$$
\mathcal{M}_{13} = \mathcal{M}_{24} =
\begin{pmatrix}
\alpha & 0 & 0 & 0 \\
0 & 0 & 0 & \alpha \\
0 & 0 & \alpha & 0 \\
0 & \alpha & 0 & 0
\end{pmatrix}
$$

$$
\mathcal{M}_{14} = \mathcal{M}_{23} =
\begin{pmatrix}
\cos \delta & 0 & 0 & i \sin \delta \\
0 & i \sin \delta & \cos \delta & 0 \\
0 & \cos \delta & i \sin \delta & 0 \\
i \sin \delta & 0 & 0 & \cos \delta
\end{pmatrix},
$$

where $\alpha = e^{-si2\pi s}$.

Interestingly, the monodromy matrices satisfy the Yang–Baxter relations,

$$
\mathcal{M}_{12}\mathcal{M}_{23}\mathcal{M}_{12} = \mathcal{M}_{23}\mathcal{M}_{12}\mathcal{M}_{23}
\tag{30.46}
$$

and

$$
\mathcal{M}_{23}\mathcal{M}_{34}\mathcal{M}_{23} = \mathcal{M}_{34}\mathcal{M}_{23}\mathcal{M}_{34}
\tag{30.47}
$$

as they should, according to the theory of braid groups. Each of these equations just expresses the fact that the two different sequences of braiding operations on each side of the equation are topologically equivalent and, therefore, yield the same result.

Now, notice that if we insert the spin value $s = 1/4$ in the expression for $\mathcal{M}_{12}$, we generate the CNOT logic gate

$$\mathcal{M}_{12} = i \begin{pmatrix} 0 & 1 & 0 & 0 \\ 1 & 0 & 0 & 0 \\ 0 & 0 & 1 & 0 \\ 0 & 0 & 0 & 1 \end{pmatrix}.$$

More logic keys may be obtained by a straightforward generalization of the previous procedure for correlation functions with a larger number of points. In order to generate the Toffoli gate, for instance, which is 8×8, we would need to consider the 6-point correlation functions. In general, for generating a gate represented by a matrix of dimension 2^n, we would have to consider $2n$-point correlation functions.

30.6 Non-Abelian Statistics in Dirac Systems in 1+1D

30.6.1 The Model

We now introduce a system of Dirac quasi-particles in one spatial dimension, assumed to be described by the massless Lagrangean density [309]

$$\mathcal{L} = i \overline{\psi} \; \partial\!\!\!/ \psi - \mathcal{V}(\psi, \overline{\psi}), \tag{30.48}$$

where $\psi = \begin{pmatrix} \psi_1 \\ \psi_2 \end{pmatrix}$ is a two-component Dirac spinor and $\mathcal{V}$ is an arbitrary potential. Our convention for the γ-matrices is $\gamma^0 = \sigma_x$, $\gamma^1 = i\sigma_y$, $\gamma^5 = \gamma^0 \gamma^1 = -\sigma_z$.

We also assume invariance under global U(1) and chiral U(1) symmetries, namely

$$\psi \to e^{i\theta} \psi$$
$$\psi \to e^{i\theta\gamma^5} \psi \tag{30.49}$$

that leads to the conservation of charge and chirality.

A Lorentz boost will act on the space-time coordinates x_μ under the vector representation of the Lorentz group, namely, $x^\mu \to \Lambda^\mu{}_\nu x^\nu$, where

$$\Lambda^\mu{}_\nu = \begin{pmatrix} \cosh\omega & -\sinh\omega \\ -\sinh\omega & \cosh\omega \end{pmatrix}, \tag{30.50}$$

in such a way that $\tanh\omega = \mathrm{v}$, where v is the relative velocity between the two reference frames connected by the boost.

The Dirac field, conversely, will transform under the spinor representation of such group, namely

$$\psi \rightarrow \left(e^{-s\omega\gamma^5}\right)\psi = \left(\begin{array}{cc} e^{s\omega} & 0 \\ 0 & e^{-s\omega} \end{array}\right)\psi \tag{30.51}$$

where s is a real parameter known as the Lorentz spin of the Dirac field. In what follows, we demonstrate that the parameter s determines how a many-particle state-vector behaves under the interchange of identical particles. It consequently characterizes the particle statistics, which may be either bosonic ($2s =$ even), fermionic ($2s =$ odd) or anyonic ($2s \neq$ integer).

From (30.50), it follows that the light-cone coordinates $u = x^0 + x^1$ and $v = x^0 - x^1$ transform as

$$u \rightarrow e^{-\omega}u \qquad v \rightarrow e^{\omega}v. \tag{30.52}$$

30.6.2 General Form of the Dirac Correlator

Symmetry Considerations

We want to determine the two-point correlation function of the Dirac field. On the basis of the symmetries of the system, one can write

$$\langle 0|\psi_i(x)\psi_j^\dagger(0)|0\rangle = \left(\begin{array}{cc} f(x) & 0 \\ 0 & g(x) \end{array}\right). \tag{30.53}$$

Using (30.51), we may infer that under a Lorentz boost this shall transform as

$$\langle 0|\psi(x)\psi^\dagger(0)|0\rangle \rightarrow \left(\begin{array}{cc} e^{s\omega} & 0 \\ 0 & e^{-s\omega} \end{array}\right)\left(\begin{array}{cc} f(x) & 0 \\ 0 & g(x) \end{array}\right)\left(\begin{array}{cc} e^{s\omega} & 0 \\ 0 & e^{-s\omega} \end{array}\right)$$

$$\langle 0|\psi(x)\psi^\dagger(0)|0\rangle \rightarrow \left(\begin{array}{cc} e^{2s\omega}f & 0 \\ 0 & e^{-2s\omega}g \end{array}\right). \tag{30.54}$$

It follows from this that the Dirac field correlator may be written in the form

$$\langle 0|\psi(x)\psi^\dagger(0)|0\rangle = \left(\begin{array}{cc} \tilde{F}(-x^2)v^{2s} & 0 \\ 0 & \tilde{G}(-x^2)u^{2s} \end{array}\right)$$

$$= \left(\begin{array}{cc} F(-x^2)\frac{v^{2s}}{(-x^2)^s} & 0 \\ 0 & G(-x^2)\frac{u^{2s}}{(-x^2)^s} \end{array}\right), \tag{30.55}$$

where the functions F, G depend on the specific form of the interaction potential $\mathcal{V}$.

Prescriptions

As we saw in Chapter 5, field operator correlation functions usually need a prescription in order to make them well-defined. This is certainly the case for the correlation function above. The Feynman prescription would consist in the replacements

$$u \to u - i\epsilon(x^0) \; ; \;\; v \to v - i\epsilon(x^0) \; ; \;\; -uv = -x^2 \to -x^2 + i\epsilon, \qquad (30.56)$$

where $\epsilon(x^0)$ is $\epsilon \times sign(x^0)$.

For the case of bosonic or fermionic fields for which $2s$ is an integer, only the functions $F(-x^2)$, $G(-x^2)$ might need a prescription in the time-like region, where $x^2 > 0$. When the Dirac field is an anyon, however, the u and v factors would need a prescription in (30.55), both in the time-like and space-like regions, because $2s$ is no longer an integer. It happens that in the space-like region $\epsilon(x^0)$ can have its sign reversed by continuous Lorentz boosts, and consequently the Feynman prescription is not well defined for a general observer. It follows that, in the case of anyonic statistics, we cannot employ Feynman's prescription. Instead, we are going to use Wightman's prescription [303], namely

$$u \to u - i\epsilon \; ; \;\; v \to v - i\epsilon \; ; \;\; -uv = -x^2 \to -x^2 + i\epsilon(x^0). \qquad (30.57)$$

Notice that $-x^2$ only needs a prescription in the time-like region, where $\epsilon(x^0)$ cannot be changed by continuous Lorentz transformations and therefore is always well defined. The prescriptions for u and v are well defined in the whole Minkowski space. In the case of anyons, therefore, we shall adopt Wightman's prescription.

Euclidean Limit

We now take the Euclidean limit of the Dirac field correlation functions. Starting from (30.55), we make the analytic continuation $x_0 \to -ix_2^E$. Introducing the complex variable $z = x_1 + ix_2^E$, we have $-x^2 \to x_E^2 \equiv |z|^2$ and we can see that the Wightman correlators (30.55) are mapped into the (Euclidean) Schwinger functions [304]

$$\langle \psi_1(x)\psi_1^\dagger(0)\rangle_S = F(x_E^2)\frac{(-z)^{2s}}{|z|^{2s}}$$

$$\langle \psi_2(x)\psi_2^\dagger(0)\rangle_S = G(x_E^2)\frac{(z^*)^{2s}}{|z|^{2s}}$$

$$\langle \psi_1(x)\psi_2^\dagger(0)\rangle_S = 0 \; ; \;\; \langle \psi_2(x)\psi_1^\dagger(0)\rangle_S = 0, \qquad (30.58)$$

which are functions of a complex variable.

Introducing the polar representation $z = |z|e^{i\text{Arg}(z)}$, we can re-write those functions as

$$\langle \psi_1(x)\psi_1^\dagger(0)\rangle_S = F(|z|^2)e^{i2s\text{Arg}(-z)} \; ; \;\; \langle \psi_2(x)\psi_2^\dagger(0)\rangle_S = G(|z|^2)e^{-i2s\text{Arg}(z)},$$

$$\qquad (30.59)$$

where we chose the cuts of the Arg functions as $-\pi \le \text{Arg}(z) < \pi$ and $0 \le \text{Arg}(-z) < 2\pi$ in such a way that we may write $\text{Arg}(-z) = \text{Arg}(z) + \pi$.

The Schwinger functions corresponding to field correlators containing the same Dirac fields as above but in a reversed order can be obtained in similar way and are given by [31, 32]

$$\langle \psi_1^\dagger(0)\psi_1(x)\rangle_S = F(|z|^2)e^{i2s\text{Arg}(z)} \quad ; \quad \langle \psi_2^\dagger(0)\psi_2(x)\rangle_S = G(|z|^2)e^{-i2s\text{Arg}(-z)}.$$

(30.60)

It follows that

$$\langle \psi_i^\dagger(0)\psi_i(x)\rangle_S = e^{-i2\pi s}\langle \psi_i(x)\psi_i^\dagger(0)\rangle_S.$$

(30.61)

The "braiding" or particle exchange operation in this one-dimensional system is a well-defined operation in the framework of the Schwinger functions. We can understand this fact in a better way by considering (5.92) and (5.88). According to these, it becomes clear that Schwinger functions of opposite arguments will correspond to Wightman functions W_+ and W_-, which, by their turn, correspond to different ordering of field operators.

30.7 Majorana Spinors with Non-Abelian Statistics

Now, in analogy to what we did for the vortices of the non-abelian Chern–Simons theory, we introduce the Majorana and anti-Majorana spinor fields $\varphi_\pm = \begin{pmatrix} \varphi_{1\pm} \\ \varphi_{2\pm} \end{pmatrix}$, where

$$\varphi_{i+} = \frac{1}{2}\left(\psi_i + \psi_i^\dagger\right) \quad ; \quad \varphi_{i-} = \frac{1}{2}\left(\psi_i - \psi_i^\dagger\right)$$

(30.62)

$$\psi_i = \varphi_{i+} + \varphi_{i-} \quad ; \quad \psi_i^\dagger = \varphi_{i+} - \varphi_{i-}.$$

(30.63)

Here, again, we are going to extract properties of the states created by the φ-fields out of their correlation functions.

30.7.1 Two-Point Functions

Let us analyze now the two-point functions of the Majorana and anti-Majorana fields built out of the Dirac field,

$$\langle 0|\varphi_{i+}(x)\varphi_{i+}(y)|0\rangle_W = -\langle 0|\varphi_{i-}(x)\varphi_{i-}(y)|0\rangle_W$$
$$= \frac{1}{4}\left(\langle 0|\psi_i(x)\psi_i^\dagger(y)|0\rangle_W + \langle 0|\psi_i^\dagger(x)\psi_i(y)|0\rangle_W\right)$$

(30.64)

and

$$\langle 0|\varphi_{i-}(x)\varphi_{i+}(y)|0\rangle_W = -\langle 0|\varphi_{i+}(x)\varphi_{i-}(y)|0\rangle_W$$
$$= \frac{1}{4}\left(\langle 0|\psi_i(x)\psi_i^\dagger(y)|0\rangle_W - \langle 0|\psi_i^\dagger(x)\psi_i(y)|0\rangle_W\right), \qquad (30.65)$$

where $i = 1, 2$, represent the two chiralities, namely right-movers and leftmovers.

Notice that for $i \neq j$, $\langle 0|\varphi_{i\pm}(x)\varphi_{j\pm}(y)|0\rangle_W$ has a trivial analytic structure in terms of the complex variables, implying that nontrivial braiding will only occur involving states for which $i = j$, that is, states with the same chirality.

Using (30.59), we obtain

$$\langle\varphi_{1+}(x)\varphi_{1+}(y)\rangle_S = -\langle\varphi_{1-}(x)\varphi_{1-}(y)\rangle_S$$
$$= \frac{1}{4} F(|x - y|^2)\left[e^{i2s\mathrm{Arg}(y-x)} + e^{i2s\mathrm{Arg}(x-y)}\right] \qquad (30.66)$$

and

$$\langle\varphi_{1-}(x)\varphi_{1+}(y)\rangle_S = -\langle 0|\varphi_{1+}(x)\varphi_{1-}(y)|0\rangle_S$$
$$= \frac{1}{4} F(|x - y|^2)\left[e^{i2s\mathrm{Arg}(y-x)} - e^{i2s\mathrm{Arg}(x-y)}\right] \qquad (30.67)$$

with similar expressions for the 2-components.

Let us investigate now the braiding properties of the above functions. These will reflect, in the Euclidean space, the particle exchange of the corresponding real-time wave functions, or equivalently, the operator commutation in the Wightman functions. For this purpose, we use the relation $\mathrm{Arg}(y - x) = \mathrm{Arg}(x - y) + \pi$ and obtain

$$\left[e^{i2s\mathrm{Arg}(x-y)} \pm e^{i2s\mathrm{Arg}(y-x)}\right] x \overset{\rightarrow}{\leftrightarrow} y$$
$$\left[e^{i2s\mathrm{Arg}(y-x)}e^{-i2\pi s} \pm e^{i2s\mathrm{Arg}(x-y)}e^{i2\pi s}\right], \qquad (30.68)$$

with analogous relations for the corresponding expressions in the φ_2 functions. We can already see the similarity with the vortex correlation function studied before in this chapter.

When the Dirac field ψ is either bosonic or fermionic, the two complex phases generated above by the braiding operation are equal, namely $e^{i2\pi s} = e^{-i2\pi s} = \pm 1$, and we conclude that

$$\langle\varphi_\pm(y)\varphi_\pm(x)\rangle_S = e^{i2\pi s}\langle\varphi_\pm(x)\varphi_\pm(y)\rangle_S, \qquad (30.69)$$

implying that $\varphi_\pm$ will also be bosonic or fermionic.

Conversely, when the Dirac field is an anyon, $e^{i2\pi s} \neq e^{-i2\pi s}$ and we obtain from (1.8)

$$\langle \varphi_{1+}(y)\varphi_{1+}(x)\rangle_S = \cos\delta \langle \varphi_{1+}(x)\varphi_{1+}(y)\rangle_S - i\sin\delta \langle \varphi_{1+}(x)\varphi_{1-}(y)\rangle_S \qquad (30.70)$$

$$\langle \varphi_{1-}(y)\varphi_{1+}(x)\rangle_S = -i\sin\delta \langle \varphi_{1+}(x)\varphi_{1+}(y)\rangle_S + \cos\delta \langle \varphi_{1+}(x)\varphi_{1-}(y)\rangle_S, \qquad (30.71)$$

where in the above expressions $\delta = 2\pi s$.

We conclude that when the Dirac field is an anyon, the $\varphi_\pm$ fields will have non-abelian braiding given by

$$\begin{pmatrix} \langle \varphi_{1+}(y)\varphi_{1+}(x)\rangle_S \\ \langle \varphi_{1-}(y)\varphi_{1+}(x)\rangle_S \end{pmatrix} = \begin{pmatrix} \cos\delta & -i\sin\delta \\ -i\sin\delta & \cos\delta \end{pmatrix} \begin{pmatrix} \langle \varphi_{1+}(x)\varphi_{1+}(y)\rangle_S \\ \langle \varphi_{1+}(x)\varphi_{1-}(y)\rangle_S \end{pmatrix}$$

$$(30.72)$$

with similar expressions for the φ_2 functions.

This is identical to the braiding matrix obtained for the vortex Majorana states in the Chern–Simons–Higgs, non-abelian theory and characterizes the corresponding states as having non-abelian statistics.

It is remarkable that the braiding properties of the gapless Dirac spinors in 1+1D are identical to those of the charged magnetic vortices of the level-two, non-abelian Chern–Simons–Higgs theory in 2+1D. We will see that this also holds true for the four-point functions, thus strongly suggesting a close relation between both systems.

30.7.2 Four-Point Functions

Here, for simplicity, we will consider just the 1-component of the ψ and φ fields, omitting the component index in order to simplify the notation.

Consider the four different correlation functions, namely

$$\langle \varphi_+(x_1)\varphi_+(x_2)\varphi_+(x_3)\varphi_+(x_4)\rangle_S = \langle \varphi_-(x_1)\varphi_-(x_2)\varphi_-(x_3)\varphi_-(x_4)\rangle_S$$
$$\langle \varphi_+(x_1)\varphi_+(x_2)\varphi_-(x_3)\varphi_-(x_4)\rangle_S = \langle \varphi_-(x_1)\varphi_-(x_2)\varphi_+(x_3)\varphi_+(x_4)\rangle_S$$
$$\langle \varphi_+(x_1)\varphi_-(x_2)\varphi_+(x_3)\varphi_-(x_4)\rangle_S = \langle \varphi_-(x_1)\varphi_+(x_2)\varphi_-(x_3)\varphi_+(x_4)\rangle_S$$
$$\langle \varphi_+(x_1)\varphi_-(x_2)\varphi_-(x_3)\varphi_+(x_4)\rangle_S = \langle \varphi_-(x_1)\varphi_+(x_2)\varphi_+(x_3)\varphi_-(x_4)\rangle_S.$$

$$(30.73)$$

They are related to the Dirac field correlation functions exactly in the same way as the ones in (30.43) are related to the ones in (30.42) [309].

Let us now study the effect of braiding in the above correlation functions. It can be shown [309] that the set of degenerate Majorana (anti-Majorana) states that corresponds to the following correlation functions

$$\begin{pmatrix} \langle \varphi_+(x_1)\varphi_+(x_2)\varphi_+(x_3)\varphi_+(x_4) \rangle_S \\ \langle \varphi_+(x_1)\varphi_+(x_2)\varphi_-(x_3)\varphi_-(x_4) \rangle_S \\ \langle \varphi_+(x_1)\varphi_-(x_2)\varphi_+(x_3)\varphi_-(x_4) \rangle_S \\ \langle \varphi_+(x_1)\varphi_-(x_2)\varphi_-(x_3)\varphi_+(x_4) \rangle_S \end{pmatrix} \tag{30.74}$$

transforms under the interchange of particles, precisely by the action of the *same* monodromy braiding matrices $\mathcal{M}_{ij}$, with $(ij) = (12), (34), (14), (23), (13), (24)$ derived in Section 30.3.

30.8 Majorana Qubits and Coherence Protection

We have seen that the interchange of identical particles exhibiting non-abelian statistics, in $2n$-particle states, produces matrices of dimension 2^n, known as the monodromy, or braiding, matrices. These act on the set of 2^n degenerate $2n$-particle states producing as the output some linear combinations of the states in the set. Linear combinations of these 2^n degenerate $2n$-particle states with non-abelian statistics are the qubits, where the information is stored in a process of quantum computation.

The particles with non-abelian statistics are, therefore, the basic building blocks of the qubits. They are, by their turn, obtained from abelian anyons in the form of self-adjoint or anti-self-adjoint combinations, which characterizes them as Majorana and anti-Majorana quantum states.

The monodromy matrices process the Majorana (anti-Majorana) qubits, producing a certain output state for a given input state. Now, for the systems considered here, whenever the spin/statistics of the basic abelian particles used to build the Majoranas is $s = 1/4$, the monodromy matrices become the basic logic gates employed in a quantum computation algorithm: NOT, CNOT, etc.

The fact that qubits are made out of quantum states of Majorana quasi-particles has profound implications because of the peculiar features of such states.

Degeneracy, Zero Gap

Degeneracy at zero energy is a first property. Suppose a field is given by

$$\varphi(t, \mathbf{x}) = \int d\mathbf{p}\, \varphi(E(\mathbf{p}), \mathbf{p}) e^{-iE(\mathbf{p})t} e^{i\mathbf{p}\cdot\mathbf{x}},$$

Then, imposing the Majorana condition in coordinate space,

$$\varphi^\dagger(t, \mathbf{x}) = \varphi(t, \mathbf{x}) \Longrightarrow \varphi^\dagger(E, \mathbf{p}) = \varphi(-E, -\mathbf{p}). \tag{30.75}$$

Imposing now the Majorana condition on the energy-momentum space, namely

$$\varphi^\dagger(E, \mathbf{p}) = \varphi(E, \mathbf{p}), \tag{30.76}$$

and using it in (30.75), we get

$$\varphi(E, \mathbf{p}) = \varphi(-E, -\mathbf{p}).$$

Assuming that the Majorana field operator creates an energy eigenstate with eigenvalue $E(\mathbf{p})$, it follows that

$$H|\varphi(E, \mathbf{p})\rangle = E|\varphi(E, \mathbf{p})\rangle$$
$$= H|\varphi(-E, -\mathbf{p})\rangle = -E|\varphi(-E, -\mathbf{p})\rangle = -E|\varphi(E, \mathbf{p})\rangle,$$
$$H|\varphi(E, \mathbf{p})\rangle = -E|\varphi(E, \mathbf{p})\rangle. \tag{30.77}$$

The first and last lines together imply that, if a Majorana state $|\varphi(E, \mathbf{p})\rangle$ is an energy eigenstate, then the energy eigenvalue must vanish: $E = 0$. It follows that all the Majorana states are degenerate. The Majorana modes are also gapless. In the case of vortices, remarkably, this happens despite the fact that the associate vortex states are gapped.

30.9 Superselecting Sectors and Coherence Robustness

According to (30.32) and (30.62), it follows that Majorana and anti-Majorana states can be written as linear combinations of states carrying opposite charges. Surprisingly, such states cannot be physical [310]. The reason is that charged states belong to superselecting sectors of the Hilbert space, which never mix in a physical state. Charge belongs to a class of observables that, despite their quantum-mechanical nature, are immune to the uncertainty principle. Indeed, one can measure the charge of an electron or of a proton one million times and the result will be always precisely the same. Consequently, a coherent combination of an electron field and its hermitean adjoint, each of which creates states with opposite charge, just cannot be physical [310]. This seldom-mentioned fact is indeed remarkable. It is precisely the reason underlying the coherence robustness of the Majorana qubits. What happens is that the above property precludes the occurrence of isolated Majorana states in the bulk, and therefore they hide in the edges, in the case of a Pfaffian quantum Hall system, or in the vortices that pierce a superconductor. Strictly speaking, in both cases they are out of the system. Nevertheless, pairs of Majorana modes can manifest in the bulk as charged states. It follows that by expressing a Dirac state as a combination of Majoranas, we can use the fact that these cannot be in the bulk to place each one of them far away from the other on the edges of the sample. This construction would naturally make the Majorana pair, or the qubit, immune to the local environmental perturbations, which are responsible for decoherence.

30.10 Overview

The non-abelian Chern–Simons–Higgs theory with level $k = 2$ in $d = 2$ spatial dimensions is the Landau–Ginzburg type field theory associated to the $\nu = 5/2$ quantum Hall system. It contains anyonic gapped charged vortex excitations, which combine in the form of Majorana and anti-Majorana states, which present non-abelian statistics. States with two such excitations form the qubits, which have their coherence protected as a consequence of the very peculiar properties of Majorana states. The monodromy or braiding matrices that result from the interchange of these excitations become, when the spin/statistics of anyon vortices is $s = 1/4$, precisely the logic gates NOT, CNOT, etc. It follows that the logic operations of a quantum computer may be performed by interchange manipulations of the degenerate Majorana excitations forming the qubits.

Interestingly, precisely the same structure can be found in a system of Dirac fields in $d = 1$ spatial dimensions, with the exception that the excitations are now gapless. This completely resembles the bulk-boundary correspondence, where we interpret the gapless one-dimensional Dirac modes as the edge excitations that correspond to the gapped vortices on the bulk.

30.11 Appendix: Commutators

Here we show the details of calculations leading to the results of Section 30.3.

$$1) \ [\mu, \Phi_M]$$

From (30.16) and (30.14), we get

$$[\mu(\mathbf{x}, t), \Phi_M] = \frac{1}{2} \frac{a}{2\pi} \mu(\mathbf{x}, t) \, n^a n^b \int d^2 y \int_{\mathbf{x}, L}^{+\infty} d\xi^k [A_k^a(\xi, t), \epsilon^{ij} F_{ij}^b(\mathbf{y})]$$

$$= \mu(\mathbf{x}, t) \frac{a}{k} \, n^a n^b \int d^2 y \int_{\mathbf{x}, L}^{+\infty} d\xi^k \partial_k^{(\xi)} \delta^{ab} \delta^2(\xi - \mathbf{y})$$

$$+ \mu(\mathbf{x}, t) \frac{a}{2\pi} \, n^a n^b \int d^2 y \int_{\mathbf{x}, L}^{+\infty} d\xi^k \epsilon^{ij} \epsilon^{bcd} [A_k^a(\xi), A_i^c(\mathbf{y}) A_j^d(\mathbf{y})].$$

$$(30.78)$$

Now, using the facts that

$$[A_k^a(\xi), A_i^c(\mathbf{y}) A_j^d(\mathbf{y})] = A_i^c(\mathbf{y})[A_k^a(\xi), A_j^d(\mathbf{y})] + [A_k^a(\xi), A_i^c(\mathbf{y})] A_j^d(\mathbf{y})$$

$$(30.79)$$

and that the first and second terms on the right-hand-side will be proportional to δ^{ad} and δ^{ac}, respectively, we see that the second term in (30.78) will be proportional to

$$\epsilon^{abc} n^a n^b = 0.$$

The first term is easily seen to yield $\frac{a}{k}\mu(\mathbf{x}, t)$, thus establishing (30.20).

$$2)\ [\sigma, Q]$$

From (30.17) and (30.12), we get

$$[\sigma(\mathbf{x}, t), Q] = \sigma(\mathbf{x}, t) b\, n^a n^b \int d^2y \int_{S_{\mathbf{x}}} d^2\xi\, \epsilon^{ij} \partial_i \epsilon^{kl} [A^a_j(\xi, t), \partial_k A^b_l(\mathbf{y}, t) -$$

$$\epsilon^{bcd} A^c_k(\mathbf{y}, t) A^d_l(\mathbf{y}, t)] \arg(\xi - \mathbf{x}) \tag{30.80}$$

The second term in the commutator again is proportional to $\epsilon^{abc} n^a n^b$ and therefore vanishes. The first term is proportional to

$$\int d^2y \int_{S_{\mathbf{x}}} d^2\xi\, \epsilon^{ij} \partial_i \partial_j \arg(\xi - \mathbf{x}) \delta^2(\xi - \mathbf{y})$$

$$= 2\pi \int d^2y \int_{S_{\mathbf{x}}} d^2\xi\, \delta^2(\xi - \mathbf{x}) \delta^2(\xi - \mathbf{y}) = 2\pi, \tag{30.81}$$

where we used (9.68). Inserting (30.81) in (30.80), we establish (30.19).

$$3)\ [\sigma, \mu]$$

Writing $\mu \equiv e^A$ and $\sigma \equiv e^B$, we have from (30.16) and (30.17)

$$[A, B] = i \frac{ab}{(2\pi)^2} n^a n^b \int d^2\xi \int_{\mathbf{x}}^{\infty} d\eta^l \arg(\xi - \mathbf{x}) \epsilon^{ij} \partial_i [A^a_l(\eta, t), A^b_j(\xi, t)]. \tag{30.82}$$

Using (30.18), it is straightforward to perform the two integrals yielding

$$[A, B] = i \frac{ab}{2\pi k} \arg(\mathbf{y} - \mathbf{x}). \tag{30.83}$$

From this, one establishes (30.21) by using (10.31).

Further Reading

E. Fradkin, *Field Theories of Condensed Matter Physics*, 2nd Edition, Cambridge University Press, Cambridge, UK (2013)

S. Sachdev, *Quantum Phase Transitions*, 2nd Edition, Cambridge University Press, Cambridge, UK (2011)

A. A. Altland and B. Simons *Condensed Matter Field Theory*, 2nd Edition, Cambridge University Press, Cambridge, UK (2010)

A. Tsvelik, *Quantum Field Theory in Condensed Matter Physics*, 2nd Edition, Cambridge University Press, Cambridge, UK (2007)

X. G. Wen, *Quantum Field Theory of Many Body Systems*, Oxford University Press, Oxford, UK (2004)

A. Gogolin, A. Nersesyan and A. Tsvelik, *Bosonization and Strongly Correlated Systems*, Cambridge University Press, Cambridge, UK (2004)

J. Zinn-Justin, *Quantum Field Theory and Critical Phenomena*, 4th Edition, Oxford University Press, Oxford, UK (2002)

D. J. Amit, *Field Theory, the Renormalization Group and Critical Phenomena*, McGraw-Hill, New York, USA (1980)

A. A. Abrikosov, L. P. Gorkov and I. E. Dzyaloshinskii, *Methods of Quantum Field Theory in Statistical Mechanics*, Prentice-Hall, Englewood Cliffs, NJ, USA (1963)

References

[1] A. Einstein, Annalen der Physik 22, 180 (1907) (in German)
[2] G. Bednorz and A. K. Müller, Z. Physik B64, 189 (1986)
[3] A. P. Drozdov, M. I. Eremets, I. A. Troyan, V. Ksenofontov and S. I. Shylin, Nature 525, 73 (2015)
[4] Y. Kamihara, T. Watanabe, M. Hirano and H. Hosono, J. Am. Chem. Soc. 130, 3296 (2008)
[5] M. Gell-Mann and F. Low, Phys. Rev. 95, 1300 (1954)
[6] C. G. Callan, Phys. Rev. D2, 1541 (1970); K. Symanzik, Comm. Math. Phys. 18, 227 (1970)
[7] G. 't Hooft, Nucl. Phys. B61, 455 (1973); S. Weinberg, Phys. Rev. D8, 3497 (1973)
[8] S. Coleman, *Aspects of Symmetry*, Cambridge University Press, Cambridge, UK (1985)
[9] J. Goldstone, A. Salam and S. Weinberg, Phys. Rev. 127, 965 (1962)
[10] P. W. Anderson, Phys. Rev. 130, 439 (1962)
[11] P. W. Higgs, Phys. Rev. Lett. 13, 508 (1964)
[12] F. Englert and R. Brout, Phys. Rev. Lett. 13, 321 (1964)
[13] G. S. Guralnik, C. R. Hagen and T. W. B. Kibble, Phys. Rev. Lett. 13, 585 (1964)
[14] D. Kastler, D. W. Robinson and J. A. Swieca, Commun. Math. Phys. 2, 108 (1966)
[15] E. C. Marino and L. H. C. M. Nunes, Nucl. Phys. B741, 404 (2006)
[16] E. C. Marino and L. H. C. M. Nunes, Nucl. Phys. B769, 275 (2007)
[17] L. Mondaini, E. C. Marino and A. A. Schmidt, J. of Phys. A42, 055401 (2009)
[18] H. B. Nielsen and P. Olesen, Nucl. Phys. B61, 45 (1973)
[19] S. W. Lovesey, *Theory of Neutron Scattering from Condensed Matter*, Clarendon Press, Oxford, UK (1986)
[20] A. A. Belavin and A. M. Polyakov, JETP Lett. 22, 245 (1975)
[21] T. Skyrme, Proc. R. Soc. London 262, 237 (1961)
[22] H. Georgi and S. Glashow, Phys. Rev. Lett. 32, 438 (1974)
[23] G. 't Hooft, Nucl. Phys. B79, 276 (1974)
[24] A. M. Polyakov, JETP Lett. 20, 194 (1975)
[25] R. Köberle and E. C. Marino, Phys. Lett. B126, 475 (1983)
[26] B. Schroer, in *The Algebraic Theory of Superselection Sectors*, D. Kastler, ed., World Scientific, Singapore, pp. 499, 587 (1990)
[27] B. Schroer, *Differential Methods in Theoretical Physics: Physics and Geometry*, L-L Chau and W. Nahm, eds., Springer, New York, p. 138 (1990)
[28] G. 't Hooft, Nucl. Phys. B138, 1 (1978)

[29] H. A. Kramers and G. H. Wannier, Phys. Rev. 60, 252 (1941)

[30] L. P. Kadanoff and H. Ceva, Phys. Rev. B3, 3918 (1971)

[31] E. C. Marino and J. A. Swieca, Nucl. Phys. B170 [FS1], 175 (1980)

[32] E. C. Marino, B. Schroer and J. A. Swieca, Nucl. Phys. B200 [FS4], 499 (1982)

[33] E. C. Marino, Nucl. Phys. B217, 413 (1983)

[34] E. C. Marino, Nucl. Phys. B230 [FS10], 149 (1984)

[35] E. C. Marino, Phys. Rev. D38, 3194 (1988)

[36] E. C. Marino and J. E. Stephany Ruiz, Phys. Rev. D39, 3690 (1989)

[37] K. Furuya and E. C. Marino, Phys. Rev. D41, 727 (1990)

[38] E. C. Marino, Int. J. Mod. Phys. A10, 4311 (1995)

[39] E. C. Marino, G. C. Marques, R. O. Ramos and J. E. Stephany Ruiz, Phys. Rev. D45, 3690 (1992)

[40] E. C. Marino, Phys. Rev. D55, 5234 (1997)

[41] E. C. Marino, Ann. of Phys. 224, 225 (1993)

[42] E. C. Marino, J. of Phys A: Math. and Gen. 39, L277 (2006)

[43] E. C. Marino, *"Dual Quantization of Solitons"* in *Applications of Statistical and Field Theory Methods to Condensed Matter*, D. Baeriswyl, A. Bishop and J. Carmelo, eds., NATO Advanced Studies Institute Series B (Physics) 218, Plenum Press, New York (1990)

[44] J. Fröhlich and P. A. Marchetti, Commun. Math. Phys. 112, 343 (1987); Europhys. Lett. 2, 933 (1986)

[45] J. Fröhlich and P. A. Marchetti, Commun. Math. Phys. 116, 127 (1988)

[46] E. Fradkin and L. Susskind, Phys. Rev. D17, 2637 (1978)

[47] J. B. Kogut, Rev. Mod. Phys. 51, 659 (1979)

[48] H. Araki, K. Hepp and D. Ruelle, Helv. Phys. Acta 35, 164 (1962)

[49] E. Bogomolnyi, Sov. J. Nuc. Phys. 24, 449 (1976); E. Bogomolnyi and M. S. Marinov, Sov. J. Nuc. Phys. 23, 355449 (1976)

[50] M. Prasad and C. Sommerfield, Phys. Rev. Lett. 35, 760 (1975)

[51] T. Kirkman and C. Zachos, Phys. Rev. D24, 999 (1981)

[52] B. Klaiber, *"The Thirring Model"* in *Boulder Lectures in Theoretical Physics (1967)*, Gordon and Breach, New York (1968), p. 141

[53] A. B. Zamolodchikov and Al. B. Zamolodchikov, Ann. Phys. (NY) 120, 253 (1979)

[54] V. E. Korepin, Commun. Math. Phys. 76, 165 (1980).

[55] W. Thirring, Ann. of Phys. 3, 91 (1968)

[56] J. Schwinger, Phys. Rev. 128, 2425 (1962)

[57] C. G. Callan, R. Dashen and D. Gross, Phys. Lett. B63, 334 (1976)

[58] R. Jackiw and C. Rebbi, Phys. Rev. Lett. 37, 172 (1976)

[59] F. Wilczek, Phys. Rev. Lett. 49, 957 (1982)

[60] A. Luther, Phys. Rep. 49, 261 (1979)

[61] J. Fröhlich and P. A. Marchetti, Lett. Math. Phys. 16, 347 (1988)

[62] S. N. Deser and A. N. Redlich, Phys. Rev. Lett. 61, 1541 (1988)

[63] M. Lüscher, Nucl. Phys. B326, 557 (1989)

[64] E. Fradkin and F. A. Schaposnik, Phys. Lett. B338, 253 (1994)

[65] F. A. Schaposnik, Phys. Lett. B356, 39 (1995)

[66] J. C. Le Guillou, C. Núñez and F. A. Schaposnik, Ann. of Phys. 251, 426 (1996)

[67] C. D. Fosco and F. A. Schaposnik, Phys. Lett. B391, 136 (1997)

[68] E. F. Moreno and F. A. Schaposnik, Phys. Rev. D88, 025033 (2013)

[69] D. G. Barci, C. D. Fosco and L. D. Oxman, Phys. Lett. B375, 267 (1996)

[70] C. P. Burgess, C. Lütken and F. Quevedo, Phys. Lett. B336, 18 (1994)

[71] E. C. Marino, Phys. Lett. B263, 63 (1991)

[72] E. C. Marino, J. of Stat. Mech.: Theory and Experiment 2017, 033103 (2017)

[73] S.-Y. Xu, C. Liu, S. K. Kushwaha, R. Sankar, J. W. Krizan, I. Belopolski, M. Neupane, G. Bian, N. Alidoust, T.-R. Chang, H.-T. Jeng, C.-Y. Huang, W.-F. Tsai, H. Lin, P. P. Shibayev, F.-C. Chou, R. J. Cava and M. Z. Hasan, Science 347, 294 (2015)

[74] S.-Y. Xu, I. Belopolski, N. Alidoust, M. Neupane, G. Bian, C. Zhang, R. Sankar, G. Chang, Z. Yuan, C.-C. Lee, S.-M. Huang, H. Zheng, J. Ma, D. S. Sanchez, B. Wang, A. Bansil, F. Chou, P. P. Shibayev, H. Lin, S. Jia and M. Z. Hasan, Science 349, 613 (2015)

[75] B. Q. Lv, H. M. Weng, B. B. Fu, X. P. Wang, H. Miao, J. Ma, P. Richard, X. C. Huang, L. X. Zhao, G. F. Chen, Z. Fang, X. Dai, T. Qian and H. Ding, Phys. Rev. X5, 031013 (2015)

[76] S. Adler, Phys. Rev. 177, 2496 (1969); J. S. Bell and R. Jackiw, N. Cimento A60, 47 (1969)

[77] G. 't Hooft, Phys. Rep. 142, 357 (1986)

[78] H. B. Nielsen and M. Ninomiya, Phys. Lett. 130, 389 (1983)

[79] D. T. Son and B. Z. Spivak, Phys. Rev. B88, 104412 (2013)

[80] A. A. Burkov, Phys. Rev. B91, 245157 (2015); A. A. Burkov, J. Phys: Cond. Matter 27, 113201 (2015)

[81] X. Fustero, R. Gambini and A. Trias, Phys. Rev. Lett. 62, 1964 (1989); R. Gambini and R. Setaro, Phys. Rev. Lett. 65, 2623 (1990); H. Fort and R. Gambini, Phys. Lett. 372, 226 (1996)

[82] I. S. Gradshteyn and I. M. Ryzhik, *Table of Integrals, Series and Products*, Academic Press, New York (2007)

[83] L. V. Keldysh, Pis'ma Zh. Eksp. Teor. Fiz. 29, 716 (1979)

[84] N. Dorey and N. E. Mavromatos, Nucl. Phys. B386, 614 (1992)

[85] E. C. Marino, Phys. Lett. B393, 382 (1997)

[86] J. Barcelos-Neto and E. C. Marino, Europhys. Lett. 57, 473 (2002)

[87] J. Barcelos-Neto and E. C. Marino, Phys. Rev. D66, 127901 (2002)

[88] S. Deser, R. Jackiw and S. Templeton, Ann. Phys. 140, 372 (1982)

[89] E. C. Marino, Nucl. Phys. B408 [FS], 551 (1993)

[90] R. L. P. G. do Amaral and E. C. Marino, J. of Phys. A25, 5183 (1992)

[91] A. O. Caldeira and A. Leggett, Ann. of Phys. 149, 374 (1983)

[92] M. E. Peskin and D. V. Schroeder, *An Introduction to Quantum Field Theory*, Westview Press, Boulder, CO, USA (1995)

[93] E. C. Marino, L. O. Nascimento, V. S. Alves, C. Morais Smith, Phys. Rev. D90, 105003 (2014)

[94] S. F. Edwards and P. W. Anderson, J. of Phys. F5, 965 (1975)

[95] P. W. Anderson, Phys. Rev. 109, 1492 (1958)

[96] N. F. Mott, Proc. Phys. Soc. London Sect. A62, 416 (1949)

[97] E. C. Marino, Phys. Rev. Lett. 55, 2991 (1985)

[98] L. D. Landau and S. I. Pekar, Zh. Eksp. Teor. Fiz. 18, 419 (1948) [in Russian]; Ukr. J. Phys. 53 (Special Issue), 71 (2008) [English]

[99] H. Fröhlich, Adv. Phys. 3, 325 (1954)

[100] R. P. Feynman, Phys. Rev. 97, 660 (1955)

[101] J. T. Devreese, *"Optical Properties of Fröhlich Polarons"* in *Polarons in Advanced Materials*, A. S. Alexandrov, ed., Springer (2007)

[102] A. S. Alexandrov, *"Superconducting Polarons and Bipolarons"* in *Polarons in Advanced Materials*, A. S. Alexandrov, ed., Springer (2007)

[103] W. P. Su, J. R. Schrieffer and A. J. Heeger, Phys. Rev. Lett. 42, 698 (1979); Phys. Rev. B22, 2099 (1980)

[104] H. Takayama, Y. R. Lin-Liu and K. Maki, Phys. Rev. B21, 2388 (1980)

[105] D. Gross and A. Neveu, Phys. Rev. D15, 3235 (1974)

[106] D. K. Campbell and A. R. Bishop, Nucl. Phys. B200 [FS4], 297 (1982); K. Fesser, D. K. Campbell and A. R. Bishop, Phys. Rev. B27, 4804 (1983)

[107] H. Yukawa, Proc. Phys. Math. Soc. of Japan 17, 48 (1935)

[108] C. M. G. Lattes, H. Muirhead, G. P. S. Occhialini and C. F. Powell, Nature 159, 694 (1947); C. M. G. Lattes, G. P. S. Occhialini and C. F. Powell, Nature 160, 453 (1947)

[109] P. G. de Gennes, *Superconductivity of Metals and Alloys*, Benjamin, New York, (1966)

[110] N. Suzuki, M. Ozaki, S. Etemad, A. J. Heeger and A. G. MacDiarmid, Phys. Rev. Lett. 45, 1209 (1980); A. Feldblum, J. Kaufman, S. Etemad, A. J. Heeger, T.-C. Chung and A. G. MacDiarmid, Phys. Rev. B 26, 819 (1982)

[111] A. Feldblum, A. J. Heeger, T.-C. Chung and A. G. Mac Diarmid, J. Chem. Phys. 77, 5114 (1982); L. Lauchlan, S. Etemad, T.-C. Chung and A. J. Heeger, Phys. Rev. B 24, 1 (1981)

[112] J. Orenstein and G. L. Baker, Phys. Rev. Lett. 49, 1043 (1982)

[113] R. Jackiw and C. Rebbi, Phys. Rev. D13, 3398 (1976)

[114] R. Jackiw and P. Rossi, Nucl. Phys. B190, 681 (1981)

[115] R. Jackiw and J. R. Schrieffer, Nucl. Phys. B190, 253 (1981)

[116] C. K. Chiang, C. R. Fincher Jr., Y. W. Park, A. J. Heeger, H. Shirakawa, E. J. Louis, S. C. Gau and A. G. MacDiarmid, Phys. Rev. Lett. 39, 1098 (1977)

[117] M. F. Atiyah and I. M. Singer, Bull. Amer. Math. Soc. 69, 422 (1963)

[118] P. Jordan and E. Wigner, Z. Phys. 47, 631 (1928).

[119] I. Affleck, Phys. Rev. Lett. 56, 2763 (1986)

[120] A. Gogolin, A. Nersesyan and A. Tsvelik, *Bosonization and Strongly Correlated Systems*, Cambridge University Press, Cambridge, UK (1998)

[121] S. Coleman, Phys. Rev. D11, 2088 (1975)

[122] S. T. Chui and P. A. Lee, Phys. Rev. Lett. 35, 315 (1975)

[123] S. Samuel, Phys. Rev. D18, 1916 (1978);

[124] J. Fröhlich and T. Spencer, J. Stat. Phys. 24, 617 (1981)

[125] J. A. Swieca, Fortschr. Phys. 25, 303 (1977); B. Schroer and T. Truong, Phys. Rev. D15, 1684 (1977)

[126] G. Benfatto, G. Gallavotti and F. Nicolò, Commun. Math. Phys. 83, 387 (1982); F. Nicolò, Commun. Math. Phys. 88, 581 (1983); F. Nicolò, J. Renn and A. Steinmann, Commun. Math. Phys. 105, 291(1986)

[127] G. Gallavotti, Rev. Mod. Phys. 57, 471 (1985)

[128] E. C. Marino, Nucl. Phys. B251 [FS13], 227 (1985)

[129] E. C. Marino, Phys. Lett. A105, 215 (1984)

[130] A. Lima-Santos and E. C. Marino, J. Stat. Phys. 55, 157 (1989)

[131] D. J. Amit, Y. Goldschmidt and G. Grinstein, J. of Phys. A: Math. Gen. 13, 585 (1980)

[132] P. di Vecchia and S. Ferrara, Nucl. Phys. B130, 93 (1977); E. Witten, Phys. Rev. D16, 299 (1977); J. Hruby, Nucl. Phys. B131, 275 (1977)

[133] A. Lima-Santos and E. C. Marino, Nucl. Phys. B336, 547 (1990)

[134] E. Witten, Nucl. Phys. B185, 513 (1981)

[135] S. Eggert, I. Affleck and M. Takahashi, Phys. Rev. Lett. 73, 332 (1994)

[136] N. Motoyama, H. Eisaki and S. Uchida, Phys. Rev. Lett. 76, 3212 (1996)

[137] I. Affleck and M. Oshikawa, Phys. Rev. B60, 1038 (1999)

[138] I. Dzyaloshinskii, J. Phys. Chem. Solids 4, 241 (1958); T. Moriya, Phys. Rev. 120, 91 (1960).

[139] F. Essler, Phys. Rev. B59, 14376 (1999)

[140] D. C. Dender, P. R. Hammar, D. H. Reich, C. Broholm and G. Aeppli, Phys. Rev. Lett. 79, 1750 (1997).

[141] L. Mondaini and E. C. Marino, J. Stat. Phys. 118, 767 (2005)

[142] J. Kondo, Progr. Th. Phys. 32, 37 (1964)

[143] H. Bethe, Z. Phys. 71, 205 (1931)

[144] N. Andrei, Phys. Rev. Lett. 45, 379 (1980)

[145] P. B. Wiegmann, Zh. Eksp. Teor. Fiz. Pis'ma Red. 31, 392 (1980)(JETP Lett. 31, 364 (1980))

[146] N. Andrei and J. H. Lowenstein, Phys. Rev. Lett. 43, 1693 (1979)

[147] N. Andrei and J. H. Lowenstein, Phys. Lett. B90, 106 (1980)

[148] N. Andrei and J. H. Lowenstein, Phys. Lett. B91, 401 (1980)

[149] N. Andrei and J. H. Lowenstein, Phys. Rev. Lett. 46, 356 (1981)

[150] N. Andrei, K. Furuya and J. H. Lowenstein, Rev. Mod. Phys. 55, 331 (1983)

[151] P. W. Anderson, G. Yuval and D. R. Hamann, Phys. Rev. B1, 4464 (1970)

[152] D. Gross and A. Neveu, Phys. Rev. D10, 323 (1974)

[153] E. Witten, Nucl. Phys. B145, 110 (1978)

[154] P. Minnhagen, Rev. Mod. Phys. 59, 1001 (1987).

[155] V. L. Berezinskii, Sov. Phys. (JETP) 32, 493 (1970)

[156] J. M. Kosterlitz and D. J. Thouless, J. Phys. C: Solid State Phys. 6, 1181 (1973).

[157] J. M. Kosterlitz, J. Phys. C: Solid State Phys. 7, 1046 (1974).

[158] J. V. José, L. P. Kadanoff, S. Kirkpatrick and D. Nelson, Phys. Rev. B16, 1217 (1977).

[159] D. J. Amit, Y. Y. Goldschmidt and G. Grinstein, J. Phys. A: Math. Gen. 13, 585 (1980).

[160] T. Giamarchi and H. J. Schulz, Phys. Rev. B39, 4620 (1989).

[161] G. Arfken, *Mathematical Methods for Physicists*, 2nd Edition, Academic Press, New York (1970)

[162] E. C. Marino and F. I. Takakura, Int. J. Mod. Phys. A12, 4155 (1997)

[163] E. Fradkin, *Field Theories of Condensed Matter Physics*, 2nd Edition, Cambridge University Press, Cambridge UK (2012)

[164] S. Sachdev, *Quantum Phase Transitions*, 2nd Edition, Cambridge University Press, Cambridge UK (2011)

[165] R. P. Feynman, Phys. Rev. 84, 108 (1951)

[166] J. R. Klauder and B. Skagerstam, *Coherent States*, World Scientific, Singapore (1985); A. Perelomov, *Generalized Coherent States and Their Applications*, Springer-Verlag, New York (1986).

[167] M. Berry, Proc. R. Soc. Lond. A392, 45 (1984).

[168] L. D. Landau and E. M. Lifshitz, Phys. Z. Sov. 8, 153 (1935); D. ter Haar, ed., *Collected Papers of L. D. Landau*, Pergamon Press, New York (1965), pp. 101–114.

[169] A. G. Abanov and Ar. Abanov, Phys. Rev. B65, 184407 (2002)

[170] P. C. Hohenberg, Phys. Rev. 158, 383 (1967); N. D. Mermin, Phys. Rev. 176, 250 (1968); N. D. Mermin and H. Wagner, Phys. Rev. Lett. 22, 1133 (1966)

[171] S. Coleman, Commun. Math. Phys. 31, 259 (1973)

[172] F. Wilczek and A. Zee, Phys. Rev. Lett. 51, 2250 (1983)

[173] X. G. Wen and A. Zee, Phys. Rev. Lett. 61, 1025 (1988)

[174] A. P. Kampf, Phys. Rep. 249, 219 (1994)

[175] E. C. Marino and L. H. C. M. Nunes, Ann. of Phys. 340, 13 (2014)

[176] Y. Nambu, G. Jona-Lasinio, Phys. Rev. 124 (1961) 246

[177] C. M. S. da Conceição and E. C. Marino, Phys. Rev. Lett. 101, 037201 (2008)

[178] C. M. S. da Conceição and E. C. Marino, Nucl. Phys. B820, 565 (2009)

[179] C. M. S. da Conceição and E. C. Marino, Phys. Rev. B80, 064422 (2009)

[180] K. Binder and P. Young, Rev. Mod. Phys. 58, 801 (1986)

[181] D. Sherrington and S. Kirkpatrick, Phys. Rev. Lett. 35, 1792 (1975)

[182] J. R. L. de Almeida and D. J. Thouless, J. Phys. A 11, 983 (1978)

[183] G. Parisi, Phys. Rev. Lett. 43, 1754 (1979)

[184] P. Kapitza, Nature, 141, 74 (1938)

[185] J. F. Allen and A. D. Misener, Nature, 141, 75 (1938)

[186] F. London, Phys. Rev. 54, 947 (1938)

[187] E. P. Gross, N. Cimento 20, 454 (1961); E. P. Gross, J. Math. Phys. 4, 195 (1963)

[188] L. P. Pitaevskii, Zh. Eksp. Teor. Fiz. 40, 646 (1961) [Sov. Phys. JETP 13, 451 (1961)]

[189] L. Onsager, N. Cimento 6, 249 (1949)

[190] F. Lund and T. Regge, Phys. Rev. D14, 1524 (1976); E. Witten, Phys. Lett. B153, 243 (1985); A. Vilenkin and T. Vachaspati, Phys. Rev. D4, 1138 (1987); R. L. Davis and E. P. S. Shellard, Phys. Lett. B214, 219 (1988)

[191] M. Kalb and P. Ramond, Phys. Rev. D9, 227 (1974)

[192] H. Fort and E. C. Marino, Int. J. Mod. Phys. A15, 2225 (2000)

[193] E. C. Marino, Phys. Rev. D53, 1001 (1996)

[194] N. Nücker, J. Fink, J. C. Fuggle, P. J. Durham and W. M. Temmerman, Phys. Rev. B 37, 5158 (1988)

[195] E. Manousakis, Rev. Mod. Phys. 63, 1 (1991)

[196] Z. X. Shen, D. S. Dessau, B. O. Wells, D. M. King, W. E. Spicer, A. J. Arko, D. Marshall, L. W. Lombardo, A. Kapitulnik, P. Dickinson, S. Doniach, J. DiCarlo, T. Loeser and C. H. Park, Phys. Rev. Lett. 70, 1553 (1993)

[197] T. Dombre and N. Read, Phys. Rev. B38, 7181 (1988)

[198] E. Fradkin and M. Stone, Phys. Rev. B38, 7215 (1988)

[199] X. G. Wen and A. Zee, Phys. Rev. Lett. 61, 1025 (1988)

[200] F. D. M. Haldane, Phys. Rev. Lett. 61, 1029 (1988)

[201] L. B. Ioffe and A. I. Larkin, Int. J. Mod. Phys. B2, 203 (1988)

[202] E. C. Marino, Phys. Lett. A263, 446 (1999)

[203] E. C. Marino and M. B. Silva Neto, Phys. Rev. B64, 092511 (2001)

[204] A. Ino, C. Kim, T. Mizokawa, Z. X. Shen, A. Fujimori, M. Takaba, K. Tamasaku, H. Eisaki and S. Uchida, J. Phys. Soc. Jpn. 68, 1496 (1999)

[205] T. Yoshida, X. J. Zhou, T. Sasagawa, W. L. Yang, P. V. Bogdanov, A. Lanzara, Z. Hussain, T. Mizokawa, A. Fujimori, H. Eisaki, Z.-X. Shen, T. Kakeshita and S. Uchida Phys. Rev. Lett. 91, 027001 (2003)

[206] A. Coste and M. Lüscher, Nucl. Phys. B323, 631 (1989)

[207] P. B. Wiegmann, Phys. Rev. Lett. 60, 821 (1988)

[208] B. I. Schraiman and E. Siggia, Phys. Rev. Lett. 61, 467 (1988); *ibid*, Phys. Rev. B42, 2485 (1990)

[209] J. P. Rodriguez, Phys. Rev. B39, 2906 (1989);*ibid*, Phys. Rev. B41, 7326 (1990)

[210] R. J. Gooding, Phys. Rev. Lett. 66, 2266 (1991); R. J. Gooding and A. Mailhot, Phys. Rev. B48, 6132 (1993)

[211] S. Haas, F. C. Zhang, F. Mila and T. M. Rice, Phys. Rev. Lett. 77, 3021 (1996)

[212] R. F. Kiefl, J. H. Brewer, I. Affleck, J. F. Carolan, P. Dosanjh, W. N. Hardy, T. Hsu, R. Kadono, J. R. Kempton, S. R. Kreitzman, Q. Li, A. H. OReilly, T. M. Riseman, P.

Schleger, P. C. E. Stamp, H. Zhou, L. P. Le, G. M. Luke, B. Sternlieb, Y. J. Uemura, H. R. Hart and K. W. Lay, Phys. Rev. Lett. 64, 2082 (1990)

[213] E. C. Marino, Phys. Rev. B61, 1588 (2000)

[214] M. Matsuda, M. Fujita, K. Yamada, R. J. Birgeneau, Y. Endoh and G. Shirane, Phys. Rev. B65, 134515 (2002)

[215] F. Borsa, P. Carretta, J. H. Cho, F. C. Chou, Q. Hu, D. C. Johnston, A. Lascialfari, D. R. Torgeson, R. J. Gooding, N. M. Salem and K. J. E. Vos, Phys. Rev. B52, 7334 (1995)

[216] A. H. Castro Neto and D. Hone, Phys. Rev. Lett. 76, 2165 (1995)

[217] V. Yu. Irkhin and A. A. Katanin, Phys. Rev. B55, 12318 (1997); *ibid* Phys. Rev. B57, 379 (1998)

[218] C. M. S. da Conceição, M. B. Silva Neto and E. C. Marino, Phys. Rev. Lett. 106, 117002 (2011)

[219] S. Raghu, X.-L. Qi, C.-X. Liu, D. J. Scalapino and S.-C. Zhang, Phys. Rev. B77, 220503 (2008)

[220] L. J. de Jongh, *Introduction to Low-Dimensional Magnetic Systems*, in *Magnetic Properties of Layered Transition Metal Compounds*, L. J. de Jongh, ed. Kluwer (1990)

[221] J. Zhao, D.-X. Yao, S. Li, T. Hong, Y. Chen, S. Chang, W. Ratcliff, J. W. Lynn, H. A. Mook, G. F. Chen, J. L. Luo, N. L. Wang, E. W. Carlson, J. Hu and P. Dai, Phys. Rev. Lett. 101, 167203 (2008)

[222] P. Richard, K. Nakayama, T. Sato, M. Neupane, Y.-M. Xu, J. H. Bowen, G. F. Chen, J. L. Luo, N. L. Wang, X. Dai, Z. Fang, H. Ding and T. Takahashi, Phys. Rev. Lett. 104, 137001 (2010)

[223] M. El Massalami, E. C. Marino and M. B. Silva Neto, J. M. Mag. Mat. 350, 30 (2014)

[224] G. F. Chen, Z. Li, J. Dong, G. Li, W. Z. Hu, X. D. Zhang, X. H. Song, P. Zheng, N. L. Wang and J. L. Luo Phys. Rev. B 78, 224512 (2008)

[225] C. Krellner, N. Caroca-Canales, A. Jesche, H. Rosner, A. Ormeci and C. Geibel Phys. Rev. B 78, 100504(R) (2008)

[226] K. von Klitzing, G. Dorda and M. Pepper, Phys. Rev. Lett. 45, 494 (1980)

[227] D. C. Tsui, H. L. Störmer and A. C. Gossard, Phys. Rev. Lett. 48, 1559 (1982)

[228] L. Landau, Z. Phys. 64, 629 (1930)

[229] M. Born and V. A. Fock, Z. Phys. 51, 165 (1928)

[230] M. V. Berry, Proc. Roy. Soc. Lond. A392, 45 (1984)

[231] D. J. Thouless, M. Kohmoto, M. P. Nightingale and M. den Nijs, Phys. Rev. Lett. 49, 405 (1982)

[232] R. Laughlin, Phys. Rev. Lett. 50, 1395 (1983)

[233] J. K. Jain, Phys. Rev. Lett. 63, 2 (1989); *ibid* Phys. Rev. B40, 8079 (1989); *ibid* Phys. Rev. B41, 7653 (1990)

[234] B. I. Halperin, Phys. Rev. Lett. 52, 1583 (1984)

[235] R. de-Picciotto, M. Reznikov, M. Heiblum, V. Umansky, G. Bunin and D. Mahalu, Nature 389, 162 (1997)

[236] D. Arovas, J. R. Schrieffer and F. Wilczek, Phys. Rev. Lett. 53, 722 (1984)

[237] F. E. Camino, W. Zhou and V. J. Goldman, Phys. Rev. B72, 075342 (2005); *ibid*, Phys. Rev. Lett. 98, 076805 (2007)

[238] Y. Aharonov and D. Bohm, Phys. Rev. 115, 485 (1959)

[239] S. C. Zhang, T. H. Hansson and S. Kivelson, Phys. Rev. Lett. 62, 82 (1989)

[240] S. C. Zhang, Int. J. Mod. Phys. B6, 25 (1992)

[241] G. Moore and N. Read, Nucl. Phys. B360, 362 (1991)

[242] E. Fradkin, C. Nayak, A. Tsvelik and F. Wilczek, Nucl. Phys. B516, 704 (1998)

[243] K. S. Novoselov, A. K. Geim, S. V. Morozov, D. Jiang, Y. Zhang, S. V. Dubonos, I. V. Grigorieva and A. A. Firsov, Science. 306, 666 (2004)

[244] K. S. Novoselov, A. K. Geim, S. V. Morozov, D. Jiang, M. I. Katsnelson, I. V. Grigorieva, S. V. Dubonos and A. A. Firsov, Nature 438, 197 (2005)

[245] A. H. Castro Neto, F. Guinea, N. M. R. Peres, K. S. Novoselov and A. K. Geim, Rev. Mod. Phys. 81, 109 (2009)

[246] V. N. Kotov, B. Uchoa, V. M. Pereira, F. Guinea and A. H. Castro Neto, Rev. Mod. Phys. 84, 1067 (2012)

[247] P. R. Wallace, Phys. Rev. 71, 622 (1947)

[248] G. W. Semenoff, Phys. Rev. Lett. 53, 2449 (1984)

[249] D. P. Di Vincenzo and E. J. Mele, Phys. Rev. B29, 1685 (1984)

[250] M. I. Katsnelson, K. S. Novoselov and A. K. Geim, Nature Phys. 2, 620 (2006)

[251] M. I. Katsnelson, Eur. Phys. J. B51, 157 (2006)

[252] N. Dombey and A. Calogeracos, Phys. Rep. 315, 41 (1999)

[253] C. W. J. Beenakker, Rev. Mod. Phys. 80, 1337 (2008)

[254] Y. Zhang, Y.-W. Tan, H. L. Stormer and P. Kim, Nature 438, 201 (2005)

[255] K. I. Bolotin, F. Ghahari, M. D. Shulman, H. L. Stormer and P. Kim, Nature 462, 196 (2009)

[256] X. Du, I. Skachko, F. Duerr, A. Luican and E. Y. Andrei, Nature 462, 192 (2009)

[257] E. C. Marino, L. O. Nascimento, V. S. Alves and C. Morais Smith, Phys. Rev. X5, 011040 (2015)

[258] C. D. Roberts and A. G. Williams, Prog. Part. Nucl. Phys. 33, 477 (1994)

[259] D. V. Khveshchenko, Phys. Rev. Lett. 87, 246802 (2001)

[260] D. V. Khveshchenko, Phys. Rev. Lett. 87, 206401 (2001)

[261] D. V. Khveshchenko and H. Leal, Nucl. Phys. B687, 323 (2004)

[262] D. V. Khveshchenko and W. F. Shively, Phys. Rev. B73, 115104 (2006)

[263] E. V. Gorbar, V. P. Gusynin, V. A. Miransky and I. A. Shovkovy, Phys. Rev. B66, 045108 (2002)

[264] E. V. Gorbar, V. P. Gusynin, V. A. Miransky and I. A. Shovkovy, Phys. Lett. A313, 472 (2003)

[265] D. V. Khveshchenko, J. Phys. Condens. Matter 21, 075303 (2009)

[266] H. Isobe and N. Nagaosa, Phys. Rev. B87, 205138 (2013)

[267] N. Menezes, V. S. Alves, E. C. Marino, L. Nascimento, L. O. Nascimento and C. Morais Smith, Phys. Rev. B95, 245138 (2017)

[268] O. Vafek and A. Vishwanath, Annual Review of Condensed Matter Physics Vol. 5, 83–112 (2014) (arXiv 1306.2272, p. 18)

[269] J. Gonzales, F. Guinea and M. A. H. Vozmediano, Nucl. Phys. B424, 595 (1994); *ibid* Phys. Rev. B59, 2474 (1999)

[270] D. C. Elias, R. V. Gorbachev, A. S. Mayorov, S. V. Morozov, A. A. Zhukov, P. Blake, L. A. Ponomarenko, I. V. Grigorieva, K. S. Novoselov, F. Guinea and A. K. Geim, Nature Phys. 7, 701 (2011)

[271] A. Coste and M. Lüscher, Nucl. Phys. B323, 631 (1989)

[272] S. Teber, Phys. Rev. D86, 025005 (2012)

[273] S. Teber, Phys. Rev. D89, 067702 (2014)

[274] A. V. Kotikov and S. Teber, Phys. Rev. D89, 065038 (2014)

[275] S. Teber and A. V. Kotikov, Europhys. Lett. 107, 57001 (2014)

[276] X. Du, I. Skachko, A. Barker and E. Y. Andrei, Nature Nanotech. 3, 491 (2008)

[277] K. Ziegler, Phys. Rev. B75, 233407 (2007)

[278] S. Coleman and B. Hill, Phys. Lett. B159, 184 (1985)

[279] K. Takeda and K. Shiraishi, Phys. Rev. B50, 14916 (1994)

[280] M. Ezawa, Phys. Rev. B 87, 155415 (2013); *ibid* Phys. Rev. Lett. 109, 055502 (2012); M. Ezawa, Phys. Rev. Lett. 110, 026603 (2013)

[281] C.-C. Liu, W. Feng and Y. Yao, Phys. Rev. Lett. 107, 076802 (2011)

[282] D. Xiao, G.-B. Liu, W. Feng, X. Xu and W. Yao, Phys. Rev. Lett. 108, 196802 (2012)

[283] B. Lalmi, H. Oughaddou, H. Enriquez, A. Kara, S. Vizzini, B. Ealet and B. Aufray, Appl. Phys. Lett. 97, 223109 (2010); P.E. Padova, C. Quaresima, C. Ottaviani, P. M. Sheverdyaeva, P. Moras, C. Carbone, D. Topwal, B. Olivieri, A. Kara, H. Oughaddou, B. Aufray and G. L. Lay, Appl. Phys. Lett. 96, 261905 (2010); B. Aufray A. Vizzini, H. Oughaddou, C. Lndri, B. Ealet and G. L. Lay, Appl. Phys. Lett. 96, 183102 (2010)

[284] S. Cahangirov, M. Topsakal, E. Aktürk, H. Şahin and S. Ciraci, Phys. Rev. Lett. 102, 236804 (2009)

[285] X.-S. Ye, Z.-G. Shao, H. Zhao, L. Yang and C.-L. Wang, Royal Soc. Chemistry Adv. 4, 21216 (2014)

[286] Y. Xu, B. Yan, H. J. Zhang, J. Wang, G. Xu, P. Tang, W. Duan and S. C. Zhang, Phys. Rev. Lett. 111, 13 (2013)

[287] L. O. Nascimento, E. C. Marino, V. S. Alves and C. Morais Smith, arXiv 1702.01573 (2017)

[288] C. Itzykson and J. B. Zuber, *Quantum Field Theory*, McGraw-Hill, New York (1980)

[289] C. L. Kane and E. J. Mele, Phys. Rev. Lett. 95, 226801 (2005)

[290] C. Nayak, S.H. Simon, A. Stern, M. Freedman and S.D. Sarma, Rev. Mod. Phys. 80, 1083 (2008)

[291] A. Yu. Kitaev, Ann. of Phys. 303, 2 (2003)

[292] J. K. Pachos, *Introduction to Topological Quantum Computation*, Cambridge University Press, Cambridge UK (2012).

[293] F. Navarro-Lérida, E. Radu and D. H. Tchrakian, Phys. Rev. D79, 65036 (2009)

[294] N. Read and D. Green, Phys. Rev. B61, 10267 (2000)

[295] D. Ivanov, Phys. Rev. Lett. 86, 268 (2001)

[296] A. Stern, F. von Oppen and E. Mariani, Phys. Rev. B70, 205338 (2004)

[297] A. Yu. Kitaev, Ann. of Phys. 321, 2 (2006)

[298] T. H. Hansson, M. Herrmanns, N. Regnault and S. Viefers, Phys. Rev. Lett. 102, 166805 (2009)

[299] E. C. Marino and J. C. Brozeguini, J. Stat. Mech.: Theory and Experiment, P09038 (2014)

[300] J. R. Woolton, V. Lahtinen, Z. Wang and J. K. Pachos, Phys. Rev B 78, 161102 (2008)

[301] H. Bombin, Phys. Rev. Lett. 105, 030403 (2010)

[302] J. R. Woolton, V. Lahtinen, B. Doucot and J. K. Pachos, Ann. Phys. 326, 2307 (2011)

[303] A. S. Wightman, Phys. Rev. 101, 860 (1956)

[304] K. Osterwalder and R. Schrader, Commun. Math. Phys. 31, 83 (1973); *ibid* Commun. Math. Phys. 42, 281 (1975)

[305] G. Moore and N. Read, Nucl. Phys. B360, 362 (1991)

[306] C. Nayak and F. Wilczek, Nucl. Phys. B470, 529 (1996)

[307] N. Read and E. Rezayi, Phys. Rev. B54, 16864 (1996)

[308] V. Gurarie and C. Nayak, Nucl. Phys. B506, 685 (1997)

[309] E. C. Marino and J. C. Brozeguini, J. Stat. Mech.: Theory and Experiment, P03011 (2015)

[310] R. F. Streater and A. S. Wightman, *PCT, Spin and Statistics, and All That*, Benjamin/Cummings, Reading, MA, USA (1964), p. 5

[311] M. Z. Hasan and C. L. Kane, Rev. Mod. Phys. 82, 3045 (2010)

[312] X.-L. Qi and S.-C. Zhang, Rev. Mod. Phys. 83, 1057 (2011)

[313] B. A. Bernevig and T. L. Hughes, *Topological Insulators and Topological Super-conductors*, Princeton University Press, Princeton, NJ, USA (2013)

[314] F. D. M. Haldane, Phys. Rev. Lett. 61, 2015 (1988)

[315] C.-Z. Chang, J. Zhang, X. Feng, J. Shen, Z. Zhang, M. Guo, K. Li, Y. Ou, P. Wei, L.-L. Wang, Z.-Q. Ji, Y. Feng, S. Ji, X. Chen, J. Jia, X. Dai, Z. Fang, S.-C. Zhang, K. He, Y. Wang, L. Lu, X.-C. Ma and Q.-K. Xue, Science 340, 167 (2013)

[316] D. Sticlet, F. Piéchon, J.-N. Fuchs, P. Kalugin and P. Simon, Phys. Rev. B85, 165456 (2012)

[317] C. L. Kane and E. J. Mele, Phys. Rev. Lett. 95, 146802 (2005)

[318] L. Fu and C. L. Kane, Phys. Rev. B76, 045302 (2007)

[319] M. Fruchart and D. Carpentier, Comptes Rendus de l'Academie des Sciences (Paris), 14, 779 (2013)

[320] B. A. Bernevig, T. L. Hughes and S.-C. Zhang, Science 314, 1757 (2006)

[321] M. König, S. Wiedmann, C. Brüne, A. Roth, H. Buhmann, L. W. Molenkamp, X.-L. Qi and S.-C. Zhang, Science 318, 766 (2007)

[322] L. Fu, C. L. Kane and E. J. Mele, Phys. Rev. Lett. 98, 106803 (2007)

[323] J. E. Moore and L. Balents, Phys. Rev. B75, 121306 (2007)

[324] R. Roy, Phys. Rev. B79, 195322 (2009)

Index

512